Environmental and Microbial Biotechnology

Series Editor

Ram Prasad, Department of Botany, Mahatma Gandhi Central University,
Motihari, Bihar, India

Innovative and novel advances in microbial biotechnology are providing great understandings in to the machineries of nature, presenting fascinating prospects to apply principles of biology to different arenas of science. Sustainable elucidations are emerging to address the concerns on improving crop productivity through microbes, depleting natural resources, environmental pollution, microbial degradation of pollutants, nanomaterials, nanotoxicity & safety issues, safety of food & agricultural products etc. Simultaneously, there is an increasing demand for natural bio-products of therapeutic and industrial significance (in the areas of healthcare, environmental remediation, microbial biotechnology). Growing awareness and an increased attention on environmental issues such as climate change, energy use, and loss of non-renewable resources have carried out a superior quality for research that provides potential solutions to these problems. Emerging microbiome approaches potentially can significantly increase agriculture productivity & human healthcare and henceforth can contribute to meet several sustainable development goals.

The main objectives have provided an impetus for research on plants and microorganisms that produce novel bio-products with variable properties and understanding their mechanisms of action at cellular and molecular level. Hence, research activities of the environmental and microbial Biotechnology are comprehensively focused up on major sectors viz., bioresources, biorefining, bioremediation of organic and inorganic pollutants, environmental risk analysis of microorganisms, environmental assessment using microbiological indicators, enzymes for environment, food & industrial applications, nanomaterials & nanotoxicity, sustainable ecobiotechnology, biofertilizer, biocontrol agents for agriculture improvement and natural products for healthcare applications.

This book series is a state-of-the-art for a wide range of scientists, researchers, students, policy makers and academician involve in understanding and implementing the knowledge on environmental and microbial biotechnology to develop biologics for proper health care to continue life in smooth and sustainable strategy without any adverse effect.

More information about this series at http://www.springer.com/series/16324

Naga Raju Maddela • Sagnik Chakraborty •
Ram Prasad
Editors

Nanotechnology for Advances in Medical Microbiology

Springer

Editors
Naga Raju Maddela
Facultad de Ciencias de la Salud y Instituto de Investigación
Universidad Técnica de Manabí
Portoviejo, Ecuador

Sagnik Chakraborty
School of Environmental Science and Safety Engineering
Jiangsu University
Zhenjiang, Jiangsu, China

Ram Prasad
Department of Botany
Mahatma Gandhi Central University
Motihari, Bihar, India

ISSN 2662-1681 ISSN 2662-169X (electronic)
Environmental and Microbial Biotechnology
ISBN 978-981-15-9915-6 ISBN 978-981-15-9916-3 (eBook)
https://doi.org/10.1007/978-981-15-9916-3

This Springer imprint is published by the registered company Springer Nature Singapore Pte Ltd.
The registered company address is: 152 Beach Road, #21-01/04 Gateway East, Singapore 189721, Singapore

Preface

The era of Nanotechnology is evolving at a very rapid pace due to its precise development, manipulation and application of nanomaterials. Evidently, nanoscience in medical field has touched many areas such as drug development, nano-tags for detecting infectious agents, nano-sensors for detection of bioterrorism agents, cantilever technology, bio-conjugated nanoparticle-based bioassay for in situ pathogen, quantification of pathogen etc. In Particular, the combination of Microbiology and Nanotechnology has strengthened the field of Science and Technology, since this combination is the most successful in providing novel solutions for protecting the health of the human and environment. Several areas of Microbiology are impacted by Nanoscience, such as visualization at the molecular-assembly levels of a process, identification of molecular recognition and self-assembly motifs as well as the assessment of these processes. Recent advances of nanotechnology in the medical microbiology includes emergence of nanoscopic vectors for presenting promising diagnostic effects, precise targeting of diseased tissues by therapeutic substances, development of nanomedicine for early detection, cure and diagnosis of cancer, disorders of central nerve system (e.g. Parkinson's and Alzheimer's diseases) lungs (e.g. Tuberculosis). The main aim of this volume is to provide recent updates in nanoscience towards medical microbiology. It can serve as a "handbook" dealing with nano techniques that evolved recently. Therefore, the present book has been designed to cover the topics related to implications of nano-strategies to combat bacterial pathogens, applications of nano techniques in Microbiology, and innovative advances in the area of Medical Microbiology. Likewise, we would like to justify the title of the proposed Volume i.e. "*Nanotechnology for Advances in Medical Microbiology*".

The chief target audience for this book are the professors and researchers who are associated with Nanoscience in Medical Microbiology. Graduate students of the aforesaid area too can be beneficiated by this book. In order to meet our objectives, we paid special attention while inviting the chapter contributors. Key contributors that have participated in this book have a solid research background, most of them are Research Professors, Scientists, Postdoctoral Research Scholars and Doctoral students, which have proven track-record internationally. Overall, the book portrays a very clear idea about the emerging nano techniques and also to direct young minds in the same path. This volume will also serve as a ready reference by practicing students, researchers of medical microbiology and medical professionals. Distinctive features of this book are lucid language, updated information, discussion on modern technologies related to the persisting field which is absent in available books in the market and clear figures and illustrations.

In order to justify the title of this volume, a wide range of topics have been included, and the details are as follows. Controlling of pathogen biofilms using green synthesized nanoparticles, present and future of visual health nanocomposites, prospective of nanomaterials in biomedical engineering, medical microbiology, and therapeutic use. Detailed information included on the preparation, characterization, and biomedical applications of Chitosan nanoparticles. Chapters have also been incorporated related to the impact of nanomaterials on HeLa cell lines, and the therapeutic role of nanoparticles in the bacteria mediated cancer. Additionally, this volume provides information on epidemiology of COVID-19 and possible role of nano techniques in dealing the coronavirus, application of nanoparticles for water disinfection, enhanced production of prodigiosin (antimicrobial, immunosuppressive drug) by nanobiotechnological approach.

Portoviejo, Manabí, Ecuador — Naga Raju Maddela
Zhenjiang, Jiangsu, China — Sagnik Chakraborty
Motihari, Bihar, India — Ram Prasad

Acknowledgements

We greatly acknowledge the support of the chapter's contributors for their valuable contribution and timely responses during this project. Without their enthusiasm and support, this volume would not be ready in the scheduled time, therefore we really appreciate their cooperation and collaboration. We also thank anonymous reviewers for their constructive criticism, which would help us in improving the quality of this book by inviting the experts to contribute the additional chapters. We greatly acknowledge the Springer Editorial and Production team for their valuable support and guidelines to complete this project. It is our honour to work with them, honestly. Finally yet importantly, we are very much thankful to the colleagues at Universidad Técnica de Manabí (Ecuador), Jiangsu University (China), and Mahatma Gandhi Central University (India) for their unconditional support and for the provision of valuable suggestions at the time of book proposal and final book preparation.

Naga Raju Maddela
Sagnik Chakraborty
Ram Prasad

Contents

Editors and Contributors

About the Editors

Naga Raju Maddela received his M.Sc. (1996–1998) and Ph.D. (2012) in Microbiology from Sri Krishnadevaraya University, Anantapuramu, India. During his doctoral program in the area of Environmental Microbiology, he investigated the effects of industrial effluents/insecticides on soil microorganisms and their biological activities and worked as a Faculty in Microbiology for 17 years, teaching undergraduate and postgraduate students. He received "Prometeo Investigator Fellowship" (2013–2015) from Secretaría de Educación Superior, Ciencia, Tecnología e Innovación (SENESCYT), Ecuador, and "Postdoctoral Fellowship" (2016–2018) from Sun Yat-sen University, China. He also received external funding from "China Postdoctoral Science Foundation" in 2017, internal funding from "Universidad Técnica de Manabí" in 2020, worked in the area of Environmental Biotechnology, participated in 19 national / international conferences, and presented research data in China, Cuba, Ecuador, India and Singapore. Currently, he is working as a full-time Professor at the Departamento de Ciencias Biológicas, Facultad de Ciencias de la Salud, Universidad Técnica de Manabí, Portoviejo, Ecuador. He has published 4 Books (Springer), 2 Chapters (InTech Open / Springer) and 42 research papers.

Sagnik Chakraborty Completed M.Sc. in Biotechnology (2006–2008) from Bangalore University and Ph.D. from National Institute of Technology Durgapur in 2014 Department of Biotechnology, India. His research work focused on the removal of the dyes by biomass-derived from waste materials and removal of the dyes from an industrial outlet of the textile plants- biomass derived from the various waste. Until now, he published forty papers according to the research findings. He has received several awards during his research career. He has mentored undergraduate and postgraduate students for their projects. Furthermore, Sagnik completed his Post-Doctoral Research 2015–2018 from Hebei University of Technology, Tianjin, China, the fund approved by the Ministry of Education China. He has received several intramural projects funds in his research domain. Moreover, he received an International Young Scientist Award from the National Natural Science Foundation of China in 2019 with grant aid for further research. Sagnik received the Brand Ambassador Award in 2019 from The Bentham Science Publisher and won several outstanding reviewing awards from Elsevier. Lately, he is engaged as Senior Researcher at Jiangsu University in the School of Environmental Science and Safety Engineering.

Ram Prasad Ph.D. is associated with Department of Botany, Mahatma Gandhi Central University, Motihari, Bihar, India. His research interest includes applied & environmental microbiology, plant-microbe-interactions, sustainable agriculture and nanobiotechnology. Dr. Prasad has more than one hundred seventy-five publications to his credit, including research papers, review articles & book chapters and five patents issued or pending, and edited or authored several books. Dr. Prasad has twelve years of teaching experience and has been awarded the Young Scientist Award & Prof. J.S. Datta Munshi Gold Medal by the International Society for Ecological Communications; FSAB fellowship by the Society for Applied Biotechnology; the American Cancer Society UICC International Fellowship for Beginning Investigators, USA; Outstanding Scientist Award in the field of Microbiology; BRICPL Science Investigator Award and Research Excellence

Award etc. He has been serving as editorial board members for several reputed journals and the series editor of Springer Nature's book series—Nanotechnology in Life Sciences. Previously, Dr. Prasad served as Assistant Professor Amity University Uttar Pradesh, India; Visiting Assistant Professor, Whiting School of Engineering, Department of Mechanical Engineering at Johns Hopkins University, Baltimore, United States and Research Associate Professor at School of Environmental Science and Engineering, Sun Yat-sen University, Guangzhou, China.

Contributors

Moulika Aerupula Centre for Biotechnology, Institute of Science & Technology, Jawaharlal Nehru Technological University Hyderabad, Hyderabad, Telangana, India

H. Abdul Jaffar Ali Department of Biotechnology, Islamia College (Autonomous), Vaniyambadi, Tamil Nadu, India

Célia G. Amorim LAQV-REQUIMTE/Departamento de Ciências Químicas, Faculdade de Farmácia, Universidade do Porto, Porto, Portugal

Alberto N. Araújo LAQV-REQUIMTE/Departamento de Ciências Químicas, Faculdade de Farmácia, Universidade do Porto, Porto, Portugal

S. Balasubramanian ICAR-Central Institute of Agricultural Engineering Regional Centre, Coimbatore, Tamil Nadu, India

Ramesh Balli Department of Genetics and Genomics, Yogi Vemana University, Kadapa, Andhra Pradesh, India

Surojit Bera Department of Microbiology, School of Bioengineering and Biosciences, Lovely Professional University, Phagwara, Punjab, India

Govindh Boddeti Department of H&S, Raghu Institute of Technology, Visakhapatnam, Andhra Pradesh, India

M. Subhosh Chandra Department of Microbiology, Yogi Vemana University, Kadapa, Andhra Pradesh, India

Jaeyeop Choi Industry 4.0 Convergence Bionics Engineering, Pukyong National University, Busan, Republic of Korea

Abhigyan Choudhury Department of Animal Science, Kazi Nazrul University, Asansol, West Bengal, India

Nabarun Chandra Das Department of Animal Science, Kazi Nazrul University, Asansol, West Bengal, India

B. S. Diwakar Department of Engineering Chemistry, SRKR Engineering College, Bhimavaram, India

Subhasish Dutta Department of Biotechnology, Haldia Institute of Technology, Haldia, West Bengal, India

Lizziane Kretli Winkelstroter Health Sciences Faculty and Master in Health Science, University of Western Sao Paulo, Sao Paulo, Brazil

Afreen fathima Centre for Biotechnology, Institute of Science & Technology, Jawaharlal Nehru Technological University Hyderabad, Hyderabad, Telangana, India

G. Flora Department of Botany, St. Mary's College (Autonomous), Thoothukudi, Tamil Nadu, India

Jaime Humberto flores Garcia Departamento de ciencias de la Enfermería, Universidad técnica de Manabí, Portoviejo, Ecuador

Gusdanis Campos Facultad de Ciencias de la Salud, Grupo de Investigación Ciencias de Laboratorio Clínico, Universidad Técnica de Manabí, Portoviejo, Ecuador

Madhureema Ghosh Department of Animal Science, Kazi Nazrul University, Asansol, West Bengal, India

Renato L. Gil LAQV-REQUIMTE/Departamento de Ciências Químicas, Faculdade de Farmácia, Universidade do Porto, Porto, Portugal

S. Gunasekaran Department of R&D, St. Peter's Institute of Higher Education and Research – SPIHER, Chennai, Tamil Nadu, India

Swapnil C. Kamble Department of Technology, Savitribai Phule Pune University, Pune, Maharashtra, India

Ajay Kumar Department of Microbiology, School of Bioengineering and Biosciences, Lovely Professional University, Phagwara, Punjab, India

Rathna Silviya Lodi International Genome Center, Jiangsu University, Zhenjiang, China

Arundathi Mesa Department of Genetics and Genomics, Yogi Vemana University, Kadapa, Andhra Pradesh, India

Rajkrishna Mondal Department of Biotechnology, Nagaland University, Dimapur, Nagaland, India

Sudip Mondal Department of Biomedical Engineering, Pukyong National University, Busan, Republic of Korea

Maria C. B. S. M. Montenegro LAQV-REQUIMTE/Departamento de Ciências Químicas, Faculdade de Farmácia, Universidade do Porto, Porto, Portugal

Suprabhat Mukherjee Department of Animal Science, Kazi Nazrul University, Asansol, West Bengal, India

Grace Sugandha Sowjanya Mythatha Department of Genetics and Genomics, Yogi Vemana University, Kadapa, Andhra Pradesh, India

Lauren Vila Naldi Health Sciences Faculty, University of Western Sao Paulo, Sao Paulo, Brazil

Junghwan Oh Department of Biomedical Engineering, Pukyong National University, Busan, Republic of Korea

Karolinny Cristiny de Oliveira Vieira Master in Health Science, University of Western Sao Paulo, Sao Paulo, Brazil

Patricia Durán Facultad de Ciencias de la Salud, Grupo de Investigación de Ciencias Visuales y Optométricas. Universidad Técnica de Manabí, Portoviejo, Ecuador

Sumin Park Industry 4.0 Convergence Bionics Engineering, Pukyong National University, Busan, Republic of Korea

Ritwik Patra Department of Animal Science, Kazi Nazrul University, Asansol, West Bengal, India

Valéria Cataneli Pereira Health Sciences Faculty and Master in Health Science, University of Western Sao Paulo, Sao Paulo, Brazil

Guhankumar Ponnusamy Qatar Environment and Energy Research Institute, Qatar Foundation, Doha, Qatar

Amit Pratush Arboreal Bio-Innovations Pvt. Ltd., Lucknow, Uttar Pradesh, India

Bellamkonda Ramesh Department of Food Technology, Vikrama Simhapuri University, Nellore, Andhra Pradesh, India

Ranjit Pabbati Centre for Biotechnology, Institute of Science & Technology, Jawaharlal Nehru Technological University Hyderabad, Hyderabad, Telangana, India

Sudheer Ravuri Center for Regenerative Sports Medicine (CRSM), Steadman Philippon Research Institute (SPRI), Vail, CO, USA

Venkateswar Reddy Kondakindi Centre for Biotechnology, Institute of Science & Technology, Jawaharlal Nehru Technological University Hyderabad, Hyderabad, Telangana, India

Venu Reddy Department of Engineering Chemistry, SRKR Engineering College, Bhimavaram, India

Bishnupada Roy Department of Chemistry, Visva-Bharati University, Santiniketan, West Bengal, India

Joan Manuel Rodríguez-Díaz Laboratorio de Análisis Químicos y Biotecnológicos, Instituto de Investigación, Universidad Técnica de Manabí, Portoviejo, Ecuador

Joyita Sarkar Institute of Chemical Technology Marathwada Campus, Jalna, Maharashtra, India

Jayaprakash Saththasivam Qatar Environment and Energy Research Institute, Qatar Foundation, Doha, Qatar

K. SenthilKannan Department of R&D, Edayathangudy G S Pillay Arts and Science College (Autonomous), Nagapattinam, India
Department of Physics, Edayathangudy G S Pillay Arts and Science College (Autonomous), Nagapattinam, India

Firdoz Shaik Schulich Faculty of Chemistry, Technion-Israel Institute of Technology, Haifa, Israel

Farhan F. Shaikh Institute of Chemical Technology Marathwada Campus, Jalna, Maharashtra, India

Silpa Somavarapu Department of Food Technology, Vikrama Simhapuri University, Nellore, Andhra Pradesh, India

Gabrielle Messias Souza Health Sciences Faculty, University of Western Sao Paulo, Sao Paulo, Brazil

Andrea Villegas Grupo de investigación en lentes de contacto, córnea y superficie ocular “Miguel Refojo”, Universidad de Valencia, Valencia, Spain

Ch. Venkatrayulu Department of Food Technology, Vikrama Simhapuri University, Nellore, Andhra Pradesh, India

M. Vimalan Department of Physics, Thirumalai Engineering College, Kanchipuram, Tamil Nadu, India

Chapter 1
Green Synthesized Nanoparticles as a Promising Strategy for Controlling Microbial Biofilm

Gabrielle Messias Souza, Karolinny Cristiny de Oliveira Vieira, Lauren Vila Naldi, Valéria Cataneli Pereira, and Lizziane Kretli Winkelstroter

Abstract Microbial biofilms are communities of cells adhered to a surface embedded with a matrix of polymeric extracellular substances. The biofilm can present one or more species of microorganisms, depending on its duration and location. It may involve Gram-positive, Gram-negative bacteria and yeast such as *Candida albicans*, Coagulase negative Staphylococcus, *Enterococcus* spp., *Klebsiela pneumoniae*, *Pseudomonas aeruginosa* and *Staphylococccus aureus*. In addition to bacteria, other cellular elements may be aggregated in the biofilm, such as platelets, for example, when the biofilm is installed on a surface bathed in blood. In the health area, there are a variety of possible situations that allow biofilm formation in medical devices directly connected to the patient or indirectly by contamination of the machines or pipes of that environment. Biofilm formation has been linked to 60% of hospital infections. In this way, components and methods that can inhibit the biofilm formation or even dissolve it have been investigated. Nanobiotechnology is an area of nanotechnology related to the creation, use and improvement of nanostructures in biotechnological processes. Among the various research fields in this area is the synthesis, characterization and application of nanoparticles with different sizes, shapes and chemical compositions. The traditional methods used for the synthesis of these nanoparticles are generally chemical methods in which toxic solvents are used and the generation of dangerous by-products can occur and involve high energy consumption. Due to these factors, there is an increasing need to develop non-toxic and environmentally friendly procedures; but with a high yield and low cost. In this context, the routes of synthesis of nanoparticles by biological

G. M. Souza · L. V. Naldi
Health Sciences Faculty, University of Western Sao Paulo, Sao Paulo, Brazil

K. C. de Oliveira Vieira
Master in Health Science, University of Western Sao Paulo, Sao Paulo, Brazil

V. C. Pereira · L. K. Winkelstroter (✉)
Health Sciences Faculty and Master in Health Science, University of Western Sao Paulo, Sao Paulo, Brazil

N. R. Maddela et al. (eds.), *Nanotechnology for Advances in Medical Microbiology*, Environmental and Microbial Biotechnology,
https://doi.org/10.1007/978-981-15-9916-3_1

systems, also known as green synthesis or biosynthesis, become quite relevant. This chapter addresses the perspectives by which green synthesis nanoparticles can be integrated as an effective method of control and prevention of microbial biofilms.

Keywords Infection · Prevention · Control · Nanotechnology

1.1 Introduction

Microbial infections are among the main public health concerns, as various bacteria, yeasts, fungi, and other pathogenic microorganisms may be resistant to drugs (Teixeira et al. 2020). Furthermore, the formation of biofilms represents a major challenge in this sense, as it makes the control of these microorganisms even more difficult (Winkelströter et al. 2016; Barzegari et al. 2020).

Thus, the development of new and effective antimicrobial agents is required, especially considering the current scenario. A promising strategy is the application of nanotechnology. The distinctive properties of the nanoscale confer impressive antimicrobial capabilities to nanomaterials which should be further explored (Reshma et al. 2017; Inamuddin et al. 2021).

Nanotechnology can be particularly advantageous in the treatment of bacterial infections. Examples include the use of nanoparticles (NPs) in antibacterial coatings for medical implants and devices to prevent infections and promote wound healing, in antibiotic delivery and bacterial detection systems, and in antibacterial vaccines (Wang et al. 2017).

The "green" synthesis of nanoparticles is based on the use of plant metabolites and natural substances. Its use is particularly rewarding as it is considered a clean, low-cost method, in addition to presenting high safety and stability (Gour and Jain 2019; Prasad 2014; Prasad et al. 2018; Srivastava et al. 2021). However, an important issue associated with the use of nanomaterials is the concern about side effects *in vivo*. Therefore, in-depth knowledge of biocompatible nanostructures intended for antimicrobial therapy is needed (Reshma et al. 2017). This chapter addresses the perspectives by which green synthesis NPs can be integrated as an effective method of control and prevention of microbial biofilms.

1.2 Microbial Biofilms

A biofilm is defined as a group of free microorganisms adhered to a surface and encrusted in a polymeric matrix produced by themselves (Winkelströter et al. 2016; Barzegari et al. 2020). Biofilms can be constituted by microorganisms of a single species or multiple species, with the formation from two or more populations of bacteria and/or fungi, yeasts, and protozoa being more common in nature (Lohse et al. 2018; Boltz et al. 2017).

Among the bacteria most commonly involved in the formation of biofilms are: *Staphylococcus epidermidis*, *Pseudomonas aeruginosa*, *Staphylococcus aureus*, *Klebsiella pneumoniae*, *Enterococcus faecalis*, *Streptococcus viridans*, *Escherichia coli*, *Proteus mirabilis*, *Methicillin* resistant *staphylococcus aureus* (MRSA), *Streptococcus mutans*, and *Gardnerella vaginalis* (Barzegari et al. 2020; Ghosh et al. 2020). Among the fungi, the following are common; *Aspergillus fumigatus* and *Candida albicans* (Sheppard and Howell 2016; Chevalier et al. 2018; Kean and Ramage 2019).

The presence of the extracellular matrix in the biofilm is a major factor in its structure, in addition to enabling the survival and protection of these microorganisms against various adverse conditions (Limoli et al. 2015; Yan and Bassler 2019). The matrix composed of extracellular polymeric substances (EPS), among which we can mention polysaccharides, proteins, and extracellular DNA (Dragoš and Kovács 2017; Yan and Bassler 2019).

The synthesis of polysaccharides can be influenced by the environment and although there are variations in their composition. Polysaccharides are responsible for the aggregation, protection, nutrition, and architecture of the biofilm (Limoli et al. 2015; Neu and Lawrence 2017).

Extracellular proteins present in biofilms are classified into secreted extracellular proteins, extracellular enzymes, and cell surface adhesins (Neu and Lawrence 2017). An example is the protein associated with biofilm (Bap), an important protein found on the surface of the film and directly related to the role of binding, migration, and formation capacity of biofilms (Colagiorgi et al. 2016; Neu and Lawrence 2017).

Extracellular DNA is produced by active secretion or controlled cell lysis. This is an element of the matrix with a significant role in the maturation of the biofilm, gene transfer, and when together with the other components, enables construction, structural integrity, and protection (Colagiorgi et al. 2016; Neu and Lawrence 2017; Kavanaugh et al. 2019).

The formation of biofilm is carried out in several phases such as initial fixation, irreversible fixation, initial formation of microcolonies, maturation, and dispersion of microorganisms, as shown in Fig. 1.1 (Winkelströter et al. 2016; Barzegari et al. 2020). The cycle of formation of a biofilm begins with the adhesion of single and/or clustered cells to a surface, where the expression of the planktonic state genes changes to the sessile state (Winkelströter et al. 2016; Gordon et al. 2017).

The reversible initial fixation is facilitated by the presence of organic residues on the surface where the biofilm will be installed. This fact is dependent on van der Waal's forces that favor interaction with the EPS. On the other hand, irreversible adhesion occurs through the interaction of extracellular substances in the polymeric matrix when adhering to microcolonies (Winkelströter et al. 2016; van Wolferen et al. 2018).

After adhering to the surface, the microorganisms bind to each other and form a monolayer biofilm composed of single cells or a multilayer biofilm formed by aggregated cells. At this stage of maturation, the structure is developed and characterized by a complex 3D architecture that favors biofilm nutrition (Winkelströter et al. 2016; van Wolferen et al. 2018).

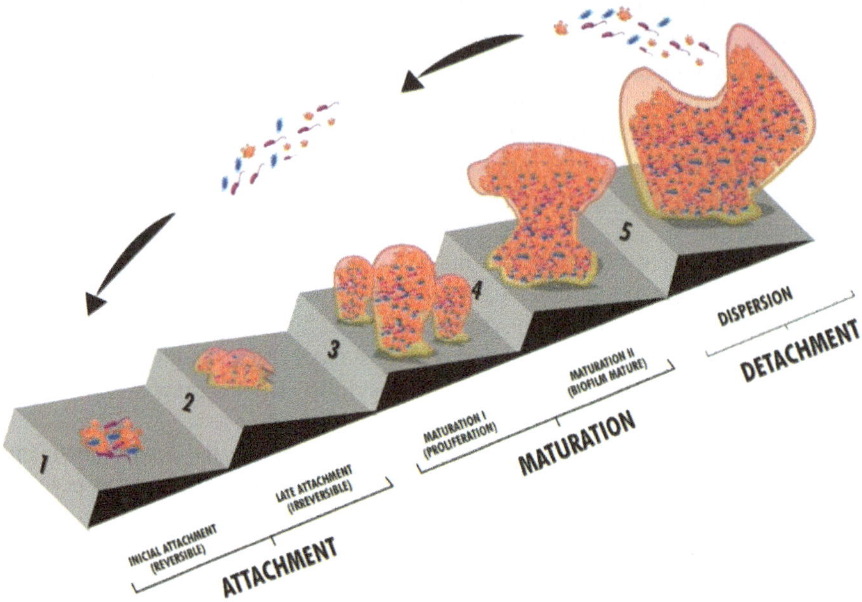

Fig. 1.1 Biofilm formation cycle with the stages of adhesion, maturation, and dispersion

The final stage is characterized by the dispersion process where microbial cells in an isolated or aggregate form detach themselves from the mature biofilm and colonize other sites to form a new biofilm. This process is mainly due to nutrient deprivation, environmental changes, and the *quorum-sensing* (QS) system between bacteria (Gordon et al. 2017; van Wolferen et al. 2018).

Some aspects can influence this process, such as environmental factors, microbial species, temperature, culture medium, the type of adhesion surface, regulation of genes to guarantee the development and maintenance of cells, architecture, and the extracellular matrix. In this way, single methods cannot be used to study biofilms (Winkelströter et al. 2016).

Biofilms can be present in several niches. The presence of microbial biofilms has been observed on biotic surfaces (natural aquatic, plant and mammalian tissues) and abiotic surfaces (artificial devices, implanted hospital devices, and biomaterials) (Boltz et al. 2017; Lohse et al. 2018).

In the clinical context, biofilms are related to contamination and systemic infections, mainly carried by contaminated implants and catheters. According to the National Institutes of Health (NIH), approximately 65% to 80% of cases of bacterial infections in humans may be related to the presence of microbial biofilms (Barzegari et al. 2020; Ghosh et al. 2020). In addition, they are also present as dental plaque, and their pathogenesis can cause cavities and gum disease (Tsui et al. 2016; Boltz et al. 2017; Daubert and Weinstein 2019). In the environment, they can be installed on the internal walls of pipes and tubes, resulting in corrosion and damage, besides generating a bad taste and smell in drinking water (Boltz et al. 2017; Di Pippo et al. 2018).

The presence of biofilms also impacts the food sector, as microbial communities aggregate on surfaces such as plastic, wood, metal, glass, soil fragments, and food products, which motivates contamination of the raw material, economic loss, short validity periods of goods, and transmission of diseases through food (Winkelströter et al. 2014; Winkelströter and Martinis 2015).

Although biofilms are most often considered pathogenic, they can be beneficial. Many of them are part of plant, animal, and human microbiomes, generating benefits in agriculture, and biotechnological applications to be used in industries, contributing to the production of fermented foods and wastewater treatment (Winkelströter et al. 2014; Velmourougane et al. 2019; Yan and Bassler 2019).

In biofilms, mutations in the genes and reduced diffusion of substances can occur, increasing their resistance to antimicrobials and making eradication more difficult (Tsui et al. 2016; Yan and Bassler 2019). Because of this, new alternatives have emerged to combat and prevent biofilms, such as plant extracts, QS inhibitory substances, industrial chemical disinfectants, probiotic products, and nanoparticles (Malafaia et al. 2018; Barzegari et al. 2020).

1.3 Methods to Prevent and Control Biofilms

The damages caused by the presence of biofilms have stimulated the research and development of new methods of elimination and prevention of this active form of life of microorganisms. In order to avoid the formation of biofilms, the following critical points must be taken into account: (1) prevent the initial connection of the microorganisms to the biofilm forming surfaces, (2) interruption of the maturation process, and (3) interference in the microorganism communication system—the QS system (Galié et al. 2018; Subhadra et al. 2018).

Some biofilm control methods act directly on these points, such as: modifications in the surface of biomaterials, such as nanoparticles with different metal oxides, nanocomposites, antimicrobial polymers, hydrogels, or liposomes; inhibition of cell signaling with lactic and citric acids; chemical treatments, such as ozone, quaternary ammonium compounds, NaOCl, and other disinfectants; enzymatic disruption strategies, such as cellulases, proteases, glycosidases, and DNAses; non-thermal plasma treatments; use of thermal processing; use of bacteriophages, such as P100; bacteriocins, such as nisin; biosurfactants, such as lichenisin or surfactin; and essential oils from plants, such as oils containing citral or carvacrol (Zabielska et al. 2016; Galié et al. 2018; Subhadra et al. 2018).

A more detailed view of the main methods of control and prevention will be detailed in the following topics:

1.3.1 *Chemical Treatments*

1.3.1.1 Disinfectants and Sanitizers

These components have a wide spectrum and the most commonly used are hydrogen peroxide, quaternary ammonium compounds, and chlorine-based products. They act both in the matrix and in the microorganisms present in biofilms, acting as important anti-biofilm agents. Although some mechanisms of action are suggested, further investigation is needed on potential toxic effects. Sodium hypochlorite is believed to act directly on the proteins that make up the biofilm matrix and inhibit the main enzymatic functions of microorganisms. Hydrogen peroxide is an oxidizing disinfectant responsible for producing free radicals that come into contact with the biofilm structures. The quaternary ammonium compounds are soluble in water and have a positive charge. Due to these characteristics, they influence the stability of the cell membrane of microorganisms, which can result in lysis (Strempel et al. 2015; Galié et al. 2018; Lineback et al. 2018).

1.3.1.2 Essential Oils

Essential oils (EOs) are volatile substances from plants that may have anti-biofilm, antibacterial, and preservative properties. Studies suggest that the antimicrobial potential of EOs is related to their action on the cell wall of microorganisms. These are a safe option compared to chemical antimicrobial agents. Several substances already have a proven anti-biofilm action against various microorganisms such as monoterpenoids (borneol, camphor, carvacrol, eucalyptol, limonene, pinene, thujone), sesquiterpenoids (caryophyllene, humulene), flavonoids (cinnamaldehyde), and other phenolic compounds. Several EOs are responsible for the regulation of the QS system, which results in the alteration of the expression of various virulence factors, including those related to the maintenance of the biofilm. EOs are recognized for inactivating microorganisms without resistance, in addition to presenting low toxicity and rapid degradation in the environment. A large number of essential oils are available (Galié et al. 2018; Correa et al. 2019; Shahid et al. 2019; Wang et al. 2019).

1.3.1.3 Bacteriocins

Bacteriocins are peptides synthesized by bacterial ribosomes in order to inhibit or eliminate other microorganisms (Strempel et al. 2015).

Bacteriocins have great relevance in the preservation of food and in the fight against pathogenic microorganisms. The bactericidal actions of bacteriocins involve the direct death of bacteria by cell membrane lysis and interruption of cellular processes, such as DNA replication, transcription, protein biosynthesis, and folding or impairment of protein functions and immunomodulatory effects, causing the stimulation of non-inflammatory immune responses in the host. Anti-biofilm effects have also been reported and include the inhibition of bacterial adhesion to surfaces, preventing the formation of new biofilms and the interference of pre-existing biofilms, in addition to great penetration capacity of formed biofilms (Strempel et al. 2015; Mathur et al. 2018; Chikindas et al. 2020).

1.3.1.4 Phagotherapy

Bacteriophages are viruses that infect bacteria. Their composition is formed by a protein capsule that surrounds a DNA genome and their structure contains several devices to contaminate the host cell (Shaffer 2019).

The use of bacteriophages is an alternative to avoid resistance to antimicrobials. The use of phages can control the biofilm EPS matrix and results in the control of biofilm proliferation. Although some bacteriophages produce exopolysaccharide depolymerases, others have low potential for access to the biofilm and demonstrate difficulty reaching bacterial cells within the biofilm (Galié et al. 2018; Shahid et al. 2019).

1.3.1.5 Enzymes

Enzymes are an important tool for biofilm control, since they have low toxicity and are biodegradable. Enzymes are currently used in conjunction with detergents (Galié et al. 2018).

Regarding the composition of the biofilm structure, organic macromolecules (proteins and polysaccharides) characterize the main target components for the action of enzymes. Thus, proteases (serine proteases, proteinase K, pepsin, and trypsin) and glycosides (amylases, dextranase, and pectinase) are the main choices for biofilm removal. Pectin methylesterase is an enzyme that reduces biofilm formation in bioreactors. Other enzymes, such as amylases, cellulases, lyases, glycosides, and DNAses, are used in industrial detergents to remove biofilms (Wang et al. 2016; Galié et al. 2018).

The association of enzymes with different methods, as with other chemical (disinfectant) or physical (ultrasound) treatments, improved the removal of biofilm (Meireles et al. 2016).

1.3.1.6 *Quorum-sensing* Inhibitors (QS)

Multiple signaling pathways are essential for biofilm formation. Among them, QS and cyclic di-GMP (cGMP) signaling are the best defined (Galié et al. 2018; Subhadra et al. 2018; Shahid et al. 2019; Wang et al. 2019).

The biosynthesis of extracellular polymeric substances and reduction in bacterial motility are caused by the high intracellular content of cGMP, favoring the formation of biofilm. Many molecules can disrupt cGMP biosynthesis, such as terpenoid saponin, compounds generated from nitric oxide, azathioprine, or sRNAs (Galié et al. 2018).

Quorum-sensing inhibitors (QSI) have been proposed as a new branch of antimicrobial agents. Several methods are applied to generate interference in bacterial *quorum-sensing* through inhibiting cell-to-cell communication, encompassing the competitive binding of inhibitors to QS receptors, enzymatic degradation of QS signals, post-transcriptional control of QS genes via sRNAs, and inhibition of biosynthesis of the QS signals (Galié et al. 2018; Shahid et al. 2019).

1.3.2 Physical Treatments

1.3.2.1 Heat

Thermal treatment of bacterial biofilm can be applied through heating, representing a non-invasive and non-antibiotic treatment category (Pijls et al. 2020).

Thermal shock is generally used to inactivate bacteria, making it a viable and efficient approach for inhibiting biofilm at accessible temperatures, such as 80 °C. This method can reduce the population density of biofilm by 5 orders of magnitude. The shock significantly inactivates bacteria in biofilms and inhibits the multiplication of microorganisms, in addition to being responsible for damaging bacterial cell membranes, making biofilms more vulnerable and significantly decreasing biomass and biovolume (Pijls et al. 2020).

1.3.2.2 Radiation

In the living organism, ionizing radiation damages the cell's DNA. Cell death is predominantly driven by double-stranded DNA breaks, separated by base pairs that are not repaired by the cell. Regarding biofilm control and prevention, there are several types of radiation, such as ultraviolet rays, gamma radiation, and ultrasound irradiation (Marjani and Khadam 2016). It is suggested that gamma irradiation can weaken the intermolecular interaction of lipopolysaccharide components, disrupt the biofilm structure, leaving it permeable to antimicrobials (Marjani and Khadam 2016).

1.3.3 Genetic Manipulation

Methods with molecular biology tools are becoming increasingly used to control biofilms. CRISPR/Cas9 technology provides modifications in specific locations of selected genes in a main step, leading to deletions, insertions, or transformations in the sequence. Gene inactivation is a powerful method for functional studies. When deletion of genes responsible for structuring the biofilm occurs, the exopolysaccharide matrix exhibits irregular density and discontinuities, demonstrating that the biofilm formation is impaired with modified physical properties (Noirot-Gros et al. 2019; Leonova and Gainetdinov 2020).

Although there are numerous methods for the control and prevention of biofilm, no method can be considered effective for all forms of existing biofilms. In view of this, it is essential that new strategies, such as the use of nanoparticles, be researched. The table below presents the advantages and disadvantages of the main methods of biofilm control and prevention described in the literature (Table 1.1).

1.4 Anti-biofilm Effect of Nanoparticles

Nanoparticles (NPs) are atomic aggregates with a diameter between 1 and 100 nm, a characteristic that guarantees a large surface area and high reactivity. Currently, NPs play an important role in the production of materials from different areas, with emphasis on the pharmaceutical and medical industry (Prasad 2014; Prasad et al. 2016). Their wide application is the result of properties related to their size, which provides a large surface area, favoring their access. Their chemical and physical properties can be improved through combination with biomolecules, a fact that provides greater stability and improves biological properties compared to their native state (Baudrimont et al. 2018; Iriarte-Mesa et al. 2020; Mylona et al. 2020).

Currently, NPs play an important role in the medical field, mainly in oncology. In addition, their potential use in combating microbial infections has been investigated. The antimicrobial effect of NPs and their ability to inhibit the formation of biofilms enables greater therapeutic efficacy, less side effects, and greater adherence to treatment (Meeker et al. 2018; Sengan et al. 2019; de Mélo Silva et al. 2020; Prasad et al. 2020).

1.4.1 Classification of Nanoparticles

NPs can be classified according to their dimensions, shapes, sizes, and the material used in their development. Their dimension is considered to be zero when the length, width, and height are fixed in a single point. They can take on a cylindrical, spherical, tubular, or spiral shape, among others, and may even be irregular. Sizes

Table 1.1 Advantages and disadvantages of the main methods of control and prevention of biofilms

Types of biofilm prevention	Methods of biofilm prevention	Advantages	Disadvantages	References
	Disinfectants	High efficiency in premature biofilms; no side or toxic effects	Resistance, limited effectiveness; high dose	Strempel et al. (2015), Galié et al. (2018) and Lineback et al. (2018)
Chemical treatment	Phagotherapy	One of the most useful anti-biofilm agents; simple and fast isolation; low cost; ecologically friendly; does not disturb the normal microbiota	Low effectiveness; release of a considerable amount of endotoxins; reduced number of phages encoding toxins; insufficient pharmacokinetic data; resistance	Galié et al. (2018), Shaffer (2019) and Shahid et al. (2019)
	Bacteriocins	High efficacy in inhibiting biofilm formation; alternative route to antimicrobials; low toxicity; ease of penetration into the biofilm; more effective than antimicrobials	Reduced efficacy in biofilms already formed; high cost; it can become ineffective as it is of protein nature; lack of information on mechanism of action	Strempel et al. (2015), Mathur et al. (2018) and Chikindas et al. (2020)
	Essential oil	Good efficacy in inhibiting biofilm formation; low cost; low toxicity; rapid degradation in the environment; high availability; low risk of resistance	Little or no ability to inhibit pre-formed biofilm	Galié et al. (2018), Correa et al. (2019), Shahid et al. (2019) and Wang et al. (2019)
	Enzymes	Biodegradable; low toxicity	High cost due to patent protection	Wang et al. (2016), Meireles et al. (2016) and Galié et al. (2018)
	Quorum-sensing inhibitors	No resistance; decreased resistance to multiple drugs; high effectiveness; obtained from natural sources	Varying potential	Galié et al. (2018), Subhadra et al. (2018), Shahid et al. (2019) and Wang et al. (2019)
Physical treatment	Heat	Easy to perform; low cost; does not damage the	Risk of resistance	Pijls et al. (2020)

(continued)

Table 1.1 (continued)

Types of biofilm prevention	Methods of biofilm prevention	Advantages	Disadvantages	References
		environment; high effectiveness; greater vulnerability; non-invasive technique		
	Radiation	Inhibition of biofilm formation; molecule passage facilitated; innovative method	Gene expression does not present a significant decrease; low effectiveness	Marjani and Khadam (2016)
Genetic manipulation	CRISPR Technique	EPS formation drastically affected; structure of the biofilm compromised	A lot of work; limited; high cost	Noirot-Gros et al. (2019) and Leonova and Gainetdinov (2020)

can vary from 1 nm to 100 nm, and the constituent material will determine the characteristic of an inorganic or organic nanoparticle (Ealia and Saravanakumar 2017).

1.4.1.1 Organic Nanoparticles

Organic NPs are biodegradable and non-toxic, and can be composed of biopolymers, liposomes, micelles, chitosan, lignin, among other biocomponents. These nanoparticles are widely used in the biomedical area, mainly in medication administration systems for ensuring their targeted distribution. There is currently a wide variety of edible organic NPs prepared from food-derived ingredients such as polysaccharides, lipids, and proteins (Ealia and Saravanakumar 2017; Azeredo et al. 2019).

1.4.1.2 Inorganic Nanoparticles

Inorganic NPs are made up of metals or metal oxides such as gold (Au), silver (Ag), copper (Cu), and zinc (Zn). All metallic components can be used to form a nanoparticle, using destructive or constructive methods. Therefore, metallic NPs can have different properties in relation to their size, high surface to volume ratio, and shape that varies from spherical to cylindrical. Metal oxide nanoparticles demonstrate greater reactivity and efficiency when compared to metallic nanoparticles, with aluminum oxide (Al_2O_3), iron oxide (Fe_2O_3), titanium oxide (TiO_2), and zinc (ZnO) being the most commonly used (Ealia and Saravanakumar 2017; Kumari et al. 2020).

Silver Nanoparticles

Silver nanoparticles (AgNPs) are the most commonly applied; it is estimated that their global production is greater than 500 t. They have wide applications, ranging from electronic food products to health-related products. However, the fact that they can be released into the environment and the possibility of having a great impact on ecosystems and especially on human health are some of their disadvantages. Silver can induce fibrosis of the bronchial epithelial cells in addition to resulting in disorders of the intestinal microbiota (Azeredo et al. 2019; Liang et al. 2020; Mylona et al. 2020).

The Ag + ions present in nanoparticles larger than 10 nm have an antimicrobial function through electrostatic interaction with negatively charged bacterial membranes. However, those smaller than 10 nm, demonstrate activity through their internalization by bacterial cells resulting in the oxidation of Ag to Ag+. It is also believed that AgNPs lead to gene expression disorders, damage to mitochondrial function, destabilization of the bacterial cell membrane, and formation of reactive oxygen species. In biofilms, the inhibition is due to its high penetrating power, however, some elements of this structure such as genetic elements, membrane proteins, and an efflux pump are able to confer resistance to silver, resulting in inhibition failure (Gholamrezazadeh et al. 2018; Azeredo et al. 2019; Meza-Villezcas et al. 2019; Meier et al. 2020; Zou et al. 2020).

Copper Nanoparticles

Copper nanoparticles (CuNPs) are used in several applications such as household, medical, industrial, and environmental products (Yadav et al. 2017). Copper is widely used in the synthesis of nanoparticles due to its natural abundance and free elemental form. CuNPs are more potent than other metal oxide nanoparticles as they induce greater cytotoxicity resulting from DNA damage. The antimicrobial activity of CuNPs is also related to the damage to the bacterial cell wall, which generates high Cu++ efflux and creates a localized ionic effect (Sengan et al. 2019; Padmavathi et al. 2020). However, its synthesis is more complex, since copper is prone to rapid oxidation, a fact that reduces the effectiveness of its production (Lotha et al. 2019; Miao et al. 2019; Sengan et al. 2019).

Regarding the anti-biofilm activity, studies suggest that copper nanoparticles have the ability to remove established biofilms through the interaction of electrical charges with the exopolysaccharide matrix. Anti-biofilm activity may also be related to inhibition of QS and genes involved in biofilm formation as seen for the species *Methyl bacterium* spp. (Seo et al. 2018).

Gold Nanoparticles

Gold nanoparticles (AuNPs) have widespread use due to their optical-electronic properties, bio-stability, and catalytic and antimicrobial activity. This nanomaterial provides a versatile surface and can easily function as a binder for the surface receptors of target cells. It can also act in biomarking, nano diagnosis, vectorization of drug molecules, radiotherapy, and transmission electron microscopy (Boda et al. 2015; Ahmed et al. 2016; Feurtet-Mazel et al. 2016; Baudrimont et al. 2018; Khan et al. 2019).

Compared to other nanomaterials, AuNPs are highly inert, being considered non-toxic due to their low chemical reactivity. However, their toxicity may change according to their oxidation state, the composition of the nanomaterial, as well as the location or type of cell exposed to AuNPs (Boda et al. 2015).

The anti-biofilm effect of AuNPs is related to strong electrostatic interactions with negatively charged bacterial membranes and by the photothermal action, capable of inhibiting the proliferation and formation of biofilms. In addition, AuNPs have a low propensity to develop microbial resistance compared to antibiotics (Ahmed et al. 2016; Habimana et al. 2018; Lu et al. 2018).

Zinc Nanoparticles

Zinc nanoparticles (ZnONPs), have low synthesis cost, excellent biocompatibility, robustness, and resistance to both corrosion and oxidation, allowing wide use in products. Although, in low concentrations, ZnONPs have no toxicity to eukaryotic cells, in the environment their interaction with natural organic materials can alter their toxicity, stabilization, agglomeration, and dissolution (Vijayakumar et al. 2015; Al-Shabib et al. 2016; Khan et al. 2016; Ouyang et al. 2017).

ZnONPs exhibit strong protein adsorption properties, which modulate cytotoxicity, metabolism, and cellular responses. Thus, this nanocomponent has a potential inhibitory effect on cancer cells and Gram-positive bacteria and fungi (Al-Shabib et al. 2016; Mehta et al. 2019).

The antimicrobial effects of ZnONPs are associated with the size of the nanoparticle, since smaller ones can more easily penetrate the cell membrane and have a large surface area, increasing the degree of antimicrobial activity. Studies suggest that upon entering the cell, ZnONPs generate reactive oxygen species, inducing apoptosis through lipid peroxidation, in addition to deactivating proteins and causing structural changes in membranes and nucleic acids (Bhuyan et al. 2015). Their anti-biofilm potential is limited when faced with mature biofilms, being potentially insufficient to eradicate them (Abdulkareem et al. 2015; Gong et al. 2019; Lim et al. 2018; Mehta et al. 2019).

1.4.1.3 Carbon-based Nanoparticles

Carbon-based nanoparticles are used in various applications in everyday life, in industry, and in technologies with multiple-walled or single-walled nanotubes, being advantageous mainly in the electronic and optical areas. Fluorescent carbon nanoparticles are used as carbonaceous emitters with superior photoluminescence properties, and have been applied in areas such as bioimaging (Gorrochategui et al. 2017; Hu et al. 2020).

These NPs have the ability to interact with cell membranes, being able to penetrate them and thus interact with the internal components of the cell. Carbon nanotubes present properties that favor the antimicrobial effect, related to the ability to generate reactive oxygen species (Li et al. 2016; Gorrochategui et al. 2017; Seo et al. 2018).

1.4.2 Methods to Synthesize Nanoparticles

1.4.2.1 Conventional

NPs synthesized conventionally come from physical and chemical methods that involve chemical reduction, electrochemistry, laser waves, and lithography, responsible for generating monodisperse particles of homogeneous composition and morphology. Conventional syntheses require several preparation steps that interfere with large-scale production, requiring the use, in most cases, of potentially hazardous materials that may remain on the surface of the nanoparticle even after its purification process (Starsich et al. 2019; Foroohimanjili et al. 2020).

For conventional synthesis of metallic NPs, for example, chemical reducing agents such as sodium hydrochloride and hydrazine are used, which have the ability to reduce metal ions. Conventional synthesis of gold nanoparticles can be performed using the Turkevich or Brust-Schifrin method. In the Turkevich technique, AuNPs are obtained by reducing hydrogen tetrachloroaurate ($HAuCl_4$) by the agent trisodium citrate, generating nanoparticles that vary between 10 and 150 nm. In the Brust-Schifrin method, particles smaller than 10 nm are obtained, using sodium borohydride ($NaHB_4$) as the reducing agent and tetraoctylammonium bromide (TOAB) as a catalyst phase transfer, thus, the anion of the gold complex ($AuCl_{4-}$) is transferred from the aqueous to organic phase through electrostatic interactions with the positively charged TOAB (Iriarte-Mesa et al. 2020).

With the progressive influence of nanoparticles in the most diverse fields of application, they began to be produced on a large scale, thus, the production method became the focus of attention. Conventional synthesis techniques are no longer widely used owing to their high cost, use of toxic substances, and that large-scale production is impaired by the low stability and monodispersion of the produced nanoparticles (Ali et al. 2019; Yew et al. 2020).

1.4.2.2 Green Synthesis of Nanoparticles Using Plants, Microorganisms, and Biopolymers

The "green" synthesis of nanoparticles is an alternative method to conventional chemical and physical methods and is based on the use of biological systems, plant extracts, and biomass. Green synthesis does not require elaborate processes for ion reduction, as occurs, for example, in the development of silver and gold nanoparticles, since in this type of synthesis biomolecules are used. The biological components that line the surface of NPs during their green synthesis, act as ecologically correct precursors for their formation, this being an important differential for their application (Islam et al. 2019).

Biological methods for synthesis of NPs provide several benefits, since they are clean, low cost methods, in addition to presenting high safety, stability, and speed during the synthesis process. In this way, there is a reduction in the potential risks for both human health and the environment, in addition to presenting better properties in terms of size, shape, biocompatibility, and stability of the nano material (Al-Shabib et al. 2016; Khan et al. 2019; Mellinas et al. 2019).

There are several biological systems for "green" synthesis of NPs such as the use of biomolecules: enzymes, amino acids, vitamins, proteins, phenolic compounds, and alkaloids (Al-Shabib et al. 2016).

Plant extracts, rich in biomolecules, can be used for the development of metallic NPs, and act as reducing and stabilizing agents during the process (Prasad 2014, 2019a, b; Prasad et al. 2018). Bacteria can also be involved in biological synthesis, such as *Pseudomonas* spp. which allows the preparation of colloidal silver (Foroohimanjili et al. 2020; Iriarte-Mesa et al. 2020; Silveira et al. 2020). Biopolymers such as Levan have promising properties due to their ability to form nanostructures in water through self-assembly, resulting in a biodegradable polymeric micelle with high biocompatibility (González-Garcinuño et al. 2019).

The great challenge in the application of this nanomaterial in *in vivo* systems is related to the concentration of administered NPs. The large volume of blood reduces the concentration of nanoparticles and disintegrates their chains, releasing the encapsulated drug in an uncontrolled manner (González-Garcinuño et al. 2019).

1.4.3 Characterization Techniques and Properties

After the synthesis of the NPs, characterization aims to verify their morphology, surface area, size, distribution, and other important parameters of nanomaterial compliance. Knowledge of these characteristics is essential for the application of the nanoparticle, in addition to ensuring better efficiency in its synthesis. Some techniques can be used for this purpose such as: UV absorption spectroscopy (UV-vis), powder X-ray diffraction (XRD), dynamic light scattering (DLS), Fourier infrared transmission spectroscopy (FTIR), dispersive energy x-ray examination

(EDAX), scanning electron microscopy (SEM), and transmission electron microscopy (TEM) (Khade et al. 2015; Muthukrishnan et al. 2015; Chaudhuri and Malodia 2017).

Spectroscopic and diffractographic techniques are indirect methods used to analyze the composition, structure, and crystalline phase of NPs. UV-vis enables the study of the arrangement of NPs by measuring optical absorption, transmittance, and reflectance, which assist in calculating band intervals and determining the photoactivity and conductance of the material. The FTIR technique identifies functional or metabolic groups present on the surface of the NPs, responsible for their reduction and stabilization. DLS analysis can estimate the size distribution in addition to quantifying the surface loads of the nanoparticles. The purity, crystalline size, geometry, and phase orientation are determined by the XRD technique with the use of diffraction patterns present in crystallographic databases. The EDX technique separates characteristic x-rays from different elements in a given energy spectrum, being used to identify the elemental composition of, for example, metallic NPs (Santhoshkumar et al. 2017; Gour and Jain 2019; Khanna et al. 2019).

To determine the size, morphological characteristics, and topography of the NPs surface, advanced electron microscopy techniques are used. The SEM technique provides three-dimensional information for direct visualization of the product, with spatial resolution of 10 to 5 nm. In this technique, the sample is digitized by electron beams and its surface characteristics are obtained from secondary electrons, thus allowing data to be obtained regarding the surface morphology as well as the dispersion of the NPs in the mass or matrix. On the other hand, the TEM technique presents a two-dimensional image resulting from the interaction of an electron beam with the sample, with a resolution higher than SEM. Thus, this technique is widely used in determining the size, shape, and number of layers. The possible combination of the SEM technique with EDX enables important elementary analysis of the nanoparticle, this being important to provide information about the metals present (Yedurkar et al. 2016; Khanna et al. 2019).

1.4.4 Application of Green Synthesis of Nanoparticles in Microbial Biofilm

Several studies have reported the use of NPs produced by biological synthesis as an alternative for the control of biofilms, due not only to their ability to eliminate microorganisms present in the structure, but also to promote the inhibition of EPS secretion by biofilm-forming bacteria (Rajkumari et al. 2017; Hasan et al. 2019; Shanmuganathan et al. 2019).

It is believed that the use of biosynthesized NPs facilitates electrostatic interactions that result in the structural rupture of the biofilm matrix. Their reduced size allows penetration into microbial cell walls leading to loss of cell viability and alteration in the biofilm cell physiology. The use of NPs also ensures controlled

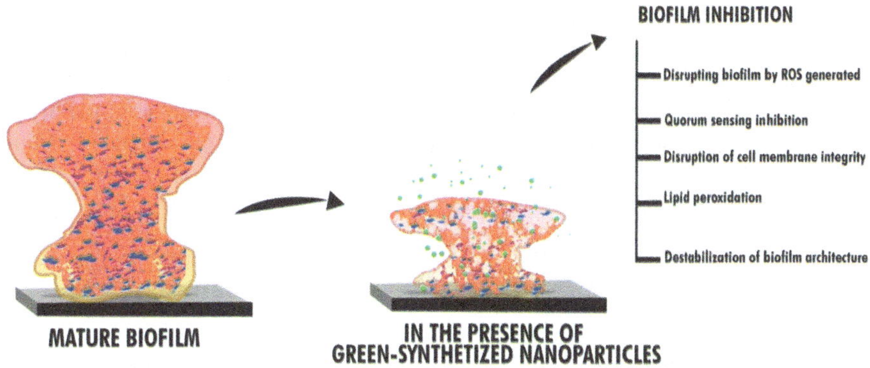

Fig. 1.2 Effects of the application of green synthesis of nanoparticles on microbial biofilms

release of the antimicrobial agent, presenting reduced toxicity and greater stability, thus also providing greater antimicrobial effects on the biofilm (Fig. 1.2) (Habimana et al. 2018; Singh et al. 2018a; Banerjee et al. 2019; Prasad et al. 2020).

The possibility of using medicinal plants and phytochemicals with antimicrobial effects on the synthesis of NPs contributes to the biofilm eradication. Many phytochemicals have the ability to alter the cell membrane and compromise the respiratory activity of cells, in addition to causing lipid peroxidation and the production of free radicals that cause deleterious effects on biofilms (Rajkumari et al. 2017; Rodríguez-Serrano et al. 2020; Ruddaraju et al. 2020).

Nanoparticles biosynthesized with polystyrene interact with polysaccharides from the biofilm matrix and through hydrophobic complexation lead to the rupture of the biofilm. Nanoparticles of biodegradable hydrogels are capable of reducing cell viability and act in the elimination of biomass, removing planktonic and sessile forms. NPs synthesized with biopolymers also act effectively in the eradication of biofilms since their polycationic groups have the ability to interact and break with the bacterial membrane resulting in the inhibition of biofilm formation. The mature biofilm can also undergo alterations caused by nanoparticles due to their elevated degree of penetration and high relationship between surface and volume (Liu et al. 2017; Banerjee et al. 2019).

In the studies of Rajkumari et al. (2017), dextrin was used for biological synthesis of silver particles, demonstrating a remarkable reduction of approximately 70% in the formation of biofilms by the pathogens *K. pneumoniae*, *P. aeruginosa*, *C. albicans,* and *S. aureus MRSA*. The study also evaluated a positive effect on the inhibition of EPS secretion, guaranteeing an important role of biosynthesized NPs in the face of infection by multi-resistant microorganisms that form biofilms. Ramalingam et al. (2019) used *avicennia marina* for biosynthesis of Fe_2O_3NPs, obtaining a percentage of 65% in the inhibition of biofilm formation at a concentration of 2 ppm for species of *P. aeruginosa* and 5 ppm for the pathogen *S. aureus*, while complete anti-biofilm activity was achieved in doses of 28 μg/mL for *E. coli* and 52 μg/mL for *P. aeruginosa*.

The great demand for innovative strategies to control and prevent the formation of biofilms has stimulated studies regarding the application of nanoparticles. The effective action of nanocomponents is remarkable due to their physical and chemical characteristics. However, improvement in some techniques is necessary, mainly in relation to large-scale production and knowledge of their mechanism of action (Zacchino et al. 2017; Wolferen et al. 2018; Shanmuganathan et al. 2019).

1.5 Other Applications of Green Synthesis of Nanoparticles

In addition to the application as an antimicrobial and anti-biofilm agent, the "green" synthesis of NPs has been directed to other applications, such as antioxidant, anti-cancer, and anticoagulant, among others (Qais et al. 2020). Rehman et al. (2020) performed the synthesis of AgNPs and titanium oxide (TiO_2) from fungi of the species *Fomes fomentarius*, and observed good antibacterial action against *S. aureus* and *E. coli*, in addition to verifying the anti-cancer activity of these particles. In human colorectal carcinoma cells, strong cytotoxic effects and cell death were observed after treatment with concentrations below 0.5 μg/mL of both studied nanoparticles, with no damage observed to the control cells. Although the mechanism that induces the apoptosis of cancer cells by these NPs is not well understood, it is suggested that AgNPs activate the apoptotic pathway with the production of oxygen free radicals, which results in anti-tumor and anti-proliferative effects (Barabadi et al. 2017; Aziz et al. 2019).

Hosseinzadeh et al. (2020), synthesized AuNPs from essential oil of *ferula persica*, which showed greater cytotoxic, apoptotic, and antiproliferative effects in colon cancer CT26 cells when compared to *Vero* cells. The morphological difference between cancerous and normal cell membranes, in terms of pore size, can result in a noticeable difference in the toxicity of NPs in cancer and normal cells, underscoring the importance of studying NPs in anti-cancer applications (Basavaraja et al. 2008).

AuNPs and AgNPs from *Olax nana* extract also presented anti-cancer activity, and the authors emphasize that the use of these same nanoparticles in imaging is promising, as it allows better visualization and outlining of the tumor design (Ovais et al. 2018). Nanoparticles can be phagocyted by cells of the liver, spleen, lungs, and bone marrow, in addition to being detected in the blood or lymphatic circulation, which can improve the diagnosis of the tumor (Brigger et al. 2012).

The NPs obtained by Ovais et al. (2018) also presented significant antibacterial and anti-leishmania activity, by presenting inhibition of amastigotes and promastigotes of *Leishmania tropica*. The use of nanotechnology in patients with leishmaniasis is promising, both in treatment and in prevention with the use of vaccines.

In addition to the antiparasitic action, the insecticidal potential of nanoparticles has also been studied (Kumar et al. 2020; Bhattacharyya et al. 2016). Chitra et al. (2015) used AgNPs synthesized from leaf extracts of *Mukia maderaspatana* and verified effective larvicidal activity against *Aedes aegypti* and *Culex quinquefasciatus* (Buhroo et al. 2017). Hajra et al. (2016) found a high mortality rate of *Aedes albopictus* when subjected to concentrations of cadmium NPs (CdNPs) synthesized from clove petal extract. A study developed by Arokiyaraj et al. (2015) presented insecticidal potential against *A. albopictusde* AgNPs obtained from *Chryasanthemum indicum* L. extract, emphasizing the importance of nanoparticles in the elimination of mosquitoes that are vectors of viruses and parasites that cause diseases in humans.

The catalytic properties of NPs are currently well described (Wang et al. 2020; Kumar et al. 2020). 4-Nitrophenol and its derivatives are considered compounds of great risk to the environment, since their use in the production of herbicides, insecticides, and synthetic dyes can harm the ecosystem as a wastewater pollutant, so that a reduction is necessary (Singh et al. 2018b). Knowing that sodium borohydride ($NaBH_4$) is used as a reducer of 4-Nitrophenol to 4-Aminophenol, Nayan et al. (2018) demonstrated a high catalytic capacity of these substances using AuNPs synthesized from *Mangifera indica*.

Another industrial pollutant of great importance is methylene blue dye, since, if ingested it can restrict oxidase enzymes, leading to central nervous system toxicity, gastrointestinal infections, and discoloration of the brain parenchyma (Kumar et al. 2020). Varadavenkatesan et al. (2019) verified the catalytic activity of AgNPs synthesized from *Ipomoea digitata* flower extract in reducing methylene blue dye, using $NaBH_4$ as a reducing agent. The results confirmed the catalytic activity of the nanoparticles, with a reduction of methylene blue in 15 minutes, indicating good perspectives of AgNPs for environmental remediation (Kumar et al. 2020).

The green synthesis of NPs is also being studied for application in food preservation. Lignocellulose is one of the most commonly used materials for packaging food products and beverages and many studies aim to modify the surface of this material for better food preservation. In studies by Bumbudsanpharoke and Ko (2018), the unbleached "kraft" (UBK) pulp of lignin and hemicellulose was used for the "green" synthesis of AuNPs, which were deposited on the surface of the UBK fiber, promoting the elimination of radicals and an antioxidant effect.

Other models of "green" nanoparticle synthesis and their applications are shown in Table 1.2. Given the various applications of NPs, the green synthesis approach opens alternative paths to the use of chemicals that are toxic to man and harmful to the environment, representing a field of study that has been extensively explored in the last decade and with a promising future.

1.6 Conclusion

The "green" synthesis of nanoparticles has been a highly attractive area of research for the past decade. Nanoparticles have been applied extensively in important areas, such as the pharmaceutical, food, and cosmetic industries. In this chapter, several applications were demonstrated, including as antimicrobial and anti-biofilm agents. However, research should be encouraged in order to observe the action of

Table 1.2 Nanoparticles obtained from green synthesis and their applications

NPs	Source	Activity	References
Ag	*Allamanda cathartica*	Antioxidant and antibacterial	Karunakaran et al. (2016)
	Tagetes erecta	Antibacterial	Padalia et al. (2015)
	Chrysanthemum indicum L.	Larvicidal and pupicidal	Arokiyaraj et al. (2015)
	Heracleum persicum	catalytic	Mohammadi et al. (2020)
	Atropa acuminata	Antioxidant, anti-inflammatory, anti-cancer, and Larvicidal	Rajput et al. (2020)
	Carum copticum	*Quorum-sensing* inhibition and bacterial biofilm	Qais et al. (2020)
Au	*Gnidia glauca*	Chemocatalytic	Ghosh et al. (2012)
	Tussilago farfara	Antibacterial and anti-cancer	Lee et al. (2019)
Cu	*Mimusops elengi*	Antibacterial; antifungal; antioxidant; thrombolytic; larvicidal; cytotoxic; heavy metal removal	Kumar et al. (2020)
	Cymbopogon Citratus	Antibacterial and anti-biofilm	Cherian et al. (2020)
Fe	*Piliostigma thonningii*	Antibacterial	Bibi et al. (2019)
Zn	*Nyctanthes arbor-tristis*	Antifungal	Jamdagni et al. (2018)
Cd	*Tagetes* sp.	Larvicidal	Hajra et al. (2016)
Ti	*Calotropis gigantean*	Acaricide	Marimuthu et al. (2013)
Mg	*Rosmarinus o_cinalis L.*	Antibacterial	Abdallah et al. (2019)

nanoparticles on an industrial scale and their effects on the environment. Concerns have also been raised about their safety and side effects for humans during therapy. Although the promising results of research in this area are increasing, it is necessary to establish guidelines to ensure the safe use of green synthesized nanomaterials to turn these new formulations into reality as anti-biofilm agents.

References

Abdallah Y, Ogunyemi SO, Abdelazez A et al (2019) The green synthesis of MgO nano-flowers using *Rosmarinus officinalis* L. (Rosemary) and the antibacterial activities against *Xanthomonasoryzae* pv. oryzae. Biomed Res Int 2019:1–8. https://doi.org/10.1155/2019/5620989

Abdulkareem EH, Memarzadeh K, Allaker RP et al (2015) Anti-biofilm activity of zinc oxide and hydroxyapatite nanoparticles asdental implant coating materials. J Dent 43(12):1462–1469. https://doi.org/10.1016/j.jdent.2015.10.010

Ahmed A, Khan AK, Anwar A et al (2016) Biofilm inhibitory effect of chlorhexidine conjugated gold nanoparticles against *Klebsiella pneumoniae*. Microb Pathog 98:50–56. https://doi.org/10.1016/j.micpath.2016.06.016

Ali J, Ali N, Wang L et al (2019) Revisiting the mechanistic pathways for bacterial mediated synthesis of noble metal nanoparticles. J Microbiol Methods 159:18–25. https://doi.org/10.1016/j.mimet.2019.02.010

Al-Shabib N, Husain F, Ahmed F et al (2016) Biogenic synthesis of Zinc oxide nanostructures from *Nigella sativa* seed: prospective role as food packaging material inhibiting broad-spectrum quorum sensing and biofilm. Sci Rep 6:36761. https://doi.org/10.1038/srep36761

Arokiyaraj S, Dinesh Kumar V, Elakya V et al (2015) Biosynthesized silver nanoparticles using floral extract of Chrysanthemum indicum L.—potential for malaria vector control. Environ Sci Pollut Res 22:9759–9765. https://doi.org/10.1007/s11356-015-4148-9

Azeredo HMC, Otoni CG, Corrêa DS et al (2019) Nanostructured antimicrobials in food packaging—recent advances. Biotechnol J 14:1900068. https://doi.org/10.1002/biot.201900068

Aziz N, Faraz M, Sherwani MA, Fatma T, Prasad R (2019) Illuminating the anticancerous efficacy of a new fungal chassis for silver nanoparticle synthesis. Front Chem 7:65. https://doi.org/10.3389/fchem.2019.00065

Banerjee D, Shivapriya PM, Gautam PK et al (2019) A review on basic biology of bacterial biofilm infections and their treatments by nanotechnology-based approaches. Proc Natl Acad Sci, India Sect B Biol Sci. https://doi.org/10.1007/s40011-018-01065-7

Barabadi H, Ovais M, Shinwari ZK, Saravanan M (2017) Anti-cancer green bionanomaterials: present status and future prospects. Green Chem Lett Rev 10:285–314. https://doi.org/10.1080/17518253.2017.1385856

Barzegari A, Kheyrolahzadeh K, Hosseiniyan Khatibi SM et al (2020) The battle of probiotics and their derivatives against biofilms. Infect Drug Resist 13:659–672. https://doi.org/10.2147/IDR.S232982

Basavaraja S, Balaji SD, Lagashetty A et al (2008) Extracellular biosynthesis of silver nanoparticles using the fungus *Fusariumsemitectum*. Mater Res Bull 43:1164–1170. https://doi.org/10.1016/j.materresbull.2007.06.020

Bhattacharyya A, Duraisamy P, Govindarajan M, Buhroo AA, Prasad R (2016) Nano-biofungicides: emerging trend in insect pest control. In: Prasad R (ed) Advances and applications through fungal nanobiotechnology. Springer International Publishing, Cham, pp 307–319

Baudrimont M, Andrei J, Mornet S et al (2018) Trophic transfer and effects of gold nanoparticles (AuNPs) in *Gammarus fossarum* from contaminated periphytic biofilm. Environ Sci Pollut Res 25:11181–11191. https://doi.org/10.1007/s11356-017-8400-3

Bhuyan T, Mishra K, Khanuja M, Prasad R, Varma A (2015) Biosynthesis of zinc oxide nanoparticles from *Azadirachta indica* for antibacterial and photocatalytic applications. Mater Sci Semicond Process 32:55–61

Bibi I, Nazar N, Ata S et al (2019) Green synthesis of iron oxide nanoparticles using pomegranate seeds extract and photocatalytic activity evaluation for the degradation of textile dye. J Mater Res Technol 8:6115–6124. https://doi.org/10.1016/j.jmrt.2019.10.006

Boda SK, Broda J, Schiefer F et al (2015) Cytotoxicity of ultrasmall gold nanoparticles on planktonic and biofilm encapsulated gram-positive staphylococci. Small 11:3183–3193. https://doi.org/10.1002/smll.201403014

Boltz J, Smets B, Rittmann J et al (2017) From biofilm ecology to reactors: a focused review. Water Sci Technol 75:1753–1760. https://doi.org/10.2166/wst.2017.061

Brigger I, Dubernet C, Couvreur P (2012) Nanoparticles in cancer therapy and diagnosis. Adv Drug Deliv Rev 64:24–36. https://doi.org/10.1016/j.addr.2012.09.006

Buhroo AA, Nisa G, Asrafuzzaman S, Prasad R, Rasheed R, Bhattacharyya A (2017) Biogenic silver nanoparticles from *Trichodesma indicum* aqueous leaf extract against *Mythimna separata* and evaluation of its larvicidal efficacy. J Plant Protect Res 57(2):194–200. https://doi.org/10.1515/jppr-2017-0026

Bumbudsanpharoke N, Ko S (2018) The green fabrication, characterization and evaluation of catalytic antioxidation of gold nanoparticle-lignocellulose composite papers for active packaging. Int J Biol Macromol 107:1782–1791. https://doi.org/10.1016/j.ijbiomac.2017.10.046

Chaudhuri SK, Malodia L (2017) Biosynthesis of zinc oxide nanoparticles using leaf extract of *Calotropis gigantea*: characterization and its evaluation on tree seedling growth in nursery stage. Appl Nanosci 7:501–512. https://doi.org/10.1007/s13204-017-0586-7

Cherian A, Saquib Q et al (2020) Cymbopogon citratus functionalized green synthesis of CuO-nanoparticles: novel prospects as antibacterial and antibiofilm agents. Biomol Ther 10:169. https://doi.org/10.3390/biom10020169

Chevalier M, Ranque S, Prêcheur I (2018) Oral fungal-bacterial biofilm models *in vitro:* a review. Med Mycol J 56:653–667. https://doi.org/10.1093/mmy/myx111

Chikindas ML, Weeks R, Drider D et al (2020) Functions and emerging applications of bacteriocins. Curr Opin Biotechnol 49:23–28. https://doi.org/10.1016/j.copbio.2017.07.011

Chitra G, Balasubramani G, Ramkumar R et al (2015) *Mukia maderaspatana* (Cucurbitaceae) extract-mediated synthesis of silver nanoparticles to control *Culex quinquefasciatus* and *Aedes aegypti* (Diptera: Culicidae). Parasitol Res 114:1407–1415. https://doi.org/10.1007/s00436-015-4320-7

Colagiorgi A, Di Ciccio P, Zanardi E et al (2016) A look inside the *Listeria monocytogenes* biofilms extracellular matrix. Microorganisms 4:22. https://doi.org/10.3390/microorganisms4030022

Correa MS, Schwambach J, Frazzon J et al (2019) Antimicrobial and antibiofilm activity of the essential oil from dried leaves of *Eucalyptus staigeriana.* Arq Inst Biol 86. https://doi.org/10.1590/1808-1657000202018

Daubert D, Weinstein B (2019) Biofilm as a risk factor in implant treatment. Periodontol 81:29–40. https://doi.org/10.1111/prd.12280

De Mélo Silva IS, do Amorim Costa Gaspar LM, Rocha AMO et al (2020) Encapsulation of red propolis in polymer nanoparticles for the destruction of pathogenic biofilms. AAPS Pharm Sci Tech 21:49. https://doi.org/10.1208/s12249-019-1576-8

Di Pippo F, Di Gregorio L, Congestri R et al (2018) Biofilm growth and control in cooling water industrial systems. FEMS Microbiol Ecol 94. https://doi.org/10.1093/femsec/fiy044

Dragoš A, Kovács ÁT (2017) The peculiar functions of the bacterial extracellular matrix. Trends Microbiol 25:257–266. https://doi.org/10.1016/j.tim.2016.12.010

Ealia SAM, Saravanakumar MP (2017) A review on the classification, characterization, synthesis of nanoparticles and their application. IOP Conf Ser Mater Sci Eng 263:032019. https://doi.org/10.1088/1757-899X/263/3/032019

Feurtet-Mazel A, Mornet S, Charron L et al (2016) Biosynthesis of gold nanoparticles by the living freshwater diatom *Eolimna minima,* a species developed in river biofilms. Environ Sci Pollut Res 23:4334–4339. https://doi.org/10.1007/s11356-015-4139-x

Foroohimanjili F, Mirzaie A, Hamdi SMM et al (2020) Antibacterial, antibiofilm, and antiquorum sensing activities of phytosynthesized silver nanoparticles fabricated from *Mespilus germanica* extract against multidrug resistance of *Klebsiella pneumoniae* clinical strains. J Basic Microbiol 60:216–230. https://doi.org/10.1002/jobm.201900511

Galié S, García-Gutiérrez C, Miguélez EM et al (2018) Biofilms in the food industry: health aspects and control methods. Front Microbiol 9. https://doi.org/10.3389/fmicb.2018.00898

Gholamrezazadeh M, Shakibaie MR, Monirzadeh F et al (2018) Effect of nano-silver, nano-copper, deconex and benzalkonium chloride on biofilm formation and expression of transcription regulatory quorum sensing gene (rh1R) in drug-resistance *Pseudomonas aeruginosa* burn isolates. Burns 44(3):700–708. https://doi.org/10.1016/j.burns.2017.10.021

Ghosh S, Patil S, Ahire M et al (2012) Gnidia glauca flower extract mediated synthesis of gold nanoparticles and evaluation of its chemocatalytic potential. J Nanobiotechnol 10:17. https://doi.org/10.1186/1477-3155-10-17

Ghosh A, Jayaraman N, Chatterji D (2020) Small-molecule inhibition of bacterial biofilm. ACS Omega 5:3108–3115. https://doi.org/10.1021/acsomega.9b03695

Gong CP, Luo Y, Pan YY (2019) Novel synthesized zinc oxide nanoparticles loaded alginate-chitosan biofilm to enhanced wound site activity and anti-septic abilities for the management of complicated abdominal wound dehiscence. J Photochem Photobiol B Biol 192:124–130. https://doi.org/10.1016/j.jphotobiol.2019.01.019

González-Garcinuño A, Tabernero A, Marcelo G et al (2019) Differences in levan nanoparticles depending on their synthesis route: microbial vs cell-free systems. Int J Biol Macromol 137:62–68. https://doi.org/10.1016/j.ijbiomac.2019.06.128

Gordon VD, Davis-Fields M, Kovach K, Rodesney CA (2017) Biofilms and mechanics: a review of experimental techniques and findings. J Phys D Appl Phys 50:223002. https://doi.org/10.1088/1361-6463/aa6b83

Gorrochategui E, Li J, Fullwood NJ et al (2017) Diet-sourced carbon-based nanoparticles induce lipid alterations in tissues of zebrafish (*Danio rerio*) with genomic hypermethylation changes in brain. Mutagenesis 32(1):91–103. https://doi.org/10.1093/mutage/gew050

Gour A, Jain NK (2019) Advances in green synthesis of nanoparticles. Artifi Cell Nanomed Biotechnol 47(1):844–851. https://doi.org/10.1080/21691401.2019.1577878

Habimana O, Zanoni M, Vitale S et al (2018) One particle, two targets: a combined action of functionalised gold nanoparticles, against *Pseudomonas fluorescens* biofilms. J Colloid Interface Sci 526:419–428. https://doi.org/10.1016/j.jcis.2018.05.014

Hajra A, Dutta S, Mondal NK (2016) Mosquito larvicidal activity of cadmium nanoparticles synthesized from petal extracts of marigold (*Tagetes* sp.) and rose (*Rosa* sp.) flower. J Parasit Dis 40:1519–1527. https://doi.org/10.1007/s12639-015-0719-4

Hasan I, Qais FA, Husain FM et al (2019) Eco-friendly green synthesis of dextrin based poly (methyl methacrylate) grafted silver nanocomposites and their antibacterial and antibiofilm efficacy against multi-drug resistance pathogens. J Clean Prod 230:1148–1155. https://doi.org/10.1016/j.jclepro.2019.05.157

Hosseinzadeh N, Shomali T, Hosseinzadeh S et al (2020) Green synthesis of gold nanoparticles by using *Ferulapersica* Willd. gum essential oil: production, characterization and in vitro anti-cancer effects. J Pharm Pharmacol 72(8):1013–1025. https://doi.org/10.1111/jphp.13274

Hu Q, Pan Y, Gong X et al (2020) A sensitivity enhanced fluorescence method for the detection of ferrocyanide ions in foodstuffs using carbon nanoparticles as sensing agentes. Food Chem 308:125590. https://doi.org/10.1016/j.foodchem.2019.125590

Inamuddin, Ahamed MI, Prasad R (2021) Advanced antimicrobial materials and applications. Springer, Singapore. ISBN 978-981-15-7098-8. https://www.springer.com/gp/book/9789811570971

Iriarte-Mesa C, López YC, Matos-Peralta Y et al (2020) Gold, silver and iron oxide nanoparticles: synthesis and bionanoconjugation strategies aimed at electrochemical applications. Top Curr Chem 378(1). https://doi.org/10.1007/s41061-019-0275-y

Islam NU, Jalil K, Shahid M (2019) Green synthesis and biological activities of gold nanoparticles functionalized with Salix alba. Arab J Chem 12(8):2914–2925. https://doi.org/10.1016/j.arabjc.2015.06.025

Jamdagni P, Khatri P, Rana JS (2018) Green synthesis of zinc oxide nanoparticles using flower extract of *Nyctanthes* arbor-tristis and their antifungal activity. J King Saud Univ Sci 30:168–175. https://doi.org/10.1016/j.jksus.2016.10.002

Karunakaran G, Jagathambal M, Gusev A et al (2016) *Allamanda cathartica* flower's aqueous extract-mediated green synthesis of silver nanoparticles with excellent antioxidant and antibacterial potential for biomedical application. MRS Commun 6:41–46. https://doi.org/10.1557/mrc.2016.2

Kavanaugh JS, Flack CE, Lister J et al (2019) Identification of extracellular DNA-binding proteins in the biofilm matrix. MBio 10:e01137-19. https://doi.org/10.1128/mBio.01137-19

Kean R, Ramage G (2019) Combined antifungal resistance and biofilm tolerance: the global threat of *Candida auris*. mSphere 4:e00458-19. https://doi.org/10.1128/mSphere.00458-19

Khade GV, Suwarnkar MB, Gavade NL et al (2015) Green synthesis of TiO_2 and its photocatalytic activity. J Mater Sci Mater Electron 26:3309–3315. https://doi.org/10.1007/s10854-015-2832-7

Khan ST, Ahmad J, Ahamed M et al (2016) Zinc oxide and titanium dioxide nanoparticles induce oxidative stress, inhibit growth, and attenuate biofilm formation activity of *Streptococcus mitis*. J Biol Inorg Chem 21:295–303. https://doi.org/10.1007/s00775-016-1339-x

Khan F, Manivasagan P, Lee J et al (2019) Fucoidan-stabilized gold nanoparticle-mediated biofilm inhibition, attenuation of virulence and motility properties in *Pseudomonas aeruginosa* PAO1 *Mar*. Drugs 17(4):E208. https://doi.org/10.3390/md17040208

Khanna P, Kaur A, Goyal D (2019) Algae-based metallic nanoparticles: synthesis, characterization and applications. J Microbiol Methods 163:105656. https://doi.org/10.1016/j.mimet.2019.105656

Kumar H, Bhardwaj K, Kuča K et al (2020) Flower-based green synthesis of metallic nanoparticles: applications beyond fragrance. Nano 10:766. https://doi.org/10.3390/nano10040766

Kumari S, Yadav BS, Yadav RB (2020) Synthesis and modification approaches for starch nanoparticles for their emerging food industrial applications: a review. Food Res Int 128:108765. https://doi.org/10.1016/j.foodres.2019.108765

Lee YJ, Song K, Cha S-H et al (2019) Sesquiterpenoids from *Tussilago farfara* flower bud extract for the eco-friendly synthesis of silver and gold nanoparticles possessing antibacterial and anticancer activities. Nano 9:819. https://doi.org/10.3390/nano9060819

Leonova EI, Gainetdinov RR (2020) CRISPR/Cas9 technology in translational biomedicine. Cell Physiol Biochem 54:354–370. https://doi.org/10.33594/000000224

Li J, Tian M, Cui L et al (2016) Low-dose carbon-based nanoparticle-induced effects in A549 lung cells determined by biospectroscopy are associated with increases in genomic methylation. Sci Rep 6:20207. https://doi.org/10.1038/srep20207

Liang Y, Zhou J, Dong Y et al (2020) Evidence for the critical role of nanoscale surface roughness on the retention and release of silver nanoparticles in porous media. Environ Pollut 258:113803. https://doi.org/10.1016/j.envpol.2019.113803

Lim JH, Jeong Y, Song S et al (2018) Penetration of an antimicrobial zinc-sugar alcohol complex into *Streptococcus mutans* biofilms. Sci Rep 8:16154. https://doi.org/10.1038/s41598-018-34366-y

Limoli DH, Jones CJ, Wozniak DJ (2015) Bacterial extracellular polysaccharides in biofilm formation and function. Microbiol Spectr 3. https://doi.org/10.1128/microbiolspec.MB-0011-2014

Lineback B, Nkemngong CA, Wu ST et al (2018) Hydrogen peroxide and sodium hypochlorite disinfectants are more effective against Staphylococcus aureus and *Pseudomonas aeruginosa* biofilms than quaternary ammonium compounds. Antimicrob Resist Infect Control 7. https://doi.org/10.1186/s13756-018-0447-5

Liu C, Guo J, Yan X et al (2017) Antimicrobial nanomaterials against biofilms: an alternative strategy. Environ Rev 25(2):225–244. https://doi.org/10.1139/er-2016-0046

Lohse MB, Gulati M, Johnson AD, Nobile CJ (2018) Development and regulation of single- and multi-species *Candida albicans* biofilms. Nat Rev Microbiol 16:19–31. https://doi.org/10.1038/nrmicro.2017.107

Lotha R, Shamprasad BR, Sundaramoorthy NS et al (2019) Biogenic phytochemicals (cassinopin and isoquercetin) capped copper nanoparticles (ISQ/CAS@CuNPs) inhibits MRSA biofilms. Microb Pathog 132:178–187. https://doi.org/10.1016/j.micpath.2019.05.005

Lu B, Lu F, Ran L et al (2018) Imidazole-molecule-capped chitosan–gold nanocomposites with enhanced antimicrobial activity for treating biofilm-related infections. J Colloid Interface Sci 531:269–281. https://doi.org/10.1016/j.jcis.2018.07.058

Malafaia CB, Jardelino ACS, Silva AG et al (2018) Effects of caatinga plant extracts in planktonic growth and biofilm formation in *Ralstonia solanacearum*. Microb Ecol 75:555–561. https://doi.org/10.1007/s00248-017-1073-0

Marimuthu S, Rahuman AA, Jayaseelan C et al (2013) Acaricidal activity of synthesized titanium dioxide nanoparticles using *Calotropisgigantea* against *Rhipicephalusmicroplus* and

Haemaphysalisbispinosa. Asian Pac J Trop Med 6:682–688. https://doi.org/10.1016/S1995-7645(13)60118-2

Marjani MFA, Khadam ZA (2016) Effect of gamma irradiation on biofilm formation of some gram-negative bacteria isolated from burn and wound infections. Int J Pharm Tech Res 9:607–613, 2455–9563

Mathur H, Field D, Rea MC et al (2018) Fighting biofilms with lantibiotics and other groups of bacteriocins. NPJ Biofilms Microbiomes 4. https://doi.org/10.1038/s41522-018-0053-6

Meeker DG, Wang T, Harrington WN et al (2018) Versatility of targeted antibiotic-loaded gold nanoconstructs for the treatment of biofilm-associated bacterial infections. Int J Hyperth 34 (2):209–219. https://doi.org/10.1016/j.envpol.2019.113816

Mehta M, Allen-Gipson D, Mohapatra S et al (2019) Study on the therapeutic index and synergistic effect of Chitosan-zinc oxide nanomicellar composites for drug-resistant bacterial biofilm inhibition. Int J Pharm 565:472–480. https://doi.org/10.1016/j.ijpharm.2019.05.003

Meier MJ, Dodge AE, Samarajeewa AD et al (2020) Soil exposed to silver nanoparticles reveals significant changes in community structure and altered microbial transcriptional profiles. Environ Pollut 258:113816. https://doi.org/10.1016/j.envpol.2019.113816

Meireles A, Borges A, Giaouris E et al (2016) The current knowledge on the application of anti-biofilm enzymes in the food industry. Food Res Int 86:140–146. https://doi.org/10.1016/j.foodres.2016.06.006

Mellinas C, Jiménez A, Garrigós MDC (2019) Microwave-assisted green synthesis and antioxidant activity of selenium nanoparticles using *Theobroma cacao* L. bean shell extract. Molecules 24 (22):4048. https://doi.org/10.3390/molecules24224048

Meza-Villezcas A, Gallego-Hernández AL, Yildiz FH et al (2019) Effect of antimicrobial nanocomposites on *Vibriocholerae* lifestyles: pellicle biofilm, planktonic and surface-attached biofilm. PLoS One 14(6):e0217869. https://doi.org/10.1371/journal.pone.0217869

Miao L, Wang P, Hou J et al (2019) Low concentrations of copper oxide nanoparticles alter microbial community structure and function of sediment biofilms. Sci Total Environ 653:705–713. https://doi.org/10.1016/j.scitotenv.2018.10.354

Mohammadi P, Heravi MM, Sadjadi S (2020) Green synthesis of Ag NPs on magnetic polyallylamine decorated extract : efficient catalyst for reduction of dyes. Sci Rep 10:1–10. https://doi.org/10.1038/s41598-020-63756-4

Muthukrishnan S, Bhakya S, Kumar TS et al (2015) Biosynthesis, characterization and antibacterial effect of plant-mediated silver nanoparticles using *Ceropegia thwaitesii*—an endemic species. Ind Crop Prod 63:119–124. https://doi.org/10.1016/j.indcrop.2014.10.022

Mylona Z, Panteris E, Kevrekidis T et al (2020) Silver nanoparticle toxicity effect on the seagrass *Halophilastipulacea*. Ecotoxicol Environ Saf 189:109925. https://doi.org/10.1016/j.ecoenv.2019.109925

Nayan V, Onteru SK, Singh D (2018) *Mangifera indica* flower extract mediated biogenic green gold nanoparticles: efficient nanocatalyst for reduction of 4-nitrophenol. Environ Prog Sustain Energy 37:283–294. https://doi.org/10.1002/ep.12669

Neu TR, Lawrence JR (2017) The extracellular matrix—an intractable part of biofilm systems. In: Flemming H-C, Neu TR, Wingender J (eds) The perfect slime: microbial extracellular polymeric substances (EPS). IWA Publishers, London, pp 25–59. https://doi.org/10.2166/9781780407425

Noirot-Gros MF, Forrester S, Malato G et al (2019) CRISPR interference to interrogate genes that control biofilm formation in *Pseudomonas fluorescens*. Sci Rep 9. https://doi.org/10.1038/s41598-019-52400-5

Ouyang K, Yu X, Zhu Y et al (2017) Effects of humic acid on the interactions between zinc oxide nanoparticles and bacterial biofilms. Environ Pollut 231(1):1104–1111. https://doi.org/10.1016/j.envpol.2017.07.003

Ovais M, Khalil AT, Raza A et al (2018) Multifunctional theranostic applications of biocompatible green-synthesized colloidal nanoparticles. Appl Microbiol Biotechnol 102:4393–4408. https://doi.org/10.1007/s00253-018-8928-2

Padalia H, Moteriya P, Chanda S (2015) Green synthesis of silver nanoparticles from marigold flower and its synergistic antimicrobial potential. Arab J Chem 8:732–741. https://doi.org/10.1016/j.arabjc.2014.11.015

Padmavathi AR, Arindam Das SMP, Priya A et al (2020) Impediment to growth and yeast-to-hyphae transition in *Candida albicans* by copper oxide nanoparticles. Biofouling 36(1):56–72. https://doi.org/10.1080/08927014.2020.1715371

Pijls BG, Sanders IMJG, Kujiper EJ et al (2020) Induction heating for eradicating *Staphylococcus epidermidis* from biofilm. Bone Joint Res 9:192–199. https://doi.org/10.1302/2046-3758.94.BJR-2019-0274.R1

Prasad R (2014) Synthesis of silver nanoparticles in photosynthetic plants. J Nanoparticles, Article ID 963961. https://doi.org/10.1155/2014/963961

Prasad R (2019a) Plant nanobionics: advances in the understanding of nanomaterials research and applications. Springer International Publishing, Cham. ISBN 978-3-030-12495-3. https://www.springer.com/gp/book/9783030124953

Prasad R (2019b) Plant nanobionics: approaches in nanoparticles biosynthesis and toxicity. Springer International Publishing, Cham. ISBN 978-3-030-16379-2. https://www.springer.com/gp/book/9783030163785

Prasad R, Pandey R, Barman I (2016) Engineering tailored nanoparticles with microbes: quo vadis. WIREs Nanomed Nanobiotechnol 8:316–330. https://doi.org/10.1002/wnan.1363

Prasad R, Jha A, Prasad K (2018) Exploring the realms of nature for nanosynthesis. Springer International Publishing, Cham. ISBN 978-3-319-99570-0. https://www.springer.com/978-3-319-99570-0

Prasad R, Siddhardha B, Dyavaiah M (2020) Nanostructures for antimicrobial and antibiofilm applications. Springer International Publishing, Cham. ISBN 978-3-030-40336-2. https://www.springer.com/gp/book/9783030403362

Qais FA, Shafiq A, Ahmad I et al (2020) Green synthesis of silver nanoparticles using Carum copticum: assessment of its quorum sensing and biofilm inhibitory potential against Gram negative bacterial pathogens. Microb Pathog 144:104172. https://doi.org/10.1016/j.micpath.2020.104172

Rajkumari J, Busi S, Vasu AC et al (2017) Facile green synthesis of baicalein fabricated gold nanoparticles and their antibiofilm activity against *Pseudomonasaeruginosa* PAO1. Microb Pathog 107:261–269. https://doi.org/10.1016/j.micpath.2017.03.044

Rajput S, Kumar D, Agrawal V (2020) Green synthesis of silver nanoparticles using Indian Belladonna extract and their potential antioxidant, anti-inflammatory, anticancer and larvicidal activities. Plant Cell Rep. https://doi.org/10.1007/s00299-020-02539-7

Ramalingam V, Dhinesh P, Sundaramahalingam S, Rajaram R (2019) Green fabrication of iron oxide nanoparticles using grey mangrove Avicennia marina for antibiofilm activity and in vitro toxicity. Surf Interfaces 15:70–77. https://doi.org/10.1016/j.surfin.2019.01.008

Rehman S, Farooq R, Jermy R et al (2020) A Wild *Fomes fomentarius* for biomediation of one pot synthesis of titanium oxide and silver nanoparticles for antibacterial and anticancer application. Biomol Ther 10:622. https://doi.org/10.3390/biom10040622

Reshma VG, Syama S, Sruthi S et al (2017) Nanoparticles with antimicrobial property. Curr Drug Metab 18(11):1040–1054.10.2174/1389200218666170925122201

Rodríguez-Serrano C, Guzmán-Moreno J, Ángeles-Chávez C et al (2020) Biosynthesis of silver nanoparticles by *Fusarium scirpi* and its potential as antimicrobial agent against uropathogenic *Escherichia coli* biofilms. PLoS One 15(3):e0230275. https://doi.org/10.1371/journal.pone.0230275

Ruddaraju LK, Pammi SVN, Guntuku GS et al (2020) A review on anti-bacterials to combat resistance: from ancient era of plants and metals to present and future perspectives of green nano technological combinations. Asian J Pharm Sci 15(1):42–59. https://doi.org/10.1016/j.ajps.2019.03.002

Santhoshkumar J, Rajeshkumar S, Kumar SV et al (2017) Phyto-assisted synthesis, characterization and applications of gold nanoparticles—a review. Biochem Biophys Rep 11:46–57. https://doi.org/10.1016/j.bbrep.2017.06.004

Sengan M, Subramaniyan SB, Prakash AS et al (2019) Effective elimination of biofilm formed with waterborne pathogens using copper nanoparticles. Microb Pathog 127:341–346. https://doi.org/10.1016/j.micpath.2018.12.025

Seo Y, Hwang J, Lee E et al (2018) Engineering copper nanoparticles synthesized on the surface of carbon nanotubes for anti-microbial and anti-biofilm applications. Nanoscale 10 (33):15529–15544. https://doi.org/10.1039/c8nr02768d

Shaffer C (2019) What are bacteriophages? https://www.news-medical.net/life-sciences/What-are-Bacteriophages.aspx

Shahid A, Rasool M, Akhter N et al (2019) Innovative strategies for the control of biofilm formation in clinical settings. In: Bacterial biofilms. IntechOpen. https://doi.org/10.5772/intechopen.89310

Shanmuganathan R, Karuppusamy I, Saravanan M et al (2019) Synthesis of silver nanoparticles and their biomedical applications—a comprehensive review. Curr Pharm Des 25. https://doi.org/10.2174/1381612825666190708185506

Sheppard DC, Howell PL (2016) Biofilm exopolysaccharides of pathogenic fungi: lessons from bacteria. J Biol Chem 291:12529–12537. https://doi.org/10.1074/jbc.R116.720995

Silveira C, Shimabuku-Biadola QL, Silva MF et al (2020) Development of an activated carbon impregnation process with iron oxide nanoparticles by green synthesis for diclofenac adsorption. Environ Sci Pollut Res 27:6088–6102. https://doi.org/10.1007/s11356-019-07329-7

Singh J, Dutta T, Kim K-H et al (2018a) 'Green' synthesis of metals and their oxide nanoparticles: applications for environmental remediation. J Nanobiotechnol 16:84. https://doi.org/10.1186/s12951-018-0408-4

Singh P, Pandit S, Beshay M et al (2018b) Anti-biofilm effects of gold and silver nanoparticles synthesized by the *Rhodiola rosea* rhizome extracts. Artif Cell Nanomed B 46(3):S886–S899. https://doi.org/10.1080/21691401.2018.1518909

Srivastava S, Usmani Z, Atanasov AG, Singh VK, Singh NP, Abdel-Azeem AM, Prasad R, Gupta G, Sharma M, Bhargava A (2021) Biological nanofactories: using living forms for metal nanoparticle synthesis. Mini-Rev Med Chem 21:245–265

Starsich FHL, Herrmann IK, Pratsinis SE (2019) Nanoparticles for biomedicine: coagulation during synthesis and applications. Ann Rev Chem Biomol Eng 10(1):155–174. https://doi.org/10.1146/annurev-chembioeng-060718-030203

Strempel N, Strehmel J, Overhage J (2015) Potential application of antimicrobial peptides in the treatment of bacterial biofilm infections. Curr Pharm Des 21:67–84. https://doi.org/10.2174/1381612820666140905124312

Subhadra B, Kim DH, Woo K et al (2018) Control of biofilm formation in healthcare: recent advances exploiting quorum-sensing interference strategies and multidrug efflux pump inhibitors. Materials (Basel) 11. https://doi.org/10.3390/ma11091676

Teixeira MC, Carbone C, Sousa MC et al (2020) Nanomedicines for the delivery of antimicrobial peptides (AMPs). Nanomaterials (Basel) 20:10(3). https://doi.org/10.3390/nano10030560

Tsui C, Kong EF, Jabra-Rizk MA (2016) Pathogenesis of *Candida albicans* biofilm. Pathog Dis 74: ftw018. https://doi.org/10.1093/femspd/ftw018

Van Wolferen M, Orell A, Albers S (2018) Archaeal biofilm formation. Nat Rev Microbiol 16:699–713. https://doi.org/10.1038/s41579-018-0058-4

Varadavenkatesan T, Selvaraj R, Vinayagam R (2019) Dye degradation and antibacterial activity of green synthesized silver nanoparticles using *Ipomoea digitata* Linn. flower extract. Int J Environ Sci Technol 16:2395–2404. https://doi.org/10.1007/s13762-018-1850-4

Velmourougane K, Prasanna R, Chawla G et al (2019) *Trichoderma-Azotobacter* biofilm inoculation improves soil nutrient availability and plant growth in wheat and cotton. J Basic Microbiol 59:632–644. https://doi.org/10.1002/jobm.201900009

Vijayakumar S, Vinoj G, Malaikozhundan B et al (2015) *Plectranthus amboinicus* leaf extract mediated synthesis of zinc oxide nanoparticles and its control of methicillin resistant *Staphylococcus aureus* biofilm and blood sucking mosquito *larvae*. Spectrochim Acta A Mol Biomol Spectrosc 137:886–891. https://doi.org/10.1016/j.saa.2014.08.064

Wang H, Wang H, Xing T et al (2016) Removal of Salmonella biofilm formed under meat processing environment by surfactant in combination with bio-enzyme. LWT—Food Sci Technol 66:298–304. https://doi.org/10.1016/j.lwt.2015.10.049

Wang L, Hu C, Shao L (2017) The antimicrobial activity of nanoparticles: present situation and prospects for the future. Int J Nanomedicine 12:1227–1249. https://doi.org/10.2147/IJN.S121956

Wang W, Li D, Huang X et al (2019) Study on antibacterial and quorum-sensing inhibition activities of *Cinnamomum camphora* leaf essential oil. Mol Ther 24:37–92. https://doi.org/10.3390/molecules24203792

Wang G, Li F, Li L et al (2020) In situ synthesis of Ag-Fe3O4 nanoparticles immobilized on pure cellulose microspheres as recyclable and biodegradable catalysts. ACS Omega. https://doi.org/10.1021/acsomega.0c00437

Winkelströter LK, Martinis ECP (2015) Different methods to quantify *Listeria monocytogenes* biofilms cells showed different profile in their viability. Braz J Microbiol 46:231–235. https://doi.org/10.1590/S1517-838220131071

Winkelströter LK, Teixeira FBR, Silva EP et al (2014) Unraveling microbial biofilms of importance for food microbiology. Microbiol Ecol 68:35–46. https://doi.org/10.1007/s00248-013-0347-4

Winkelströter LK, dos Reis-Teixeira FB, Lameu GS, Martinis ECP (2016) Understanding *Listeria monocytogenes* biofilms: perspectives into mechanisms of adaptation and regulation under stress conditions. In: Stress and environmental regulation of gene expression and adaptation in bacteria. Wiley-Blackwell, p 1274–1278. https://doi.org/10.1002/9781119004813.ch123

Wolferen MV, Orell A, Albers S (2018) Archaeal biofilm formation. Nat Rev Microbiol 16:699–713. https://doi.org/10.1038/s41579-018-0058-4

Yadav L, Tripathi RM, Prasad R, Pudake RN, Mittal J (2017) Antibacterial activity of Cu nanoparticles against *E. coli*, *Staphylococcus aureus* and *Pseudomonas aeruginosa*. Nano Biomed Eng 9(1):9–14

Yan J, Bassler BL (2019) Surviving as a community: antibiotic tolerance and persistence in bacterial biofilms. Cell Host Microbe 26:15–21. https://doi.org/10.1016/j.chom.2019.06.002

Yedurkar S, Maurya C, Mahanwar P (2016) Biosynthesis of zinc oxide nanoparticles using ixora coccinea leaf extract—a green approach. Open J Synthes Theo Appl 5(1). https://doi.org/10.4236/ojsta.2016.51001

Yew YP, Shameli K, Miyake M et al (2020) Green biosynthesis of superparamagnetic magnetite Fe3O4 nanoparticles and biomedical applications in targeted anticancer drug delivery system: a review. Arab J Chem 13(1):2287–2308. https://doi.org/10.1016/j.arabjc.2018.04.013

Zabielska J, Tyfa A, Kunicka-Styczynska A (2016) Methods for eradication of the biofilms formed by opportunistic pathogens using novel techniques—a review. Folia Biologica et Oecologica 12:26–37. https://doi.org/10.1515/fobio-2016-0003

Zacchino SA, Butassi E, Cordisco E, Svetaz L (2017) Hybrid combinations containing natural products and antimicrobial drugs that interfere with bacterial and fungal biofilms. Phytomedicine 37(15):14–25. https://doi.org/10.1016/j.phymed.2017.10.021

Zou X, Li P, Wang X et al (2020) Silver nanoparticle and Ag+−induced shifts of microbial communities in natural brackish waters: are they more pronounced under oxic conditions than anoxic conditions? Environ Pollut 258:113686. https://doi.org/10.1016/j.envpol.2019.113686

Chapter 2
Visual Health Nanocomposites: Present and Future

Patricia Durán, Andrea Villegas, and Gusdanis Campos

Abstract Background: In recent years, the therapeutics for eye diseases as retinopathy diabetic, dry eye, glaucoma and contact lenses fabrication are including nanocompounds as ocular drug delivery systems, biomarkers for detection of eye diseases, copper and silver ions that has been incorporated by means of the nanotechnology for the production of less toxic and more effective alternatives. Several compounds such as silver nanoparticles, turmeric-based nanocomposites, neovascularization blockers and nanoparticles extracted from the plant *Costus pictus* among others have been researched for this purpose. Those compounds are some of the promising research efforts that are studying and will contribute to preventing vitreous hemorrhages and ocular manifestations that affect large populations, being one of the main causes of blindness in the world. Although, nanotechnology has a lot of contributions which are advantageous for several medicinal areas.

Objective: To carry out a bibliographic review of the latest advances and clinical or toxics findings in nanotechnology for the therapeutic treatment for visual health.

Method: The review was undertaken in the next databases: Google Scholar, Science Direct, Scopus, Elsevier, Springer and PubMed using the keywords: nanotechnology in diabetic retinopathy, nanotechnology and visual health. 75 papers

P. Durán (✉)
Grupo de Investigación de Ciencias Visuales y Optométricas, Facultad de Ciencias de la Salud, Universidad Técnica de Manabí, Portoviejo, Ecuador
e-mail: julia.duran@utm.edu.ec

A. Villegas
Grupo de Investigación en Lentes de Contacto, Córnea y Superficie Ocular "Miguel Refojo", Universidad de Valencia, Valencia, Spain

Instituto de Posgrados, Grupo de Investigación de Ciencias Visuales y Optométricas, Universidad Técnica de Manabí, Portoviejo, Ecuador

G. Campos
Grupo de Investigación Ciencias de Laboratorio Clínico, Facultad de Ciencias de la Salud, Universidad Técnica de Manabí, Portoviejo, Ecuador
e-mail: gusdanis.campos@utm.edu.ec

N. R. Maddela et al. (eds.), *Nanotechnology for Advances in Medical Microbiology*, Environmental and Microbial Biotechnology,
https://doi.org/10.1007/978-981-15-9916-3_2

were reviewed with the keywords "nanotechnology", "visual health", "diabetic retinopathy", "glaucoma" and "dry eye". The Boolean Connector used was "and".

Results: The advantage of nanotechnology at the ocular level is that it allows in the treatment and pharmacology to create nanocomposites that allow them to pass through the delicate ocular tissues and reach the posterior segment of the eye, where many traditional drugs fail. The first nanocomposites were the anti-neovascularization agents such as bevacizumab, atorvastatin, and recently they are producing longer release compounds based on nanotechnology as curcumin silver nanoparticles, extract of insulin Plant (*Costus pictus*) leaves and its silver nanoparticle are some of those nano compounds that promise to be an alternative way to treat diabetic retinopathy.

Conclusion: Many nanotechnology-based drugs have been obtained from plant extracts although they are used in treatment applications at the systemic level, recently some have begun to be used at the ocular level in nanocomposite presentations. Nanotechnology is an alternative to the control of diabetic retinopathy, which affects a large part of the population. The use of drug delivery systems based on nanoparticles as liposomes, dendrimers, cationic nanoemulsions, polymeric nanoparticles pointing out the advantages of the proposed nanomedicines to target this ocular disease.

Keywords Nanocomposites and diabetic retinopathy · Liposomes · Ophthalmic drugs · Diabetes mellitus · Nanoparticles · Nanotechnology · Ocular drug delivery · Retinal neovascularization · Toxicology

2.1 Introduction

In recent years, nanotechnology offers new alternatives for the diagnosis and treatment of diabetic retinopathy, dry eye, glaucoma, and retina diseases (Jo et al. 2011; Raju and Goldberg 2008). The first nano compounds was applied for ocular drugs, nanoparticles for avoid ocular infections and recently for diabetes as anti-neovascularization agents such as bevacizumab, atorvastatin, the curcumin silver nanoparticles, and the extract of insulin plant (*Costus pictus*) leaves and its silver nanoparticle are some of those nano compounds that promise to be an alternative way to treat diabetic retinopathy and other ocular alterations (Aruna et al. 2014). Some studies have shown that curcumin reduces the inflammation and delays or prevents obesity-induced insulin resistance and associated complications, including atherosclerosis and immune mediated liver disease. Unfortunately, dietary curcumin is poorly absorbed by the digestive system and undergoes glucuronidation and excretion rather than being released into the serum and systemically distributed. New improved methods developed with nanoparticles and lipid/liposome formulations that increase absorption and bioavailability of curcumin. Development and refinement of these technologies will enable cell-directed targeting of curcumin and improved therapeutic outcome (Maradana et al. 2013).

The nanocomposites reduce the side effects of traditional drugs and have been used for more than two decades in pharmaceutical formulations, making them more hydrophilic. Drugs at the ocular level can be absorbed in the conjunctiva through the blood vessels and from there go to the choroid and the vitreous, therefore very small particles are required for better absorption and distribution in the tissues (Card and Magnuson 2011; Rupenthal 2020).

When there is an alteration in the retina, the product of a systemic damage caused by diabetes diagnosed in a patient, there is a reduction in the retinal vascular flow which produces ischemia and leads to a decrease in oxygenation at the cellular level in the retina. The retina is the tissue that consumes more oxygen in relation to weight, throughout the human body. Causing a breaking the hemato/retinal barrier and producing plasma exudates and lipids in retinal cells which ends up producing an edema in the whole tissue affected. In diabetic retinopathy the small capillaries of the retina present rupture and obstruction so hemorrhages appear, edema (accumulation abnormal fluid) and important areas of ischemia (total lack of blood) retinal that will carry over time and if they are not treated well, to the loss progressive vision (Campos et al. 2017). The dependence type of diabetes suffered by the patient, the age of the same, the time of evolution since the diagnosis of the disease for the first time base, metabolic controls, and other pathologies such as arterial hypertension or hyperlipidemias, impact directly on the degree of severity of diabetic retinopathy (Farkaš and De Leeuw 2020; Ghafoorianfar et al. 2020).

A complete assessment of the fundus of the eye, supported in addition to complementary exams such as fluorescein angiography and/or tomography of optical coherence is important for the early detection. The diabetic retinopathy does not produce symptoms in the early stages of its evolution and when it manifests, it is usually in a phase advanced. Hence the importance of its early detection with the background examination of the eye, under pupillary dilation. Among the clinical manifestations, the changes are mild, with the presence of microaneurysms, microhemorrhages or subclinical macular edema. Once the pathology starts its progression, it is common to observe abnormal vessels, with irregular calibers, the greater amount of microaneurysms and exudates in the posterior pole. Macular edema is the initial cause of the progressive decrease in central visual acuity. In the phase of proliferative diabetic detinopathy, ischemia is evidenced, the retina compensates for its hypoxia by forming new vessels. These new vessels (neovessels) are abnormal, fragile so they break easily producing hemorrhages (intraretinal, subhyaloid or vitreous), which ends up being the main cause of central and peripheral retinal detachments. Traditional treatment, as photocoagulation laser, can be initially indicated (Samuel et al. 2011). In cases where there is an extensive hemorrhage inside the eye, by bleeding from the retinal vessels, is performed a surgical procedure called vitrectomy, eliminating the blood that is located in the sub vitreous spaces, allowing the use of laser in the retina, minimizing the risk of continuing with visual loss. Other treatments alternatives, include therapy with stem cells, where their effectiveness is not highly proven, and is also found antiangiogenic therapy, which its main objective is to attack the endothelial growth factor, decreasing the likelihood of new vessel formation (Xu et al. 2011).

2.2 State of Art of Nanotechnology Applied to Visual Health

The nanotechnology and the artificial intelligence (AI), also called computational intelligence, at the early stages was displayed by machines, and at these time health technologies are based on it. In computer science, an ideal "intelligent" device is a flexible rational agent that perceives its environment and performs actions that maximize its chances of success in some objective or task. The term artificial intelligence applies when a machine imitates the "cognitive" functions that humans associate with other human minds, such as "learning" and "solving problems". As devices become increasingly capable, a technology that was once thought to require intelligence is removed from the definition (Samuel et al. 2011). For example, optical character recognition is no longer perceived as an example of "artificial intelligence" becoming a conventional technology. Technological advances still classified as artificial intelligence are systems capable of playing chess and self-handling. It is to improve it at the surgical room or to avoid malpractices for ethical and moral purposes (Rhoads 2020).

In 1956, John McCarthy coined the term "artificial intelligence", and defined it as: "the science and ingenuity of making intelligent machines, especially intelligent computer programs"(Durán Ospina et al. 2014). For Nils John Nilsson, there are four fundamental pillars on which artificial intelligence is based on:

A search of the required state in the set of rules produced by the possible actions. Genetic algorithms (analogous to the process of evolution of DNA strings). The nanotechnology and the artificial neural networks (comparable to the physical functioning of the brain of animals and humans). Reasoning through a formal logic similar to human thought (Zerfass et al. 2020). There are also different types of perceptions and actions that can be obtained and produced, respectively, by physical sensors and mechanical sensors in machines, electrical or optical pulses in computers, as well as by inputs and outputs of software and its software environment. Several examples are in the area of systems control, automatic planning, the ability to respond to diagnostics and consumer queries, writing recognition, speech recognition, and pattern recognition. The systems based on artificial intelligence are currently part of the routine in fields such as economics, medicine, engineering, and the military, and have been used in a variety of software applications, strategy games, such as computer chess, and other video games (Ting et al. 2020).

Ophthalmology, within medicine, has always been characterized as one of the fields that uses the most significant technology for diagnosis of pathologies and innovation in new surgical techniques. For decades, the incursion of refractive surgery, assessment of optical coherence tomography, new designs of sophisticated surgical microscopes and the incorporation of intraocular lenses with multifocal designs have allowed improving the quality of life of patients. However, some of these advances still have very high costs and are not accessible to all users. Recently, artificial intelligence has been incorporated into ophthalmology operating rooms to provide the surgeon with new diagnostic and treatment alternatives for different surgeries. At first glance, stem cells, the implementation of artificial corneas in

Table 2.1 Applications of nanotechnology in visual health

Applied	Nanotechnology in visual Health
Glaucoma	• Nanocompounds for new ocular drug delivery • Implant to detect PIO early
Ocular Antibiotics	• Copper, cobalt, gold, polyethilenglicol (PEG), graphene and silver nanoparticles • Nanocarriers
Retina Conjunctiva Ocular Pharmacology Diagnosis and treatment	• Retina implants • Pharmacology Nanocompounds • Intravitreal injections as liposomes encapsulating an angiogenesis inhibitor to avoid neovascularization • Nanocompounds against COVID-19 on eye surface • Develop new molecules as nanosuspensions, nanospheres, liposomes, and microemulsions • New biomarkers for detect deseases and improve treatments

biopolymers and new retinal implants that in one look like science fiction, but today they have become tools for ophthalmologists and visual health professionals (Durán Ospina 2013). Nanotechnology and its applications in medicine require multidisciplinary teams among ophthalmologists, researchers, biomedical engineers, molecular biologists, robotics engineers, research managers and even financial and marketing experts to make these developments go beyond an operating room and patents can be achieved to improve patients life quality (Fangueiro et al. 2015).

Otherwise, in the recent years, the stem cell technology has recently become an ally of nanotechnology many advances have been made in the stem cells field for eye surgery purposes, especially at the level of minor conditions, such as myopia (Sheng et al. 2016). However, getting back those who do not have it, either from birth or from a later problem, remains a difficult obstacle to overcome. Stem cells have much to offer in these investigations, as they can transform into different cell types, so it is not unreasonable to regenerate damaged eye tissues with their help. Precisely this is what a group of scientists, who have used "Isogenic pairs of wild type and mutant induced pluripotent stem cell (iPSC)," based on one of the most recent advances in stem cells, have thought to obtain all components of the ocular tissues. Although at the moment they have only done so in rabbits, the results are very hopeful for the extrapolation to humans (Chichagova et al. 2020). Although they can be obtained from many sources, the most common are embryonic stem cells, and therefore have a large number of ethical disadvantages by those who oppose the use of research embryos (Prasad et al. 2019).

Two hundred fifty genes to date have been identified or associated with the development of retinal diseases, and numerous more have been implicated in diverse corneal genetic dystrophies. A genetic condition affects vision and also can significantly impact the patient quality of life. Gene therapy seeks to slow the progression of these diseases by treating the underlying etiology at the level of the genome (Lane et al. 2020). The Table 2.1 summarize the applications of nanotechnology in visual health.

This shows that nanotechnology will not only be an application science in the fields of engineering, agriculture and medicine and more specifically in ophthalmology and visual health.

People affected with ocular diseases will significantly increase over the next decades, and, consequently, a substantial increase in health costs is expected. Diabetic retinopathy is the most common chronic complication of diabetes. The treatment of eye diseases affecting the posterior segment, such as diabetic retinopathy, is quite challenging due to the anatomy, physiology, and biochemistry of the eye. Therefore, the development of new therapeutics for posterior eye diseases has been a major focus of pharmaceutical research in the area of vision sciences. Several nanosystems already offer efficient solutions for ophthalmological conditions, targeting internal eye tissues, as the retina, and many novel products are expected to appear hereafter (Weng et al. 2017), and Syed (2017) developed fluocinolone acetonide intravitreal implant 0.19 mg is a useful option for the treatment of DME in these patients. This review provides an insight into nanoparticle-based solutions for therapies directed to the posterior segment of the eye diseases, particularly diabetic retinopathy, the present scenarios, and the demands and expectations for the future or toxic effects (Jo et al. 2011).

Therefore, it is important to carry out a good review and vigilance strategy regarding not only the benefits but also the possible toxic effects or interactions with other drugs. Nanotoxicology represents a new and growing research area in toxicology. It deals with the assessment of the toxicological properties of nanoparticles (NPs) with the intention of determining whether (and to what extent) they pose an environmental or societal threat (Jo et al. 2011). Inherent properties of NPs (including size, shape, surface area, surface charge, crystal structure, coating, and solubility/dissolution), as well as environmental factors (such as temperature, pH, ionic strength, salinity, and organic matter), collectively influence NP behavior, fate and transport, and ultimately toxicity (Durán Ospina and Zapata 2016; Prasad et al. 2019). The mechanisms underlying the toxicity of nanomaterials (NMs) have recently been studied extensively (Prasad 2019). Reactive oxygen species (ROS) toxicity represents one such mechanism. An overproduction of ROS induces oxidative stress, resulting in an inability of the cells to maintain normal physiological redox-regulated functions. In context, this chapter includes topics pertaining to chemical and physical properties of NMs and characterization for proper toxicological evaluation, exposure, and environmental fate and transport and ecological and genotoxic effects (Campos et al. 2017). The purpose of this chapter is to review the available research pertaining specifically to NMs in the aquatic environment (implants, aquatic invertebrates, and fish) and their use in biomarker studies. Artificial intelligence and nanotechnology are closely intertwined to allow image visualization and simulated models required for the construction of the nanocarriers and nano compounds (Xu et al. 2011).

Those Clustered Regularly Interspaced Short Palindromic Repeats (CRISPR-Cas) therapies in ophthalmology must be incorporated in the targeting of autosomal dominant pathogenic alleles as well as the knockdown of specific wild-type genes that produce pathogenic phenotypes in particular locations, such as vascular

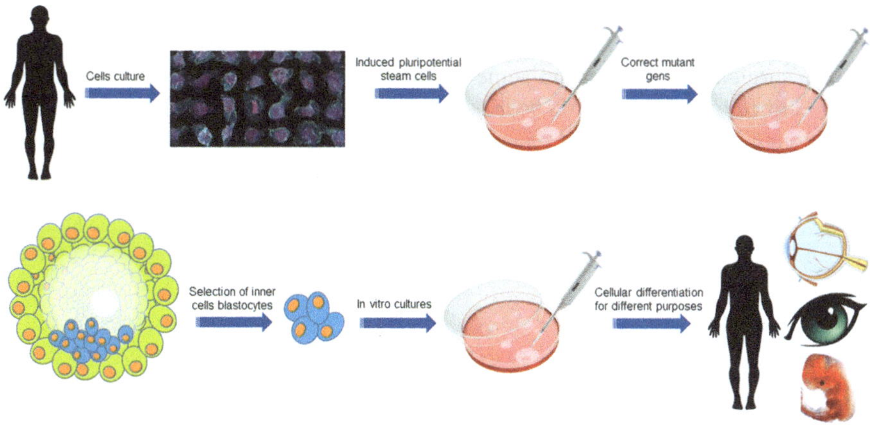

Fig. 2.1 Schematic of the steam cells culture

endothelial growth factor (VEGF) in causing neovascularization of the choroid. Do not require a donor DNA sequence for homologous recombination, reducing the size requirements of the delivery vector. Further development to differentiate CRISPR-Cas modified iPSCs into retinal, corneal and other cell types, as well as methods for autologous transplantation, could make sight-saving advances in numerous genetic diseases of the eye possible (DiCarlo et al. 2017).

Genome surgery: the process of selection of somatic cells from patients, reprogramming to iPSC and implanting through autologous transplantation. The generation of embryonic stem cells by a selection of inner cell mass from a blastocyst, culture, and then reprogramming/genome editing for different purposes. See Fig. 2.1.

The CRISPR and CRISPR-associated systems(Cas) represent one of the powerful tools for studying diseases through the creation of model organisms generated by targeted modification and by the correction of disease mutations for therapeutic purposes. CRISPR-Cas systems have been applied successfully to the visual sciences and study of ophthalmic disease. As a modification of zebrafish and mammalian models of eye development and disease, to the correction of pathogenic mutations in patient-derived stem cells. Recent advances in CRISPR-Cas delivery and optimization boast improved functionality that continues to enhance genome-engineering applications in the eye. This review provides a synopsis of the latest implementations of CRISPR-Cas tools in the field of ophthalmology. It meant for many years, the search for a way to obtain large quantities of these cells without having to use human embryos. Finally, in 2006, this process was successful with the discovery of induced pluripotency stem cells known as iPS, Takahashi and Yamanaka obtained those cells, and to generate iPS, those researchers took cells from adult tissues, which induced the expression of four genes capable of de-differentiating and transforming them into stem cells, similar to those found in embryos (Chichagova et al. 2020). It marked the beginning of a new era in the field of regenerative medicine, since they offered a range of immense possibilities and, also, successfully surpassed the barrier of ethics (Peng et al. 2017).

The surgical procedures and some techniques currently used for ocular surgery are described below, which has been coming in for decades to benefit patients. Some of them are already used, and others are still in the experimental phase before they can be marketed or approved by bioethics committees and implemented. The progress has been classified according to the most used eye surgery.

2.2.1 Laser Eye Surgery

Of the ocular surgical techniques that have incorporated nanotechnologies and artificial intelligence techniques, is Laser assisted in Situ Keratomileusis (LASIK) computerized surgery has been describing as one of the eldest, with the incorporation of mathematical algorithms for the postoperative prediction and although not known as artificial intelligence, have been the most accurate calculations involving mathematical formulas. The recent inclusion of robotic surgery and monitors inside operating rooms are some of the challenges for eye surgery shortly. A group of equations that have contributed mainly concerning the previous ones is the inclusion of more predictive values. Olsen was the one that in addition to predictive values of axial length and keratometry, added the preoperative anterior camera deep (ACD) and the crystalline thickness. Later, Holladay developed a new formula, Holladay II, increasing the number of predictive values to seven (axial length, keratometry, ACD, white- white, crystalline thickness, preoperative refraction, and patient age) (Srivannaboon et al. 2013).

Although the author before the development of the formula Holladay II, developed strategies to improve clinical results by adding diopters to the power of the lens calculated with the formula Holladay I, the publication of his new formula was a before and after in terms of said formula, especially in short eyes, offering the possibility to improve refractive results. Indeed, authors such as Fenzl (Fenzl et al. 1998) state that with this formula 90% of patients can be achieved within a range of (positive and negative Dioptrias) ± 1D of the desired refraction and 100% in the field of ± 2D.

Another of the fourth generation formulas for LASIK surgery was the one developed by Haigis et al., which uses the axial length, keratometry, and depth of the anterior chamber for the calculation of intraocular power. Three constants characterize the difference of this formula: constant provided by the manufacturer (a_0), associated to the ACD (a_1) and constant (a_2) associated with the axial length and calculated by regression methods using data from multiple surgeons (Whang et al. 2020). Another essential feature of this formula is its usefulness in the calculation after refractive surgery, since for the estimation of the effective lens position (ELP), it does not use the keratometry, is the one known with Haigis-L (Kim et al. 2015). Corneal essential to think about, the conceptions of regulation, around the world, and know all the new technologies to research is demonstrating that the incorporation of exogenous recombinant human stromal cell-derived factor-1 alpha (SDF-1 alpha) with thermosensitive chitosan-gelatin hydrogel (CHI hydrogel) accelerated corneal epithelium reconstruction with more native structural and functional properties and increased local expression of growth factors that are essential for corneal epithelium repair. The mechanism by which the exogenous SDF-alpha promotes corneal repair may involve in inducing proliferation and migration of chemokine stromal-cell derived factor-1 known as CXCR4-expressing in limbal epithelial stem cells (LESCs) and mesenchymal stem cells cells (MSCs) to the injury site via the SDF-1/CXCR4 chemokine axis. Therefore, SDF-1 alpha/CHI hydrogel complexes could be a benefit to the future application. As a biosecurity way, to prevent pre and post-surgery infections, novel hydrogels as a chitosan/β-glycerol phosphate solution can be administered as an eye drop that will form a transparent dressing on the front of the eye. The dressing will deliver a sustained release of antibiotics, killing bacteria, but not harming the cells in the cornea (Tang et al. 2017).

For precision surgery, an early detection of ectasia the tomographic and biomechanical index, which combines scheimpflug-based corneal tomography and biomechanics for enhancing ectasia detection. The detection of mild or subclinical forms of ecstatic corneal diseases has gained relevance because these cases are at high risk for developing iatrogenic progressive ectasia (keratectasia) after corneal laser vision correction. Ectasia progression occurs due to the biomechanical decompensation of the corneal stroma, which is related to the preoperative predisposition or biomechanical status of the cornea and to the structural impact from the surgical procedure itself and after surgery. The effect from the procedure may be evaluated using parameters such as the residual stromal bed and the percent of tissue altered. Although different artificial intelligence methods are available, the random forest method provided the most efficient strategy for developing the TBI (Ambrósio et al. 2017). As for any machine learning method, it is fundamental to include a cross-validation method to infer or presume external validity of the model. The current study chose the leave-one-out cross-validation (LOOCV). This method increases computational time and complexity, but also significantly increases the reliability or robustness of the model in classifying new data. A possible study for assessing ectasia susceptibility involves the analysis of the preoperative state of cases that developed ectasia after laser vision correction along with the surgical parameters that represent the impact from surgery on the cornea. Another possible approach is to integrate finite element simulations with the corneal structural and shape analysis. Also, adding longitudinal study for a retrospective evaluation of patients who progressed to clinical ectasia would further improve criteria to define such a group as simulation models very useful also in nanotechnology for design nanocompounds.

2.2.2 *Cataract Surgery*

As a field of regenerative medicine and nanotechnology, the repair and regeneration of tissues using endogenous stem cells represent an ultimate goal. Currently, the only treatment for cataracts, the leading cause of blindness worldwide, is to extract the cataractous lens and implant an artificial intraocular lens. However, this procedure poses notable risks of complications; they isolate lens epithelial stem/progenitor cells (LECs) in mammals and show that Pax6 and Bmi1 are required for LEC renewal. A new surgical technique of cataract removal that preserves endogenous LECs and achieves functional lens regeneration in rabbits and macaques, as well as in human infants with cataracts was proved on China. The method differs conceptually from current practice, as it preserves endogenous LECs and their natural environment maximally, and regenerates lenses with visual function. This approach demonstrates a novel treatment strategy for cataracts and provides a new paradigm for tissue regeneration using endogenous stem cells (Sharma et al. 2016). Researchers have developed a surgical technique with stem cells to eliminate the cataract. The method developed by scientists at Sun Yat-sen University (China) and the University of California, San Diego (USA) removes the cataract from the inside of the lens through a small incision. This structure is coated with epithelial crystalline stem cells, which usually repair the damage. Before the tests, scientists estimated that this procedure would regenerate the lens. The team reported that tests with rabbits and monkeys were successful, so the method was then tested in 12 children. Otherwise, the intraoperative optical coherence tomography (iOCT) is a noninvasive imaging modality that provides real-time dynamic feedback of the various surgical steps. The use of iOCT as an aid to decision-making has been successfully reported in cases undergoing keratoplasty, implantable collamer lens (ICL) implantation as well as cataract surgery. iOCT helps to assess the graft-host relationship in penetrating keratoplasty, to detect subclinical big bubbles and guide layer by layer stromal dissection in cases of deep anterior lamellar keratoplasty. It acts as a guide during crucial surgical steps in endothelial keratoplasty, right from the scoring of the descemet membrane to ensuring graft apposition at the end of surgery. The morphological features of the corneal incision in phacoemulsification may be assessed cells (Ehlers et al. 2014).

The iOCT is a useful tool in assessing the status of the posterior capsule and may help identify preexisting posterior capsular defects during cataract surgery in various clinical scenarios such as posterior polar cataract, traumatic cataract, and vitrectomized eyes. It allows on-table assessment of the Implantable Collamer lens (ICL) vault and potentially facilitates the exchange of ICL in the same sitting in extremes of the vault. Ocular surface disorders such as ocular surface squamous neoplasia, pterygium, and dermoid may find an application for iOCT, wherein an iOCT-guided stromal dissection will ensure adequate depth of dissection. Further technological advancements may allow for automatic centration and tracking and address the present limitation of instrument-induced shadowing. Intraoperative OCT is a useful tool in assessing the status of the posterior capsule during cataract surgery in various clinical scenarios. In cases with posterior polar cataract, it may help detect cases with an actual posterior capsular defect.

The tools exposed before may allow the surgeon to exercise extra caution in such cases, thus reducing the incidence of complications. It may also help ascertain the posterior capsule status in cases with traumatic cataract. Patients with posterior segment pathology often develop cataract during treatment, which may be associated with the iatrogenic posterior capsular defect. Intraoperative Optical coherence tomography (OCT) aids in decision-making in various anterior segment surgeries and has the potential to decrease surgical time as well as postoperative complications. Technological advancements have led to the replacement of handheld and microscope-mounted OCT devices with microscope-integrated iOCT that seamlessly integrates image acquisition with the various surgical steps. A prototype automated stereo vision surgical instrument tracking system developed automatically centers the iOCT scan-field on the surgical instrument tip and allows for continuous visualization of instrument-tissue interactions over a 2500 mm^2 field. Widespread use of this technology will further enhance the advantages of iOCT. One of the significant limitations of microscope-integrated iOCT is the shadowing induced by the surgical instruments, obscuring the underlying cross-sectional view. Further advances technology may help overcome these obstacles shortly and enhance the safety of various surgical procedures (Ehlers et al. 2014).

2.2.3 Glaucoma Surgery

Stem cells derived from tissues of healthy adults such as skin, brain, bone marrow, nasal mucosa and are recently used to glaucoma treatment for their high potential to protect the optic nerve from further damage and slow the progression of vision loss due to glaucoma. It can replace the ocular tissues and include regeneration of trabecular meshwork. Also, it can be implanted inside the eye to be functional and establish working connections with the brain. The challenge for researchers and experts in nanotechnology and ophthalmologists now is to reliably differentiate into the specific ocular tissues that are damaged in glaucoma and following for any side effect (Squillaro et al. 2018; Fangueiro et al. 2015).

To avoid the fibrosis and hence capsule formation around the glaucoma implants, two of the main reasons for glaucoma implant failure. To address these issues, researchers have designed a microfluidic meshwork and tested its biocompatibility in a rabbit eye model. The amount of fibrosis elicited by the microfluidic meshwork was compared to the amount obtained by the plate of the conventional glaucoma drainage device. The microfluidic meshwork got minimal fibrosis and capsule formation after 3-months implantation in a rabbit model. It provides promising evidence to aid in the future development of a new glaucoma drainage implant that will elicit minimal scar formation and provide better long-term surgical outcomes. The drainage devices were fabricated using photolithography techniques similar to those presented previously. The fabrication was done on silicon wafers with a nickel-releasing layer. Briefly, microchannel walls were patterned with negative photoresist SU-8, and the microchannels were formed by sacrificial photoresist (LOR 5A and AZ1505, Microchem, Westborough, MA). The meshwork had an overall area of 7 mm × 7 mm and a grid period of 100 μm. The thickness of the meshwork was four μm. The microfluidic channels had outer diameters of 20 μm and inner diameters of 8 μm. These parameters were determined according to finite element simulations to provide sufficient AH outflow (2 μL/min at 10 mmHg). After being released from the substrate, the meshworks were washed and stored in a buffer solution before autoclave and implantation (Amoozgar et al. 2017), as glaucoma drugs, the atorvastatin as a statin based on solid lipid nanoparticles has been studying as eye drops for the treatment of the age-related macular degeneration (AMD) (Yadav et al. 2020).

2.2.4 Vitreo Surgery

Since they were first derived more than three decades ago, embryonic stem cells have been proposed as a source of replacement cells in regenerative medicine, and this science includes the nanotechnology field. Their plasticity and unlimited capacity for self-renewal raises concerns about their safety, including tumor formation ability, potential immune rejection, and the risk of differentiating into unwanted cell types. The authors report the medium-term to the long-term safety of cells derived from human embryonic stem cells (hESC) transplanted into patients. The human embryonic stem cells hESC and hESC-derived retinal pigment epithelium cells were generated as previously described. Briefly, vials of hESC-MA09 were thawed, expanded, and differentiated into pigmented retinal pigment epithelium patches by the current good manufacturing practices. The hESC-retinal pigment epithelium cells were assessed for safety and characterized for retinal-pigment-epithelium-specific attributes at various times. Vials of cryopreserved hESC-retinal pigment epithelium cells were thawed, formulated, Gramstained, and delivered to the operating room. Pars plana vitrectomy, including the surgical induction of posterior vitreous separation from the optic nerve anteriorly to the posterior border of the vitreous base, was done in the eye with the worst vision. 150 μL of retinal pigment epithelium was injected through a MedOne PolyTip Cannula 23/38 or 25/38 (MedOne Surgical, Sarasota, FL, USA) to deliver the targeted dose of viable retinal pigment epithelium cells into the subretinal space in sites with a preselected transition zone (the area between atrophic photoreceptor, retinal pigment epithelium, and choriocapillaris and fairly healthy post-equatorial retina) as the centre is assessed with autofluorescence and optical coherence tomography imaging. Transplantation sites were chosen carefully based on the presence of native, albeit compromised, retinal pigment epithelium and similarly compromised overlying photoreceptors to optimize the chances of transplant integration and potential for photoreceptor cell rescue. Three dose cohorts were treated for each disorder: cohort 150,000 cells (three patients with Stargardt's macular dystrophy and three with age-related macular degeneration); cohort 2, 100,000 cells (three patients with Stargardt's macular dystrophy and three with age-related macular degeneration); and cohort 3, 150,000 cells (three patients with Stargardt's macular dystrophy and three with age-related macular degeneration). The oral-systemic immunosuppression regimen included tacrolimus and mycophenolate mofetil 1 week before the surgical procedure and continued for 12 weeks. Researchers found that hESC-derived cells were well tolerated for up to 37 months after transplantation in individuals with atrophic age-related macular degeneration and Stargardt's macular dystrophy. So far, in the two clinical trials, there were no serious adverse safety signals attributed to the transplanted cells. Potential safety concerns about the use of hESC in people, including the possibility of teratoma formation, immune reactions, and the risk of cells differentiating into unwanted ectopic cell types were not noted. According to literature reports, teratoma formation was expected to arise within the first few months after transplantation, but this was not the case in our patients who have been followed up for a median of 22 months. To the best of our knowledge, this is the first report of the results of medium- term to long-term safety and tolerability after transplantation of cells derived from pluripotent stem cells in individuals with any disease (Schwartz et al. 2015).

On the other hand, the intravitreal implant of fluocinolone acetonide 0.19 mg (ILUVIEN) is one of the recent corticosteroids implants for the treatment of diabetic macular edema (DME).

2.2.5 Retina Surgery

The artificial soft retina developed used rigid and hard material. In 2017, a Colombia student of Post-Doctoral study at Oxford University at the Chemistry Department can successfully obtain a biological artificial retina tissue developed in the laboratory. It will be a revolutionary innovation for bionic implants and future surgeries and a new alternative for patients with retinosis pigmentaria. They are using light into electrical signals that travel through the nervous system, triggering a response from the brain, ultimately building a picture of the scene on. This new development of a new synthetic, double-layered retina which closely mimics the natural human retinal process immerse in soft water droplets (hydrogels) and biological cell membrane proteins. Designed like a camera where the cells act as pixels, detecting, and reacting to light to create a greyscale image (Schild 2020). See Fig. 2.2.

In 2015, Steven Schwartz and colleagues reported the results of treatment of 18 patients with advanced dry age-related macular degeneration and Stargardt's disease (juvenile macular degeneration) using stem cells. Transplantation of stem-cell derived retinal pigment epithelial cells cannot restore vision but might be able to prevent further disease progression; low vision is one of the challenges for surgeons in recent years. With a micro fundus perimeter, which allows visualization of the retina and where a person fixes a target, they identified that the eyes that improved were unable to move the fixation target out of the scotoma onto seeing the retina at

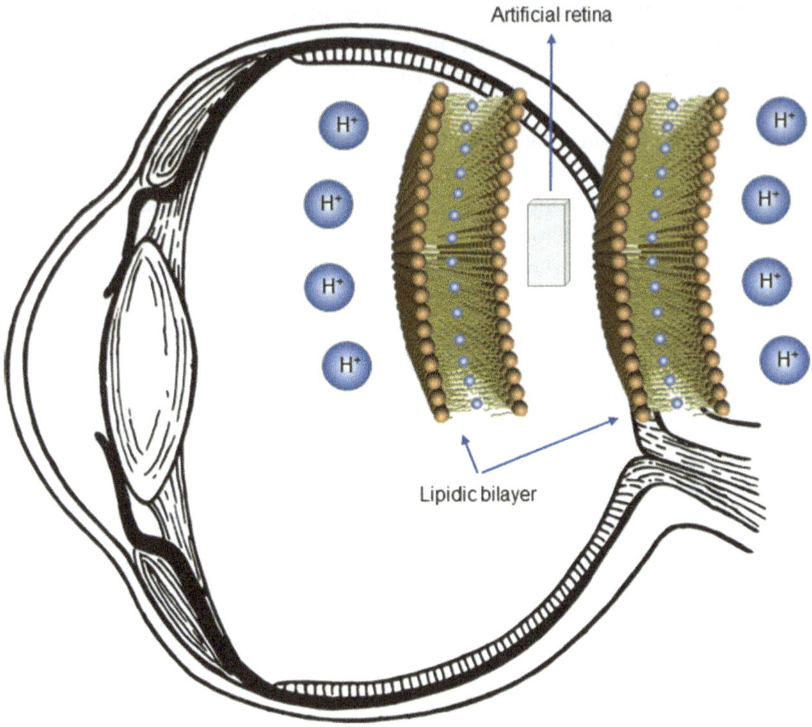

Fig. 2.2 Schematic retina implant

baseline. This improvement of vision in the worse eye inevitably comes into play when interest focuses on repeated testing or administration of treatment to that eye pre and postsurgical.

Otherwise, to creating an automatic system to locate and segment the optic nerve head (ONH) in eye fundus photographic images using genetic algorithms, the diagnosis and surgical procedures will have more certain pictures of the optical nerve. It is called "Domain knowledge," and it is used to create a set of images' guide. Initially, using an eye fundus color image as input, a collection of hypothesis points was obtained that exhibited geometric properties and intensity levels similar to the ONH contour pixels. Next, a genetic algorithm was used to find an ellipse containing the maximum number of hypothesis points in an offset of its perimeter, considering some constraints. The curve thus obtained is the approximation to the ONH. This study was performed with the database belonging to the Ophthalmology Service at Miguel Servet Hospital, Saragossa (Spain). Researchers report that those algorithms can reinforce when it is applied to a different image base to predict the changes in the contour of the papilla and see pixels with geometric characteristics and intensity levels using MATLAB software (Schwartz et al. 2015).

2.2.6 Oculoplasty

Various techniques of autologous fat grafting have been described and are becoming increasingly popular as a primary surgical procedure and as an adjunct to other processes; for example, the use of periocular fat grafting and fillers in their approach to rejuvenation of the periocular aesthetic unit and management of cicatricial ectropion. The surgical techniques suggest a real benefit from combining autologous fat grafting with traditional lifting techniques as a substitute for using a posterior spacer graft in the management of the retracted lower eyelid after blepharoplasty. The method consists of autologous fat harvesting that was performed using a Coleman-type method. Fat grafting was performed in combination with various lifting procedures, as a substitute to a posterior spacer graft, to function as a scaffold that provides additional vertical support to the eyelid. The fat harvesting and preparation are performed under intravenous propofol anesthesia with supplemental local anesthesia and nano compounds will be a good alternative for better penetration. Two percent of Xylocaine with 1:100,000 epinephrine was injected around the lower eyelids and in the submental area with a 30-gauge needle. The tumescent anesthetic agent was infiltrated in the outer thigh or suprapubic area where liposuction was performed. A 2-hole Coleman harvesting cannula with a blunt tip was attached to a 10-mL Luer- Lok syringe, enabling manual aspiration of the fat. Fatwas manually filtered and then placed into 3-mL syringes and then placed upright in a sterile bowl to allow separation of the fat. Once the fat settled, the infra- and supernatant were decanted out, leaving the pure yellow fat. The purified fat was transferred to 1-mL syringes via a Luer-Lok connector (Al Fayez 2013).

The principal role of fat grafting in the setting of post blepharoplasty ectropion is to push/stretch the anterior and middle lamellae up across the entire horizontal length of the retracted lower eyelid while the posterior lamella is being released. At the same time, retractor release and lateral tightening necessary to elevate further and stabilize the retracted eyelid. Fat stem cells may also improve internal scarring and reduce further cicatrix formation. This technique provides significant improvement in eyelid position while avoiding the use of a posterior lamellar spacer graft. The authors have not used posterior lamella spacers in the surgical management of blepharoplasty-induced lower eyelid retraction for some years. In conclusion, the authors' experience suggests that for management of blepharoplasty-induced lower eyelid retraction, autologous fat grafting plays a useful role in optimizing both the functional and aesthetic rehabilitation, in combination with standard lifting techniques. It is the first case series demonstrating the use of autologous fat grafting as an adjunct procedure for the correction of post blepharoplasty lower eyelid retraction.

Otherwise, the use of monoclonal antibodies has been described in the treatment of orbital vascular lesions, lymphoma, and squamous cell carcinoma. Inflammatory conditions treated with monoclonal antibodies include thyroid eye disease, IgG4 disease, and granulomatosis with polyangiitis. Immunotherapy with checkpoint inhibitors has also found applications to orbital disease. Use of small molecule inhibitors has been described in the treatment of basal cell carcinoma, squamous cell carcinoma, and Erdheim–Chester disease. There are many orbital, lacrimal, and eyelid side effects of molecular target agents (MTAs) with which the oculoplastic surgeon should be familiar, including hypertrichosis, edema, and orbital and eyelid inflammation. Small molecule inhibitors (SMIs) block a metabolic or enzymatic step in the target cell, which halts the growth of the cell. As opposed to monoclonal antibodies, small molecule inhibitors can often cross the blood–eye, and blood-brain barrier. The name of the molecule describes the targeted metabolic or enzymatic step. Ending in -tinib,—zomoib,—ciclib, or -parib denotes a tyrosine kinase, proteasome, cyclin-dependent kinase, or poly-ADP-ribose polymerase inhibitor, respectively. Thus, ibrutinib, an inhibitor of Bruton's tyrosine kinase, (BTK) is a tyrosine kinase inhibitor (-tinib). Straddling oculoplastic surgery and external disease and success of the treatment of vernal keratoconjunctivitis and mucous membrane pemphigoid has been investigated with monoclonal antibodies. Omalizumab, an anti-IgE antibody that binds to circulating IgE has been reported to be effective in the treatment of severe refractory vernal keratoconjunctivitis. Treating refractory mucous membrane pemphigoid with nanocomposites of rituximab has also shown promises for recovery postsurgical inflammation (Allen 2017).

This are recognized companies designing clinical trials for advanced dispositions to include those new technologies based on nanotechnologies to the surgeons service: Stargardt disease, including Advanced Cell Technology®, Acucela®, Alcon (Novartis)®, Alexion®, BioMarin®, Cell Cure®, Genentech (Roche) ®, Intrexon®, Janssen®, Pfizer®, Stem cells®, and Sucampo® and the new innovations ones which are working on patents and improve new techniques to be implemented in surgical environments. One of the innovations was "Argus II," one of the recently approved retina implants has a miniature video camera located in the eyeglasses of the patient who captures a scene. The patient carries a minicomputer that receives and processes the video. These instructions are transmitted wirelessly to the retinal implant. Once there the chip converts the signals into small pulses of electricity that bypass the damaged photoreceptors of the macula and directly stimulate the remaining cells of the retina, which transmit the information through the optic nerve, to the brain creating the perception of patterns of light. A micro camera housed in the patient's glasses captures the images, and these are sent to a small computer that the patient carries on, where they are processed and transformed into instructions. These are transmitted wirelessly to the antenna of the retinal implant. These impulses stimulate healthy cells that remain in the retina and send information to the brain through the optic nerve to create the perception of light patterns, which patients learn to interpret (Schaffrath et al. 2019).

In summary, "Argus II" works as a macular implant that is attached to an external high definition camera and a processor that stimulates the inner retina and ends up generating a visual stimulus in the visual pathways and improves the patient's vision. The user of this chip should wear glasses, which have the camera inserted, and a mini computer on top, which receives the scenes that the camera captures. The computer system transmits the information wirelessly to the implant, and the chip converts the signals into tiny pulses of electricity that stimulate the retina and create

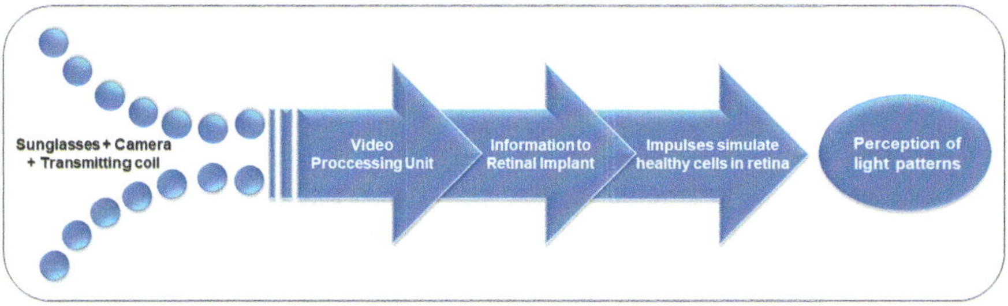

Fig. 2.3 Functional process of Argus II retinal implant

patterns of light. The dispositive in 2017, cost $US 100.000 approximately. See Fig. 2.3.

2.3 Present and Future of Eye Surgery: Nanotechnology, Genomic and Nanorobotics

A recent project named WET: Whole eye transplantation will be the most ambitious. Blindness afflicts ~39 million people worldwide. Retinal ganglion cells are unable to regenerate, making this condition irreversible in many cases. Whole-eye transplantation (WET) provides the opportunity to replace diseased retinal ganglion cells, as well as the entire optical system and surrounding facial tissue, if necessary. Recent success in face transplantation demonstrates that this may be a promising treatment for what has been to this time an incurable condition. An animal model for WET must be established to further enhance our knowledge of nerve regeneration, immunosuppression, and technical aspects of surgery. A systematic review of the literature was performed to evaluate studies describing animal models for WET. In the majority of published research, WET can result in the recovery of vision in cold-blooded vertebrates. There are a few instances in which mammalian WET models demonstrate survival of the transplanted tissue following neurovascular anastomosis and the ability to maintain brief electroretinogram activity in the new host. Mammalian animal models for WET will be the future research for translation to human eye transplantation. Human WET holds promise to restore vision to patients who have blindness. Significant research has been conducted to achieve this goal. However, it was seen as science fiction some decades ago, the first mention of eye transplantation in humans occurred as early as 1885 *Revue Générale d'Ophthalmology*. Dr. Chibret removed the staphylomatous and buphthalmic eye of a 17-year-old girl and placed a rabbit. But what if it is true is that science and advances are promising for these patients. Table 2.2, Summarize the recent innovation of nanotechnology and artificial intelligence for eye surgery.

Table 2.2 Innovation in nanotechnology and artificial intelligence for eye surgery

Innovation	Eye surgery	Purpose
Steam Cells	Cataract, Cornea, Vitreo, Oculoplastic, Glaucoma	Tissue regeneration (Tang et al. 2017)
Algorithms for precision measurements	LASIK, Cataract, Keratoplasty	Application of mathematical algorithms for precise surgeries (Sharma et al. 2016)
iOCT	Retina, Glaucoma	Glaucoma, retina (Ehlers et al. 2014)
Retina Implant and micro-electronic memristors (MEMS)	Retina	Restoration of vision through electrical signals (Schaffrath et al. 2019)
CRISPR-Cas systems	Reconstruction of ocular tissues	Regenerative ophthalmology (Lane et al. 2020)

2.4 Conclusions

Advances in nanotechnology, stem cells and the incorporation of robotics and nanotechnology continue to revolutionize surgical techniques and present a challenge to surgeons and institutions that train future visual health professionals to incorporate these new technologies, create protocols and establish mechanisms so that the transfer of technology can reduce costs to benefit patients. Shortly, blindness in the world will be a subject of research and social inclusion, so that these new patents and developments can reach from public policies and foundations people who need it. It requires teams of professionals such as ophthalmologists, engineers, molecular biologists and professionals who in one way or another know about research management to help communities to make these new devices known and can be used by those who need them.

The training of new ophthalmologist surgeons should be viewed from a global perspective not only with medicine, but even their preparation should incorporate engineering, artificial intelligence, robotics, nanotechnology and learn to make stem cell cultures. In developing countries, these devices are not accessible because of the vast sums of money involved in establishing research centers and doctoral training at this level. But you can think of the help of fellows and support through governments to finance these innovative talents. The ophthalmic industry is also changing and incorporating this new equipment, standardizing tests, and making the transition from animal models to humans. Bioethics committees must also be trained to allow and be aware of any changes that may or may not be beneficial to making significant adjustments.

Acknowledgments The authors want to thank the researchers of those reviews, who have made these scientific advances possible to solve visual problems. Also, to the Universidad Técnica de Manabí, Portoviejo, Ecuador.

Conflict of Interest The authors declare to have no conflict of interest.

References

Al Fayez MF (2013) Limbal-Conjunctival vs Conjunctival autograft transplant for recurrent Pterygia: a prospective randomized controlled trial. Arch Ophthalmol 131(1):11–16

Allen RC (2017) Molecularly targeted agents in Oculoplastic surgery. Curr Opin Ophthalmol 28 (5):485–492

Ambrósio R et al (2017) Integration of Scheimpflug-based corneal tomography and biomechanical assessments for enhancing Ectasia detection. J Refract Surg 33(7):434–443

Amoozgar B, Wei X, Hui Lee J, Bloomer M, Zhao Z, Coh P, Han Y (2017) A novel flexible microfluidic meshwork to reduce fibrosis in glaucoma surgery. PLoS One 12(3):e0172556. https://doi.org/10.1371/journal.pone.0172556

Aruna A et al (2014) Comparative anti-diabetic effect of Methanolic extract of insulin. Indo Am J Pharm Res 4(07):3217–3230

Campos EJ, Campos A, Martins J, Ambrósio AF (2017) Opening eyes to Nanomedicine: where we are, challenges and expectations on Nanotherapy for diabetic retinopathy. Nanomedicine 13 (6):2101–2113. https://doi.org/10.1016/j.nano.2017.04.008

Card JW, Magnuson BA (2011) A review of the efficacy and safety of nanoparticle-based Oral insulin delivery systems. Am J Physiol Gastrointest Liver Physiol 301(6):956–967

Chichagova V et al (2020) Human IPSC differentiation to retinal Organoids in response to IGF1 and BMP4 activation is line- and method-dependent. Stem Cells 38(2):195–201

DiCarlo JE et al (2017) CRISPR-Cas genome surgery in ophthalmology. Transl Vis Sci Technol 6 (3):13

Durán Ospina P (2013) Are nanotechnology and robotics alternatives for therapeutic and Theragnostic ophthalmic applications Technologies for eye Care Services. J Ocul Dis Ther 1:24–32

Durán Ospina P, Zapata PC (2016) Send orders for reprints to Reprints@benthamscience.Ae nanotechnology composites and visual health Patient's quality of life. Pharm Nanotechnol 4:1–10

Durán Ospina P, Díaz MC, Plaza JP (2014) A review in innovation in ocular prostheses and visual implants: new biomaterials and Neuro-implants is the challenge for the visual care. J Ocul Dis Ther 2:9–16

Ehlers JP, Kaiser PK, Srivastava SK (2014) Intraoperative optical coherence tomography using the RESCAN 700: preliminary results from the DISCOVER study. Br J Ophthalmol 98 (10):1329–1332

Fangueiro JF, Silva AM, Garcia ML, Souto EB (2015) Current nanotechnology approaches for the treatment and Management of Diabetic Retinopathy. Eur J Pharm Biopharm 95:307–322. https://doi.org/10.1016/j.ejpb.2014.12.023

Farkaš B, De Leeuw NH (2020) Towards a morphology of cobalt nanoparticles: size and strain effects. Nanotechnology 31(19):195711

Fenzl RE, Gills JP, Cherchio M (1998) Refractive and visual outcome of hyperopic cataract cases operated on before and after implementation of the Holladay II formula. Ophthalmology 105 (9):1759–1764. https://doi.org/10.1016/S0161-6420(98)99050-9

Ghafoorianfar S, Ghorani-Azam A, Mohajeri SA, Farzin D (2020) Efficiency of nanoparticles for treatment of ocular infections: systematic literature review. J Drug Deliv Sci Technol 57:101765. https://doi.org/10.1016/j.jddst.2020.101765

Jo DH, Lee TG, Kim JH (2011) Nanotechnology and Nanotoxicology in retinopathy. Int J Mol Sci 12(11):8288–8301

Kim DH, Kim MK, Wee WR (2015) Estimation of intraocular Lens power calculation after myopic corneal refractive surgery: using corneal height in anterior segment optical coherence tomography. Korean J Ophthalmol 29(3):195–202

Lane A et al (2020) Modeling and rescue of RP2 retinitis Pigmentosa using IPSC-derived retinal Organoids. Stem Cell Reports 15(1):67–79. https://doi.org/10.1016/j.stemcr.2020.05.007

Maradana MR, Thomas R, O'Sullivan BJ (2013) Targeted delivery of Curcumin for treating type 2 diabetes. Mol Nutr Food Res 57(9):1550–1556

Peng YQ, Tang LS, Yoshida S, Ye DZ (2017) Applications of CRISPR/Cas9 in retinal degenerative diseases. Int J Ophthalmol 10(4):646–651

Prasad R (2019) Plant nanobionics: approaches in nanoparticles biosynthesis and toxicity. Springer International Publishing, Cham. ISBN 978-3-030-16379-2. https://www.springer.com/gp/book/9783030163785

Prasad R, Kumar V, Kumar M, Choudhary D (2019) Nanobiotechnology in bioformulations. Springer International Publishing, Cham. ISBN 978-3-030-17061-5. https://www.springer.com/gp/book/9783030170608

Raju HB, Goldberg JL (2008) Nanotechnology for ocular therapeutics and tissue repair. Expert Rev Ophthalmol 3(4):431–436

Rhoads DD (2020) Computer vision and artificial intelligence are emerging diagnostic tools for the clinical microbiologist. J Clin Microbiol 58(6). https://doi.org/10.1128/JCM.02053-19

Rupenthal ID (2020) Nanomedicine approaches in ocular drug delivery. https://www.oznanomed.org/wp-content/uploads/2020/01/abstract_Ilva-Rupenthal.pdf. Accessed 06 February 2021

Samuel D, Bharali D, Mousa SA (2011) The role of nanotechnology in diabetes treatment: current and future perspectives. Int J Nanotechnol 8(1–2):53–65

Schaffrath K et al (2019) One-year safety and performance assessment of the Argus II retinal prosthesis: a Postapproval study. JAMA Ophthalmol 137(8):896–902

Schild VR (2020) Light-activated synthetic tissues [PhD thesis]. University of Oxford. https://ora.ox.ac.uk/objects/uuid:c611246d-0d9d-47d3-9592-608703f6eedf. Accessed 06 February 2021

Schwartz SD et al (2015) Human embryonic stem cell-derived retinal pigment epithelium in patients with age-related macular degeneration and Stargardt's macular dystrophy: follow-up of two open-label phase 1/2 studies. Lancet 385(9967):509–516. https://doi.org/10.1016/S0140-6736(14)61376-3

Sharma N, Aron N, Kakkar P, Titiyal JS (2016) Continuous intraoperative OCT guided Management of Post-Deep Anterior Lamellar Keratoplasty Descemet's membrane detachment. Saudi J Ophthalmol 30(2):133–136. https://doi.org/10.1016/j.sjopt.2016.01.001

Sheng L, Christopher M, Melnick A (2016) 乳鼠心肌提取 HHS public access. Physiol Behav 176 (1):100–106

Squillaro T et al (2018) Concise review: the effect of low-dose ionizing radiation on stem cell biology: a contribution to radiation risk. Stem Cells 36(8):1146–1153

Srivannaboon S et al (2013) Accuracy of Holladay 2 formula using IOLMaster parameters in the absence of Lens thickness value. Graefes Arch Clin Exp Ophthalmol 251(11):2563–2567

Syed YY (2017) Fluocinolone acetonide intravitreal implant 0.19 mg (ILUVIEN®): a review in diabetic macular edema. Drugs 77(5):575–583. https://doi.org/10.1007/s40265-017-0722-4

Tang Q et al (2017) Thermosensitive chitosan-based hydrogels releasing stromal cell derived Factor-1 alpha recruit MSC for corneal epithelium regeneration. Acta Biomater 61:101–113. https://doi.org/10.1016/j.actbio.2017.08.001

Ting DSW et al (2020) Artificial intelligence, the internet of things, and virtual clinics: ophthalmology at the digital translation forefront. Lancet Digit Health 2(1):e8–e9. https://doi.org/10.1016/S2589-7500(19)30217-1

Weng Y et al (2017) Nanotechnology-based strategies for treatment of ocular disease. Acta Pharm Sin B 7(3):281–291. https://doi.org/10.1016/j.apsb.2016.09.001

Whang WJ, Yoo YS, Kim HS, Yoon G (2020) Influence of corneal spherical aberration on prediction error of the Haigis-L formula. Sci Rep 10(1):1–8

Xu Y, Zhang L, Chen G, Chen P (2011) Thinking on the application of nanotechnology in the mechanism research on the traditional Chinese medicine diagnosis and treatment of diabetes mellitus. J Phys Conf Ser 276:1

Yadav M, Schiavone N, Guzman-Aranguez A, Giansanti F, Papucci L, Perez de Lara MJ, Singh M, Kaur IP (2020) Correction to: atorvastatin-loaded solid lipid nanoparticles as eye drops: proposed treatment option for age-related macular degeneration (AMD). Drug Deliv Transl Res 10(5):1531. https://doi.org/10.1007/s13346-020-00777-6

Zerfass A, Hagelstein J, Tench R (2020) Artificial intelligence in communication management: a cross-National Study on adoption and knowledge, impact, challenges and risks. J Commun Manag 24(4):1–16

Chapter 3
Applications of Nanomaterials in Biomedical Engineering

Ranjit Pabbati, Venkateswar Reddy Kondakindi, and Firdoz Shaik

Abstract In recent years, we have witnessed the revolutionary changes brought by nanotechnology in many fields. Nanomaterials are the building blocks of nanotechnology. Nanomaterials exhibit unique properties which are absent in bulk materials. Among nanomaterials, noble metal nanoparticles (NMNs) have been attracted due to their enchanting optical and chemical properties. At present, NMNs have enormous applications including catalysis, biosensors, bioimaging, theranostics, antimicrobial, cosmetics and medical appliances, and many others. In this book chapter, we have discussed briefly about the current scenario of nanotechnology, synthetic strategies of nanoparticles including top-down, bottom-up and biological approaches, characterization techniques, optical properties and different applications of nanoparticles in medical biotechnology.

Keywords Nanotechnology · Nanoparticles · Synthetic strategies · LSPR · Biomedical

3.1 Introduction and Historical Background

Nanotechnology has considered as the revolutionary technology of the twenty first century. It has brought revolutionary changes in large sectors including catalysis (Shaik et al. 2014), electronics (Tapio et al. 2016), energy (Xu et al. 2020; Yang et al. 2020), environmental (Liu 2006; Prasad et al. 2017; Shash et al. 2019), biomedical

---

Ranjit Pabbati and Venkateswar Reddy Kondakindi have been contributed equally to this chapter.

R. Pabbati · V. R. Kondakindi
Centre for Biotechnology, Institute of science & Technology, Jawaharlal Nehru Technological University, Hyderabad, Telangana, India

F. Shaik (✉)
Schulich Faculty of Chemistry Technion, Israel Institute of Technology, Haifa, Israel

N. R. Maddela et al. (eds.), *Nanotechnology for Advances in Medical Microbiology*, Environmental and Microbial Biotechnology,
https://doi.org/10.1007/978-981-15-9916-3_3

(Li et al. 2019a) and many other fields (Mitrano et al. 2014; Yetisen et al. 2016; Thangadurai et al. 2020a, b). Any technology done on a nanoscale regime which has real practical applications is normally considered as nanotechnology (2000). Nanomaterials are the building blocks of nanotechnology. A lot of curiosity has created towards nanomaterials from the talk given by Richard Feynman in 1959 on "There's plenty of room at the bottom", at the annual meeting of the American Physical Society (Feynman 2011). He discussed on tuning the things at small scale to its extreme limits, and consist of several solid assumptions including information on tiny scale, comparing the size of various biological systems, creation of miniaturization by evaporation, creation of tiny machines, and manipulating atoms one by one the way we need (Feynman 2011). In 1974, Taniguchi used the term "Nanotechnology" for the first time to describe the precision engineering in semiconductor processing in his scientific publication (Taniguchi 1974). However, the real interest towards nanotechnology and nanoscience has developed with the invention of the scanning tunnelling microscope by Binning and Rohrer in 1986 (Gerber et al. 1986). Eventually, it was helpful for the improvement of atomic force microscope, invented by Calvin Quate and Christoph Gerber, which provided better atomic resolution of nearly 3 orders higher magnitude than that of the diffraction limit of the optical microscope (Binnig et al. 1986). In 1985, Kroto et al. discovered the fullerene a ball like carbon material of just 1nm diameter in size (Kroto et al. 1985). Later in 1986, Eric Drexler discussed on molecular nanotechnology in his book "Engine of Creation" which brought many nanotechnology things closely to the layman outside of the scientific community (Eric Drexler 1986). Since then, several important discoveries had been taken place including discovery of carbon nanotubes (Iijima 1991), nano-imprint technology (Chou et al. 1995), investigation on photocatalytic activity of TiO_2 nanosheets (Sasaki et al. 1996) and many other things. In 2001, USA has established National Nanotechnology Initiative (NNI) which primarily investigates and conducts research on nanotechnology. Nanotechnology has developed into a huge distinctive area and several scientific journals (Nature Nanotechnology, Nano letters, Journal of Nanoscience and Nanotechnology, ACS applied Nanomaterials, Journal of Nanoparticle research and many others) have introduced which record exclusively on the advancement of nanomaterials and various applications. It is not exaggerated to say that there is no technology revolutionary without nanotechnology in the twenty first century. It has brought tremendous advances in many fields including medical biotechnology.

3.2 Nanomaterials

Nanotechnology has built on the building blocks of nanomaterials. The Greek word "nano" is derived from the word "dwarf" meaning reduction in size or small (Melo et al. 2013). There is no specific definition to nanomaterials, usually any material whose size is in the range of 1-100 nm at least in one of its dimensions is normally considered as a nanomaterial. The nanometre (nm) scale is typically defined as one

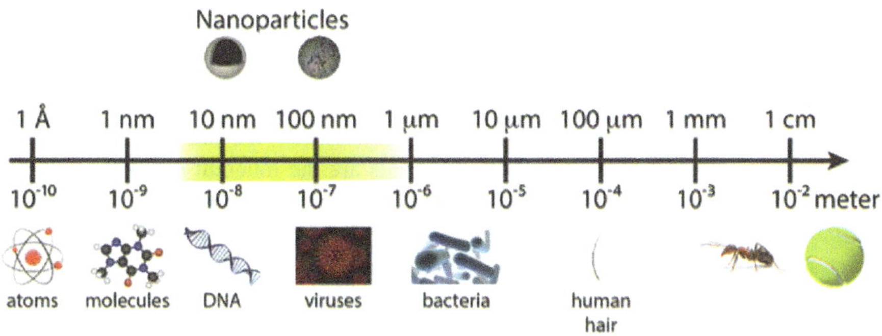

Fig. 3.1 Comparison of different sizes of materials with nanoscale dimension. Modified from Bloemen 2015

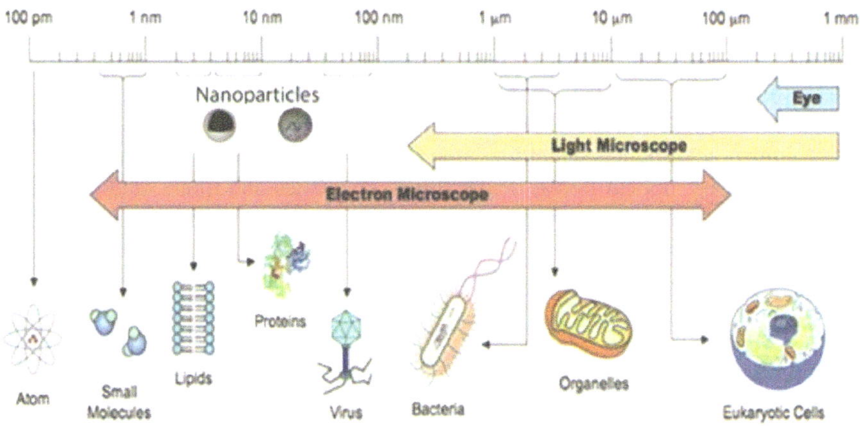

Fig. 3.2 Visibility range of electron microscope, light microscope and human eye. Modified from Fiona Eadie (2017)

billionth of a meter (10^{-9} m) which usually consists of 10 to 10^5 atoms. Different sizes of materials are compared with nanoscale dimension as shown in Fig. 3.1. Nanomaterials act as the bridging materials between microparticles and atoms. Nanomaterials exhibit unique properties compared with bulk materials owing to quantum effects.

The size of nanomaterials is in the order of nanometre dimensions which limit its visibility with human eye. Different types of electronic microscopes are normally employed to see different types of nanomaterials. The visibility range of different materials using electron and optical microscopes, and human eye is compared as shown in Fig. 3.2.

Michael Faraday (Faraday 1857) reported the synthesis of colloidal gold nanoparticles in 1857, since then there is tremendous improvements on synthesis of different types of nanomaterials. Nanomaterials are fabricated using different methods and these methods are mainly classified into two subcategories top-down and bottom-up. A detailed information on various methods for the synthesis of nanomaterials is provided under section "synthesis of nanomaterials". Though many nanomaterials are engineered using different methods, yet some nanomaterials are found naturally, for example: seashells, skeleton, DNA and RNA molecules, some viruses, proteins, enzymes. Nanomaterials are the inherent of unique optical, electrical, mechanical, physical and chemical, and electronic properties which are normally absent in micro or bulk materials (Yang et al. 2015). We are exploiting the different novel properties of nanomaterials for various applications. Several different types of nanomaterials are used for commercial applications form several years or decades. Different types of commercial products including optical fibres (Pawar and Kale 2019), wrinkle-free textiles (von Goetz et al. 2013), biosensors (Wongkaew et al. 2019), sunscreen (Contado and Pagnoni 2008), paints and varnishes (Zhu et al. 2019), cosmetics (Auffan et al. 2010) and electronics (Li et al. 2019b) are making use of nanomaterials. The thin nanocoatings of different nanocomposites are used in diversified commercial products including sports equipment, automobiles, windows and bicycles (Rao et al. 2018). TiO_2 based nanocoatings are used in beverages glass bottles to protect from UV light damaging, self-cleaning windows, sunscreens and cosmetics (Auffan et al. 2010). Nanosilica particles are used in everyday consumer and commercial products including non-diary coffee creamer, automobile tires, catalyst supports, optical fibres, dental fillers, cosmetics and filters, and nano-clay composites are used in longer-lasting tennis balls (Guo et al. 2018).

3.2.1 *Classification of Nanomaterials*

Nanomaterials have different classifications based on dimensions, shape and type of materials. Richard W. Siegel classified nanomaterials based on the dimensions into zero-dimensional, single-dimensional, two-dimensional and three-dimensional nanostructures (Fig. 3.3) (Sajanlal et al. 2011). Based on nature of the materials, nanomaterials are generally classified into organic and inorganic nanoparticles. Based on shape they are classified into isotropic and anisotropic nanoparticles. Different examples for organic and inorganic nanomaterials are shown in Fig. 3.3. Organic nanomaterials include dendrimer, liposome, polymeric micelle, nanocapsule and nanosphere, whereas inorganic nanoparticles include mesoporous silica nanoparticles, carbon nanotubes, iron oxide nanoparticles, gold nanoparticles and quantum dot (Richards et al. 2017).

3.2.2 *Properties of Nanomaterials*

Nanomaterials exhibit different unique properties owing to the existence of quantum effects which lead to different optical, magnetic, electrical, mechanical and various

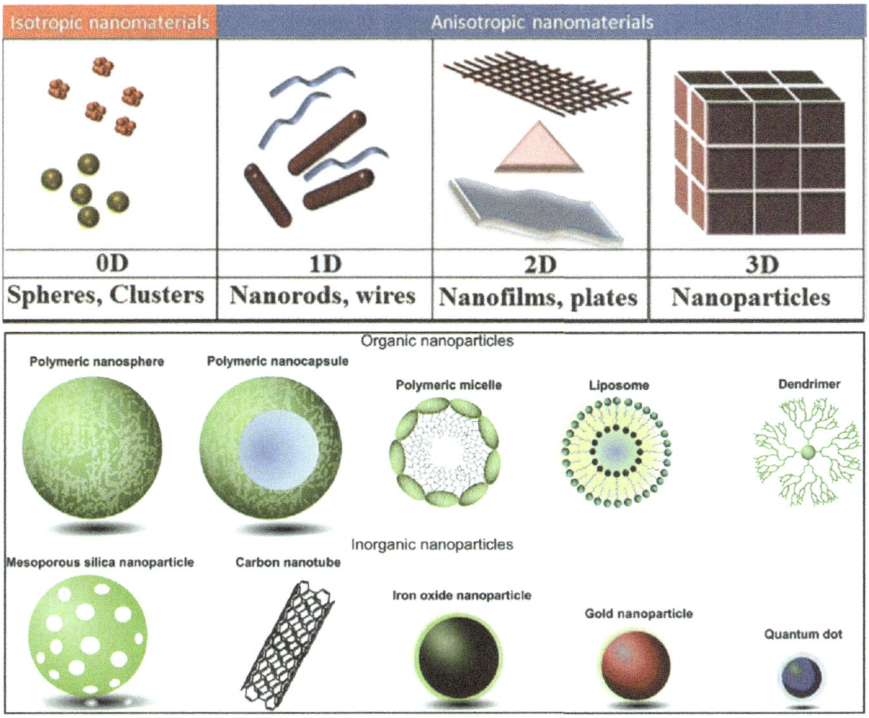

Fig. 3.3 Classification of nanomaterials based on dimensions (top panel), figure taken from Sajanlal et al. 2011 with permission and nature of materials (bottom panel). Figure taken from Richards et al. 2017 with permission

physicochemical properties. Among them optical and chemical properties are fascinating which can be tuned finely by controlling the size, morphology and composition of nanomaterials.

3.2.3 Optical Properties

Noble metal nanoparticles (NMPs) attracted tremendous interest among various nanomaterials owing to their unique optical and catalytic properties (Srivastava et al. 2021). Au, Ag, Pt and Pd are considered as noble metal nanoparticles. NMPs show different interaction properties with light. When light is incident on the NMPs, localized surface plasmon resonance (LSPR) is generated on the surface of nanoparticle at specific frequency of light radiation and this is responsible to enhance several orders of magnitude of various light–matter interactions including large increment in selective absorption and scattering of light and electric field (EF) enhancements on the surface of nanoparticles (Polo et al. 2018a). Nanoparticles

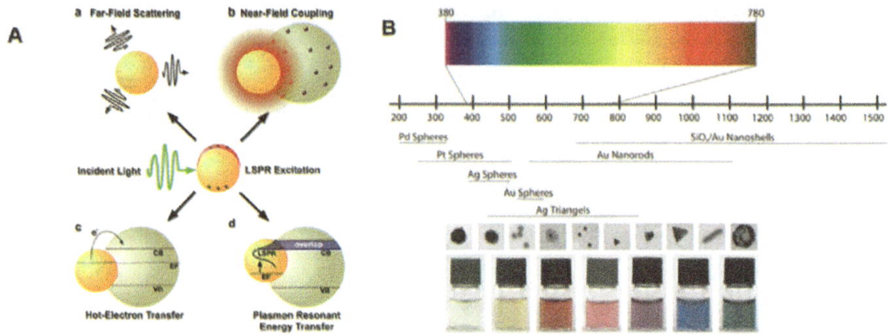

Fig. 3.4 (A) Schematic illustration for the interaction of light with plasmonic NPs. Figure taken from Erwin et al. 2016 with permission (B) The optical properties are highly dependent on the size, shape and composition of NPs. Figure taken from Tsuzuki 2009 with permission

which exhibit LSPR phenomenon with the interaction of light are known as plasmonic nanoparticles.

LSPR is responsible for the generation of different types of mechanisms on the surface of nanoparticles such as far-field scattering, near-field coupling, hot-electron transfer and plasmon resonant energy transfer (Fig. 3.4A) (Erwin et al. 2016). LSPR can be modify easily by tailoring the size, morphology and composition of various NMPs. Different morphologies and compositions of NMPs (Au, Ag, Pt, Pd) absorb and emit different frequencies of electromagnetic (EM) radiations as shown in Fig. 3.4B (Tsuzuki 2009). The interaction properties of various nanomaterials with light have emerged into a large distinct field known as nanoplasmonics (Polo et al. 2018b). Plasmonic nanomaterials can able to absorb and scatter specific wavelengths of EM with molar coefficients higher than several orders of magnitude compared with molar coefficients values of conventional organic fluorophores (Scarabelli et al. 2017).

Plasmonic nanomaterials are used in various biomedical applications (bioplasmonics) including biosensors, photothermal therapy (PTT), two-photon luminescence (TPL) imaging, optoacoustic imaging (OI) and theranostics applications (Wong et al. 2020). NMPs also exhibit unique catalytic properties owing to enhanced surface area to volume ratio (S/V), quantum size effects, interfacial and surface chemistry (Scarabelli et al. 2017).

3.3 Synthetic Strategies for Nanomaterials

The synthesis of nanomaterials is mainly classified into two major approaches such as "top-down approach" and "bottom-approach". In top-down approaches, a bulk material is break into small pieces, these small pieces are grinded, crushed and moulded into small size nanomaterials, whereas in bottom approaches,

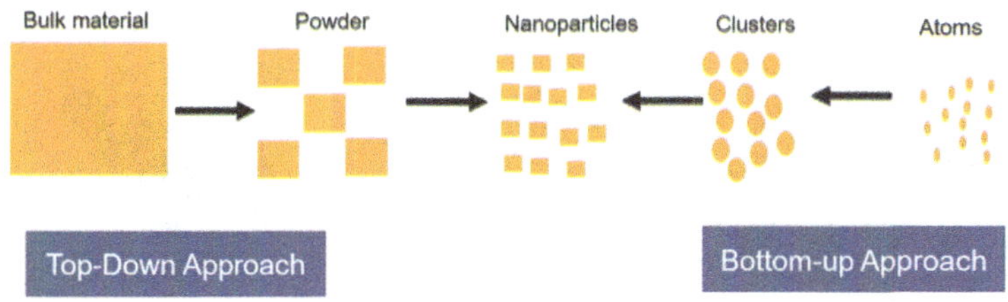

Fig. 3.5 Schematic illustration of the synthetic strategies of nanomaterials

nanomaterials are built from the systematic assemble of atoms/molecules using different types of capping ligands/surfactants (Fig. 3.5).

The major advantage of top-down approaches is that the nanomaterials are synthesized in large quantity in a short span of time, whereas in bottom-up approaches, homogeneous nanomaterials with monodisperse size, shape and composition can be synthesized. The synthetic strategies for the NMPs can also be classified into physical, chemical and biological, depending upon the type of approach used for the synthesis of NMPs. A large variety of nanoparticles such as nanospheres, nanowires, nanocubes, nanorods, nanostars, nanoprisms, nanoplates, nano-hollow spheres, nanoflowers, nanotubes, nanospheres, octahedral and tetrahedral nanoparticles are synthesized using different synthetic strategies for various applications (Jain et al. 2008). The different synthetic strategies using top-down and bottom-up approaches for the fabrication of Ag nanoparticles as a model system are shown in Fig. 3.6 (Mukherji et al. 2018). These synthetic strategies can also be employed for the synthesis of various NMPs.

3.3.1 Top-Down Synthetic Strategies

Top-down synthetic methods are mainly based on the concept of breaking down the large materials to desired nanometre scale dimensions. These synthetic methods make use of conventional workshop or microfabrication methods for milling, cutting, moulding and carving of the large size materials to specific size and shape. The commonly used methods based on the top-down strategies for the synthesis of various types of nanoparticles are discussed below.

3.3.1.1 Pyrolysis

Pyrolysis method is most commonly used for the synthesis of noble metal nanoparticles (Zhu et al. 2015). This method is facile and most commonly employed industrially viable production of nanoparticles in powder form (Kim et al. 2002). In this approach, a vaporous precursor (liquid or gas) is allowed to pass within the orifice at high pressure followed by burning which leads to ash production. The solid

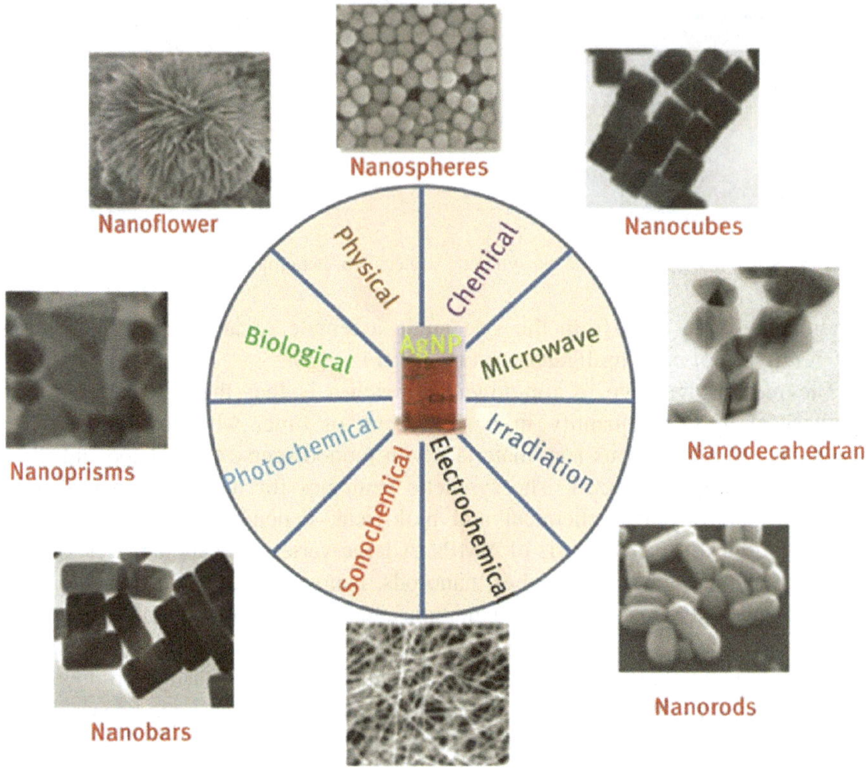

Fig. 3.6 Schematic illustration for the synthesis of Ag NPs using different synthetic strategies as a model system. Figure taken from Mukherji et al. 2018 with permission

ash is process further to recover the nanoparticles. The major drawbacks of this method are (i) large aggregates or agglomerates with non-uniform size nanoparticles are formed and (ii) requires enormous amounts of energy to maintain high pressure and temperature which are its major limitations.

3.3.1.2 Attrition (Milling)

The process of grinding the macro and micro-scale materials to make nano-size particles in a ball mill is known as attrition or milling. This method has been commonly used in the ceramic processing and powder metallurgy-based industries for many years. The basic principle in the attrition process depends on the amount of energy transfer to sample from the milling media. During this process, kinetic energy is transferred from balls to sample which helps in reduction of grain size of the macro-scale materials. This process involves various parameters such as type of mill, milling temperature, milling surrounding environment, duration of milling, milling intensity, and the amount of powder used for milling purpose, which shows greater impact on the physical properties of nanoparticles including size and composition. In order to improve these limitations, different types of milling units have been developed for various applications including shaker mills, tumbler mills, vibratory mills, attrition mills, planetary mills, etc. These devices have different operating conditions along with different efficiency milling capacities and have additional accessories such as to control heat and isolate different sizes of nanoparticles. Liquid assisted grinding has several advantages compared with "dry milling" mechano-chemical process, including greater time efficiency, reduction in wastage of materials and energy, which results in improved quality of nanoparticles (Xu et al. 2015).

This process is very simple and highly advantageous for the large-scale production of nanoparticles. It is very easy to prepare different alloys or nanocomposites using this method compared with conventional nanoparticles synthetic methods like chemical reduction, sonochemical, and electrochemical, etc. It also has disadvantages like formation of disordered crystal structures during the milling process. In addition, the sample is contaminated with Fe and minor amounts of N_2 and O_2 which usually comes from the milling tools and environmental conditions, surface defects in addition to the generation of internal stress. The presence of surface defects greatly influences the physicochemical properties of nanoparticles due to high aspect ratio.

3.3.1.3 Micropatterning

This technique is generally known as "art of miniaturization of patterns" and commonly used in nanoelectronics field (Chen et al. 2009). It also widely used in bioengineering, cell biology and biomaterials engineering areas. Photolithographic techniques and several different techniques have been developed in micropatterning process for the last several years. These include E-beam lithography, soft-lithography, scanning lithography, nano-imprint lithography, colloidal lithography and scanning probe lithography, and many others. The fundamental concept of this technique is usage of various EM radiations such as light (UV-Visible), X-rays, electrons, ions, with well controlled focused beam to remove and design specific nano-size structures on a precursor material known as "resist" and forming systematic nano-design patterns (Carrico et al. 2007).

3.3.2 *Bottom-Up Synthetic Strategies*

In this process, nanoparticles are constructed from atoms and molecules based on their chemical properties in which they are built together slowly in the presence of capping ligands/surfactants. The concept of this method is based on the molecular self-assembly and/or molecular recognition. Capping ligands are used to control the growth of nanoparticles, several different types of capping ligands are available for the synthesis of nanoparticles (Shaik 2020). In contrast to top-down approach, homogeneous nanoparticles can be synthesized using this process. The size, shape and composition of NMPs can be tailored very easily with perfect crystallographic structures for various applications are synthesized using bottom-up approach. The common bottom-up synthetic strategies are described below along with their advantages and demerits.

3.3.2.1 Chemical Reduction Method

This method is most commonly used for the synthesis of NMPs. In this approach, reduction of an ionic salt (metal ions) takes place using different reducing agents in the presence of capping ligands/stabilizing agents. Various types of reducing agents such as NaOH, $NaBH_4$, hydrogen, alcohols, carbon monoxide, hydrazine are most commonly used for the synthesis of NMPs.

Lee-Meisel and Creighton are the most commonly used solution-based methods for the synthesis of NMPs (Evanoff Jr and Chumanov 2005). The Lee-Meisel method is generally used for the synthesis of Ag nanoparticles, in this approach Ag nitrate and sulphate salts are utilized as metal precursor salts which are reduced by using H_2, sodium citrate and $NaBH_4$ as reducing agents at various temperature conditions (Lee and Meisel 1982). The size and shape of Ag nanoparticles can be tuned easily varying the pH of the reaction conditions. Different morphologies of Ag nanoparticles were obtained using this approach. For lower pH conditions (5.5-11.1), triangle and polygon shapes of Ag nanoparticles were observed due to slow reduction kinetics rates, whereas for high pH values spheres and rods were obtained due to fast reduction kinetics of Ag nitrate. The morphologies of Ag nanoparticles can be tuned easily by monitoring the pH of the reaction conditions (Evanoff Jr and Chumanov 2005).

In a similar approach, Au nanoparticles are synthesized by reduction of Au^{+3} ions using different reducing agents. In 1951, Turkevich et al. demonstrated the synthesis of Au nanoparticles of size 20 nm using citrate mediated reduction of aqueous $HAuCl_4$ solution (Turkevich et al. 1951). This method is known as Turkevich method. TEM images of Au nanoparticles of different sizes such as 10, 20, 50 and 100 nm synthesized using Turkevich method are shown in Fig. 3.7. Later, this method was modified by the Frens in 1970s.

In the early 1990s, the Brust method for the synthesis of Au nanoparticles was developed by Brust and Schiffrin (Brust et al. 1994). The Turkevich method is based on the synthesis of Au nanoparticles in aqueous solution, whereas Brust method is based on the synthesis of Au nanoparticles in organic solvent. Though several methods were developed for the synthesis of Au nanoparticles, all those methods were considered as the variants of original methods developed by Turkevich and Brust. Zhang et al. reported on the synthesis of platinum nanocubes using general chemical reduction method (Zhang and Fang 2009). Pd nanoparticles are also synthesized using chemical reduction method (Ganesan et al. 2007).

The chemical reduction method has several disadvantages though it is considered as a facile conventional method for the synthesis of nanoparticles including NMPs.

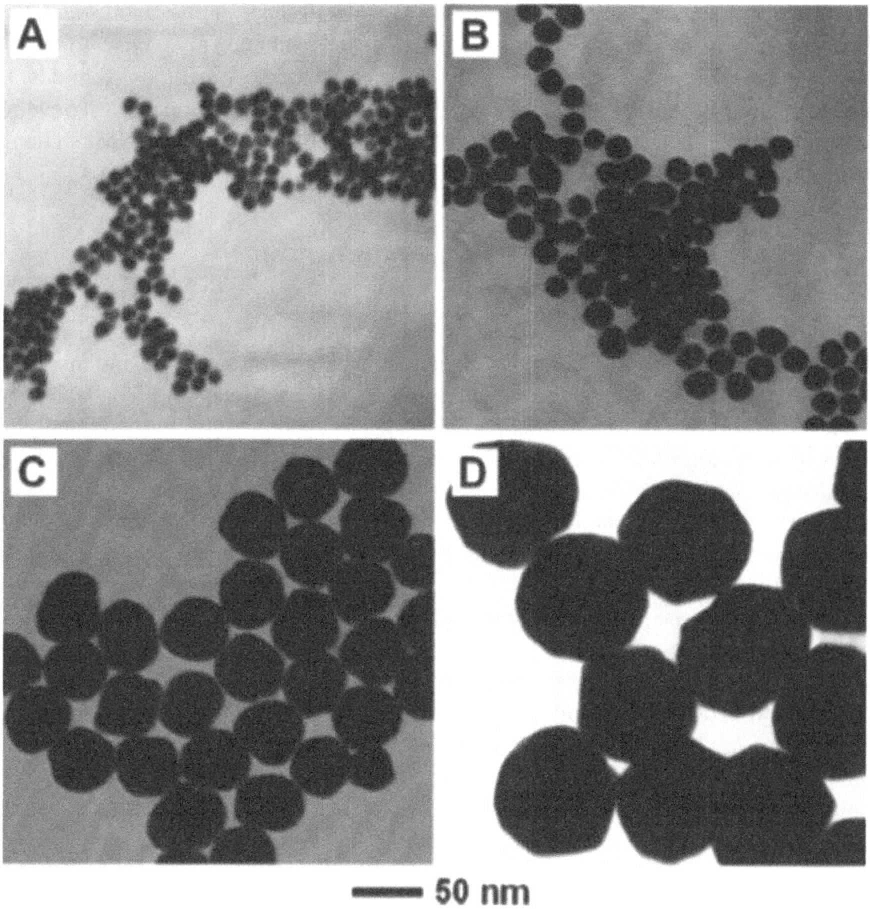

Fig. 3.7 TEM images of synthesis of Au nanoparticles with different sizes prepared using Turkevich method: (A) 10, (B) 20, (C) 50 and (D) 100 nm. Figure taken from Njoki et al. 2007 with permission

This synthetic method normally requires very high temperature and pressure conditions, and longer time periods to complete the synthesis of nanoparticles. In addition, most of toxic chemicals are used during the reaction which may show adverse effects on environment. The presence of capping ligands on the surface of nanoparticles restrains their usage in biomedical applications (Shaik et al. 2017).

3.3.2.2 Electrochemical Method

Reetz and Helbig are the pioneers for the synthesis of metal nanoparticles using electrochemical method (Reetz and Helbig 1994). In their method, metal nanoparticles were synthesized by dissolving metal sheet at the anode and the intermediate metal salt formed get reduced at the cathode, give raise to metal nanoparticles. In the last one decade, several electrochemical methods were developed to synthesize different shapes, aspect ratios and facets of nanoparticles. Several electrochemical methods were introduced for the synthesis of various NMPs (Jiang et al. 2020).

3.3.2.3 Microemulsion Method

In the early 1980s, this method was first introduced for the synthesis of Pt, Pd and Rh nanoparticles. In this approach, salts and reducing agents are dispersed and mixed in the presence of surfactants in two different emulsions (water-in-oil or water-in-water) for the synthesis of nanoparticles. The micelles were formed, and the Brownian motion of the micelles is responsible for inter-micellar collisions which results in the mixing of reactants followed by nucleation for the synthesis of nanoparticles. This method has become so popular for synthesis of various types of nanoparticles in water-in-oil or oil-in-water microemulsions (Najjar 2012). The highly advantage of this method is that thermodynamically stable and monodisperse nanoparticles can be synthesized very easily.

3.3.2.4 Microwave Method

In this method, microwave irradiation is used for the synthesis of NMPs using metal salts as precursors in the presence of polymer surfactants solution. It is a facile one pot approach used for the synthesis of metal nanoparticles. The size, shape and composition of NMPs can be tuned easily by variation and standardization of reaction conditions. There are several good review articles which can give comprehensive information on synthesis of MNPs using microwave method (Zhu and Chen 2014).

3.3.2.5 Laser Ablation

Laser ablation is a very flexible method commonly used for the synthesis of nanomaterials in which the laser light is used to peel the material from a solid surface. Laser light with low flux power is irradiated on the material which absorbs the laser energy and gets heated in which material is vaporized or sublimated. Laser wavelength and the optical properties of the material play key role to decide on the amount of material get removed from the solid surface. In recent years, this method has become an alternative potential method to chemical reduction method for the synthesis of nanoparticles. The size and shape of NMPs can be tuned easily and the absence of removal of excess reagents from the synthesis reaction is the main advantage of this method (Amendola et al. 2006). Mafune et al. reported on the synthesis of stable nanoparticles on the metal plate using sodium dodecyl sulphate surfactant (Mafuné et al. 2000). Several different types of ligand-free NMPs were synthesized using this approach. This method has potential for large-scale production of NMPs with tunable size, shape and composition (Arce et al. 2017).

3.3.2.6 Green Synthesis Methods

The conventional chemical reduction methods mainly involve the usage of toxic chemicals and reducing agents for the synthesis of NMPs, whereas in green synthesis methods the NMPs are synthesized using mild solvents and non-toxic reducing agents. Biomolecules such as starch, proteins and lipids are used as capping agents for synthesis of different types of nanoparticles (Li et al. 2017).

3.3.2.7 Biological Methods

In this method, different types of microorganisms are exploited for the synthesis of NMPs (Prasad et al. 2016; Srivastava et al. 2021). Microorganisms act as potential nanofactories which have capability to synthesize NMPs without using any toxic chemical precursors, reducing agents and avoiding requirement of high energy and sophisticated tools. In recent years, microorganisms, including bacteria (actinomycetes), yeast and fungus have been studied extensively for the intra- and extracellular production of nanoparticles (Prasad 2016, 2017, 2019a, b). Microorganisms contain several enzymes which help in absorbing different heavy metals and detoxify them, and act as reducing agents to reduce metal salts to metal nanoparticles (Prasad et al. 2018a, b; Prasad and Aranda 2018). Several biological protocols have been reported on the synthesis of NMPs using bacterial biomass, supernatant and plant tissues (Fig. 3.8) (Singh et al. 2016a). Among the various methods, extracellular production of nanoparticles has gain great attention due to absence of downstream processing steps such as sonication for the cell wall breakage, and centrifugation for isolation and purification of nanoparticles. In addition, several different types of proteins, enzymes and metal-resistant genes act as reducing agents. These act as natural capping agents for the synthesis of monodisperse nanoparticles with greater stability.

Microorganisms such as bacteria, fungi and yeast are most commonly used for the synthesis of NMPs (Table 3.1). Among bacteria species like *Bhargavaea indica* (Singh et al. 2015b), *Weissella oryzae* (Singh et al. 2016e), *Bacillus methylotrophicus* (Wang et al. 2016) and *Brevibacterium frigoritolerans* (Singh et al. 2015c) are most commonly explored for the synthesis of Au and Ag nanoparticles. Various genera of microorganisms including *Corynebacterium, Lactobacillus, Bacillus, Pseudomonas, Klebsiella, Escherichia, Enterobacter, Aeromonas, Rhodococcus, Brevibacterium, Streptomyces, Trichoderma, Desulfovibrio, Sargassum, Shewanella, Plectonema boryanum, Rhodopseudomonas, Pyrobaculum* and others have been explored for the synthesis of metal nanoparticles (Li et al. 2011).

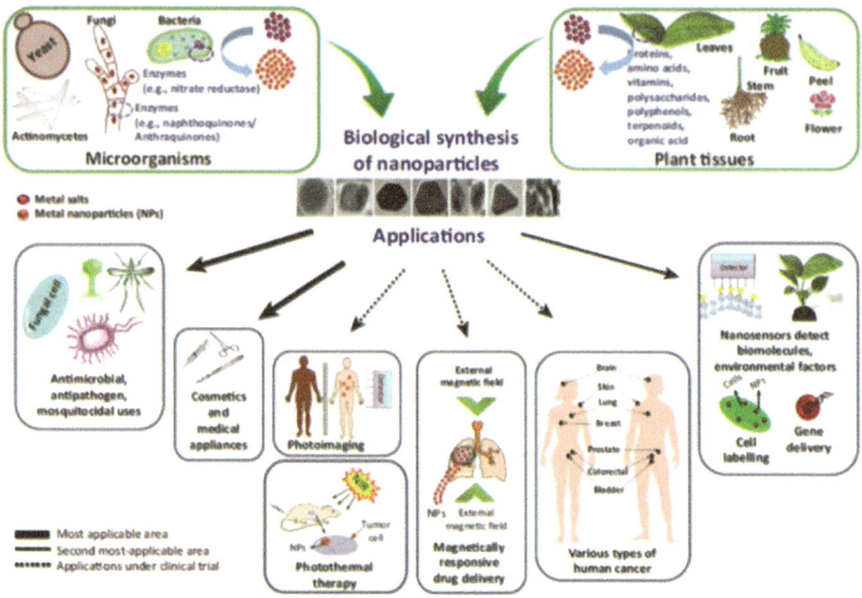

Fig. 3.8 Schematic illustration for the biological synthesis of nanoparticles and their applications in biomedical and environmental. Figure taken from Singh et al. 2016a with permission

The synthesis of NMPs by fungi is known as mycosynthesis that attracted many advantages compared to bacterial synthesis, since the production of nanoparticles in mycosynthesis is straight forward and it is easy to downstream the nanoparticles. Thus, it is considered as a cost-effective and efficient strategy for the synthesis of nanoparticles (Alghuthaymi et al. 2015; Abdel-Aziz et al. 2018). In addition, fungi show more tolerant towards the uptake and competences for metals, particularly in terms of high affinity of metal salts towards the wall-binding of fungus which leads to high production of nanoparticles. Fungal enzymes and electron shuttle quinones play major role in the synthesis of NMPs. Enzymes such as reductase enzymes from Penicillium species and *Fusarium oxysporum*, and nitrate reductase were found to have significant role and the mechanism is like the synthesis of nanoparticles in bacteria (Kumar et al. 2007; Prasad et al. 2016).

Plants are also used for the synthesis of NMPs and the applications of plants in nanotechnology are known as phytonanotechnology. Phytonanotechnology has several advantages such as simple process, cost-effective, eco-friendly, biocompatible and scalability for the synthesis of nanoparticles using water as reducing agent (Prasad 2014, 2019c, d). Nanoparticles which are produced from plants are non-toxic and can use for biomedical and environmental applications. Recently, the roots and leaf extracts of *Panax ginseng* medicinal plant are successfully used for the synthesis of nanoparticles, which show the potential of medicinal plants towards the synthesis of nanoparticles (Singh et al. 2016c). In addition, several plant roots, leaves, stems and their extracts are also used for the synthesis of nanoparticles

Table 3.1 Synthesis of nanoparticles from microorganisms

Microorganism	Extracellular/ Intracellular	Type of Nanoparticle	Shapes	Size (nm)	References
Pseudomonas deceptionensis	Extracellular	Ag	Spherical	10	(Jo et al. 2016)
Bacillus methylotrophicus	Extracellular	Ag	Spherical	30	(Wang et al. 2016)
Bhargavaea indica	Extracellular	Ag and Au	Spherical	30	(Singh et al. 2016d)
Bacillus Amyloliquefaciens	Extracellular	CdS	Cubic/ hexagonal	4	(Singh et al. 2011)
Bacillus pumilus	Extracellular	Ag	Spherical/ Tri angular	90	(Elbeshehy et al. 2015)
Listeria monocytogenes	-	Ag	Anisotropic	Non-uniform shapes and sizes	(Soni and Prakash 2015)
Neurospora crassa	Intra and extracellular	Ag, Au, Au-Ag alloy	Spherical	>100	(Castro-Longoria et al. 2011)
Yarrowia lipolytica NCYC 789	Extracellular	Ag	Spherical	15	(Apte et al. 2013)
Rhodosporidium diobovatum	Intracellular	Lead	Non-Uniform	2-5	(Seshadri et al. 2011)
Candida utilis NCIM 3469	Extracellular	Ag	Spherical	20-80	(Waghmare et al. 2015)

Table 3.2 Synthesis of nanoparticles from plants

Plants	Plant parts for Extraction	Type of Nanoparticle	Shapes	Size (nm)	References
Sargassum algae	Alga	Pd	Octahedral	10	(Momeni and Nabipour 2015)
Panax ginseng	Root	Ag and Au	Spherical	40	(Singh et al. 2015a)
Red ginseng	Root	Ag	Spherical	30	(Singh et al. 2016b)
Cymbopogon citratus	Leaves	Au	Spherical	50	(Murugan et al. 2015)
Azadirachta indica	Leaves	Ag	Spherical/Tri angular	90	(Poopathi et al. 2015)
Cocos nucifera	Leaves	Lead	Spherical	47	(Elango and Roopan 2015)
Catharanthus roseus	Leaves	Pd	Spherical	40	(Kalaiselvi et al. 2015)
Pistacia atlantica	Seeds	Ag	Spherical	27	(Sadeghi et al. 2015)
Banana	Peel	CdS	Spherical	2	(Zhou et al. 2014)

(Table 3.2). The presence of various proteins, vitamins, organic acids, amino acids and secondary metabolites such as polysaccharides, flavonoids, polyphenols, terpenoids, heterocyclic compounds in plant products plays significant role for the reduction of metal salts, which can act as reducing and capping agents for the synthesis of nanoparticles. It has been demonstrated that the *Corallina officinalis* extract consisting of hydroxyl functional group from polyphenols and the carbonyl group from proteins could assist in forming and stabilizing gold nanoparticles (El-Kassas and El-Sheekh 2014). In addition, *Murraya koenigii leaf* extract has used for the synthesis of Ag and Au nanoparticles (Philip et al. 2011). Several different types of mechanisms are reported for the synthesis of nanoparticles using different plant leaf extracts.

Though there are several advantages for biological synthesis of nanoparticles, the major challenge is monodispersity of nanoparticles. It is a challenging task to obtain monodisperse nanoparticles using biological agents. There are several recent reports which rationally designed for fine tuning the size and shape of nanoparticles. The environmental conditions, pH, temperature and the nutrient supplements in the growth medium play a key role in controlling the size and shape of nanoparticles. The optimal growth of microorganisms needs to be maintained for the better synthesis of nanoparticles. pH is the most influential factor on the growth of nanoparticles, different nanoparticles can be synthesized by tuning the pH conditions. For example, Gurunathan et al. (Gurunathan et al. 2009) reported that the most Ag nanoparticles are synthesized at pH 10. Among fungi, different pH conditions such as acidic pH (for *Fusarium acuminatum*), (Banu and Balasubramanian 2014) pH 6 (for *Penicillium fellutanum*) (Gurunathan et al. 2009) and alkaline pH (*Isaria fumosorosea*) (Banu and Balasubramanian 2014) are the optimal pH conditions for the synthesis of nanoparticles. In plants also the pH conditions affect the yield and morphology of nanoparticles. For example, in *Avena sativa* extract, many small size Au nanoparticles were formed at pH 3 and 4, whereas, at pH 2, nanoparticle aggregation takes place (Sathishkumar et al. 2010). Thus, acidic pH conditions promote aggregation of nanoparticles during their synthesis using plants. Thus, environmental conditions, pH and temperature need to optimize for the biological synthesis of nanoparticles.

3.3.3 Characterization Techniques of Nanoparticles

The physicochemical properties of NMPs are mainly depend on the size, shape and composition of nanoparticles (Table 3.3). There are several techniques to characterize size, shape, phase constitution, microcrystal structure and composition of nanoparticles. Developing novel instruments and their usage for characterization of nanoparticles is the most topic in the area of nanotechnology. The most common techniques used for the characterization of nanoparticles are electron microscopes (Transmission electron microscope (TEM) and Scanning electron microscope (SEM)), atomic force microscopy (AFM), confocal microscope, dynamic light

Table 3.3 Characterization techniques of nanoparticles

Parameter of Nanoparticles	Technique
Particle size	Transmission electron microscopy (TEM), dynamic light scattering (DLS), laser diffractometry,
Optical properties	Ultraviolet-visible spectroscopy (UV-vis.), fluorescence spectroscopy
Crystal structure	X-ray diffraction (XRD)
Chemical analysis	FT-IR, EDS, scanning electron microscopy (SEM), ion spectroscopy and thermogravimetric analysis (TGA)

scattering (DLS), X-ray diffraction (XRD), energy dispersive spectroscopy (EDS), Fourier transform infrared spectroscopy (FT-IR), ICPMS, ultra-visible light spectroscopy (Uv-vis), thermogravimetric analysis (TGA) and fluorescence microscope. The recent techniques such as nanoparticle tracking analysis (NTA) and isoelectric focussing electrophoresis (IEF) mainly detect the Brownian motion and size distribution of nanoparticles, respectively, which help to monitor the tracking of nanoparticles.

3.3.4 Applications of Plasmonic Nanoparticles

In recent years, we have witnessed tremendous applications of plasmonic nanomaterials in various fields. In particular, employing the optical properties of plasmonic nanomaterials in real-life applications including biology and medicine is generally known as bioplasmonics (Prasad et al. 2020).

Bioplasmonics is one of the most attractive research areas in the field of bio-nanotechnology, Mirkin group pioneers the usage of plasmonic Au nanoparticles for the fabrication of calorimetric biosensors based on the distance dependent optical properties of nanomaterials, which triggers the usage of different plasmonic nanomaterials for biological applications (Storhoff et al. 1998). This field has grown tremendously and several clinical trials under progress and several commercial products are developed including HEATSENS, EXICURE, AuroLase therapy and many others. When the light is incident on the plasmonic nanomaterials, LSPR is generated and this is responsible for the generation of unique optical properties in plasmonic nanomaterials (the details regarding the optical properties of plasmonic nanomaterials are discussed above in the section "optical properties of nanomaterials"). We are exploiting those unique optical properties of plasmonic nanomaterials in biomedical applications. The different plasmonic responses are mainly used for numerous applications including biosensors, optoacoustic imaging, photothermal therapy (PTT), photovoltaics, two-photon luminescence (TPL) imaging and many others as shown in Fig. 3.9 (Polo et al. 2018a).

Several varieties of plasmonic nanomaterials probes are developed particularly for biological applications for fabrication of biosensors, and light-based therapies

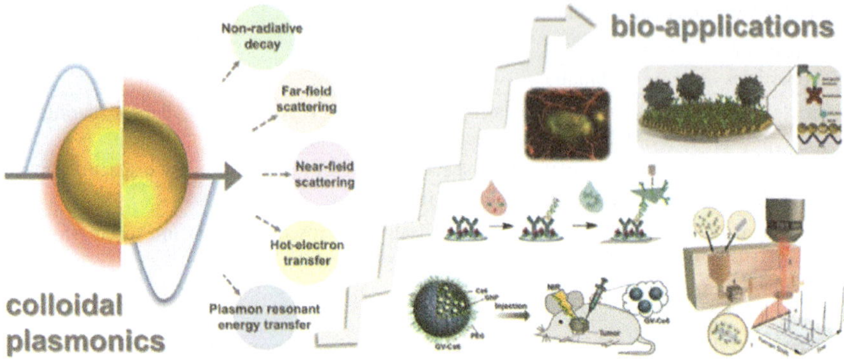

Fig. 3.9 Schematic illustration for different effects of interaction of light with plasmonic NPs and their bioapplications. Figure taken from Polo et al. 2018a with permission

and imaging. NMNs dimensions have similar size compared with several cell components, which make them ideal candidates for monitoring molecular interactions and reactions. These plasmonic NMNs behave as nanoantennas or nanotransducers, which help to remotely control the photoactivation reactions that trigger different responses including nanoheating, control binding and aggregation, or to help in enhanced signals (fluorescence, IR-absorption spectroscopy and surface-enhanced Raman techniques). The usage of nanophotonics for theranostics applications is shown in Fig. 3.10 (Conde et al. 2012).

To utilize plasmonic NMNs for bioapplications need to fulfil specific criteria based on the applications. For example, for applications in photothermal therapy, drug delivery or intracellular sensing requires of specific size, and biocompatibility. However, for sensing applications the optimization of field enhancements and homogeneity of plasmonic nanomaterials are the key parameters apart from the size and toxicity (Polo et al. 2018a).

The LSPR of plasmonic NMNs is mainly depend on the size and composition, thus tuning the size and composition is one of the intense research areas for synthesis of different plasmonic nanomaterials. At first tuning the composition of plasmonic nanomaterials looks the good option, but there are only several nanoparticles which can show plasmonic properties such as (Au, Ag, Cu and its chalcogenides, Al, Rh, Ru, In, Na). However, Rh, Ru and In metals are only active in UV region (Ren et al. 2003), Al and Na show less stability towards oxidation (Ross and Schatz 2014), and Ag, Cu and its chalcogenides are toxic in nature, which limits the availability of different types of plasmonic nanomaterials for bioapplications. Tuning the size and plasmonic coupling of plasmonic nanoparticles are also the key parameters to control the plasmonic properties of the NMNs. The general plasmonic nanomaterials for common bioapplications are shown in Table 3.4.

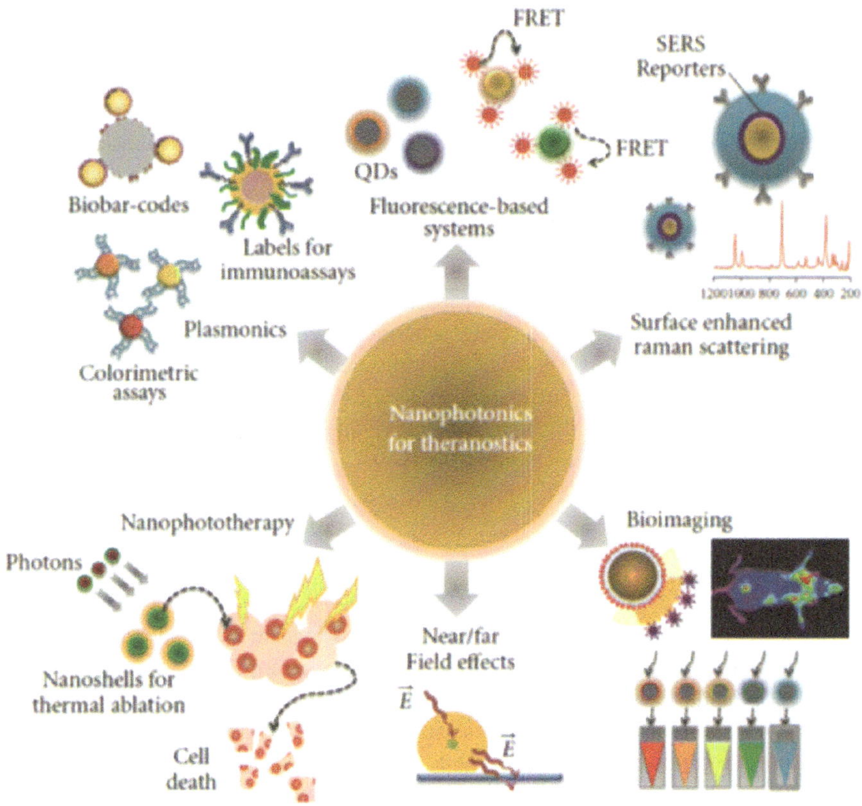

Fig. 3.10 Schematic illustration for the usage of nanophotonics for various bioapplications. Figure taken from Conde et al. 2012 with permission

Table 3.4 The most common bioapplications of different plasmonic nanomaterials

Plasmonic Probes	Dimensions[a]	LSPR (nm)	Bioapplications	References
Au nanorods	l=20-100; t= 8- 30	600-1500	PTT, TPL, SERS	(Nikoobakht and El-Sayed 2003)
Au nanostars	d=30-100	600-900	PTT, SERS	(Chen et al. 2013)
Au nanoshells	d = 50 -120; t = 2 -30	600-1000	PTT, SERS	(Hirsch et al. 2003)
Au nanocages	e = 20 -100	600-1000	PTT, TPL, SERS	(Au et al. 2010)
Au nanoprisms	e = 50 - 200; t = 8 - 20	800-1500	PTT and thermal sensing	(Polo et al. 2013)
Au NPs	d = 10- 100	510-550	Colorimetric, SERS	(Elghanian et al. 1997)
Ag NPs	d = 10 -100	400-600	Colorimetric, SERS	(Mock et al. 2003)

[a] length (l), diameter (d), edge length (e) and thickness (t)

3.3.4.1 Diagnostic Applications of Plasmonic Nanomaterials

In the recent years, different types of plasmonic sensors are fabricated in which the final output is based on the optical read-out, the basic principle is that plasmonic nanomaterials enhance different light–matter interactions by several orders (Shaik et al. 2018). The optical read out-system is mainly work depending on the changes occurring in aggregation state of nanoparticles and refractive index of the surrounding medium, and change in the colour for the techniques like ELISA.

The development of LSPR on NP surfaces induces EF enhancements, in turn which bring lot of changes in optical properties mainly in fluorescence quenching or enhancements, SERS, and their electrochemical activity, act as transducers for binding events which help for detecting biological analytes at low concentration level. The high selectivity of target molecules using label-free assays makes plasmonic NPs as ideal candidates for the fabrication of different types of sensors in environmental and biomedical areas.

3.3.4.2 Sensing

Biosensors are fabricated using different types of plasmonic nanoparticles. The unique feature of localizing the EF at the plasmonic nanoparticle surface helps for the development of various types of biosensors that have potential to detect analyte of interest in complex biological scenario (Luan et al. 2018). Many LSPR based biosensors have improved a lot owing to the increased absorption and scattering phenomenon. This is also exploited for the development of SERS-based detection of biological analytes. The mechanism of biosensors for detection of analytes is different, they are mainly work on the change in refractive index, colour change, change in fluorescence signals and enhanced SERS signals are the key parameters which are indirectly measured to detect the sample of interest (biomarkers, bacteria, virus, toxins, etc.). Several label-free biosensors are also developed for the detection of virus, bacteria and toxins, etc. Au and Ag based plasmonic nanoparticles are mainly used in several LSPR based biosensors for recognizing different types of biomolecules and biomarkers related to specific diseases (Nusz et al. 2009). To enhance the sensitivity of biosensors, the size, shape and compositions of NPs are tuned and used in the fabrication of biosensors.

The NPs surface functionalization is the key parameter for enabling the identification of different analytes in biological complex media, since NPs are exposed to whole blood, serum and plasma which may lead to unwanted interactions. The functionalization of several types of biomolecules on the surface of NPs avoids unspecific interactions. Several immune assays are developed based on tuning the aggregation of plasmonic NPs. The combination of plasmonic NPs with enzyme-linked immunosorbent assay (ELISA) is called plasma ELISA, widely employed to detect several analytes. Inci et al. developed the plasmonic based biosensor for the detection of HIV virus in the unprocessed blood. The biosensor is based on the

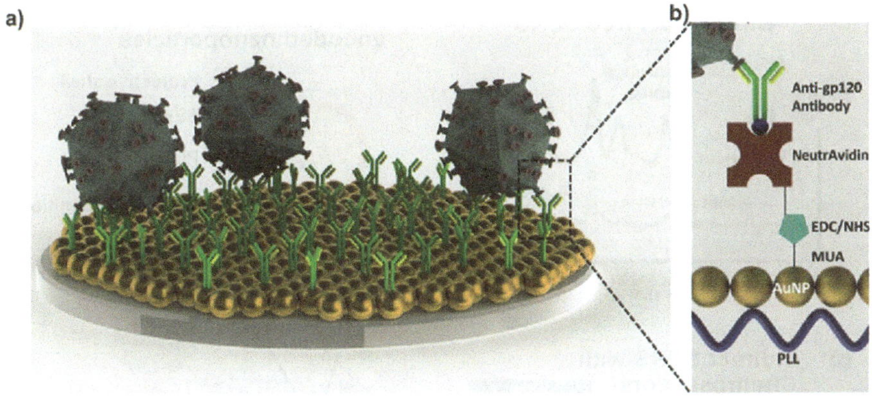

Fig. 3.11 A biosensor based on the functionalization of Au NPs with anti-gp120 antibody is developed for the detection of virus based on the change of LSPR. Figure taken from Inci et al. 2013 with permission

plasmonic Au NPs linked with antibodies to selectively capture rapidly emerging various types of HIV as shown in Fig. 3.11 (Inci et al. 2013).

In recent years, the critical developments in nanofabrication and instrumentation have taken place and the SERS has brought many changes towards the fabrication of biosensors for real-life applications. SERS-based sensing strategies are mainly divided into two types: direct vs. indirect based SERS. In direct SERS method, the analyte of interest is detected directly and it gives the characteristic SERS spectra for the target analytes. To achieve this, the analyte needs to adsorb on the plasmonic substrate and to achieve this, analyte should show affinity towards the plasmonic nanomaterials. For example, Guerrini et al. fabricated a positive charged Ag NPs to which negatively charged DNA adsorb on the surface of Ag NPs. Electrostatic adhesion between the negatively charged phosphate groups of DNA with the positively charged Ag NPs further induces the trapping of DNA molecules between the Ag NPs, which act as a highly active SERS cluster. This help to improve SERS signals which can able to detect DNA at very low concentration. Fig. 3.12a represents the schematic illustration for the detection of double-stranded DNA interaction with the chemotherapeutic agent of Cisplatin (Masetti et al. 2015). The direct SERS approach provides the intrinsic SERS spectrum of the analyte related to chemical-specific interactions. However, this method is not suitable to detect the analytes in complex biological media. In this case, different chemical species co-adsorb with the analyte of interest and gives inaccurate information.

The indirect SERS approaches can overcome the issues and difficulties arise with direct SERS measurements. In indirect SERS measurements, the analysis and detection of analyte of interest can be monitored by measuring the vibrational responses of a chemosensor or a SERS reporter. Fig. 3.12b represents an example for the real time identification of nitric oxide (NO) in the living cell using a 4-aminobenzenthiol (ABT) as a chemoreceptor (Rivera_Gil et al. 2013). ABT is

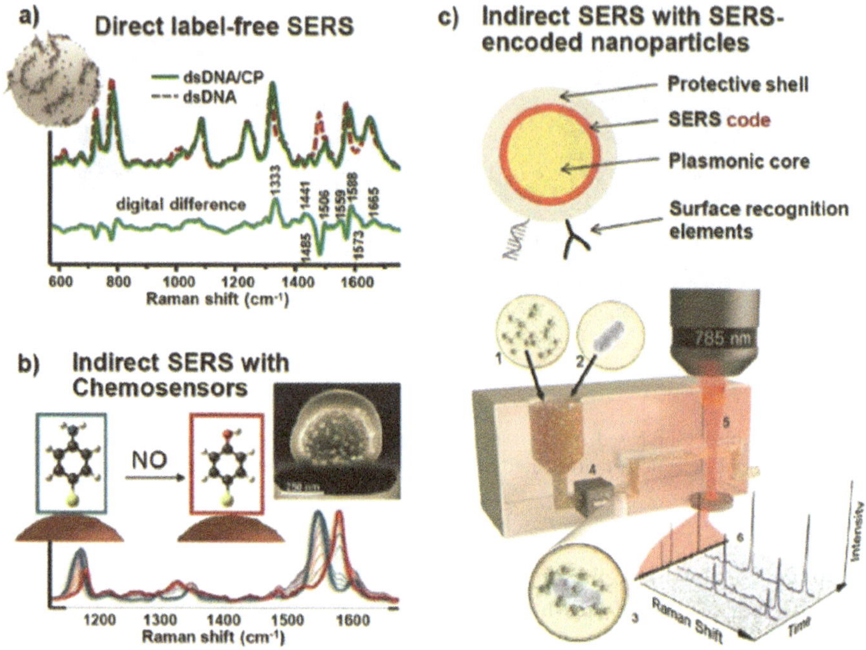

Fig. 3.12 Schematic illustration for different SERS-based sensing schemes. (a) Direct label- free SERS detection of double-stranded DNA (dsDNA). Figure taken from Masetti et al. 2015 with permission (b) Indirect SERS with chemosensors for the detection of nitric oxide (NO) in living cells. Figure taken from Rivera_Gil et al. 2013 with permission (c) Indirect SERS with SERS-encoded NPs for detection of bacteria in biofluids. Figure taken from Pazos-Perez et al. 2016 with permission

covalently attached to the Au nanoparticles coated in a mesoporous silica. NO penetrates into the mesoporous shell reacts with ABT through diazotization process, which alters the SERS spectra of the chemosensor. The extent of alterations of these chemosensor profile SERS spectra can be correlated quantitatively with NO. The other indirect SERS approach is identification of analyte based on SERS-encoded NPs (SEPS) other than chemosensors. The encoded NPs typically consist of plasmonic core labelled with a SERS code stabilized with inert shell which is usually consist of silica, PEG or some other materials. The outer surface of the inert shell is conjugated with bio-recognition elements (aptamers, oligonucleotides, antibodies, etc.) which improve the selectivity affinity towards the analyte of interest. For example, Alvarez-Puebla et al. integrated SEPS with microfluids to fabricate a device which operates based on optical detection of the microorganisms for the quick detection and quantification of pathogens in large volume of biofluids as shown in Fig. 3.12c (Pazos-Perez et al. 2016). For the detection of analyte of interest in infected sample, SEPs encoded with specific biological elements for the detection of specific pathogens are mixed with the sample and these SEPs bound in large number at the surface of the pathogen membrane. The sample is irradiated with a

785 laser to detect SERS signals, as analyte of interest passes it generates huge amount of SERS signals as a result of large amount of SEPs present on the surface of pathogen membrane. The SERs code provides an unique spectral fingerprint for specific pathogen, which has potential to measure in complex sample consisting of different pathogens.

3.3.4.3 Bioimaging

Plasmonic NPs have large scattering effects when interact with light and this phenomenon is exploited for studying the NP-cell interaction studies in dark field microscopy. For example, Au NP of size 5 nm is used for a single molecule imaging in a live cell and for detection of cancerous cells by conjugated NPs with epidermal growth factor (EGF) (Leduc et al. 2013). These EGF receptors present in larger amount in cancerous cell. The mechanism is that Au NP of 5 nm will exhibit less scattering effect when they are inside the cells, but when they attached with EGF receptors on the cell membranes in which inter NPs greatly enhances the scattering effects. The light emitting NPs have improved the biomolecular imaging field, to which the multi-photon resonance microscopy has greatly contributed. A large variety of plasmonic nanomaterials which can absorb near infrared absorption (NIR) radiation have been used as TPL (two-photon luminescence imaging) contrast agents. Several types of multifunctional plasmonic nanocomposites have been developed for the multimodal imaging (Li et al. 2018). Nanoparticles are functionalized with other molecules such as dyes, or NPs or semiconductor NPs, upconverting NPs are used for enhanced contrast agents for imaging. For example, a dye integrated with plasmonic NP in a polymer or silica shell can enhance the fluorescence signal up to 20-fold high (Sotiriou et al. 2012). The size, shape and composition of Au NPs can be tuned to have similar excitations of fluorescence molecules in NIR region. Several different types of plasmonic NPs have been reported as contrast agents in optical imaging (OI) and non-invasive bioimaging applications.

For example, Leduc et al. used 5 nm Au nanoparticles functionalized with camelid antibodies that have capability to recognize widely used green fluorescent proteins (GFPs). These act as probes for imaging a single molecule in a living cell and in vitro using photothermal imaging shown in Fig. 3.13 (Leduc et al. 2013).

3.4 Therapeutic Applications

The physicochemical properties of plasmonic NPs are widely used in therapeutic applications mainly based on photothermal heating or heat triggered to release drug from plasmonic nanocarriers.

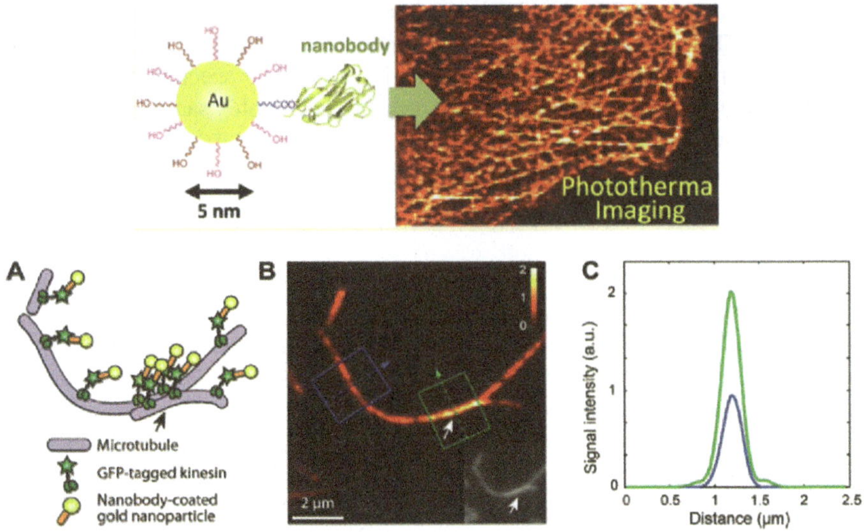

Fig. 3.13 (a) Nanobody functionalized Au NPs as an optical probe for single molecule tracking in live cells by photothermal imaging. (b) NB-Au-NPs target kinesin-GFPs in vitro. Figure taken from Leduc et al. 2013 with permission

3.4.1 Photothermal Therapy (PTT)

The photothermal response from plasmonic NPs has been exploited for many biomedical applications. In this therapy, plasmonic NPs generate heat which is higher than the physiological temperature (few degrees above 37 °C) and it is known as hyperthermia. The elevated temperature can be used to kill cancerous cells, since the cancerous cells are highly susceptible to heating. The nanoheating is localized one that the heating is confined within the tumour cells and do not spread to the surrounding healthy cells. The tumour cells can be inactivated with the slight increase in temperature prior to cancer treatments using radio or chemotherapy. Au and Ag nanoparticles are widely used for PTT applications. The optical properties of these plasmonic NPs can be monitored to achieve high near-IR (near infrared) absorption region by tuning the size, shape and composition of Au NPs. Halas and coworkers first reported the PTT in 2003 using silica-Au nanoshells with silica core of size 110 nm and a 10 nm thick Au shell (Pham et al. 2002). Many other shapes of Au NPs such as nanocages, nanoprisms, nanostars and nanorods are used for PTT (Pérez-Hernández et al. 2015).

Several distinct responses from the excitation of NPs by light have been reported inside the cell such as ablation, apoptosis and necrosis. These have different mechanisms such as generation of bubbles, reactive oxygen species (ROS), elevated temperatures and lysosomal membrane disruption for treatment using in PTT. Au nanoprisms that specifically induce apoptosis in mouse embryonic fibroblast cells transformed with the SV40 virus are shown in Fig. 3.14b. Au nanocages of size

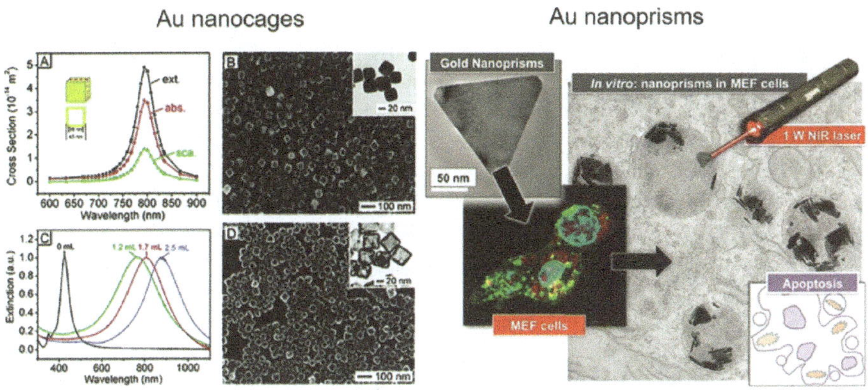

Fig. 3.14 Au nanocages and nanoprisms as model systems for photothermal applications. Figure taken from Chen et al. 2007 and Pérez-Hernández et al. 2015 with permission

45 nm in edge have been developed which has strong absorption in near-IR region and used as photothermal agents for cancer treatment Fig. 3.14a. Apart from this, encapsulated cargo vehicles are fabricated using polymers and plasmonic NPs such as plasmonic liposomes and lipid nanocapsules which act as drug carrying vehicles when excited using appropriate light.

3.4.1.1 Theranostics Nanosystems

In the year 2001, John Funkhouser has introduced the term theranostics (meaning a large bag of Therapeutics and Diagnostics). In general, theranostics refers to the comprehensive effort that combined the therapy and diagnostic in a single system. The concept of nanotheranostics was introduced with the emergence of nanotechnology and nanoparticles. The nanotheranostics has the potential to provide non-invasive imaging, targeting and treatment at the diseased sites without showing major impacts on the surrounding healthy cells. The theranostics nanomaterials can be made of using different chemical moieties such as targeting for specific biomolecular binding, diagnostics agents and polymer coating or matrix which gives colloidal stability and surface engineering bio-conjugation. The high surface area to volume ratio, small size, in addition to unique optical and electronics properties are the added advantage for the fabrication of nanotheranostics system (Chen et al. 2017). The complex nanotheranostics system can signal (diagnostics) and deliver drugs to specific cancerous cells and necessary target sites to perform the therapy. The more advantage is after controlled release of encapsulated drug molecules to the targeted sites, the nanoparticles can be safely degraded and excreted from body. Several different types of plasmonic NPs have been used as the one of the important components in the fabrication of theranostics system (Li et al. 2014). For example, Liu et al. reported on the fabrication of plasmonic NPs as multi-mode theranostic agents (Lin et al. 2013). The plasmonic nanocomposites consist of Au NPs which act

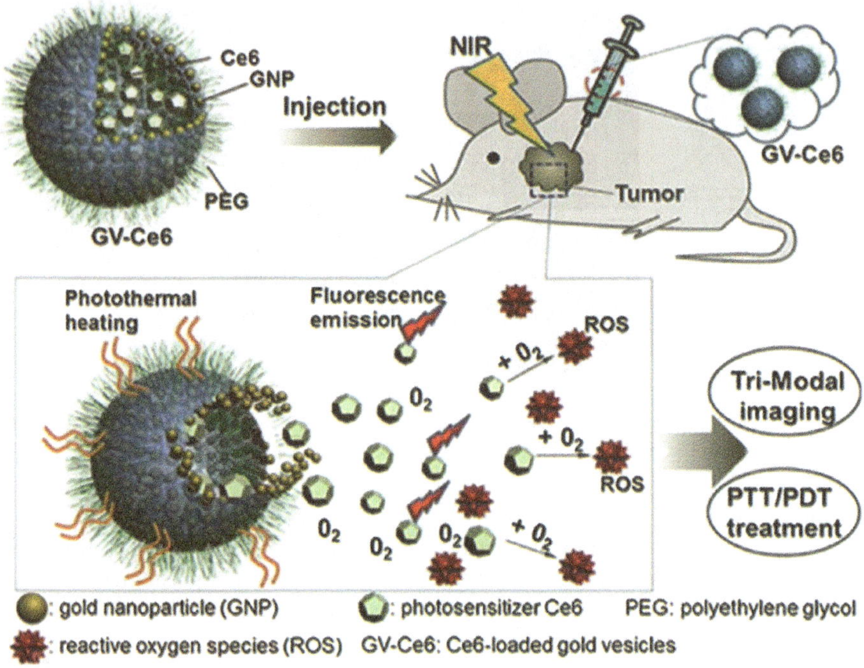

Fig. 3.15 Theranostics NPs: Plasmonic Au vesicles (GVs) loaded with photosensitizer (Ce6) which provide trimodal imaging capabilities consisting of fluorescence, thermal and OI, and PTT/PDT for cancer therapy. Figure taken from Lin et al. 2013 with permission

as combination of PTT, drug-delivery and imaging systems with encapsulated Ce6 which may act as a PDT and/or fluorescence agent (Fig. 3.15) which act as a perfect model system for theranostics application.

3.4.1.2 Antibacterial Activities

Plasmonic nanomaterials such as Au, Ag and Cu exhibited excellent antibacterial activities (Seil and Webster 2012). In the last decade the usage of Ag nanoparticles as antibacterial and antiseptic applications has been increased tremendously. Ag NPs show the best bactericidal activity among the other plasmonic nanomaterials (Ulaeto et al. 2020). Different sizes of Ag NPs show different bactericidal properties and its activity increases with the decrease in size of Ag NPs, and it also shows antiviral properties which are highly dependent on size and stabilizing agents (Sharma et al. 2009). The antimicrobial effects of Ag, Au and Cu show different mechanisms, including binding to microbial DNA, direct membrane rupture, and interacting with sulphydryl activities of several enzymes (Aziz et al. 2014, 2015, 2016). These NPs are also responsible for the generation of ROS which in turn cause enzyme and lipid oxidation. Thomas et al. investigated the effect of antimicrobial activity of the formation of Ag NPs from $AgNO_3$ which is entrapped inside the poly (acrylamide-co-acrylic acid) matrices (Thomas et al. 2007). For such systems, the antimicrobial activity will be enhanced with the increase of concentration of $AgNO_3$. In a similar approach, Ozay et al. reported on the synthesis of composite particles using emulsion polymerization of 4-vinylpyridine. These hydrogels with the incorporation of Ag and Cu NPs exhibited antibacterial activity on *B. subtilis*, *S. aureus*, *P. aeruginosa* and *E. Coli* (Ozay et al. 2010).

Zan et al. reported on the synthesis of poly vinyl alcohol (PVP) hydrogels films on biodegradable poly (L-lactic acid) (PLLA), integrated with Ag NPs and this hybrid composite shows potent antimicrobial activity towards *E. coli*, and reduced HeLa cell adhesion(Zan et al. 2010). In addition, Ag-Au hybrid systems are also used as antimicrobial systems. Dos Santos et al (dos Santos et al. 2012) reported on the fabrication of citrated capped Ag-Au alloy NPs and integrated with conventional antibiotics and tested for antimicrobial activity. The presence of those NPs in the hybrid systems decreases the requirement of the amount of conventional antibiotics needed for antimicrobial effect.

Several metal oxides related NPs including ZnO, TiO_2, Fe_2O_3/Fe_3O_4, CeO_2, ZnS and CdSe displayed antimicrobial activity (Valencia et al. 2020). Among these metal oxide nanoparticles, extensive research is conducted on TiO_2 NPs towards antimicrobial activity owing to photocatalytic nature. The photocatalysis nature of TiO_2 NPs depends on many factors such as crystallite size, morphology, crystal structure, caused UV-induced generation of hydroxyl radicals which has potential to oxidize essential bacterial components such as proteins, nucleic acids, lipids and polysaccharides. Chen et al. (Chen and Chen 2010) fabricated nanohybrid consisting of incorporated TiO_2 nanoparticles in polyurethane, and found efficient antimicrobial effect against *S. aureus, E. coli and P. aeruginosa.* In similar manner, different compositions of different metal oxide-based NPs act as antimicrobial activities.

3.5 Conclusions

Nanotechnology has emerged into an independent and advanced technology which has brought revolutionary changes in the twenty first century. Nanoparticles (1-100 nm) the building blocks of nanotechnology exhibited unique optical and chemical properties which we are exploited for different applications. The optical and chemical properties are highly dependent of size, shape and composition of nanoparticles. The synthetic strategies act as powerful tools to control the growth of nanoparticles and have created enormous nanoparticles with different properties. The availability of advanced instrumentation has further advanced to investigate the different properties of nanoparticles. Plasmonic nanoparticles are widely used in biomedical applications as bioimaging agents, antimicrobial agents, biosensors and mainly as theranostics. Nanotechnology has potential to address many of the global challenges facing today our society and can bring novel solutions to these challenges to improve the quality of life.

However, there are few areas to address, there is a great need to develop novel synthetic strategies to tune size, shape and compositions of nanoparticles easily which indeed helpful to create advanced nanohybrids, need to improve characterization techniques to study the physicochemical properties of nanoparticles easily and to explore usage of nanoparticles in diverse applications. In addition, monodispersity, stability, large-scale production, biocompatibility of the various nanoparticles need to improve. We strongly believe with the continuous research efforts that can improve the applications of nanoparticles in biology and medicine to next level.

References

Abdel-Aziz SM, Prasad R, Hamed AA, Abdelraof M (2018) Fungal nanoparticles: a novel tool for a green biotechnology? In: Prasad R, Kumar V, Kumar M, Wang S (eds) Fungal nanobionics: principles and applications. Springer Singapore Pte Ltd., Singapore, pp 61–87

Alghuthaymi MA, Almoammar H, Rai M, Said-Galiev E, Abd-Elsalam KA (2015) Myconanoparticles: synthesis and their role in phytopathogens management. Biotechnology & Biotechnological Equipment 29:221–236

Amendola V, Polizzi S, Meneghetti M (2006) Laser ablation synthesis of gold nanoparticles in organic solvents. J Phys Chem B 110:7232–7237

Apte M, Sambre D, Gaikawad S, Joshi S, Bankar A, Kumar AR, Zinjarde S (2013) Psychrotrophic yeast yarrowia lipolytica NCYC 789 mediates the synthesis of antimicrobial silver nanoparticles via cell-associated melanin. AMB Express 3:32

Arce VB, Santillán JMJ, Muñetón Arboleda D, Muraca D, Scaffardi LB, Schinca DC (2017) Characterization and stability of silver nanoparticles in starch solution obtained by femtosecond laser ablation and salt reduction. J Phys Chem C 121:10501–10513. https://doi.org/10.1021/acs.jpcc.6b12384

Au L, Zhang Q, Cobley CM, Gidding M, Schwartz AG, Chen J, Xia Y (2010) Quantifying the cellular uptake of antibody-conjugated au nanocages by two-photon microscopy and inductively coupled plasma mass spectrometry. ACS Nano 4:35–42. https://doi.org/10.1021/nn901392m

Auffan M, Pedeutour M, Rose J, Masion A, Ziarelli F, Borschneck D, Chaneac C, Botta C, Chaurand P, Labille J, Bottero J-Y (2010) Structural degradation at the surface of a tio2-based nanomaterial used in cosmetics. Environ Sci Technol 44:2689–2694. https://doi.org/10.1021/es903757q

Aziz N, Fatma T, Varma A, Prasad R (2014) Biogenic synthesis of silver nanoparticles using *Scenedesmus abundans* and evaluation of their antibacterial activity. J Nanoparticles, Article ID 689419. https://doi.org/10.1155/2014/689419

Aziz N, Faraz M, Pandey R, Sakir M, Fatma T, Varma A, Barman I, Prasad R (2015) Facile algae-derived route to biogenic silver nanoparticles: synthesis, antibacterial and photocatalytic properties. Langmuir 31:11605––11612. https://doi.org/10.1021/acs.langmuir.5b03081

Aziz N, Pandey R, Barman I, Prasad R (2016) Leveraging the attributes of *Mucor hiemalis*-derived silver nanoparticles for a synergistic broad-spectrum antimicrobial platform. Front Microbiol 7:1984. https://doi.org/10.3389/fmicb.2016.01984

Banu AN, Balasubramanian C (2014) Optimization and synthesis of silver nanoparticles using Isaria fumosorosea against human vector mosquitoes. Parasitol Res 113:3843–3851

Binnig G, Quate CF, Gerber C (1986) Atomic force microscope. Phys Rev Lett 56:930–933. https://doi.org/10.1103/PhysRevLett.56.930

Bloemen M (2015) Immunomagnetic separation of bacteria by iron oxide nanoparticles

Brust M, Walker M, Bethell D, Schiffrin DJ, Whyman R (1994) Synthesis of thiol-derivatised gold nanoparticles in a two-phase liquid–liquid system. J Chem Soc Chem Commun 1:801–802
Carrico IS, Maskarinec SA, Heilshorn SC, Mock ML, Liu JC, Nowatzki PJ, Franck C, Ravichandran G, Tirrell DA (2007) Lithographic patterning of photoreactive cell-adhesive proteins. J Am Chem Soc 129:4874–4875. https://doi.org/10.1021/ja070200b
Castro-Longoria E, Vilchis-Nestor AR, Avalos-Borja M (2011) Biosynthesis of silver, gold and bimetallic nanoparticles using the filamentous fungus neurospora crassa. Colloids Surf B Biointerfaces 83:42–48
Chen H, Zhang W, Zhu G, Xie J, Chen X (2017) Rethinking cancer nanotheranostics. Nature Reviews Materials 2:17024. https://doi.org/10.1038/natrevmats.2017.24
Chen H, Zhang X, Dai S, Ma Y, Cui S, S Achilefu YG (2013) Multifunctional gold nanostar conjugates for tumor imaging and combined photothermal and chemo-therapy. Theranostics 3:633–649. https://doi.org/10.7150/thno.6630
Chen J, Wang D, Xi J, Au L, Siekkinen A, Warsen A, Li Z-Y, Zhang H, Xia Y, Li X (2007) Immuno gold nanocages with tailored optical properties for targeted photothermal destruction of cancer cells. Nano Lett 7:1318–1322. https://doi.org/10.1021/nl070345g
Chen J, Mela P, Moller M, Lensen MC (2009) Microcontact deprinting: a technique to pattern gold nanoparticles. ACS Nano 3:1451–1456
Chen W-J, Chen Y-C (2010) Fe3o4/tio2 core/shell magnetic nanoparticle-based photokilling of pathogenic bacteria. Nanomedicine 5:1585–1593. https://doi.org/10.2217/nnm.10.92
Chou SY, Krauss PR, Renstrom PJ (1995) Imprint of sub-25 nm vias and trenches in polymers. Appl Phys Lett 67:3114–3116
Conde J, Rosa J, Lima JC, Baptista PV (2012) Nanophotonics for molecular diagnostics and therapy applications. International Journal of Photoenergy 2012:619530. https://doi.org/10.1155/2012/619530
Contado C, Pagnoni A (2008) Tio2 in commercial sunscreen lotion: flow field-flow fractionation and ICP-AES together for size analysis. Anal Chem 80:7594–7608. https://doi.org/10.1021/ac8012626
dos Santos MM, Queiroz MJ, Baptista PV (2012) Enhancement of antibiotic effect via gold: silver-alloy nanoparticles. J Nanopart Res 14:859. https://doi.org/10.1007/s11051-012-0859-8
Elango G, Roopan SM (2015) Green synthesis, spectroscopic investigation and photocatalytic activity of lead nanoparticles. Spectrochim Acta A Mol Biomol Spectrosc 139:367–373
Elbeshehy EK, Elazzazy AM, Aggelis G (2015) Silver nanoparticles synthesis mediated by new isolates of bacillus spp., nanoparticle characterization and their activity against bean yellow mosaic virus and human pathogens. Front Microbiol 6:453
Elghanian R, Storhoff JJ, Mucic RC, Letsinger RL, Mirkin CA (1997) Selective colorimetric detection of polynucleotides based on the distance-dependent optical properties of gold nanoparticles. Science 277:1078–1081. https://doi.org/10.1126/science.277.5329.1078
El-Kassas HY, El-Sheekh MM (2014) Cytotoxic activity of biosynthesized gold nanoparticles with an extract of the red seaweed corallina officinalis on the MCF-7 human breast cancer cell line. Asian Pac J Cancer Prev 15:4311–4317
Eric Drexler K (1986) Engines of creation: the coming era of nanotechnology. Doubleday, New York
Erwin WR, Zarick HF, Talbert EM, Bardhan R (2016) Light trapping in mesoporous solar cells with plasmonic nanostructures. Energ Environ Sci 9:1577–1601
Evanoff DD Jr, Chumanov G (2005) Synthesis and optical properties of silver nanoparticles and arrays. Chem Phys Chem 6:1221–1231
Faraday M (1857) On the relations of gold and other metals to light. Proceedings of the Royal Society of London 1:356–361
Feynman RP (2011) There's plenty of room at the bottom. Resonance 16:890. https://doi.org/10.1007/s12045-011-0109-x
Eadie F (2017) Crazy about cristation and fasciation. https://gardendrum.com/2017/01/08/crazy-about-cristation-or-fascinated-by-fasciation/#comments. Accessed 18 Jan 2017

Ganesan M, Freemantle RG, Obare SO (2007) Monodisperse thioether-stabilized palladium nanoparticles: synthesis, characterization, and reactivity. Chem Mater 19:3464–3471. https://doi.org/10.1021/cm062655q

Gerber C, Binnig G, Fechs H, Marti O, Rohrer H (1986) Scanning tunneling microscope combined with a scanning electron microscope. In: Scanning tunneling microscopy. Springer, London, pp 79–82

Guo P, Chen L, Wang J, Geng Z, Lu Z, Zhang G (2018) Enhanced tribological performance of aminated nano-silica modified graphene oxide as water-based lubricant additive. ACS Applied Nano Materials 1:6444–6453. https://doi.org/10.1021/acsanm.8b01653

Gurunathan S, Kalishwaralal K, Vaidyanathan R, Venkataraman D, Pandian SRK, Muniyandi J, Hariharan N, Eom SH (2009) Biosynthesis, purification and characterization of silver nanoparticles using Escherichia coli. Colloids Surf B Biointerfaces 74:328–335

Hirsch LR, Stafford RJ, Bankson JA, Sershen SR, Rivera B, Price RE, Hazle JD, Halas NJ, West JL (2003) Nanoshell-mediated near-infrared thermal therapy of tumors under magnetic resonance guidance. Proc Natl Acad Sci 100:13549–13554. https://doi.org/10.1073/pnas.2232479100

Iijima S (1991) Helical microtubules of graphitic carbon. Nature 354:56–58. https://doi.org/10.1038/354056a0

Inci F, Tokel O, Wang S, UA Gurkan ST, Kuritzkes DR, Demirci U (2013) Nanoplasmonic quantitative detection of intact viruses from unprocessed whole blood. ACS Nano 7:4733–4745. https://doi.org/10.1021/nn3036232

Jain PK, Huang X, El-Sayed IH, El-Sayed MA (2008) Noble metals on the nanoscale: optical and photothermal properties and some applications in imaging, sensing, biology, and medicine. Acc Chem Res 41:1578–1586. https://doi.org/10.1021/ar7002804

Jiang L, Santiago I, Foord J (2020) High-yield electrochemical synthesis of silver nanoparticles by enzyme-modified boron-doped diamond electrodes. Langmuir 36:6089–6094. https://doi.org/10.1021/acs.langmuir.0c00375

Jo JH, Singh P, Kim YJ, Wang C, Mathiyalagan R, Jin C-G, Yang DC (2016) Pseudomonas deceptionensis dc5-mediated synthesis of extracellular silver nanoparticles. Artificial Cells, Nanomedicine, and Biotechnology 44:1576–1581

Kalaiselvi A, Roopan SM, Madhumitha G, Ramalingam C, Elango G (2015) Synthesis and characterization of palladium nanoparticles using catharanthus roseus leaf extract and its application in the photo-catalytic degradation. Spectrochim Acta A Mol Biomol Spectrosc 135:116–119

Kim S, Liu B, Zachariah M (2002) Synthesis of nanoporous metal oxide particles by a new inorganic matrix spray pyrolysis method. Chem Mater 14:2889–2899

Kroto HW, Heath JR, O'Brien SC, Curl RF, Smalley RE (1985) C60: Buckminsterfullerene. Nature 318:162–163. https://doi.org/10.1038/318162a0

Kumar SA, MK Abyaneh SG, Kulkarni SK, Pasricha R, Ahmad A, Khan M (2007) Nitrate reductase-mediated synthesis of silver nanoparticles from agno 3. Biotechnol Lett 29:439–445

Leduc C, Si S, Gautier J, Soto-Ribeiro M, Wehrle-Haller B, Gautreau A, Giannone G, Cognet L, Lounis B (2013) A highly specific gold nanoprobe for live-cell single-molecule imaging. Nano Lett 13:1489–1494. https://doi.org/10.1021/nl304561g

Lee P, Meisel D (1982) Adsorption and surface-enhanced Raman of dyes on silver and gold sols. J Phys Chem 86:3391–3395

Li C, Zhou L, Yang H, Lv R, Tian P, Li X, Zhang Y, Chen Z, Lin F (2017) Self-assembled exopolysaccharide nanoparticles for bioremediation and green synthesis of noble metal nanoparticles. ACS Appl Mater Interfaces 9:22808–22818. https://doi.org/10.1021/acsami.7b02908

Li J, Cai P, Shalviri A, Henderson JT, He C, Foltz WD, Prasad P, Brodersen PM, Chen Y, DaCosta R, Rauth AM, Wu XY (2014) A multifunctional polymeric nanotheranostic system delivers doxorubicin and imaging agents across the blood–brain barrier targeting brain metastases of breast cancer. ACS Nano 8:9925–9940. https://doi.org/10.1021/nn501069c

Li L, Xing H, Zhang J, Lu Y (2019a) Functional DNA molecules enable selective and stimuli-responsive nanoparticles for biomedical applications. Acc Chem Res 52:2415–2426. https://doi.org/10.1021/acs.accounts.9b00167
Li Q, Nan K, Le Floch P, Lin Z, Sheng H, Blum TS, Liu J (2019b) Cyborg organoids: implantation of nanoelectronics via organogenesis for tissue-wide electrophysiology. Nano Lett 19:5781–5789. https://doi.org/10.1021/acs.nanolett.9b02512
Li X, H Xu, Z-S Chen, G Chen (2011) Biosynthesis of nanoparticles by microorganisms and their applications. Journal of Nanomaterials 2011
Li Z, Hu Y, Miao Z, Xu H, Li C, Zhao Y, Li Z, Chang M, Ma Z, Sun Y, Besenbacher F, Huang P, Yu M (2018) Dual-stimuli responsive bismuth nanoraspberries for multimodal imaging and combined cancer therapy. Nano Lett 18:6778–6788. https://doi.org/10.1021/acs.nanolett.8b02639
Lin J, Wang S, Huang P, Wang Z, Chen S, Niu G, Li W, He J, Cui D, Lu G, Chen X, Nie Z (2013) Photosensitizer-loaded gold vesicles with strong plasmonic coupling effect for imaging-guided photothermal/photodynamic therapy. ACS Nano 7:5320–5329. https://doi.org/10.1021/nn4011686
Liu W-T (2006) Nanoparticles and their biological and environmental applications. J Biosci Bioeng 102:1–7. https://doi.org/10.1263/jbb.102.1
Luan J, Xu T, Cashin J, Morrissey JJ, Kharasch ED, Singamaneni S (2018) Environmental stability of plasmonic biosensors based on natural versus artificial antibody. Anal Chem 90:7880–7887. https://doi.org/10.1021/acs.analchem.7b05470
Mafuné F, Kohno J-y, Takeda Y, Kondow T, Sawabe H (2000) Formation and size control of silver nanoparticles by laser ablation in aqueous solution. J Phys Chem B 104:9111–9117. https://doi.org/10.1021/jp001336y
Masetti M, Xie H-n, Krpetić Ž, Recanatini M, Alvarez-Puebla RA, Guerrini L (2015) Revealing DNA interactions with exogenous agents by surface-enhanced Raman scattering. J Am Chem Soc 137:469–476. https://doi.org/10.1021/ja511398w
Melo MA, Guedes SF, Xu HH, Rodrigues LK (2013) Nanotechnology-based restorative materials for dental caries management. Trends Biotechnol 31:459–467
Mitrano DM, Rimmele E, Wichser A, Erni R, Height M, Nowack B (2014) Presence of nanoparticles in wash water from conventional silver and nano-silver textiles. ACS Nano 8:7208–7219. https://doi.org/10.1021/nn502228w
Mock JJ, Smith DR, Schultz S (2003) Local refractive index dependence of plasmon resonance spectra from individual nanoparticles. Nano Lett 3:485–491. https://doi.org/10.1021/nl0340475
Momeni S, Nabipour I (2015) A simple green synthesis of palladium nanoparticles with Sargassum alga and their electrocatalytic activities towards hydrogen peroxide. Appl Biochem Biotechnol 176:1937–1949
Mukherji S, Bharti S, Shukla G, Mukherji S (2018) Synthesis and characterization of size- and shape-controlled silver nanoparticles. Physical Sciences Reviews 4:20170082. https://doi.org/10.1515/psr-2017-0082
Murugan K, Benelli G, Panneerselvam C, Subramaniam J, Jeyalalitha T, Dinesh D, Nicoletti M, Hwang J-S, Suresh U, Madhiyazhagan P (2015) Cymbopogon citratus-synthesized gold nanoparticles boost the predation efficiency of copepod mesocyclops aspericornis against malaria and dengue mosquitoes. Exp Parasitol 153:129–138
Najjar R (2012) Microemulsions: an introduction to properties and applications. BoD–Books on Demand.
National nanotechnology initiative (2000) Leading to the next industrial revolution. Microscale Thermophysical Engineering 4:205–212. https://doi.org/10.1080/10893950050148160
Nikoobakht B, El-Sayed MA (2003) Surface-enhanced Raman scattering studies on aggregated gold nanorods. Chem A Eur J 107:3372–3378. https://doi.org/10.1021/jp026770+
Njoki PN, Lim I-IS, Mott D, Park H-Y, Khan B, Mishra S, Sujakumar R, Luo J, Zhong C-J (2007) Size correlation of optical and spectroscopic properties for gold nanoparticles. J Phys Chem C 111:14664–14669

Nusz GJ, Curry AC, Marinakos SM, Wax A, Chilkoti A (2009) Rational selection of gold nanorod geometry for label-free plasmonic biosensors. ACS Nano 3:795–806. https://doi.org/10.1021/nn8006465

Ozay O, Akcali A, Otkun MT, Silan C, Aktas N, Sahiner N (2010) P(4-vp) based nanoparticles and composites with dual action as antimicrobial materials. Colloids Surf B Biointerfaces 79:460–466. https://doi.org/10.1016/j.colsurfb.2010.05.013

Pawar D, Kale SN (2019) A review on nanomaterial-modified optical fiber sensors for gases, vapors and ions. Microchimica Acta 186:253. https://doi.org/10.1007/s00604-019-3351-7

Pazos-Perez N, Pazos E, Catala C, Mir-Simon B, Gómez-de Pedro S, Sagales J, Villanueva C, Vila J, Soriano A, García de Abajo FJ, Alvarez-Puebla RA (2016) Ultrasensitive multiplex optical quantification of bacteria in large samples of biofluids. Sci Rep 6:29014. https://doi.org/10.1038/srep29014

Pérez-Hernández M, del Pino P, Mitchell SG, Moros M, Stepien G, Pelaz B, Parak WJ, Gálvez EM, Pardo J, de la Fuente JM (2015) Dissecting the molecular mechanism of apoptosis during photothermal therapy using gold nanoprisms. ACS Nano 9:52–61. https://doi.org/10.1021/nn505468v

Pham T, Jackson JB, Halas NJ, Lee TR (2002) Preparation and characterization of gold nanoshells coated with self-assembled monolayers. Langmuir 18:4915–4920. https://doi.org/10.1021/la015561y

Philip D, Unni C, Aromal SA, Vidhu V (2011) Murraya koenigii leaf-assisted rapid green synthesis of silver and gold nanoparticles. Spectrochim Acta A Mol Biomol Spectrosc 78:899–904

Polo E, Poupard MFN, Guerrini L, Taboada P, Pelaz B, Alvarez-Puebla RA, del Pino P (2018a) Colloidal bioplasmonics. Nano Today 20:58–73. https://doi.org/10.1016/j.nantod.2018.04.003

Polo E, Poupard MFN, Guerrini L, Taboada P, Pelaz B, Alvarez-Puebla RA, del Pino P (2018b) Colloidal bioplasmonics. Nano Today 20:58–73

Polo E, del Pino P, Pelaz B, Grazu V, de la Fuente JM (2013) Plasmonic-driven thermal sensing: ultralow detection of cancer markers. Chem Commun 49:3676–3678. https://doi.org/10.1039/C3CC39112D

Poopathi S, De Britto LJ, Praba VL, Mani C, Praveen M (2015) Synthesis of silver nanoparticles from azadirachta indica—a most effective method for mosquito control. Environ Sci Pollut Res 22:2956–2963

Prasad R (2014) Synthesis of silver nanoparticles in photosynthetic plants. J Nanoparticles, Article ID 963961. https://doi.org/10.1155/2014/963961

Prasad R (2016) Advances and applications through fungal nanobiotechnology. Springer International Publishing, Cham. ISBN 978-3-319-42989-2

Prasad R (2017) Fungal nanotechnology: applications in agriculture, industry, and medicine. Springer Nature Singapore Pte Ltd., Singapore. ISBN 978-3-319-68423-9

Prasad R (2019a) Microbial nanobionics: basic research and applications. Springer International Publishing, Cham. ISBN 978-3-030-16534-5. https://www.springer.com/gp/book/9783030165338

Prasad R (2019b) Microbial nanobionics: state of art. Springer International Publishing, Cham. ISBN 978-3-030-16383-9. https://www.springer.com/gp/book/9783030163822

Prasad R (2019c) Plant nanobionics: advances in the understanding of nanomaterials research and applications. Springer International Publishing, Cham. ISBN 978-3-030-12495-3. https://www.springer.com/gp/book/9783030124953

Prasad R (2019d) Plant nanobionics: approaches in nanoparticles biosynthesis and toxicity. Springer International Publishing, Cham. ISBN 978-3-030-16379-2. https://www.springer.com/gp/book/9783030163785

Prasad R, Aranda E (2018) Approaches in bioremediation. Springer International Publishing, Cham. https://www.springer.com/de/book/9783030023683

Prasad R, Pandey R, Barman I (2016) Engineering tailored nanoparticles with microbes: quo vadis. WIREs Nanomed Nanobiotechnol 8:316–330. https://doi.org/10.1002/wnan.1363

Prasad R, Kumar V, Kumar M (2017) Nanotechnology: food and environmental paradigm. Springer Nature Singapore Pte Ltd., Singapore. ISBN 978-981-10-4678-0

Prasad R, Kumar V, Kumar M, Wang S (2018a) Fungal nanobionics: principles and applications. Springer Nature Singapore Pte Ltd., Singapore. ISBN 978-981-10-8666-3. https://www.springer.com/gb/book/9789811086656

Prasad R, Jha A, Prasad K (2018b) Exploring the realms of nature for nanosynthesis. Springer International Publishing, Cham. ISBN 978-3-319-99570-0. https://www.springer.com/978-3-319-99570-0

Prasad R, Siddhardha B, Dyavaiah M (2020) Nanostructures for antimicrobial and antibiofilm applications. Springer International Publishing, Cham. ISBN 978-3-030-40336-2. https://www.springer.com/gp/book/9783030403362

Rao R, Pint CL, Islam AE, Weatherup RS, Hofmann S, Meshot ER, Wu F, Zhou C, Dee N, Amama PB, Carpena-Nuñez J, Shi W, Plata DL, Penev ES, Yakobson BI, Balbuena PB, Bichara C, Futaba DN, Noda S, Shin H, Kim KS, Simard B, Mirri F, Pasquali M, Fornasiero F, Kauppinen EI, Arnold M, Cola BA, Nikolaev P, Arepalli S, Cheng H-M, Zakharov DN, Stach EA, Zhang J, Wei F, Terrones M, Geohegan DB, Maruyama B, Maruyama S, Li Y, Adams WW, Hart AJ (2018) Carbon nanotubes and related nanomaterials: critical advances and challenges for synthesis toward mainstream commercial applications. ACS Nano 12:11756–11784. https://doi.org/10.1021/acsnano.8b06511

Reetz MT, Helbig W (1994) Size-selective synthesis of nanostructured transition metal clusters. J Am Chem Soc 116:7401–7402

Ren B, Lin X-F, Yang Z-L, Liu G-K, Aroca RF, Mao B-W, Tian Z-Q (2003) Surface-enhanced Raman scattering in the ultraviolet spectral region: UV-SERS on rhodium and ruthenium electrodes. J Am Chem Soc 125:9598–9599. https://doi.org/10.1021/ja035541d

Richards DA, Maruani A, Chudasama V (2017) Antibody fragments as nanoparticle targeting ligands: a step in the right direction. Chem Sci 8:63–77. https://doi.org/10.1039/C6SC02403C

Rivera_Gil P, Vazquez-Vazquez C, Giannini V, Callao MP, Parak WJ, Correa-Duarte MA, Alvarez-Puebla RA (2013) Plasmonic nanoprobes for real-time optical monitoring of nitric oxide inside living cells. Angew Chem Int Ed 52:13694–13698. https://doi.org/10.1002/anie.201306390

Ross MB, Schatz GC (2014) Aluminum and indium plasmonic nanoantennas in the ultraviolet. J Phys Chem C 118:12506–12514. https://doi.org/10.1021/jp503323u

Sadeghi B, Rostami A, Momeni S (2015) Facile green synthesis of silver nanoparticles using seed aqueous extract of pistacia atlantica and its antibacterial activity. Spectrochim Acta A Mol Biomol Spectrosc 134:326–332

Sajanlal PR, Sreeprasad TS, Samal AK, Pradeep T (2011) Anisotropic nanomaterials: structure, growth, assembly, and functions. Nano Reviews 2:5883

Sasaki T, Watanabe M, Hashizume H, Yamada H, Nakazawa H (1996) Macromolecule-like aspects for a colloidal suspension of an exfoliated titanate. Pairwise association of nanosheets and dynamic reassembling process initiated from it. J Am Chem Soc 118:8329–8335. https://doi.org/10.1021/ja960073b

Sathishkumar M, Sneha K, Yun Y-S (2010) Immobilization of silver nanoparticles synthesized using curcuma longa tuber powder and extract on cotton cloth for bactericidal activity. Bioresour Technol 101:7958–7965

Scarabelli L, Hamon C, Liz-Marzán LM (2017) Design and fabrication of plasmonic nanomaterials based on gold nanorod supercrystals. Chem Mater 29:15–25. https://doi.org/10.1021/acs.chemmater.6b02439

Seil JT, Webster TJ (2012) Antimicrobial applications of nanotechnology: methods and literature. Int J Nanomedicine 7:2767–2781. https://doi.org/10.2147/IJN.S24805

Seshadri S, Saranya K, Kowshik M (2011) Green synthesis of lead sulfide nanoparticles by the lead resistant marine yeast, rhodosporidium diobovatum. Biotechnol Prog 27:1464–1469

Shaik F (2020) Ligand-free yolk-shell nanoparticles: synthesis and catalytic applications. Chem Nano Mat 1:1

Shaik F, Peer I, Jain PK, Amirav L (2018) Plasmon-enhanced multicarrier photocatalysis. Nano Lett 18:4370–4376. https://doi.org/10.1021/acs.nanolett.8b01392

Shaik F, Zhang W, Niu W (2017) A generalized method for the synthesis of ligand-free M@SiO2 (M = Ag, Au, Pd, Pt) yolk–shell nanoparticles. Langmuir 33:3281–3286. https://doi.org/10.1021/acs.langmuir.7b00141

Shaik F, Zhang W, Niu W, Lu X (2014) Volume-confined synthesis of ligand-free gold nanoparticles with tailored sizes for enhanced catalytic activity. Chemical Physics Letters 613:95–99. https://doi.org/10.1016/j.cplett.2014.08.005

Sharma VK, Yngard RA, Lin Y (2009) Silver nanoparticles: green synthesis and their antimicrobial activities. Adv Colloid Interface Sci 145:83–96. https://doi.org/10.1016/j.cis.2008.09.002

Shash S, Ramanan VV, Prasad R (2019) Sustainable green technologies for environmental management. Springer, Singapore. ISBN 978-981-13-2772-8. https://www.springer.com/la/book/9789811327711

Singh BR, Dwivedi S, Al-Khedhairy AA, Musarrat J (2011) Synthesis of stable cadmium sulfide nanoparticles using surfactin produced by bacillus amyloliquifaciens strain ksu-109. Colloids Surf B Biointerfaces 85:207–213

Singh P, Kim Y, Yang D (2015a) A strategic approach for rapid synthesis of gold and silver nanoparticles by Panax ginseng leaves. Artif Cells Nanomed Biotechnol 44:1949–1957

Singh P, Kim YJ, Wang C, Mathiyalagan R, Yang DC (2016c) The development of a green approach for the biosynthesis of silver and gold nanoparticles by using Panax ginseng root extract, and their biological applications. Artificial Cells, Nanomedicine, and Biotechnology 44:1150–1157

Singh P, Kim YJ, Wang C, Mathiyalagan R, Yang DC (2016d) Microbial synthesis of flower-shaped gold nanoparticles. Artificial Cells, Nanomedicine, and Biotechnology 44:1469–1474

Singh P, Kim YJ, Wang C, Mathiyalagan R, Yang DC (2016e) Weissella oryzae dc6-facilitated green synthesis of silver nanoparticles and their antimicrobial potential. Artificial Cells, Nanomedicine, and Biotechnology 44:1569–1575

Singh P, Kim YJ, Wang C, Mathiyalagan R, El-Agamy Farh M, Yang DC (2016b) Biogenic silver and gold nanoparticles synthesized using red ginseng root extract, and their applications. Artificial Cells, Nanomedicine, and Biotechnology 44:811–816

Singh P, Kim Y-J, Zhang D, Yang D-C (2016a) Biological synthesis of nanoparticles from plants and microorganisms. Trends Biotechnol 34:588–599

Singh P, Kim YJ, Singh H, Wang C, Hwang KH, Farh ME-A, Yang DC (2015c) Biosynthesis, characterization, and antimicrobial applications of silver nanoparticles. Int J Nanomedicine 10:2567

Singh P, Kim YJ, Singh H, Mathiyalagan R, Wang C, Yang DC (2015b) Biosynthesis of anisotropic silver nanoparticles by bhargavaea indica and their synergistic effect with antibiotics against pathogenic microorganisms. Journal of Nanomaterials 2015:1

Soni N, Prakash S (2015) Antimicrobial and mosquitocidal activity of microbial synthesized silver nanoparticles. Parasitol Res 114:1023–1030

Sotiriou GA, Franco D, Poulikakos D, Ferrari A (2012) Optically stable biocompatible flame-made sio2-coated y2o3:Tb3+ nanophosphors for cell imaging. ACS Nano 6:3888–3897. https://doi.org/10.1021/nn205035p

Srivastava S, Usmani Z, Atanasov AG, Singh VK, Singh NP, Abdel-Azeem AM, Prasad R, Gupta G, Sharma M, Bhargava A (2021) Biological nanofactories: using living forms for metal nanoparticle synthesis. Mini-Rev Med Chem 21(2):245–265

Storhoff JJ, Elghanian R, Mucic RC, Mirkin CA, Letsinger RL (1998) One-pot colorimetric differentiation of polynucleotides with single base imperfections using gold nanoparticle probes. J Am Chem Soc 120:1959–1964. https://doi.org/10.1021/ja972332i

Taniguchi N (1974) On the basic concept of nano-technology proceedings of the international conference on production engineering Tokyo part ii japan society of precision engineering

Tapio K, Leppiniemi J, Shen B, Hytönen VP, Fritzsche W, Toppari JJ (2016) Toward single electron nanoelectronics using self-assembled DNA structure. Nano Lett 16:6780–6786. https://doi.org/10.1021/acs.nanolett.6b02378

Thangadurai D, Sangeetha J, Prasad R (2020a) Nanotechnology for food, agriculture, and environment. Springer International Publishing, Cham. ISBN 978-3-030-31937-3. https://www.springer.com/gp/book/9783030319373

Thangadurai D, Sangeetha J, Prasad R (2020b) Functional bionanomaterials. Springer International Publishing, Cham. ISBN 978-3-030-41464-1. https://www.springer.com/gp/book/9783030414634

Thomas V, Yallapu MM, Sreedhar B, Bajpai SK (2007) A versatile strategy to fabricate hydrogel–silver nanocomposites and investigation of their antimicrobial activity. J Colloid Interface Sci 315:389–395. https://doi.org/10.1016/j.jcis.2007.06.068

Tsuzuki T (2009) Commercial scale production of inorganic nanoparticles. International Journal of Nanotechnology 6:567–578

Turkevich J, Stevenson PC, Hillier J (1951) A study of the nucleation and growth processes in the synthesis of colloidal gold. Discuss Faraday Soc 11:55–75

Ulaeto SB, Mathew GM, Pancrecious JK, Nair JB, Rajan TPD, Maiti KK, Pai BC (2020) Biogenic Ag nanoparticles from neem extract: their structural evaluation and antimicrobial effects against Pseudomonas nitroreducens and Aspergillus unguis (nii 08123). ACS Biomater Sci Eng 6:235–245. https://doi.org/10.1021/acsbiomaterials.9b01257

Valencia L, Kumar S, Nomena EM, Salazar-Alvarez G, Mathew AP (2020) In-situ growth of metal oxide nanoparticles on cellulose nanofibrils for dye removal and antimicrobial applications. ACS Applied Nano Materials 3:7172–7181. https://doi.org/10.1021/acsanm.0c01511

von Goetz N, Lorenz C, Windler L, Nowack B, Heuberger M, Hungerbühler K (2013) Migration of ag- and tio2-(nano)particles from textiles into artificial sweat under physical stress: experiments and exposure modeling. Environ Sci Technol 47:9979–9987. https://doi.org/10.1021/es304329w

Waghmare SR, Mulla MN, Marathe SR, Sonawane KD (2015) Ecofriendly production of silver nanoparticles using candida utilis and its mechanistic action against pathogenic microorganisms. 3 Biotech 5:33–38

Wang C, Kim YJ, Singh P, Mathiyalagan R, Jin Y, Yang DC (2016) Green synthesis of silver nanoparticles by bacillus methylotrophicus, and their antimicrobial activity. Artificial Cells, Nanomedicine, and Biotechnology 44:1127–1132

Wong XY, Sena-Torralba A, Álvarez-Diduk R, Muthoosamy K, Merkoçi A (2020) Nanomaterials for nanotheranostics: tuning their properties according to disease needs. ACS Nano 14:2585–2627. https://doi.org/10.1021/acsnano.9b08133

Wongkaew N, Simsek M, Griesche C, Baeumner AJ (2019) Functional nanomaterials and nanostructures enhancing electrochemical biosensors and lab-on-a-chip performances: recent progress, applications, and future perspective. Chem Rev 119:120–194. https://doi.org/10.1021/acs.chemrev.8b00172

Xu C, De S, Balu AM, Ojeda M, Luque R (2015) Mechanochemical synthesis of advanced nanomaterials for catalytic applications. Chem Commun 51:6698–6713

Xu M, Shaik F, Jiang B, Zheng J (2020) FeCoNiP alloy nanoparticles integrated on vertically aligned phosphorus-doped single-walled carbon nanotubes for overall water splitting. J Electrochem Soc 167:102515. https://doi.org/10.1149/1945-7111/ab9e3d

Yang L, Shaik F, Pang F, Zhang W (2020) PdAgCu alloy nanoparticles integrated on three-dimensional nanoporous CuO for efficient electrocatalytic nitrogen reduction under ambient conditions. Langmuir 36:5112–5117. https://doi.org/10.1021/acs.langmuir.0c00228

Yang X, Yang M, Pang B, Vara M, Xia Y (2015) Gold nanomaterials at work in biomedicine. Chem Rev 115:10410–10488. https://doi.org/10.1021/acs.chemrev.5b00193

Yetisen AK, Qu H, Manbachi A, Butt H, Dokmeci MR, Hinestroza JP, Skorobogatiy M, Khademhosseini A, Yun SH (2016) Nanotechnology in textiles. ACS Nano 10:3042–3068. https://doi.org/10.1021/acsnano.5b08176

Zan X, Kozlov M, McCarthy TJ, Su Z (2010) Covalently attached, silver-doped poly(vinyl alcohol) hydrogel films on poly(l-lactic acid). Biomacromolecules 11:1082–1088. https://doi.org/10.1021/bm100048q

Zhang J, Fang J (2009) A general strategy for preparation of Pt 3d-transition metal (Co, Fe, Ni) nanocubes. J Am Chem Soc 131:18543–18547

Zhou GJ, Li SH, Zhang YC, Fu YZ (2014) Biosynthesis of CdS nanoparticles in banana peel extract. J Nanosci Nanotechnol 14:4437–4442. https://doi.org/10.1166/jnn.2014.8259

Zhu C, Du D, Eychmüller A, Lin Y (2015) Engineering ordered and nonordered porous noble metal nanostructures: synthesis, assembly, and their applications in electrochemistry. Chem Rev 115:8896–8943

Zhu J, Li X, Zhang Y, Wang J, Cao Y, Camaiti M, Wei B (2019) Dual functionalities of few-layered boron nitrides in the design and implementation of Ca(OH)2 nanomaterials toward an efficient wall painting fireproofing and consolidation. ACS Appl Mater Interfaces 11:11792–11799. https://doi.org/10.1021/acsami.9b00826

Zhu Y-J, Chen F (2014) Microwave-assisted preparation of inorganic nanostructures in liquid phase. Chem Rev 114:6462–6555

Chapter 4
Nanofibers in Medical Microbiology

Renato L. Gil, Célia G. Amorim, Joan Manuel Rodríguez-Díaz, Alberto N. Araújo, and Maria C. B. S. M. Montenegro

Abstract Nanotechnology is an emergent field of research and innovation applied to nanoelectronics, nanomaterials, nanobiology, nanomedicine, among other areas. Particularly in medicine it has been used to improve the disease diagnosis through novel instrumentation as well as the treatment efficacy through the encapsulation and controlled release of bioactive compounds. Overall, nanotechnology can provide novel tools and techniques to biomedical research, namely in proteomics, genetics and regenerative medicine. Polymer nanofibers are one of the nanostructures that have been used in nanomedicine, especially those fabricated by electrospinning. The development of drug delivery systems, wound dressing materials, tissue regeneration and disease diagnosis are the most representative examples about the use of electrospun nanofibers. Their use in medical microbiology has mostly contributed to an improvement in the treatment of microbial infections through the increase of response capacity of active pharmaceutical ingredients but also into the reduction of side effects. In this chapter, a comprehensive overview about the manufacturing techniques and the most common polymers used for nanofibers production is discussed. The relation between the usefulness of these nanofibers with their composition, their size, and orientation is pointed out. Their medical applications on microbiology, namely the local administration of antibiotic drugs in infected areas, the development of wound dressing materials to improve the healing process and the possibility to raise or replace biological functions without organ transplantation by

R. L. Gil · C. G. Amorim (✉) · A. N. Araújo · M. C. B. S. M. Montenegro
LAQV-REQUIMTE/Departamento de Ciências Químicas, Faculdade de Farmácia, Universidade do Porto, Porto, Portugal
e-mail: camorim@ff.up.pt; anaraujo@ff.up.pt; mcbranco@ff.up.pt

J. M. Rodríguez-Díaz
Laboratorio de Análisis Químicos y Biotecnológicos, Instituto de Investigación, Universidad Técnica de Manabí, Portoviejo, Ecuador

Departamento de Procesos Químicos, Facultad de Ciencias Matemáticas, Físicas y Químicas, Universidad Técnica de Manabí, Portoviejo, Ecuador

Programa de Pós-graduação em Engenharia Química, Universidade Federal da Paraíba, João Pessoa, Brazil

N. R. Maddela et al. (eds.), *Nanotechnology for Advances in Medical Microbiology*, Environmental and Microbial Biotechnology,
https://doi.org/10.1007/978-981-15-9916-3_4

the use of artificial materials were analysed covering the developments accomplished over the last 10 years. It is noticed a lack of information about the formation of biofilms on the surface of nanofibrous scaffolds. Moreover, more studies must be done considering the nanoscale morphology and the type of polymer used, which seems to be crucial to prevent infections by microorganisms across medical applications.

Keywords Nanosciences · Medical microbiology · Nanofibers · Biomedical engineering · Drug delivery · Wound dressing · Tissue engineering

4.1 Introduction

Nanoscience and nanoengineering are emerging and promising interdisciplinary research areas from nanotechnology. The prefix 'nano' derives from the Greek νᾶνος (Latin nanus), meaning "dwarf". The General Conference on Weights and Measures officially endorsed the usage of nano as a standard prefix in 1960 and depicts one thousand millionths of a meter (10^{-9} m) (Vert et al. 2012). The technology operated at nanoscale dimensions (1–100 nm) in material, physical, chemical, biological, and environmental sciences (Fig. 4.1) makes this content so important as the industrial revolution (Barhoum et al. 2019; Thangadurai et al. 2020a, b).

The scientific story of nanotechnology applied to nanoelectronics, nanomaterials, nanobiology, nanomedicine, etc. started in the middle of the last century. Richard Feynman, an American physicist, and Nobel Prize laureate, introduced the concept of nanotechnology in 1959, giving the first lecture about this subject "There's Plenty of Room at the Bottom" (Feynman 1960), at California Institute of Technology at the American Physical Society meeting.

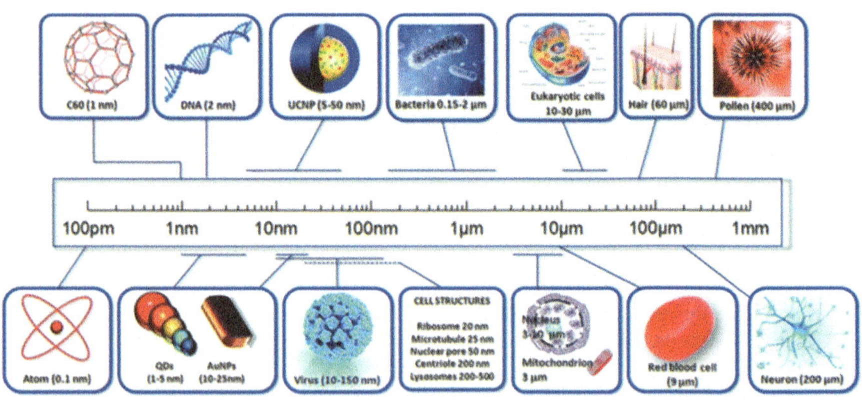

Fig. 4.1 Examples of nanoparticles sizes. (Reproduced from Gnach et al. (2015) with permission from The Royal Society of Chemistry)

Particularly in medicine, nanotechnology has been used in three main areas (Jain 2019), which includes: (a) rapid diagnosis, auxiliary treatment and transport of active substances; (b) scientific research in molecular medicine like genetics and proteomics through the production of modified microorganisms; (c) regenerative medicine by the creation of biomaterials and nanostructures with potential clinical implications (Peran et al. 2013). In the last few years, promising therapies and diagnosis devices resorting to nanoparticles were used to face the increasing incidences of cancer, cardiovascular, respiratory, musculoskeletal, and neurodegenerative diseases (Patra et al. 2018). The first generation of nanoparticle-based therapy comprised lipid systems like liposomes and micelles that could include gold or magnetic nanoparticles to change membrane fluidity for drug delivery system (Park et al. 2006). Diferent types of nanomaterials are frequently considered in designing the target-specific drug delivery systems namely in drugs with poor water solubility and less absorption ability (Mira et al. 2020; Ye et al. 2020); and they stay in the blood circulatory system for an extended period of time with an easier penetration in the tissue system, facilitating the uptake of the drugs by cells. Therefore, the increase of drug delivery efficiency in the target location is possible. They have been also referred concerning their nanostructure on the biological effects. Several parameters such as size, zeta potential, surface chemistry, mechanical property, shape and other inherent biophysical/chemical characteristics contribute to enhance drug delivery effectiveness (Venkataraman et al. 2011; Prasad et al. 2017). However, toxicity and exposure data, combined with therapeutic properties, have been contributed to the risk assessment evaluation on the use of nanomaterials in consumer products, improving its regulation by governmental organizations (FDA 2012; Hardy et al. 2018; SCCS 2020; Tsuji et al. 2006).

The most current types of nanomaterials used in medical biology are carbon-based materials, metal-based materials, dendrimers, composites, polymers, etc. (Azonano 2017; Gubala et al. 2018). Carbon-based materials are composed mainly of carbon and can be prepared with different shapes: hollow spheres, ellipsoids or tubes; metal-based materials include quantum dots, nanogold, nanosilver and metal oxides, such as titanium dioxide; dendrimers are nanosized polymers built from branched units, whose surface can be tailored to perform specific chemical functions; composites combine different nanomaterials or different sizes of bulk-type materials. Polymers are another type of nanostructured materials that can be prepared in different morphologies such as nanofibers, nanoparticles, nanowires, nanospheres, and others (Lu et al. 2017). They have been used in medical care, such as drug delivery, tissue engineering, wound dressing, diagnosis systems, etc. Among them, nanofibers are highlighted once they have been proposed to support many developments guided to therapeutic solutions such as scaffolds structures, functionalization for tissue engineering and regenerative medicine; delivery systems for drugs/proteins/genes; bioactive wound dressings; membranes for different medical applications including filtration and dialysis; and biosensing for disease diagnostics and/or prognosis.

Fibers are a bioinspired nanostructure in the way that we recognize from either continuous filaments or elongated objects. The history of fiber production by

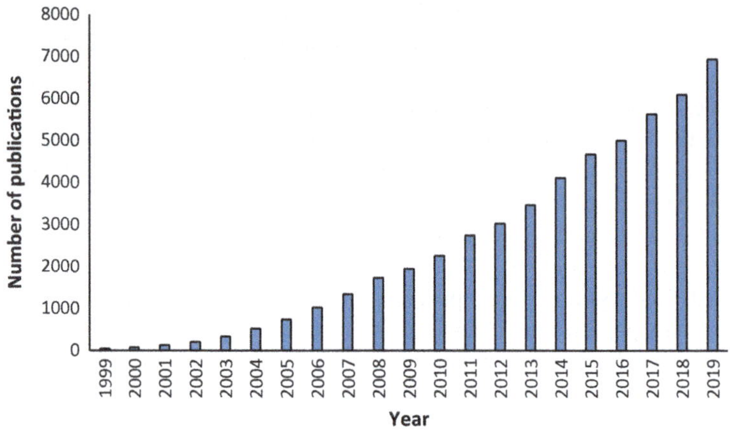

Fig. 4.2 Evolution of the number of scientific publications, since 1999 until now, using "nanofibers" as keyword. (Source: Web of Science database, June 2020)

humanity, such as wild silk threads is dated to c. 2450–2000 BCE. The spindle for the manufacture of cotton and wool fibers used in the production of clothing was invented around 1300 BCE. This practice slowly evolved into the textile industry in the 1880s (Xue et al. 2019). The history of synthetic nanofiber technology started in the late 1500s with William Gilbert work (Burgess 2012). This English physician observed that when a piece of charged amber was brought near a spherical drop of water on a dry surface, the amber "pulls the nearest parts out of their position and draws it up into a cone." (Barhoum et al. 2019). This was the first reported case of electrospraying, which was later renamed into electrospinning.

The electrospinning process for nanofibers production was recognized and registered as a patent in 1902 (Cooley 1902). Later in 1914 John Zeleny studied the behaviour of fluid droplets at the end of metal capillaries that formed fine fibers, which contributed to the development of the mathematical model explaining the behaviour of fluids under electrostatic forces (Tucker et al. 2012). Until 1995 electrospinning process was employed for over three decades in industries to produce various products, namely spun and tubular products, nonwoven fabrics, grafts, etc. (Doshi and Reneker 1995). For further details, the reader shall check the work published by Tucker et al. (2012) and Xue et al. (2019). Since that time, the term electrospinning becomes more popular and the number of publications has been increasing exponentially every year (Fig. 4.2).

Nanofibers can be prepared from natural or synthetic polymers that result in different mechanical properties and applications. The most common materials used are organic polymers in the form of either solution or melt. Synthetic polymers include poly(vinyl alcohol) (PVA), poly(lactic acid) (PLA), polycaprolactone (PCL), polyurethane (PU), poly(lactic-co-glycolic acid) (PLGA), poly (vinylpyrrolidone) (PVP), polyethylene oxide (PEO), poly(ethylene-co-vinylacetate) (PEVA), etc. (Sharma et al. 2015). Their main advantages are the low-cost, tailored architecture, controllable degradation rate, and mass production (Nair and Laurencin 2007). Natural polymers include collagen, silk fibroin (SF), heparin, keratin, gelatin, proteins, and polysaccharides such as cellulose acetate (CA), pectin, chitosan, and alginate (Sharma et al. 2015). These natural polymers are highlighted by their biocompatibility, which induces low antigenic response. However, their price and poor water stability hinder their use in nanofibers production. Nonetheless, these natural polymers could be combined with synthetic polymers to fabricate hybrid nanofibers with better mechanical properties (Homaeigohar and Boccaccini 2020).

The rheological properties of the polymer, dictated by its molecular weight, and its electrical properties determine its chain entanglement. For instance, lowering the molecular weight tends to generate beads rather than fibers. All the molecules that can have self-assemble and generate enough chain entanglement can be used for nanofibers production. Overall, lower viscosity and lower density of surface charges favour the formation of beaded nanofibers, whereas reduction of surface tension makes the beads disappear gradually (Xue et al. 2019). A variety of compounds can also be combined to produce composite nanofibers structures to have improved and inimitable physical, chemical, and biological properties (Polini and Yang 2017).

Interestingly, the choice of the polymer and the technique used for nanofibers production is based on its final application. Particularly in nanomedicine, the physico-chemical properties of the used polymer should be carefully evaluated since they affect the biocompatibility, cytotoxicity, and physical features (diameter, shape, and form) of the final nanofibers. In the present chapter, we focus on the use of nanofibers in biomedical applications, highlighting their importance in medical microbiology. PCL, PLGA, and gelatine have been used in the fabrication of nanofibers for the delivery of antibiotics and wound dressing materials with antibacterial properties. Parallelly, SF, PLA, and CA have been used for tissue engineering and drug delivery systems, while PEVA, PVA, and chitosan have been proposed, as promising polymers, for drug delivery applications (Table 4.1).

4.2 Manufacturing Techniques

There are several techniques to make nanofibers, which can be divided into non-electrospinning and electrospinning techniques (Fig. 4.3). The first group includes drawing techniques, spinneret-based tunable engineered parameter method, phase separation, self-assembly, template synthesis, freeze-drying synthesis, and interfacial polymerization of nanofibers according (Alghoraibi and Alomari 2019). The second group includes electrospinning, which is the most commonly used technique to generate nanofibers for research applications and has also demonstrated the most promising results in terms of biomedical applications. Different structures such as hollow, flat, and ribbon-shaped, with controlled, inter- and intrafiber porosity can be fabricated depending on the application purposes. Many advantages for the use of the electrospinning method have been referred such as the straightforward

Table 4.1 Summary of representative examples about the use of nanofibers in biomedical applications, particularly drug delivery systems, wound dressing materials and tissue engineering

Application	Polymer	Fabrication	Modification	Usefulness	Ref.
Drug delivery					
	CA	Co-electrospinning	Mix paclitaxel with polymer solution	Encapsulation and release of paclitaxel for cancer treatment	Y. Liu et al. (2020)
	CA	Co-electrospinning	Mix thymol with polymer solution	Encapsulation and release of thymol for bacterial infection treatment	Y. J. Chen et al. (2020a)
	CA and PVP	Coaxial electrospinning	Mix emodin with PVP solution	Encapsulation and release of emodin for MRSA treatment	Ye et al. (2020)
	PCL	Electrospinning	Mix micro-RNAs with PCL solution	Encapsulation and release of micro-RNAs for osteogenesis	Tahmasebi et al. (2020)
	PCL	Electrospinning	Mix tretinoin with polymer solution	Encapsulation and release of tretinoin for acne treatment	Khoshbakht et al. (2020)
	PCL	Forcespinning	Mix doxorubicin and functionalized carbon nano-onions with polymer solution	Encapsulation and release of doxorubicin for cancer treatment	Mamidi et al. (2020)
	PCL and PVA	Coaxial electrospinning	Mix doxorubicin with PVA solution	Encapsulation and release of doxorubicin for cancer treatment	Yan et al. (2020)
	PCL, PLA and gelatin	Blended-electrospinning	Mix tetracyclin with polymer solution	Encapsulation and release of tetracycline for bacterial infection treatment (periodontal disease)	Shahi et al. (2017)
	PEO and chitosan	Electrospinning	Mix spores of bacterial strain with the polymer solution	Encapsulation and release of probiotics for bacterial infection treatment (periodontal disease)	Zupancic et al. (2018)
	PEO and SF	Co-axial electrospinning	Gelatin nanospheres loaded with vancomycin	Encapsulation and release of vancomycin for bacterial infection treatment	Song et al. (2017)
	PLGA	Electrospinning	Mix linezolid with polymer solution	Encapsulation and release of linezolid for MRSA treatment	Eren Boncu et al. (2020)

	Poly(methyl vinyl ether-*alt*-maleic anhydride): acid and ester	Electrospinning	Mix antibiotics with polymer solution	Encapsulation and release of antibiotics for bacterial infection treatment	Mira et al. (2020)
	PVA and chitosan	Blended-electrospinning	Mix functionalized graphene oxide-curcumin particles with polymer solution	Encapsulation and release of curcumin for bacterial infection and cancer treatment	Sedghi et al. (2017)
	SF and gelatin	Electrospinning	Mix ceftazidime with polymer solution	Encapsulation and release ceftazidime for bacterial infection treatment	Safdari et al. (2016)
Wound dressing					
	Chitosan	Drawing	–	Antimicrobial effectiveness against *Clostridioides difficile*	Shahini Shams Abadi et al. (2020)
	Chitosan	Electrospinning	Mix silver nanoparticles with polymer solution	Antimicrobial effectiveness for wound healing	Lee et al. (2014)
	Gelatin	Electrospinning	Mix *Centella asiatica* extract with polymer solution	Effectiveness for wound healing (fibroblast activity and antimicrobial properties)	Yao et al. (2017)
	Oxidized cellulose and polyethylene glycol	Freeze-drying	Mix zinc oxide with polymer solution	Antimicrobial effectiveness and control bleeding for wound healing	Shefa et al. 2019)
	PCL	Electrospinning	Mix mussel adhesive protein (*r*fp-1) and DOPA-Fe(III) complexes with polymer solution	Effectiveness for wound healing (cicatrisation)	Kim et al. (2015)
	PLGA	Electrospinning	Mix *Aloe vera* extract and recombinant human epidermal growth factor with polymer solution	Effectiveness for wound healing (fibroblast activity)	
	Polyacrylonitrile and moringa extract	Electrospinning	–	Antimicrobial effectiveness for wound healing	Fayemi et al. (2018)

(continued)

Table 4.1 (continued)

Application	Polymer	Fabrication	Modification	Usefulness	Ref.
	PU CA	Blended-electrospinning	Mix reduced graphene oxide/silver with PU solution Mix curcumin with CA solution	Antimicrobial effectiveness for wound healing	Esmaeili et al. (2020)
	PVA and chitosan	Electrospinning	Mix *Bidens pilosa* extract with polymer solution	Antimicrobial effectiveness for wound healing	Kegere et al. (2019)
	PVA and sodium alginate	Electrospinning	Mix zinc oxide with polymer solution	Antimicrobial effectiveness for wound healing	Shalumon et al. (2011)
	PVA, chitosan and honey	Electrospinning	Mix HPCS-BV/PS1 bacteriophage with polymer solution	Antimicrobial effectiveness for wound healing	Sarhan and Azzazy (2017)
Tissue engineering					
	Gelatin	Electrospinning	Mix FGF-2 with polymer solution	Cell attachment and proliferation	Lee et al. (2016)
	Gliadin	Dry-spinning	–	Cell attachment and proliferation	N. Reddy and Yang (2008)
	PCL	Electrospinning	Surface modification with graphene-oxide	Skeletal muscle cells scaffold	Uehara et al. (2020)
	PCL	Electrospinning	–	Bone cells scaffold	(Tahmasebi et al. (2020)
	PCL and polyethylene terephthalate	Blended-electrospinning	–	Artificial vessels	Nejad et al. (2020)
	PCL and bovine serum albumin	Blended electrospinning	Mix nerve growth factor with polymer solution	Regeneration of nervous system cells	Valmikinathan et al. (2009)
	PCL and MXene	Electrospinning	–	Bone cells scaffold	Awasthi et al. (2020)
	PEO and casein	Co-electrospinning	Mix silver nanoparticles with polymer solution	Cell attachment and proliferation	Selvaraj et al. (2018)
	PEO and soy protein	Co-electrospinning	–	Cell attachment and proliferation	Ramji and Shah (2014)

	PLA and CA	Electrospinning and freeze-drying	–	Bone cells scaffold	J. Chen et al. (2020b)
	PLGA and bovine serum albumin	Emulsion electrospinning	Mix vascular endothelial growth factor with polymer solution	Cell attachment and proliferation	Rosa et al. (2017)
	SF and PLA	Electrospinning	Load hydroxyapatite	Bone cells scaffold	Y. Gao et al. (2018)
	SF and nylon 6	Blended-electrospinning	Surface modification with lysozyme and collagen	Potential therapy for pelvic organ prolapse	Yuan et al. (2020)

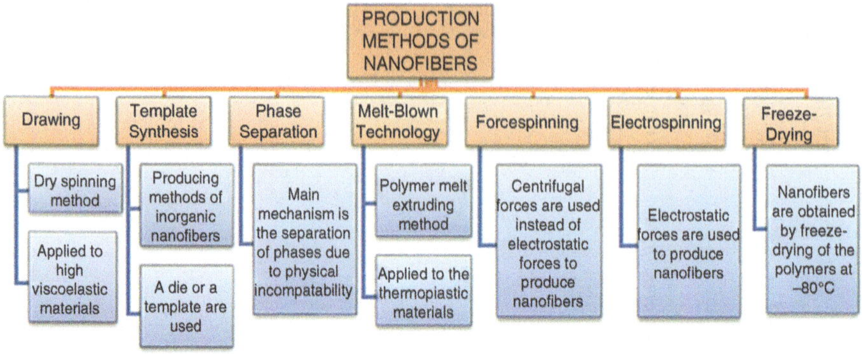

Fig. 4.3 Schematic representation of nanofibers production methods. (Reproduced from Yildiz et al. (2020). Copyright © 2020 with permission from Elsevier)

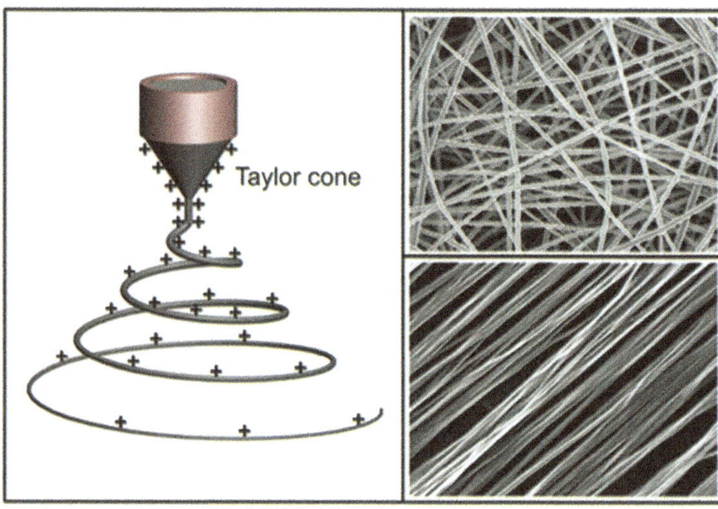

Fig. 4.4 Electrospinning technique for nanofibers production. (Reprinted with permission from Xue et al. (2019). Copyright © 2019 American Chemical Society)

setup, the ability to mass-produce continuous nanofibers from various polymers, and the capability to generate ultrathin fibers with controllable diameters, compositions, and orientations (Al-Enizi et al. 2018; X. Z. Gao et al. 2019; W. J. Lu et al. 2014).

Electrospinning technique was firstly implemented with a simple apparatus for nanofibers production. A syringe containing polymer solution or melted polymer, with a charged metallic needle tip, a grounded or oppositely charged collector, and a high-voltage power supply, are the main constituents. The electrospinning process can be divided into four stages:

(a) charging of the liquid droplet and formation of Taylor cone or cone-shaped jet (Fig. 4.4); the equilibrium reached between the electrostatic charges (resulted by the application of an electric field) and the surface tension of the solution, is responsible for the Taylor cone formation. Up this voltage point, the jet is ejected from the tip of the Taylor cone; (b) extension of the charged jet along a straight line; (c) thinning of the jet in the presence of an electric field and growth of electrical bending instability (also known as whipping instability); (d) solidification and collection of the jet as solid fiber(s) on a grounded collector (Xue et al. 2019). The jet passes through a region of instability, where the solution's solvents evaporate and are deposited onto the substrate as a fiber (Kitto et al. 2019).

Nowadays, there are available electrospinning systems with needleless technology for a high-throughput production for pilot scale and industrial-scale production (Ske 2020). In this new electrospinning platform, a rotating cylindrical spinneret with the application-specific rotating collectors makes possible the fabrication of many fibrous architectures with various fiber diameter and orientation. The fiber's properties are determined by the applied voltage, the flow rate, and the distance between the spinneret and the collector (Xue et al. 2019).

As previously mentioned organic polymers can be applied to solution electrospinning or melt electrospinning. The most used one is the solution electrospinning where the solid nanofibers are formed after solvent evaporation and deposited on the collector (Xue et al. 2019). Biocompatible and biodegradable synthetic polymers, such as PCL and PLGA, have been directly electrospun into nanofibers and further explored as scaffolds for biomedical applications (Eren Boncu et al. 2020; Tahmasebi et al. 2020). On the other hand, natural biopolymers, such as SF, fibrinogens, dextran, casein, chitosan, alginate, collagen, and gelatin, have been electrospun into nanofibers from their solutions.

The structure and morphology of the nanofibers produced by solution electrospinning are determined by the polymer (ex: molecular weight), by the solvent (suitable for polymer dissolution), processing parameters, and atmospheric conditions. Some polymers such as polyethylene and polypropylene are not suitable for solution electrospinning due to their low solubility in solvents. In those situations, fibers are produced by melting electrospinning resorting to hot circulating fluids, electrical heating tape, or laser (Xue et al. 2019).

The nanofibers obtained by electrospinning can have meticulous orientations and dimensions, which makes this technique a powerful fiber manufacturing tool. The possibility to fabricate nanofibers with controlled/uncontrolled structure and ordered/non ordered orientation empathize their potential and acceptability for biomedical applications, which require well-arranged alignment and special structures (Greiner and Wendorff 2007).

Nanofibers use is increasing in several medical areas as already described. Particularly in medical microbiology, two important aspects of their use should be noticed. The first is related to their development and application in tissue engineering and the strategies to avoid biofilm formation that can affect their medical viability. The second is related with their use as a drug delivery system for antibiotics that help to optimize the therapy, reduce the concentrations, and reduce or eliminate the lateral effects of the drug. The use of wound dressing based on nanofibers has also been considered concerning microbiological aspects, namely the antibacterial and anti-inflammatory properties of some polymers or the incorporation of bioactive substances for that purpose.

4.3 Medical Applications on Microbiology

As already mentioned nanofibers have been widely used in biomedical applications. Their attractive properties such as the ratio of high surface area and volume as well as the high porosity with small pores enhance their use because of the similar characteristics with the extracellular matrix. This novel morphology emphasize cell behaviour and offers promising support for cells, which is quite attractive for tissue engineering purposes (Al-Enizi et al. 2018; X. Z. Gao et al. 2019; W. J. Lu et al. 2014; Yoshimoto et al. 2003). A plethora of other biomedical applications using nanofibers have been reported, especially drug delivery systems (Abid et al. 2019), biological wound dressings (Homaeigohar and Boccaccini 2020) and medical diagnosis (Rezaei and Mahmoudifard 2019). Particularly in medical microbiology, the controlled release of active pharmaceutical ingredients for bacterial infection therapy, wound healing materials incorporating antimicrobial substances and the fabrication of effective biosensing systems for disease diagnosis are the representative examples of these types of applications (Fig. 4.5). Nonetheless, nanofibers can also be used for early detection, cure, and diagnosis of cancer but this subject will not be

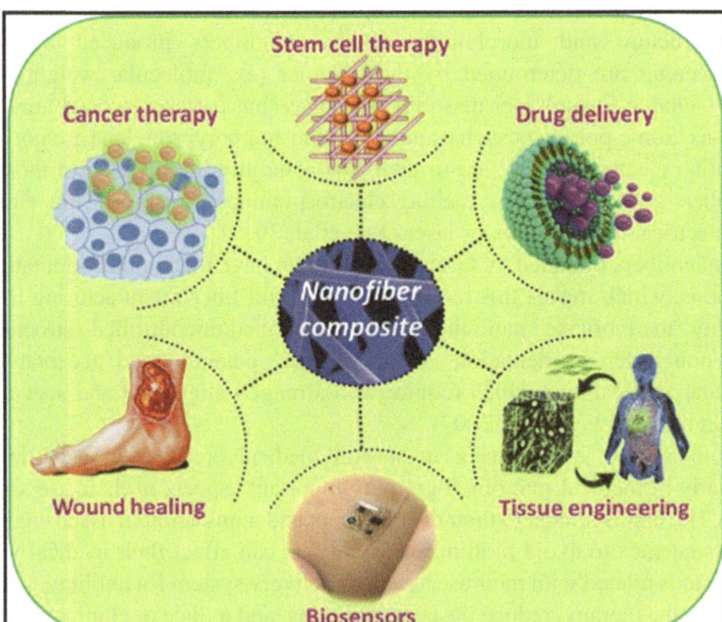

Fig. 4.5 Key of biomedical applications of nanofiber composites (Bacakova et al. 2020)

described in this chapter. For further details, the reader shall check the work published by Abid et al. (2019).

4.3.1 Drug Delivery Systems

Drug delivery systems refer to a technical system that comprehensively regulates the distribution of drugs in living organisms in space, time, and dosage (Ghosh and Murthy 2006; Prasad et al. 2017). It uses controlled-release materials that allow a slow and continuous release of the drug in the human body, maintaining its effective concentration for a long period. An ideal delivery system should release drug molecules in a controlled manner, preserving their bioactivity, optimizing their effectiveness, and reducing their side effects (Bongio et al. 2010; Mottaghitalab et al. 2015). Therefore, the development of novel and effective drug delivery systems for the controlled release of bioactive molecules is of critical importance in field medicine, particularly in medical microbiology. The recent advances in nanomaterials research opened new possibilities, allowing the production of numerous nanostructured carriers including liposomes, dendrimers, and nanoparticles (Mody et al. 2014) as well as electrospun nanofibers (Cleeton et al. 2019). The latter has been used for drug administration purposes due to their high surface-to-volume ratio. This feature increases the efficiency of the encapsulation and the stability of the drug release, as well as the ease of processing and the cost-benefit ratio (Ulubayram et al. 2015).

The selection of polymers is very important in the fabrication of electrospun nanofibers for drug delivery systems since their features like water solubility, polarity, and molecular weight would determine the properties of the final fiber (Ye et al. 2020). One of the polymers that attracted attention was chitosan due to its biodegradability, biocompatibility, and safety characteristics. However, electrospinning pure chitosan is very difficult due to the repulsive forces between its ionic groups (Desai et al. 2008). Meanwhile, PVP is a synthetic hydrophilic polymer with well spinnable and mechanical properties that can be used alone or combined with other polymers to fabricate electrospinning fibers (Ye et al. 2020).

Another important aspect is related to the solvent system, which is fundamental for the optimization of electrospinning by altering the rheological and electrostatic properties of the polymer solutions. This feature was recently demonstrated after the encapsulation of linezolid, a new generation antibiotic against methicillin-resistant *Staphylococcus aureus* (MRSA) infections, into electrospun PLGA nanofibers using different solvents (Fig. 4.6). The researchers found that mono solvent solutions were not spinnable but solvents combined at different ratios allowed the production of nanofibers (Eren Boncu et al. 2020). In this work, the best combination was attained with dichloromethane and dimethylformamide at 50:50 ratio (v/v), providing a controlled release of linezolid for 13 days.

The mechanism and the amount of drug-loaded inside the nanofibers is also a key factor for the development of highly effective nanofibers as drug delivery systems.

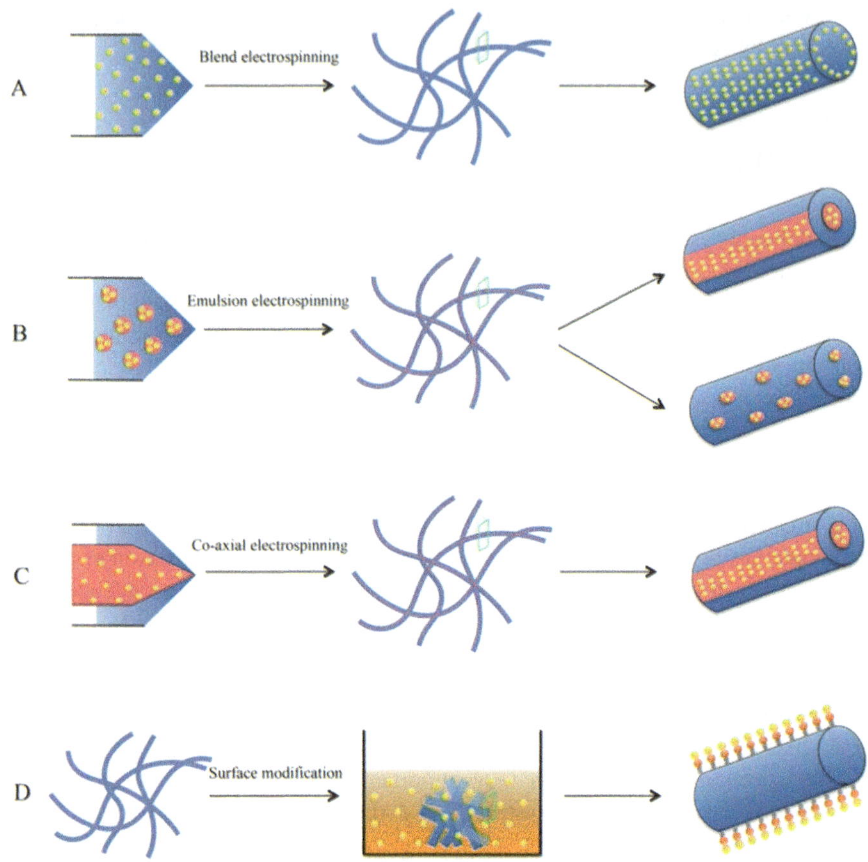

Fig. 4.6 Schematic illustration of drug incorporation strategies into nanofibers. (**a**) Blend electrospinning, where drugs and polymers are co-dissolved in solvents to be spun; (**b**) emulsion electrospinning, where drug solutions are emulsified into immiscible polymer solutions, followed by spinning; (**c**) co-axial electrospinning, where drug and polymer solutions are separately spun through two concentric nozzles; (**d**) post-immobilization, where drugs are conjugated onto fabricated nanofiber matrices through physical or chemical interaction. (Reproduced from (J. Wang and Windbergs 2017). Copyright © 2017 with permission from Elsevier)

There are several ways to incorporate the drugs inside the fibers including blended electrospinning, emulsion electrospinning, co-axial electrospinning, and surface modification (Fig. 4.6) (Wang and Windbergs 2017). Drug molecules can be directly embedded into the polymer fiber matrix or be attached to the fiber surface. Nonetheless, the mechanism release of the drug from the electrospun fibers is also an important aspect, which can be affected by different factors such as the drug content, drug distribution in nanofibers and fiber diameter distribution (Al-Enizi et al. 2018; Verreck et al. 2003).

The first study related to the use of electrospun fibers as drug delivery system came out in 2002, in which tetracycline hydrochloride, an antibiotic used for

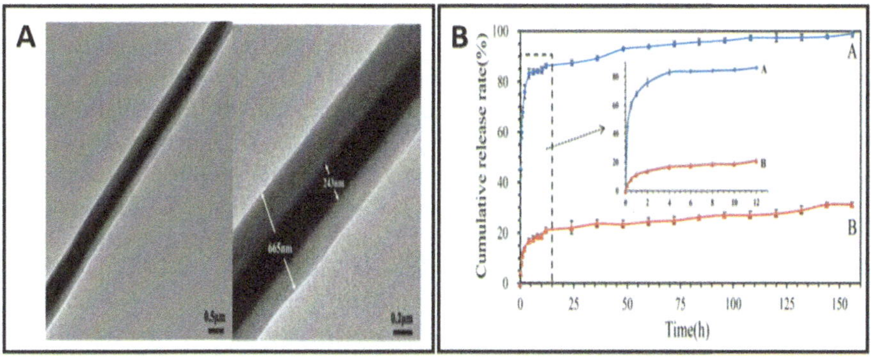

Fig. 4.7 Electrospun nanofibers used for emodin encapsulation and release. (**a**) Transmission electron microscopy image of core-shell structure nanofiber containing emodin. (**b**) Cumulative release curves of emodin from nanofibers and raw emodin. (**a**) Nanofibers containing emodin, (**b**) raw emodin. The insert profiles show the release behavior of nanofibers and raw emodin during the initial 12 h. (Ye et al. 2020)

periodontal disease treatment, was loaded into PLA, PEVA, and PLA/PEVA blend (50:50, w/w) nanofibers (Kenawy et al. 2002). The results pointed out that the release profile was dependent on the nature of the used polymer and the amount of the loaded drug. The best results showed a smooth release of the drug over about 5 days using electrospun PEVA and PLA/PEVA (50:50) mats. In recent years, electrospinning has been used to the synthesis of polymer nanofibers capable of sustained release of several pharmaceutical drugs such as antibiotics (Y. J. Chen et al. 2020a; Shahi et al. 2017; Song et al. 2017; Y. N. Yang et al. 2020) and anticancer substances (Y. Liu et al. 2020; Sedghi et al. 2017; Yan et al. 2020). The most recent and representative examples are summarized in Table 4.1, which included ceftazidime, vancomycin, emodin, doxorubicin, and paclitaxel drugs.

Although the techniques for incorporating drugs into nanofibers have proven to be somewhat effective, it was identified a fast pharmaceutical drug dissolution (or bioactive substance) on electrospun nanofibers surface, which is known as burst release (Al-Enizi et al. 2018). To circumvent this drawback, core-shell composite nanofibers can be fabricated using co-axial electrospinning, in which two miscible or immiscible materials can be electrospun together (Huang et al. 2006). Using this approach, drugs can be coated by an outer shell of a polymeric layer to enhance the release of these materials. An interesting study reported the variation of the physical and chemical properties of the core and shell solutions to control the release of a hydrophilic drug, metoclopramide hydrochloride (Tiwari et al. 2010). The results clearly indicated a more controllable release pattern with the core-shell fibers using either PLC, PLA, and PLGA comparing to the respective monolithic fibers.

On the other hand, the use of core-shell nanofibers can simplify the use of pharmaceutical drugs with poor solubility in water and low oral bioavailability. The co-axial electrospinning nanofibers with emodin drug encapsulated in the hygroscopic cellulose acetate sheath is a representative example (Fig. 4.7a) (Ye et al. 2020). The results showed a biphasic drug release profile of emodin with an initial rapid release followed by a slower sustained release (Fig. 4.7b). In another work, co-axial electrospinning was used to prepare core-shell nanofibers containing amoxicillin trihydrate loaded in SF as core and PVA in the shell (Ojah et al. 2019). The fabricated nanofibers showed a biphasic drug profile with prolonged antibacterial activity against the Gram-negative *Escherichia coli (E. coli) and the Gram-positive Staphylococcus aureus (S. aureus)* bacteria. A similar approach was described in the fabrication of PVA/PCL core-shell nanofibers, in which PVA and PCL formed the core and shell layers, respectively (Yan et al. 2020).

The remarkable progress in this field of nanotechnology, particularly in nanomaterials science, conducted on the development of new drug delivery systems based on electrospun nanofibers. The recent research trends refer to the incorporation of nanoparticles into polymer nanofibers to develop new medical treatment strategies. An interesting example refers to the incorporation of vancomycin-loaded gelatine nanospheres into SF nanofibers through a colloidal electrospinning technique (Song et al. 2017). The results showed a more sustained release of vancomycin from nanofibrous membranes fortified with gelatine B nanospheres than nanosphere-free membranes. This strategy provides an extended antibacterial effect against *S. aureus*. Additionally, SF/graphene oxide nanofibers were fabricated by blended electrospinning, and the antibacterial activity was further evaluated (Wang et al. 2018). The results showed an increase of the inhibitory effect against *S. aureus* and *E. coli* which was enhanced by the addition of graphene oxide.

The combination of natural or synthetic drugs with electrospun mats for the development of drug delivery systems has been proved to be an attractive alternative to eliminate microorganisms. The major problems like low solubility and poor bioavailability of the drugs can be overcome by their incorporation into nanofibrous scaffolds.

These excellent outcomes of nanofibers as drug delivery, combined with their mechanical properties leverage their use as wound dressing materials.

4.3.2 Wound Dressing

Recently, electrospun nanofibers have been used as materials for wound healing purposes, which is a physiological process that occurs normally as a response to tissue damage to maintain its functionality and integrity (Monaco and Lawrence 2003). An ideal polymer for wound dressing should be noncytotoxic, biodegradable, hemocompatible, easy to remove, and capable of maintaining the moisture content over the wound surface (Wang et al. 2007). Another relevant characteristic should be their impermeability to bacteria, avoiding the formation of biofilms at the surface that could infect the wound. In general, the excellent properties of nanofibers accomplish these criteria, particularly the high oxygen porosity and the similar texture to the natural extracellular matrix in the skin, which largely contributed to

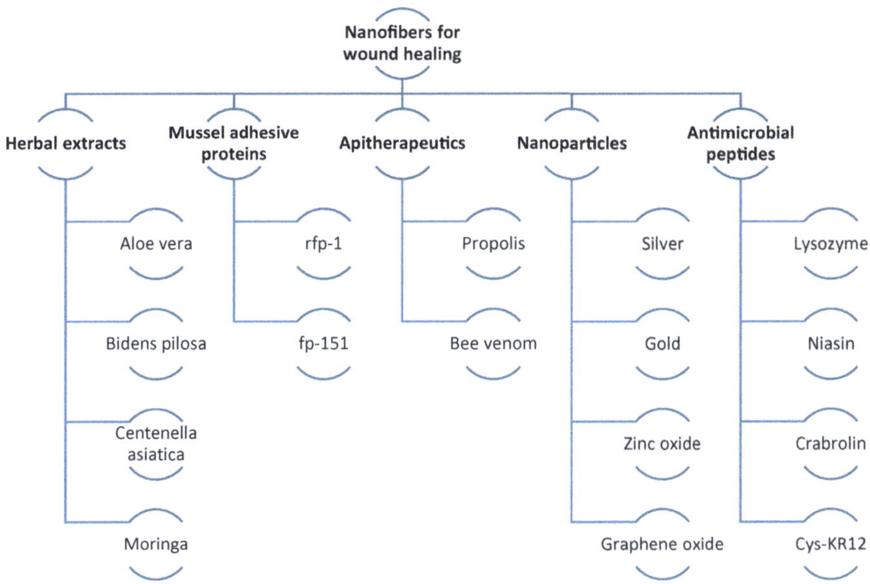

Fig. 4.8 Schematic application of electrospun nanofibers as wound healing materials

the development of new medical materials (Fayemi et al. 2018). Polymer nanofibers can be used by themselves without any additives or in combination with bioactive plant extracts or synthetic materials used to improve their efficiency (Table 4.1). Mussel adhesive proteins (Kim et al. 2015) and honey-derived products (Sarhan and Azzazy 2017) could also be used to fabricate nanofibers with enhanced wound healing capability. Some of the representative examples of materials incorporated into electrospun nanofibers are displayed in Fig. 4.8.

Among the natural herbal extracts, there are several examples stated in the scientific literature. In one case, the well-known antioxidant properties of *Aloe vera* were enhanced by its incorporation into electrospun PLGA nanofibrous membrane, stimulating the proliferation and activity of fibroblasts (Garcia-Orue et al. 2017). The wound-healing effect was even boosted by loading the nanofibers with a recombinant human epidermal growth factor, which acted as a mediator in the healing process. Electrospun nanofibers synthetized from PVA and chitosan blends loaded with extract of *Bidens pilosa* were also referred (Kegere et al. 2019). The results showed an increase in the antibacterial activity of the composite nanofibers against *S. aureus* and *E. coli*. When compared with either herbal extract or chitosan alone, suggesting their potential use as wound healing material. The herbal extract of the traditional *Centella asiatica* was also incorporated into gelatin nanofibers using electrospinning technique, promoting fibroblast proliferation and collagen synthesis, which enhanced the wound healing process (Yao et al. 2017). Other examples in the literature refer to moringa leaf extracts with antimicrobial properties incorporated into polyacrylonitrile electrospun nanofibers (Fayemi et al. 2018). Chemical substances obtained from herbal extracts can also be used such as the salvianolic acid B (*Salvia miltiorrhiza*) and bromelain (*Ananas comosu*) that were co-loaded into electrospun core-shell nanofibers (Shoba et al. 2017). In this case, the bromolein was first released from PVA/gelatin shell to remove the formed eschar as a result of its proteolytic activity while the salvianolic acid was loaded into the core of PCL to be released later and promote the formation of new blood vessels.

The recent advances in nanomaterials have drawn the attention of researchers to the development of novel wound dressings. An interesting example describes the incorporation of different contents of silver nanoparticles into chitosan nanofibers by electrospinning process (Lee et al. 2014). The composite nanofibers showed a higher degree of effectiveness against *P. aeruginosa* and MRSA when compared to the pure chitosan nanofibers. Additionally, as the number of nanoparticles increased in the fibers, its antibacterial properties were enhanced against both microorganisms. In another work, gold nanoparticles modified with 6-aminopenicillanic acid as antibacterial active ingredients were loaded into PCL/gelatin nanofibers by co-electrospinning to fabricate biocompatible wound dressings against multidrug-resistant bacteria (X. Yang et al. 2017). Zinc oxide nanoparticles were also used to produce composite scaffolds based on oxidized cellulose and the synthetic polyethylene glycol polymer (Shefa et al. 2019). The nanofiber composite was obtained by the freeze-drying method and showed biodegradable and biocompatible characteristics. The incorporation of zinc oxide enhanced the blood/fibrin coot formation and antibacterial properties, proving the potential of the proposed scaffold to be used as a hemostatic agent while inhibiting the spreading of bacterial infections. Other applications for cutaneous wound healing reported the use of graphene oxide to enhance the antibacterial activities of nanofibers, such as the SF/graphene oxide-blended nanofibers (Wang et al. 2018) and the scaffolds based in PU and CA as polymers blended with a reduced graphene oxide/silver nanocomposite and the natural polyphenolic compound called curcumin (Esmaeili et al. 2020).

Antimicrobial peptides are other group of effective broad-spectrum antibiotics that can be used for the treatment of both Gram negative and Gram positive bacteria (K. V. R. Reddy et al. 2004; Inamuddin et al. 2021). Accordingly, these short peptides have been incorporated in electrospun nanofibers for the fabrication of wound dressings. An interesting example describes the loading of the Crabrolin, a synthetic peptide, into PCL electrospun nanofibers. An extended linear release profile of the antimicrobial peptide was observed (Eriksen et al. 2013). In a different work, lysozyme and niasin were incorporated into an electrospun nanofibrous scaffold synthesized with a blend of water soluble polymers, PVA and poly(acrylic acid) and showed a stronger antibacterial activity against *S. aureus*, proving the great potential to treat infected wounds (Amariei et al. 2018). Lysozyme was loaded into electrospun nanofibrous mat based in chitosan-ethylenediaminetetraacetic acid blended with PVA (Charernsriwilaiwat et al. 2012). In animal wound healing studies, lysozyme loaded nanofibers exhibited an accelerated healing rate compared to the control (gauze) (Fig. 4.9). In another study, a synthetic single chain peptide derived from the proline-rich antimicrobial peptide dimer A3-APO was loaded into PVA electrospun nanofibers (Sebe et al. 2016). The results indicate an improvement

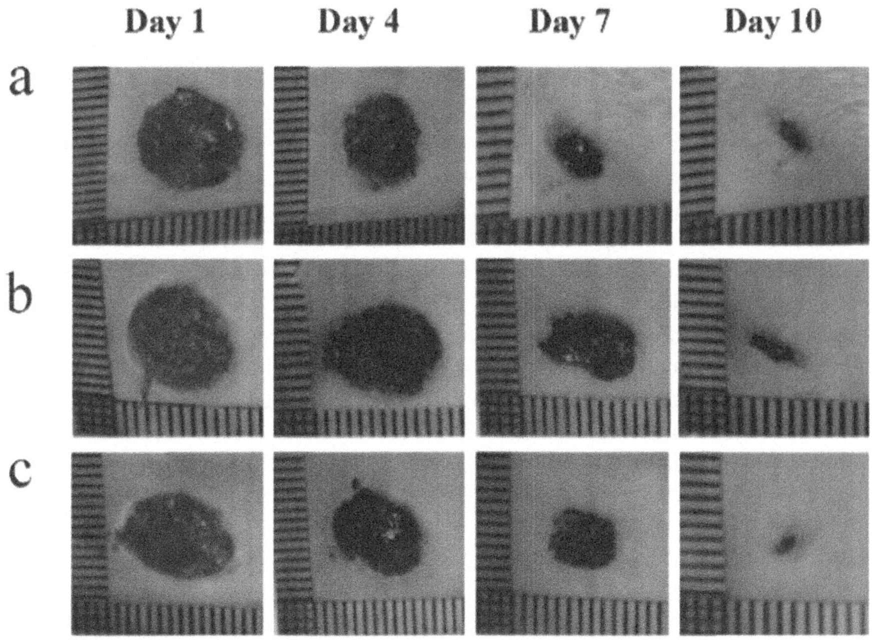

Fig. 4.9 Presence of wound healing at 1, 4, 7, and 10 days after adding (**a**) 30% lysozyme loaded CS–EDTA/PVA nanofiber mats, (**b**) gauze (−ve control), and (**c**) commercial antibacterial gauze dressing (Sofra-tulle®) (+ve control). (Reproduced from (Charernsriwilaiwat et al. 2012). Copyright © 2012 with permission from Elsevier)

in the appearance of wounds treated with APO-loaded nanofibers, reducing significantly the wound size and its bacterial presence. Other examples reported the incorporation of the synthetic Cys-KR12 into SF electrospun nanofibers (Song et al. 2016) and the co-loading of pam3CSK4 peptide and vitamin D3 into core-shell nanofibrous (S. Chen et al. 2017a).

As previously described, electrospun nanofibers have been used successfully to develop new solutions for wound healing agents. Natural and synthetic polymers have been used so far, including chitosan, SF, PVA, PU, PLA, and PCL, among others. Parallelly, several antibacterial materials such as bioactive molecules and nanoparticles were already incorporated into nanofibrous mats to prevent wound infection. The versatility of polymers and the tremendous research in nanomaterials, namely in bioactive nanoparticles will certainly contribute to the appearance of novel and cost-effective wound dressings.

4.3.3 Tissue Engineering

Tissue engineering is an interesting research field for the development of functional replacements for damaged tissues namely for musculoskeletal (including bone, cartilage, ligament, and skeletal muscle), skin, vascular and neural tissues (O'Brien 2011; Vasita and Katti 2006). It involves the use of a biocompatible scaffold capable of mimic the biological function and structure of the natural extracellular matrix, which is essential for the formation of new viable tissue. The similarity of electrospun nanofibers with the extracellular matrix offers a promising approach to develop fibrous scaffolds for cells growing, migration, and seeding. In recent years, different artificial polymeric scaffolds have been developed for bone and vascular tissue engineering, among others (Al-Enizi et al. 2018; W. J. Lu et al. 2014). Some representative examples are summarized in Table 4.1, including the formation of nanofibrous scaffolds for cell attachment and proliferation as well as the fabrication of nanofibers able to incorporate and deliver different growth factors. Nonetheless, the formation of biofilms on the surface of such nanofibrous scaffolds may be a ubiquitous problem (Kurtz and Schiffman 2018), leading to bacterial infections at the implanted place and severe health problems. To reduce the need for the use of antibacterial agents, anti-fouling nanofibers to repel microbes and proteins for as long as possible, delaying their binding to solid material is one of the areas with great impact on tissue engineering. Electrospun nanofiber mats will keep having a vital role in controlling the local interactions with microorganisms across medical applications (Kurtz and Schiffman 2018).

To understand the interaction between bacteria and the polymer, Fabrizio De Cesare recently studied the effect of the nanofiber diameter on the attachment of bacteria (De Cesare et al. 2019), using electrospun PCL nanofibers and *Burkholderia terrícola* bacteria cells as models. The results showed that bacteria attached preferentially to nanofibers with ≈100 nm diameter. Thus, materials could be created to reduce or prevent the adhesion of bacteria by adjusting the size of the fibers, acting as well as antimicrobials for the biocontrol of diseases induced by pathogens in medicine (De Cesare et al. 2019).

Nonetheless, not only in the area of microbiology these nanofiber compounds have attracted interest. Various biodegradable and biocompatible polymers, including either natural and/or synthetic polymers have been tested in the nanofibers production. An interesting example refers to the natural polymer chitin and its derivative, chitosan, for the fabrication of nanofibrous scaffolds applied mainly in bone tissue engineering (Tao et al. 2020). Similarly, a composite scaffold made of chitosan, alginate, collagen, and hydroxyapatite was fabricated by electrospinning to decrease the collagen solubility at the implanted place (Yu et al. 2013). The nanofibers were applied in studies in vitro showing a reduction of the collagen decomposition by 35% in 10 days, which contributed to the cell spreading, attachment, mineralization, and proliferation. Synthetic polymers have been also used for nanofibers production, such as the attractive study describing the synthesis of a tubular nanofibrous structure with a 6 mm internal diameter by blended electrospinning of polyethylene terephthalate and PCL (Nejad et al. 2020). The suitable mechanical properties of the attained nanofibers, combined with the biocompatibility of both polymers, make this material a good candidate for artificial vessel replacement.

Natural polymers can also be combined with synthetic polymers to produce hybrid nanofibers with better mechanical properties. An example of this valuable strategy was demonstrated by combining SF, an animal-derived protein, whose use has been hindered by the poor mechanical properties and the high biodegradation rate, with the synthetic polymer Nylon 6, which has excellent mechanical strength and a low decomposition rate (Yuan et al. 2020). These composite nanofibrous mats were loaded with lysozyme and collagen. The results showed a high potential for clinical therapy of pelvic organ prolapse. Among the natural polymers, peptides and proteins have drawn the attention of researchers, particularly in the field of tissue engineering. The great advantage relies on the fact that they are basic components of cells and have many vital functions in the body, which provides a high degree of compatibility (Yildiz et al. 2020). Besides, they are biodegradable materials and thus can be easily removed from the body without surgical intervention. Different peptide/protein nanofibers have been described in the literature using either plant or animal-derived proteins. Plant-derived proteins used in nanofibers synthesis include zein (Zhang et al. 2015), soy protein (Ramji and Shah 2014), and gluten (N. Reddy and Yang 2008) while animal-derived proteins can be listed as casein (Selvaraj et al. 2018), SF (Y. Chen et al. 2017b; Y. Gao et al. 2018), bovine serum albumin (Valmikinathan et al. 2009; Won et al. 2012), elastin (Machado et al. 2013), fibrinogen (Z. Liu et al. 2015), keratin (Rajabi et al. 2020) and gelatin (Aldana and Abraham 2017), among others. The effect of nanofibers based on peptide/protein scaffolds has been demonstrated in various in vivo studies but this material is not available on the market, yet (Yildiz et al. 2020).

Another interesting approach for tissue engineering involves the use of growth factors alone or in combination with cells or other biochemical materials for cellular reparation (Laurencin et al. 1999). These growth factors usually behave as mediators of basic cell functions through their binding to specific receptors (Anitua et al. 2012). However, due to their instability, they have a short half-life time, losing their effectiveness in a short period. Therefore, there is an urgent demand to develop new methods capable of delivering these growth factors, maintaining their stability and activity. Regarding this aspect, electrospun nanofibers have great potential due to the ability to incorporate and deliver these growth factors (Shin et al. 2012). An interesting example refers to the use of electrospun gelatin nanofibers to incorporate the fibroblast growth factor 2 (FGF-2), providing the increased cell proliferation when compared with control nanofibers (Lee et al. 2016). In another study, an electrospun core-shell nanofiber scaffold loaded with osteogenic enhancer fibroblast growth factors, FGF-18 and FGF-2, was fabricated for repairing bone defects. The core was based on PEO loaded with the FGF-2 and the FGF-18, which was first incorporated into bioactive glass nanospheres for a long-term delivery, while the outer shell was synthesized from PCL (Kang et al. 2015). The in vitro cellular studies showed significant stimulation of cell proliferation and the induction of cellular mineralization. Histological images at 6 weeks were examined after H&E and Masson's trichrome staining (Fig. 4.10), showing the formation of new bone with similar morphology to old bone, particularly observed in the scaffold loaded with FGF18/FGF2.

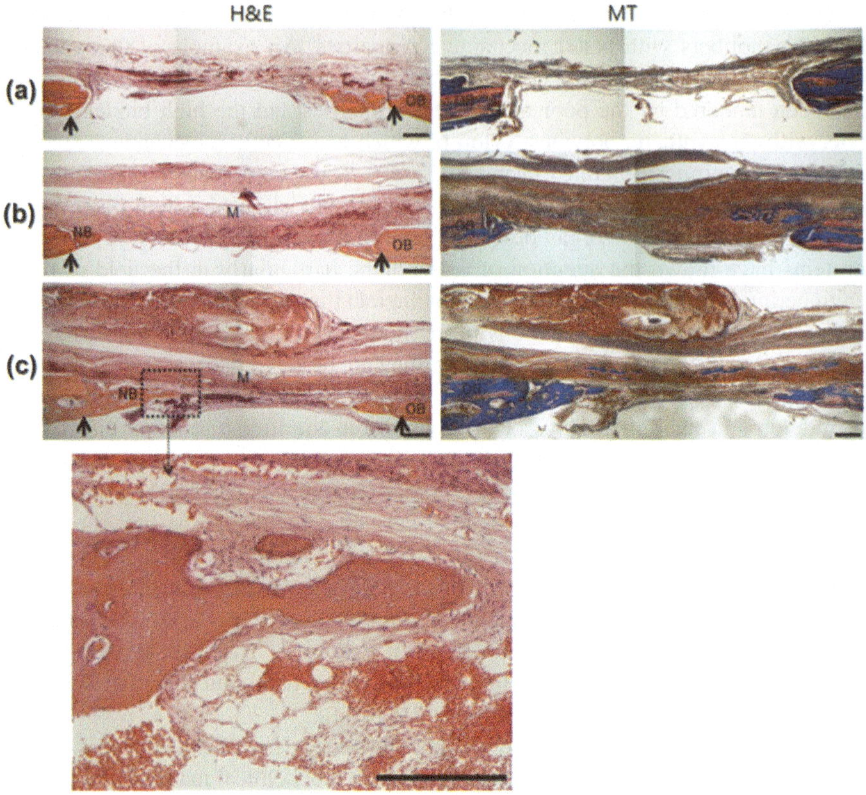

Fig. 4.10 Histological images of the harvested samples at 6 weeks, optically examined after H&E (left column) and Masson's trichrome staining (right column): (**a**) blank control, (**b**) scaffold only, and (**c**) FGF2/FGF18-scaffold. Enlarged images of the FGF2/FGF18-loaded scaffold group in the region of new bone formation revealed an immature woven bone structure with osteocytes, bone lining cells and vessel forming cells. Dashed lines indicate defect margins. OB, old bone; NB, new bone. Scale bars in (**a**–**c**): 500 μm; and in the enlarged image: 100 μm. (Reproduced from Kang et al. (2015). Copyright © 2015 with permission from Elsevier)

Vascular endothelial growth factor, involved in angiogenesis, was also incorporated into electrospun nanofibers (Rosa et al. 2017). The fabricated scaffolds based on PGLA provided a sustained release of the growth factor over 6 days, enhancing the cell adhesion without any remarkable toxicity. In another case, co-axial electrospinning was used to encapsulate a recombinant form of this vascular growth factor and PEO in the core of nanofibers, while a mixture of PCL and PEG were used to synthetize the shell (Zigdon-Giladi et al. 2017). The in vitro cellular studies demonstrated an endothelial cell migration 80-fold higher with the loaded nanofibers when compared to the scaffold without growth factor. Studies in mouses confirmed that nanofibers loaded with the vascular growth factor can stimulate cell migration into the scaffold within three days and significantly enhanced blood vessels formation within 14 days, whereas control scaffolds contained few vessels. The incorporation and release of nerve growth factor (Whitehead et al. 2018), platelet derived growth factor (Bertoncelj et al. 2014) and connective tissue growth factor using different nanofibers are other representative examples (Xu et al. 2019).

Nowadays, electrospun nanofibers have been used to develop new strategies in tissue engineering area. The results achieved so far are at embrionary stage but the replacement of bone and vascular tissue in animals was already accomplished. An important aspect of the use of nanofibers, although the biocompatibility and noncytotoxic features, is related to the formation of biofilms at the polymer's surface, which can lead to severe infections. Further work should be performed to understand this behaviour and overcome this drawback.

4.4 Future Perspectives

The importance of nanoscale science and technology in different fields is remarkable. The development of various nanostructured materials has been improved medical care not only from the perspective of early disease diagnosis but also in novel therapeutic strategies. Particularly in microbiology, nanomedicine has been contributed to improving the treatment efficiency on microbial infections through the increase of response capacity of active pharmaceutical ingredients but also into the reduction of side effects.

A plethora of nanostructures has been used for medical biology applications, in which nanofibers represent an attractive and powerful approach. The excellent properties of nanofibers and the versatility of available polymers, either natural or synthetic, highlight their use in medicine.

In this chapter, the use of nanofibers for the local administration of antibiotic drugs in infected areas by microorganisms is described. The development of wound dressing materials based on nanofibers has also been described to improve the healing process, by incorporating bioactive substances capable of eliminate bacteria and reduce the inflammation.

Tissue engineering based on nanofibers have an excellent potential to create functional materials or technologies that mimics the structural organization of natural tissues. The possibility to raise or replace biological functions without organ transplantation by the use of artificial materials that enable the seeding and proliferation of cells into the material structure is a present goal. The potential of these nanostructured materials is highlighted by the functional and mechanical properties of electrospun nanofibers similar to the extracellular matrix. However, it is scarce the data about the formation of biofilms on the surface of such nanofibrous scaffolds, considering the nanoscale morphology and the type of polymer used. These studies are crucial to controlling the local interactions with microorganisms across medical applications.

Overall, nanofibers show to be a reliable and promising approach in nanomedicine, particularly in medical microbiology.

Acknowledgments This work was supported by the National Funds through FCT/MCTES—Portuguese Foundation for Science and Technology within the scope of the project UIDB/50006/2020. Renato Gil thanks FCT (Fundação para a Ciência e Tecnologia) and ESF (European Social Fund) through POCH (Programa Operacional Capital Humano) for his PhD grant SFRH/BD/131504/2017.

References

Abid S, Hussain T, Raza ZA, Nazir A (2019) Current applications of electrospun polymeric nanofibers in cancer therapy. Mater Sci Eng C Mater Biol Appl 97:966–977. https://doi.org/10.1016/j.msec.2018.12.105

Aldana AA, Abraham GA (2017) Current advances in electrospun gelatin-based scaffolds for tissue engineering applications. Int J Pharm 523(2):441–453. https://doi.org/10.1016/j.ijpharm.2016.09.044

Al-Enizi A, Zagho M, Elzatahry A (2018) Polymer-based electrospun nanofibers for biomedical applications. Nanomaterials 8(4):259. https://doi.org/10.3390/nano8040259

Alghoraibi I, Alomari S (2019) Handbook of nanofibers. Springer

Amariei G, Kokol V, Boltes K, Letón P, Rosal R (2018) Incorporation of antimicrobial peptides on electrospun nanofibres for biomedical applications. RSC Adv 8(49):28013–28023. https://doi.org/10.1039/C8RA03861A

Anitua E, Alonso R, Girbau C, Aguirre JJ, Muruzabal F, Orive G (2012) Antibacterial effect of plasma rich in growth factors (PRGF®-Endoret®) against Staphylococcus aureus and Staphylococcus epidermidis strains. Clin Exp Dermatol 37(6):652–657. https://doi.org/10.1111/j.1365-2230.2011.04303.x

Awasthi GP, Maharjan B, Shrestha S, Bhattarai DP, Yoon D, Park CH, Kim CS (2020) Synthesis, characterizations, and biocompatibility evaluation of polycaprolactone-MXene electrospun fibers. Colloids Surf A 586:124282. https://doi.org/10.1016/j.colsurfa.2019.124282

Azonano (2017) Classification of nanomaterials, the four main types of intentionally produced nanomaterials. https://www.azonano.com/article.aspx?ArticleID=1872

Bacakova L, Pajorova J, Tomkova M, Matejka R, Broz A, Stepanovska J et al (2020) Applications of nanocellulose/nanocarbon composites: focus on biotechnology and medicine. Nanomaterials 10(2):196. https://doi.org/10.3390/nano10020196

Barhoum A, Rasouli R, Yousefzadeh M, Rahier H, Bechelany M (2019) Handbook of nanofibers. Springer

Bertoncelj V, Pelipenko J, Kristl J, Jeras M, Cukjati M, Kocbek P (2014) Development and bioevaluation of nanofibers with blood-derived growth factors for dermal wound healing. Eur J Pharm Biopharm 88(1):64–74. https://doi.org/10.1016/j.ejpb.2014.06.001

Bongio M, van den Beucken JJJP, Leeuwenburgh SCG, Jansen JA (2010) Development of bone substitute materials: from 'biocompatible' to 'instructive'. J Mater Chem 20(40):8747–8759. https://doi.org/10.1039/C0JM00795A

Burgess R (2012) Understanding nanomedicine an introductory textbook. Pan Stanford Publishing, Singapore

Charernsriwilaiwat N, Opanasopit P, Rojanarata T, Ngawhirunpat T (2012) Lysozyme-loaded, electrospun chitosan-based nanofiber mats for wound healing. Int J Pharm 427(2):379–384. https://doi.org/10.1016/j.ijpharm.2012.02.010

Chen S, Ge L, Gombart AF, Shuler FD, Carlson MA, Reilly DA, Xie J (2017a) Nanofiber-based sutures induce endogenous antimicrobial peptide. Nanomedicine 12(21):2597–2609. https://doi.org/10.2217/nnm-2017-0161

Chen Y, Yang W, Wang W, Zhang M, Li M (2017b) Bombyx mori silk fibroin scaffolds with Antheraea pernyi silk fibroin micro/nano fibers for promoting EA. hy926 cell proliferation. Materials (Basel) 10(10):1153. https://doi.org/10.3390/ma10101153
Chen Y, Qiu YY, Chen WBF, Wei QF (2020a) Electrospun thymol-loaded porous cellulose acetate fibers with potential biomedical applications. Mater Sci Eng C 109:110536. https://doi.org/10.1016/j.msec.2019.110536
Chen J, Zhang TH, Hua WK, Li PY, Wang XF (2020b) 3D porous poly(lactic acid)/regenerated cellulose composite scaffolds based on electrospun nanofibers for biomineralization. Colloids Surf A 585:124048. https://doi.org/10.1016/j.colsurfa.2019.124048
Cleeton C, Keirouz A, Chen XF, Radacsi N (2019) Electrospun nanofibers for drug delivery and biosensing. ACS Biomater Sci Eng 5(9):4183–4205. https://doi.org/10.1021/acsbiomaterials.9b00853
Cooley JF (1902) Apparatus for electrically dispersing fluids. Google Patents US692631A
De Cesare F, Di Mattia E, Zussman E, Macagnano A (2019) A study on the dependence of bacteria adhesion on the polymer nanofibre diameter. Environ Sci Nano 6(3):778–797. https://doi.org/10.1039/C8EN01237G
Desai K, Kit K, Li J, Zivanovic S (2008) Morphological and surface properties of electrospun chitosan nanofibers. Biomacromolecules 9(3):1000–1006. https://doi.org/10.1021/bm701017z
Doshi J, Reneker DH (1995) Electrospinning process and applications of electrospun fibers. J Electrost 35(2–3):151–160. https://doi.org/10.1016/0304-3886(95)00041-8
Eren Boncu T, Ozdemir N, Uskudar Guclu A (2020) Electrospinning of linezolid loaded PLGA nanofibers: effect of solvents on its spinnability, drug delivery, mechanical properties, and antibacterial activities. Drug Dev Ind Pharm 46(1):109–121. https://doi.org/10.1080/03639045.2019.1706550
Eriksen THB, Skovsen E, Fojan P (2013) Release of antimicrobial peptides from electrospun nanofibres as a drug delivery system. J Biomed Nanotechnol 9(3):492–498. https://doi.org/10.1166/jbn.2013.1553
Esmaeili E, Eslami-Arshaghi T, Hosseinzadeh S, Elahirad E, Jamalpoor Z, Hatamie S, Soleimani M (2020) The biomedical potential of cellulose acetate/polyurethane nanofibrous mats containing reduced graphene oxide/silver nanocomposites and curcumin: antimicrobial performance and cutaneous wound healing. Int J Biol Macromol 152:418–427. https://doi.org/10.1016/j.ijbiomac.2020.02.295
Fayemi OE, Ekennia AC, Katata-Seru L, Ebokaiwe AP, Ijomone OM, Onwudiwe DC, Ebenso EE (2018) Antimicrobial and wound healing properties of polyacrylonitrile-moringa extract nanofibers. Acs Omega 3(5):4791–4797. https://doi.org/10.1021/acsomega.7b01981
FDA (2012) FDA's approach to regulation of nanotechnology products. https://www.fda.gov/science-research/nanotechnology-programs-fda/fdas-approach-regulation-nanotechnology-products
Feynman RP (1960) There's plenty of room at the bottom. Eng Sci 23(5):22–36
Gao Y, Shao W, Qian W, He J, Zhou Y, Qi K et al (2018) Biomineralized poly (l-lactic-co-glycolic acid)-tussah silk fibroin nanofiber fabric with hierarchical architecture as a scaffold for bone tissue engineering. Mater Sci Eng C 84:195–207. https://doi.org/10.1016/j.msec.2017.11.047
Gao XZ, Han SY, Zhang RH, Liu GT, Wu J (2019) Progress in electrospun composite nanofibers: composition, performance and applications for tissue engineering. J Mater Chem B 7 (45):7075–7089. https://doi.org/10.1039/c9tb01730e
Garcia-Orue I, Gainza G, Gutierrez FB, Aguirre JJ, Evora C, Pedraz JL et al (2017) Novel nanofibrous dressings containing rhEGF and Aloe vera for wound healing applications. Int J Pharm 523(2):556–566. https://doi.org/10.1016/j.ijpharm.2016.11.006
Ghosh PK, Murthy RS (2006) Microemulsions: a potential drug delivery system. Curr Drug Deliv 3 (2):167–180. https://doi.org/10.2174/156720106776359168
Gnach A, Lipinski T, Bednarkiewicz A, Rybka J, Capobianco JA (2015) Upconverting nanoparticles: assessing the toxicity. Chem Soc Rev 44(6):1561–1584. https://doi.org/10.1039/c4cs00177j

Greiner A, Wendorff JH (2007) Electrospinning: a fascinating method for the preparation of ultrathin fibres. Angew Chem Int Ed 46(30):5670–5703. https://doi.org/10.1002/anie.200604646

Gubala V, Johnston LJ, Liu ZW, Krug H, Moore CJ, Ober CK et al (2018) Engineered nanomaterials and human health: part 1. Preparation, functionalization and characterization (IUPAC technical report). Pure Appl Chem 90(8):1283–1324. https://doi.org/10.1515/pac-2017-0101

Hardy A, Benford D, Halldorsson T, Jeger MJ, Knutsen HK, More S et al (2018) Guidance on risk assessment of the application of nanoscience and nanotechnologies in the food and feed chain: part 1, human and animal health. EFSA J 16(7):5327. https://doi.org/10.2903/j.efsa.2018.5327

Homaeigohar S, Boccaccini AR (2020) Antibacterial biohybrid nanofibers for wound dressings. Acta Biomater 107:25–49. https://doi.org/10.1016/j.actbio.2020.02.022

Huang ZM, He CL, Yang A, Zhang Y, Han XJ, Yin J, Wu Q (2006) Encapsulating drugs in biodegradable ultrafine fibers through co-axial electrospinning. J Biomed Mater Res A 77 (1):169–179. https://doi.org/10.1002/jbm.a.30564

Inamuddin, Ahamed MI, Prasad R (2021) Advanced antimicrobial materials and applications. Springer, Singapore. ISBN 978-981-15-7098-8. https://www.springer.com/gp/book/9789811570971

Jain KK (2019) The handbook of nanomedicine. Humana Press and Springer, Totowa, NJ

Kang MS, Kim J-H, Singh RK, Jang J-H, Kim H-W (2015) Therapeutic-designed electrospun bone scaffolds: mesoporous bioactive nanocarriers in hollow fiber composites to sequentially deliver dual growth factors. Acta Biomater 16:103–116. https://doi.org/10.1016/j.actbio.2014.12.028

Kegere J, Ouf A, Siam R, Mamdouh W (2019) Fabrication of poly(vinyl alcohol)/chitosan/bidens pilosa composite electrospun nanofibers with enhanced antibacterial activities. Acs Omega 4 (5):8778–8785. https://doi.org/10.1021/acsomega.9b00204

Kenawy E-R, Bowlin GL, Mansfield K, Layman J, Simpson DG, Sanders EH, Wnek GE (2002) Release of tetracycline hydrochloride from electrospun poly(ethylene-co-vinylacetate), poly (lactic acid), and a blend. J Control Release 81(1):57–64. https://doi.org/10.1016/S0168-3659(02)00041-X

Khoshbakht S, Asghari-Sana F, Fathi-Azarbayjani A, Sharifi Y (2020) Fabrication and characterization of tretinoin-loaded nanofiber for topical skin delivery. Biomater Res 24:8. https://doi.org/10.1186/s40824-020-00186-3

Kim BJ, Kim S, Oh DX, Masic A, Cha HJ, Hwang DS (2015) Mussel-inspired adhesive protein-based electrospun nanofibers reinforced by Fe(iii)–DOPA complexation. J Mater Chem B 3 (1):112–118. https://doi.org/10.1039/C4TB01496K

Kitto T, Bodart-Le Guen C, Rossetti N, Cicoira F (2019) 25 - processing and patterning of conducting polymers for flexible, stretchable, and biomedical electronics. In: Ostroverkhova O (ed) Handbook of organic materials for electronic and photonic devices, 2nd edn. Woodhead Publishing, pp 817–842

Kurtz IS, Schiffman JD (2018) Current and emerging approaches to engineer antibacterial and antifouling electrospun nanofibers. Materials (Basel) 11(7):1059. https://doi.org/10.3390/ma11071059

Laurencin CT, Ambrosio AMA, Borden MD, Cooper JA, J. (1999) Tissue engineering: orthopedic applications. Annu Rev Biomed Eng 1(1):19–46. https://doi.org/10.1146/annurev.bioeng.1.1.19

Lee SJ, Heo DN, Moon JH, Ko WK, Lee JB, Bae MS et al (2014) Electrospun chitosan nanofibers with controlled levels of silver nanoparticles. Preparation, characterization and antibacterial activity. Carbohydr Polym 111:530–537. https://doi.org/10.1016/j.carbpol.2014.04.026

Lee H, Lim S, Birajdar MS, Lee S-H, Park H (2016) Fabrication of FGF-2 immobilized electrospun gelatin nanofibers for tissue engineering. Int J Biol Macromol 93:1559–1566. https://doi.org/10.1016/j.ijbiomac.2016.07.041

Liu Z, Li S, Su L, Sun K, Wu X, Wu F et al (2015) Novel superhydrophilic poly(l-lactic acid-co-ε-caprolactone)/fibrinogen electrospun patch for rat abdominal wall reconstruction. J Biomater Appl 30(2):230–238. https://doi.org/10.1177/0885328215577732

Liu Y, Wang Q, Lu Y, Deng HB, Zhou X (2020) Synergistic enhancement of cytotoxicity against cancer cells by incorporation of rectorite into the paclitaxel immobilized cellulose acetate nanofibers. Int J Biol Macromol 152:672–680. https://doi.org/10.1016/j.ijbiomac.2020.02.184

Lu WJ, Sun JS, Jiang XY (2014) Recent advances in electrospinning technology and biomedical applications of electrospun fibers. J Mater Chem B 2(17):2369–2380. https://doi.org/10.1039/c3tb21478h

Lu Y, Shah KW, Xu JW (2017) Synthesis, morphologies and building applications of nanostructured polymers. Polymers 9(10):506. https://doi.org/10.3390/polym9100506

Machado R, da Costa A, Sencadas V, Garcia-Arévalo C, Costa CM, Padrão J et al (2013) Electrospun silk-elastin-like fibre mats for tissue engineering applications. Biomed Mater 8 (6):065009. https://doi.org/10.1088/1748-6041/8/6/065009

Mamidi N, Zuniga AE, Villela-Castrejon J (2020) Engineering and evaluation of forcespun functionalized carbon nano-onions reinforced poly (epsilon-caprolactone) composite nanofibers for pH-responsive drug release. Mater Sci Eng C 112:110928. https://doi.org/10.1016/j.msec.2020.110928

Mira A, Sainz-Urruela C, Codina H, Jenkins SI, Rodriguez-Diaz JC, Mallavia R, Falco A (2020) Physico-chemically distinct nanomaterials synthesized from derivates of a poly(anhydride) diversify the spectrum of loadable antibiotics. Nanomaterials (Basel) 10(3):486. https://doi.org/10.3390/nano10030486

Mody N, Tekade RK, Mehra NK, Chopdey P, Jain NK (2014) Dendrimer, liposomes, carbon nanotubes and PLGA nanoparticles: one platform assessment of drug delivery potential. AAPS PharmSciTech 15(2):388–399. https://doi.org/10.1208/s12249-014-0073-3

Monaco JL, Lawrence WT (2003) Acute wound healing an overview. Clin Plast Surg 30:1), 1–1),12. https://doi.org/10.1016/s0094-1298(02)00070-6

Mottaghitalab F, Farokhi M, Shokrgozar MA, Atyabi F, Hosseinkhani H (2015) Silk fibroin nanoparticle as a novel drug delivery system. J Control Release 206:161–176. https://doi.org/10.1016/j.jconrel.2015.03.020

Nair LS, Laurencin CT (2007) Biodegradable polymers as biomaterials. Prog Polym Sci 32 (8):762–798. https://doi.org/10.1016/j.progpolymsci.2007.05.017

Nejad MR, Yousefzadeh M, Solouk A (2020) Electrospun PET/PCL small diameter nanofibrous conduit for biomedical application. Mater Sci Eng C 110:110692. https://doi.org/10.1016/j.msec.2020.110692

O'Brien FJ (2011) Biomaterials & scaffolds for tissue engineering. Mater Today 14(3):88–95. https://doi.org/10.1016/S1369-7021(11)70058-X

Ojah N, Saikia D, Gogoi D, Baishya P, Ahmed GA, Ramteke A, Choudhury AJ (2019) Surface modification of core-shell silk/PVA nanofibers by oxygen dielectric barrier discharge plasma: studies of physico-chemical properties and drug release behavior. Appl Surf Sci 475:219–229. https://doi.org/10.1016/j.apsusc.2018.12.270

Park SH, Oh SG, Mun JY, Han SS (2006) Loading of gold nanoparticles inside the DPPC bilayers of liposome and their effects on membrane fluidities. Colloids Surf B 48(2):112–118. https://doi.org/10.1016/j.colsurfb.2006.01.006

Patra JK, Das G, Fraceto LF, Campos EVR, Rodriguez-Torres MDP, Acosta-Torres LS et al (2018) Nano based drug delivery systems: recent developments and future prospects. J Nanobiotechnol 16:71. https://doi.org/10.1186/s12951-018-0392-8

Peran M, Garcia MA, Lopez-Ruiz E, Jimenez G, Marchal JA (2013) How can nanotechnology help to repair the body? Advances in cardiac, skin, bone, cartilage and nerve tissue regeneration. Materials 6(4):1333–1359. https://doi.org/10.3390/ma6041333

Polini A, Yang F (2017) Physicochemical characterization of nanofiber composites. In: Nanofiber composites for biomedical application. Elsevier, pp 97–115

Prasad R, Pandey R, Varma A, Barman I (2017) Polymer based nanoparticles for drug delivery systems and cancer therapeutics. In: Kharkwal H, Janaswamy S (eds) Natural polymers for drug delivery. CAB International, London, pp 53–70

Rajabi M, Ali A, McConnell M, Cabral J (2020) Keratinous materials: structures and functions in biomedical applications. Mater Sci Eng C 110:110612. https://doi.org/10.1016/j.msec.2019.110612

Ramji K, Shah RN (2014) Electrospun soy protein nanofiber scaffolds for tissue regeneration. J Biomater Appl 29(3):411–422. https://doi.org/10.1177/0885328214530765

Reddy N, Yang Y (2008) Self-crosslinked gliadin fibers with high strength and water stability for potential medical applications. J Mater Sci Mater Med 19(5):2055–2061. https://doi.org/10.1007/s10856-007-3294-0

Reddy KVR, Yedery RD, Aranha C (2004) Antimicrobial peptides: premises and promises. Int J Antimicrob Agents 24(6):536–547. https://doi.org/10.1016/j.ijantimicag.2004.09.005

Rezaei Z, Mahmoudifard M (2019) Pivotal role of electrospun nanofibers in microfluidic diagnostic systems—a review. J Mater Chem B 7(30):4602–4619. https://doi.org/10.1039/c9tb00682f

Rosa AR, Steffens D, Santi B, Quintiliano K, Steffen N, Pilger DA, Pranke P (2017) Development of VEGF-loaded PLGA matrices in association with mesenchymal stem cells for tissue engineering. Braz J Med Biol Res 50(9):e5648. https://doi.org/10.1590/1414-431x20175648

Safdari M, Shakiba E, Kiaie SH, Fattahi A (2016) Preparation and characterization of Ceftazidime loaded electrospun silk fibroin/gelatin mat for wound dressing. Fibers Polym 17(5):744–750. https://doi.org/10.1007/s12221-016-5822-3

Sarhan WA, Azzazy HM (2017) Apitherapeutics and phage-loaded nanofibers as wound dressings with enhanced wound healing and antibacterial activity. Nanomedicine 12(17):2055–2067. https://doi.org/10.2217/nnm-2017-0151

SCCS (2020) The SCCS guidance on the safety assessment of nanomaterials in cosmetics comment. Regul Toxicol Pharmacol 112:104611. https://doi.org/10.1016/j.yrtph.2020.104611

Sebe I, Ostorhazi E, Fekete A, Kovacs KN, Zelko R, Kovalszky I et al (2016) Polyvinyl alcohol nanofiber formulation of the designer antimicrobial peptide APO sterilizes acinetobacter baumannii-infected skin wounds in mice. Amino Acids 48(1):203–211. https://doi.org/10.1007/s00726-015-2080-4

Sedghi R, Shaabani A, Mohammadi Z, Samadi FY, Isaei E (2017) Biocompatible electrospinning chitosan nanofibers: a novel delivery system with superior local cancer therapy. Carbohydr Polym 159:1–10. https://doi.org/10.1016/j.carbpol.2016.12.011

Selvaraj S, Thangam R, Fathima NN (2018) Electrospinning of casein nanofibers with silver nanoparticles for potential biomedical applications. Int J Biol Macromol 120:1674–1681. https://doi.org/10.1016/j.ijbiomac.2018.09.177

Shahi RG, Albuquerque MTP, Munchow EA, Blanchard SB, Gregory RL, Bottino MC (2017) Novel bioactive tetracycline-containing electrospun polymer fibers as a potential antibacterial dental implant coating. Odontology 105(3):354–363. https://doi.org/10.1007/s10266-016-0268-z

Shahini Shams Abadi M, Mirzaei E, Bazargani A, Gholipour A, Heidari H, Hadi N (2020) Antibacterial activity and mechanism of action of chitosan nanofibers against toxigenic Clostridioides (clostridium) difficile isolates. Ann Ig 32(1):72–80. https://doi.org/10.7416/ai.2020.2332

Shalumon KT, Anulekha KH, Nair SV, Nair SV, Chennazhi KP, Jayakumar R (2011) Sodium alginate/poly(vinyl alcohol)/nano ZnO composite nanofibers for antibacterial wound dressings. Int J Biol Macromol 49(3):247–254. https://doi.org/10.1016/j.ijbiomac.2011.04.005

Sharma J, Lizu M, Stewart M, Zygula K, Lu Y, Chauhan R et al (2015) Multifunctional nanofibers towards active biomedical therapeutics. Polymers 7(2):186–219. https://doi.org/10.3390/polym7020186

Shefa AA, Taz M, Hossain M, Kim YS, Lee SY, Lee BT (2019) Investigation of efficiency of a novel, zinc oxide loaded TEMPO-oxidized cellulose nanofiber based hemostat for topical bleeding. Int J Biol Macromol 126:786–795. https://doi.org/10.1016/j.ijbiomac.2018.12.079

Shin SH, Purevdorj O, Castano O, Planell JA, Kim HW (2012) A short review: recent advances in electrospinning for bone tissue regeneration. J Tissue Eng 3(1):2041731412443530. https://doi.org/10.1177/2041731412443530
Shoba E, Lakra R, Syamala Kiran M, Korrapati PS (2017) Fabrication of core–shell nanofibers for controlled delivery of bromelain and salvianolic acid B for skin regeneration in wound therapeutics. Biomed Mater 12(3):035005. https://doi.org/10.1088/1748-605x/aa6684
Ske (2020) Electrospinning system—needleless version. http://www.ske.it/index.php/product/electrospinning-system-needleless-electrospinning/
Song DW, Kim SH, Kim HH, Lee KH, Ki CS, Park YH (2016) Multi-biofunction of antimicrobial peptide-immobilized silk fibroin nanofiber membrane: implications for wound healing. Acta Biomater 39:146–155. https://doi.org/10.1016/j.actbio.2016.05.008
Song J, Klymov A, Shao J, Zhang Y, Ji W, Kolwijck E et al (2017) Electrospun nanofibrous silk fibroin membranes containing gelatin nanospheres for controlled delivery of biomolecules. Adv Healthc Mater 6(14):1700014. https://doi.org/10.1002/adhm.201700014
Tahmasebi A, Enderami SE, Saburi E, Islami M, Yaslianifard S, Mahabadi JA et al (2020) MicroRNA-incorporated electrospun nanofibers improve osteogenic differentiation of human-induced pluripotent stem cells. J Biomed Mater Res A 108(2):377–386. https://doi.org/10.1002/jbm.a.36824
Tao FH, Cheng YX, Shi XW, Zheng HF, Du YM, Xiang W, Deng HB (2020) Applications of chitin and chitosan nanofibers in bone regenerative engineering. Carbohydr Polym 230:115658. https://doi.org/10.1016/j.carbpol.2019.115658
Thangadurai D, Sangeetha J, Prasad R (2020a) Nanotechnology for food, agriculture, and environment. Springer International Publishing, Cham. ISBN 978-3-030-31937-3. https://www.springer.com/gp/book/9783030319373
Thangadurai D, Sangeetha J, Prasad R (2020b) Functional bionanomaterials. Springer International Publishing, Cham. ISBN 978-3-030-41464-1. https://www.springer.com/gp/book/9783030414634
Tiwari SK, Tzezana R, Zussman E, Venkatraman SS (2010) Optimizing partition-controlled drug release from electrospun core–shell fibers. Int J Pharm 392(1):209–217. https://doi.org/10.1016/j.ijpharm.2010.03.021
Tsuji JS, Maynard AD, Howard PC, James JT, Lam CW, Warheit DB, Santamaria AB (2006) Research strategies for safety evaluation of nanomaterials, part IV: risk assessment of nanoparticles. Toxicol Sci 89(1):42–50. https://doi.org/10.1093/toxsci/kfi339
Tucker N, Stanger JJ, Staiger MP, Razzaq H, Hofman K (2012) The history of the science and technology of electrospinning from 1600 to 1995. J Eng Fibers Fabr 7:63–73. https://doi.org/10.1177/155892501200702S10
Uehara TM, Paino IMM, Santos FA, Scagion VP, Correa DS, Zucolotto V (2020) Fabrication of random and aligned electrospun nanofibers containing graphene oxide for skeletal muscle cells scaffold. Polym Adv Technol 31(6):1437–1443. https://doi.org/10.1002/pat.4874
Ulubayram K, Calamak S, Shahbazi R, Eroglu I (2015) Nanofibers based antibacterial drug design, delivery and applications. Curr Pharm Des 21(15):1930–1943. https://doi.org/10.2174/1381612821666150302151804
Valmikinathan CM, Defroda S, Yu X (2009) Polycaprolactone and bovine serum albumin based nanofibers for controlled release of nerve growth factor. Biomacromolecules 10(5):1084–1089. https://doi.org/10.1021/bm8012499
Vasita R, Katti DS (2006) Nanofibers and their applications in tissue engineering. Int J Nanomed 1 (1):15–30. https://doi.org/10.2147/nano.2006.1.1.15
Venkataraman S, Hedrick JL, Ong ZY, Yang C, Ee PLR, Hammond PT, Yang YY (2011) The effects of polymeric nanostructure shape on drug delivery. Adv Drug Deliv Rev 63 (14–15):1228–1246. https://doi.org/10.1016/j.addr.2011.06.016
Verreck G, Chun I, Peeters J, Rosenblatt J, Brewster ME (2003) Preparation and characterization of Nanofibers containing amorphous drug dispersions generated by electrostatic spinning. Pharm Res 20(5):810–817. https://doi.org/10.1023/A:1023450006281

Vert M, Doi Y, Hellwich KH, Hess M, Hodge P, Kubisa P et al (2012) Terminology for biorelated polymers and applications (IUPAC recommendations 2012). Pure Appl Chem 84(2):377–408. https://doi.org/10.1351/pac-rec-10-12-04

Wang J, Windbergs M (2017) Functional electrospun fibers for the treatment of human skin wounds. Eur J Pharm Biopharm 119:283–299. https://doi.org/10.1016/j.ejpb.2017.07.001

Wang Z-G, Wan L-S, Xu Z-K (2007) Surface engineerings of polyacrylonitrile-based asymmetric membranes towards biomedical applications: an overview. J Membr Sci 304(1):8–23. https://doi.org/10.1016/j.memsci.2007.05.012

Wang S-D, Ma Q, Wang K, Chen H-W (2018) Improving antibacterial activity and biocompatibility of bioinspired electrospinning silk fibroin nanofibers modified by graphene oxide. Acs Omega 3(1):406–413. https://doi.org/10.1021/acsomega.7b01210

Whitehead TJ, Avila COC, Sundararaghavan HG (2018) Combining growth factor releasing microspheres within aligned nanofibers enhances neurite outgrowth. J Biomed Mater Res A 106(1):17–25. https://doi.org/10.1002/jbm.a.36204

Won JJ, Nirmala R, Navamathavan R, Kim HY (2012) Electrospun core–shell nanofibers from homogeneous solution of poly(vinyl alcohol)/bovine serum albumin. Int J Biol Macromol 50 (5):1292–1298. https://doi.org/10.1016/j.ijbiomac.2012.04.007

Xu R, Zhang Z, Toftdal MS, Møller AC, Dagnaes-Hansen F, Dong M et al (2019) Synchronous delivery of hydroxyapatite and connective tissue growth factor derived osteoinductive peptide enhanced osteogenesis. J Control Release 301:129–139. https://doi.org/10.1016/j.jconrel.2019.02.037

Xue JJ, Wu T, Dai YQ, Xia YN (2019) Electrospinning and electrospun nanofibers: methods, materials, and applications. Chem Rev 119(8):5298–5415. https://doi.org/10.1021/acs.chemrev.8b00593

Yan EY, Jiang JY, Yang XY, Fan LQ, Wang YW, An QL et al (2020) pH-sensitive core-shell electrospun nanofibers based on polyvinyl alcohol/polycaprolactone as a potential drug delivery system for the chemotherapy against cervical cancer. J Drug Deliv Sci Technol 55:101455. https://doi.org/10.1016/j.jddst.2019.101455

Yang X, Yang J, Wang L, Ran B, Jia Y, Zhang L et al (2017) Pharmaceutical intermediate-modified gold nanoparticles: against multidrug-resistant Bacteria and wound-healing application via an electrospun scaffold. ACS Nano 11(6):5737–5745. https://doi.org/10.1021/acsnano.7b01240

Yang YN, Lu KY, Wang P, Ho YC, Tsai ML, Mi FL (2020) Development of bacterial cellulose/chitin multi-nanofibers based smart films containing natural active microspheres and nanoparticles formed in situ. Carbohydr Polym 228:115370. https://doi.org/10.1016/j.carbpol.2019.115370

Yao C-H, Yeh J-Y, Chen Y-S, Li M-H, Huang C-H (2017) Wound-healing effect of electrospun gelatin nanofibres containing centella asiatica extract in a rat model. J Tissue Eng Regen Med 11 (3):905–915. https://doi.org/10.1002/term.1992

Ye P, Wei S, Luo C, Wang Q, Li A, Wei F (2020) Long-term effect against methicillin-resistant Staphylococcus aureus of emodin released from coaxial electrospinning nanofiber membranes with a biphasic profile. Biomol Ther 10(3):362. https://doi.org/10.3390/biom10030362

Yildiz A, Kara AA, Acarturk F (2020) Peptide-protein based nanofibers in pharmaceutical and biomedical applications. Int J Biol Macromol 148:1084–1097. https://doi.org/10.1016/j.ijbiomac.2019.12.275

Yoshimoto H, Shin YM, Terai H, Vacanti JP (2003) A biodegradable nanofiber scaffold by electrospinning and its potential for bone tissue engineering. Biomaterials 24(12):2077–2082. https://doi.org/10.1016/S0142-9612(02)00635-X

Yu C-C, Chang J-J, Lee Y-H, Lin Y-C, Wu M-H, Yang M-C, Chien C-T (2013) Electrospun scaffolds composing of alginate, chitosan, collagen and hydroxyapatite for applying in bone tissue engineering. Mater Lett 93:133–136. https://doi.org/10.1016/j.matlet.2012.11.040

Yuan MQ, Dai FF, Li D, Fan YQ, Xiang W, Tao FH et al (2020) Lysozyme/collagen multilayers layer-by-layer deposited nanofibers with enhanced biocompatibility and antibacterial activity. Mater Sci Eng C 112:110868. https://doi.org/10.1016/j.msec.2020.110868

Zhang Y, Cui L, Che X, Zhang H, Shi N, Li C et al (2015) Zein-based films and their usage for controlled delivery: origin, classes and current landscape. J Control Release 206:206–219. https://doi.org/10.1016/j.jconrel.2015.03.030

Zigdon-Giladi H, Khutaba A, Elimelech R, Machtei EE, Srouji S (2017) VEGF release from a polymeric nanofiber scaffold for improved angiogenesis. J Biomed Mater Res A 105 (10):2712–2721. https://doi.org/10.1002/jbm.a.36127

Zupancic S, Rijavec T, Lapanje A, Petelin M, Kristl J, Kocbek P (2018) Nanofibers with incorporated autochthonous bacteria as potential probiotics for local treatment of periodontal disease. Biomacromolecules 19(11):4299–4306. https://doi.org/10.1021/acs.biomac.8b01181

Chapter 5
Extraction of Silver Nanoparticles (Ag-NPs) by Green Synthesis from Aqueous Extract of Seaweeds and Their Consequences on HeLa Cell Line and Their Utility on Soil by Spectroscopic Tools

K. SenthilKannan, G. Flora, S. Gunasekaran, H. Abdul Jaffar Ali, M. Vimalan, and S. Balasubramanian

Abstract Seaweed extracts have been used to synthesize silver nanoparticles (Ag-NPs) as they offer many benefits and utilities. The use of biological materials in nanoparticles synthesis is preferred as they are eco-friendly and compatible for pharmaceutical and other biomedical and therapeutic applications. As the number of applications of NPs increases, assessment of the risks posed by NPs is important and it has developed into an area of research itself. Metallic nanoparticles have different applications such as in electronics, catalysts and photonic applications. Silver metal has toxicity against a wide range of micro organisms especially silver nanoparticles which has promising antimicrobial properties and other bio utilities. Silver nanoparticles are effective as anti-inflammatory, anti-angiognesis, antiviral and

K. SenthilKannan (✉)
Department of R&D, Edayathangudy G S Pillay Arts and Science College (Autonomous), Nagapattinam, India

Department of Physics, Edayathangudy G S Pillay Arts and Science College (Autonomous), Nagapattinam, India

G. Flora
Department of Botany, St. Mary's College (Autonomous), Thoothukudi, India

S. Gunasekaran
Department of R&D, St. Peter's Institute of Higher Education and Research—SPIHER, Chennai, India

H. A. J. Ali
Department of Biotechnology, Islamia College (Autonomous), Vaniyambadi, India

M. Vimalan
Department of Physics, Thirumalai Engineering College, Kanchipuram, India

S. Balasubramanian
Principal Scientist, ICAR-Central Institute of Agricultural Engineering Regional Centre, Coimbatore, India

N. R. Maddela et al. (eds.), *Nanotechnology for Advances in Medical Microbiology*, Environmental and Microbial Biotechnology,
https://doi.org/10.1007/978-981-15-9916-3_5

anti-platelet activity against cancer cells. The Ag-NPs synthesized were from the aqueous extracts of seaweeds and were characterized by using spectral analysis as a perspective tool. The effect of the synthesized silver nanoparticles on the isolated soil microbes were assessed by different assays such as Calorimetric chowder assay and Agar diffusion assay. The effect of synthesized Ag-NPs on earthworm was also studied by examining its morphology and histology. The toxicity of the combined seaweed-AgNPs against HeLa carcinoma cells has been analyzed. Also the Ag-NPs are studied with Fluorescence—FL activity with band gap in eV.

Keywords Silver NPs · Green synthesis · Seaweed · HeLa cell line · Soil · Spectroscopic tools

5.1 Synthesis and Characterization of Silver Nanoparticles

Silver nanoparticles have been broadly utilized for the time of the point of reference hardly many years in different applications because of their notable adequacy in biomedical (Cao et al. 2010), electronic (Mohan et al. 2007), catalysis (Feng et al. 2011) and optical applications (Hayward et al. 2000). Specifically, the extraordinary antimicrobial properties of Ag-NPs have prompted the advancement of a wide assortment of nano silver things, including nanosilver-covered injury dressings, prophylactic gadgets, vigilant instruments, and embeds (Lohse and Murphy 2012; You et al. 2012; Aziz et al. 2016). Aside from these antimicrobial exercises, Ag-NPs are likewise known to have antifungal, calming, antiviral, hostile to angiogenesis and against platelet properties. Also, later improvements have seen Ag-NPs utilized in room shower, backdrop gloves, clothing cleanser, and divider paint definitions just as in the material business for garments producing. The current examination depicts a solitary advance, green, and fast combination of silver nanoparticles (Ag-NPs) arranged by organic (green) strategies utilizing kelp concentrates of *Rhodymenia palmata, Gracilaria corticata, Hypnea musciformis, Sargassum tenerrimum, Stoechospermum marginatum and Dictyota dichotoma.*

Watery arrangement of 1 mM silver presented to fluid marine growth concentrates of *Rhodymenia palmata, Gracilaria corticata, Hypnea musciformis, Sargassum tenerrimum, Stoechospermum marginatum* and *Dictyota dichotoma* a noticeable shading change from straightforward, light earthy colored to dim earthy colored in 5 minutes at room temperature. It indicates the pattern of silver nanoparticles which was confirmed by spectral analysis. The amalgamations of silver nanoparticles were observed at regular intervals through UV visible spectroscopy. The shading change deduction is one of least demanding realized procedures to affirm the nanoparticles union as there is decline in size with an expansion in excitation of external surface electrons known as surface plasmon reverberation (Ding et al. 2015).

The movement of the response prompting the change of Ag^+ from $AgNO_3$ to decreased nano one by watery concentrate of *Rhp* (*Rhodymenia palmata*), *Grc*

(*Gracilaria corticata*), *Hym* (*Hypnea musciformis*), *Srt* (*Sargassum tenerrimum*), *Stm* (*Stoechospermum marginatum*) and *Dtd* (*Dictyota dichotoma*). Silver nanoparticles were checked by watching the shading change and absorbance maxima top in the scope of around 416 nm. The apex showed a surface plasmon reverberation (SPR), which has just been recorded for different metal nanoparticles which went from 2 to 100 nm in size (Henglein 1993; Ravindra and Rajasab 2014). The state of the band was even proposing uniform dispersal of nanopartcles (Travan et al. 2009). Absorbance maxima pinnacle of *Grc*-AgNPs, *Hym*-AgNPs (*Gracilaria corticata, Hypnea musciformis* ocean growth blended silver nanoparticles) is in the scope of 375 nm affirming the development of silver nanoparticle (Baia and Simon 2007; Ayman et al. 2014). The recurrence and width of the surface plasmon assimilation relies upon the size and state of the metal nanoparticles just as on the dielectric steadiness of the metal itself and the encompassing medium (Mukherjee et al. 2002).

5.2 XRD

X-beam or X-ray diffraction (XRD) of silver nanoparticles delivered utilizing *Gracilaria corticata* and *Hypnea musciformis* are appeared in the Figs. 5.1 and 5.2. Various Bragg's reflections with 2θ estimations of 38, 44, 6 and 77 which coordinate to the (111), (200), (220) and (311) arrangements of grid planes are experiential which are recorded to the face-focused cubic structures for silver. The widened Bragg's pinnacle shows the development of nano crystals (Theivasanthi and Alagar 2010; Zargar et al. 2011). The XRD pattern of the pinnacles showed that the silver nanoparticles blended by *Gracilaria corticata* and *Hypnea musciformis* are translucent in nature and a portion of the unassigned acmes were additionally watched, it might be expected to the less bimolecular of settling operators, catalysts or proteins in the marine growth.

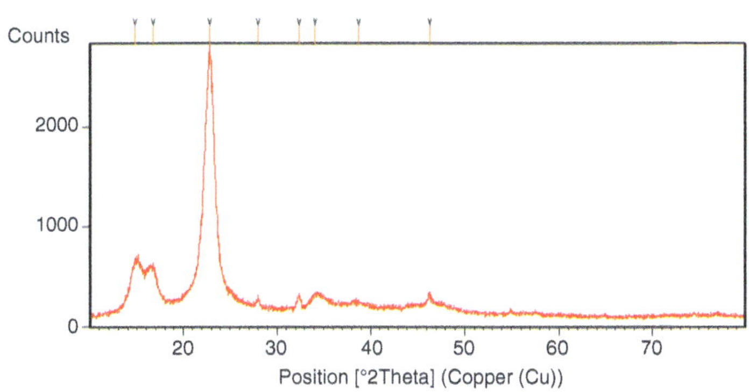

Fig. 5.1 XRD diffraction patterns recorded from drop coated films of silver nanoparticles on *Grc*-AgNPs

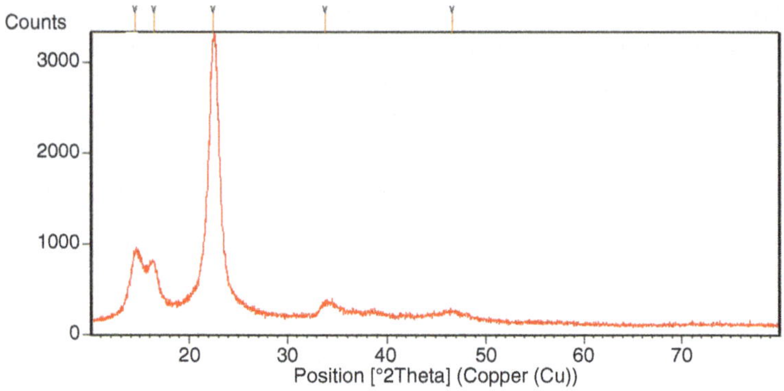

Fig. 5.2 XRD diffraction patterns recorded from drop coated films of silver nanoparticles on *Hym*-AgNPs. *Grc*-AgNPs-*Gracilaria corticata*-AgNPs. *Hym-AgNPs—Hypnea musciformis*-AgNPs

5.3 Antioxidant

Antioxidant association of thorough going phenols, tannins, flavonoids, tocopherol and terpenoids contants of watery concentrates of *Rhodymenia palmata, Gracilaria corticata, Hypnea musciformis, Sargassum tenerrimum, Stoechospermum marginatum* and *Dictyota dichotoma*, *Rhp*-AgNP (Silver nano incorporated by Rhp), *Grc*-AgNP (Silver nano orchestrated by Grc) and *Hym*-AgNP (Silver nano combined by Hym), *Srt*-AgNP (Silver nano integrated by Srt), *Stm*-AgNP (Silver nano blended by Stm) and *Dtd*-AgNP (Silver nano combined by Dtd) introduced, showed in Fig. 5.3 that these synthetic substance were prevalently higher in fluid concentrate of *Rhodymenia palmata, Gracilaria corticata, Hypnea musciformis, Sargassum tenerrimum, Stoechospermum marginatum* and *Dictyota dichotoma* contrasted with the ocean growth diminished AgNPs (*Rhp*-AgNPs, *Grc*-AgNPs, *Hym*-AgNPs, *Srt*-AgNPs, *Stm*-AgNPs and *Dtd*-AgNPs). Yet, terpenoid substance of *Grc*-AgNPs and *Rhp*-AgNPs tocopherol content were higher than the fluid concentrate.

In the current examination, cancer prevention agent limit saw on the nearness of all out phenol, tannin, flavonoid, tocopherol and terpenoid content. The significant measure of cancer prevention agents in the watery concentrate of *Rhodymenia palmata, Gracilaria corticata, Hypnea musciformis, Sargassum tenerrimum, Stoechospermum marginatum* and *Dictyota dichotomawere* noted. Polyphenolic mixes are normal cell reinforcements which are found for the most part in kelp (Moon and Shibamoto 2009). Polyphenolics contain lessening properties as hydrogen or electron giving specialists, in this way observed as cell reinforcements. So our outcome uncovered that the concentrate of marine kelp *Rhodymenia palmata, Gracilaria corticata, Hypnea musciformis, Sargassum tenerrimum, Stoechospermum marginatum* and *Dictyota dichotoma*, were fit for creating Ag nanoparticles extracellular and these nano particles are very steady in arrangement due to topping likely by the polyphenols present in the concentrate.

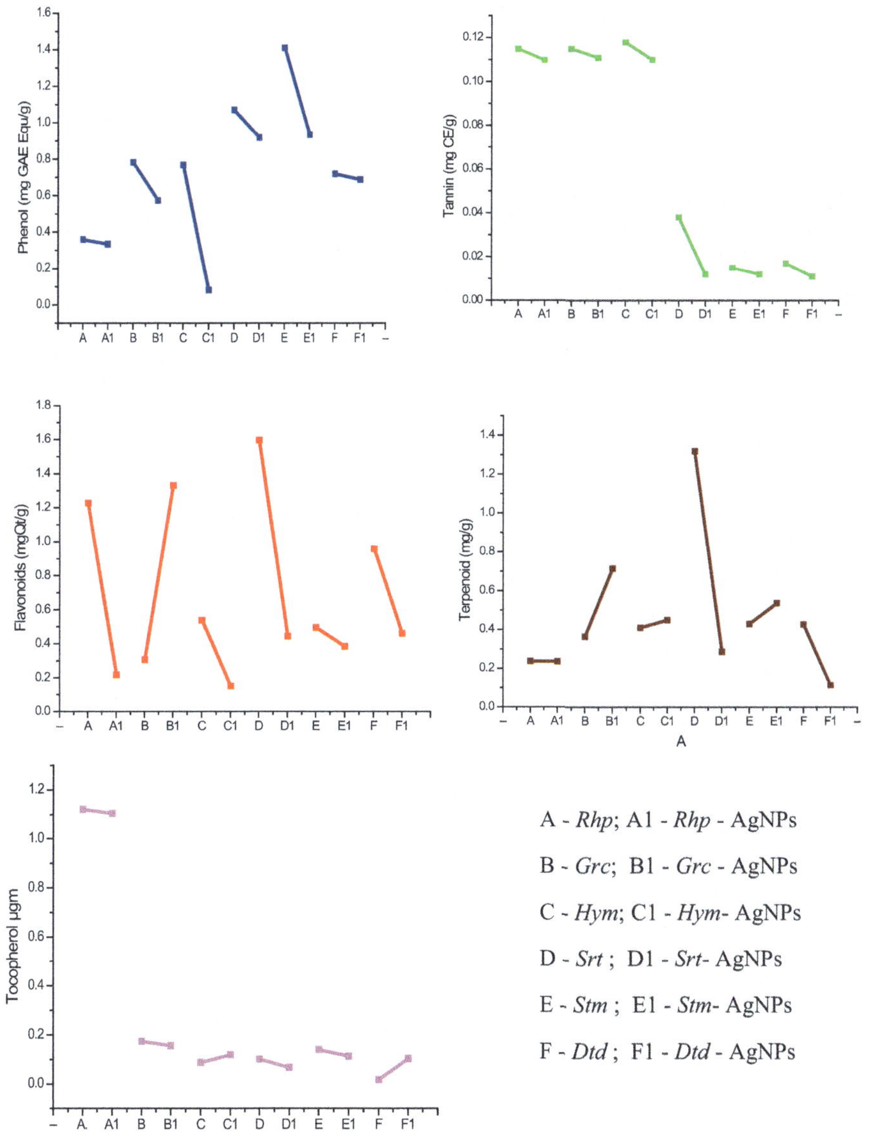

Fig. 5.3 Comparison of amount of antioxidant contents in seaweeds and seaweed synthesised silver nanoparticles

5.4 In-Vitro Cytotoxicity Investigation of Hym—AgNPs on HeLa Cell Line

Nanoparticles (NPs) of various size and physicochemical properties have been acquainted with numerous fields of life and biomedical sciences in the course of the most recent decade (Oberdorster et al. 2005). In this regard, NPs opened another period in biomedical sciences and have been utilized explicitly as quality or operator bearers, and in sedate plan, change of therapeutics, naming of fluorescents, and tissue building (Tan et al. 2007; Yoon et al. 2007; Kreuter and Gelperina 2008; Su et al. 2008; Hackenberg et al. 2010). Among the NPs, silver nanoparticles (AgNPs) have gotten consideration for their antimicrobial exercises (Cho et al. 2005; Kim et al. 2006; Prasad et al. 2011, 2012; Prasad and Swamy 2013; Prasad 2014; Joshi et al. 2018) and have been utilized for various purposes including the assembling of disinfectants, shampoos, antiperspirants, humidifiers, wound dressings, and different material items (Ahmed et al. 2008; Johnston et al. 2010; Zanette et al. 2011) they have likewise been utilized as a covering for different implantable gadgets, for example, catheters, heart valves, and inserts (Chen and Schluesener 2008; Chaloupka et al. 2010). Regardless of their advantages, there has been not kidding worry about the conceivable symptoms of AgNPs. Past investigations detailed that AgNPs actuated genotoxicity and cytotoxicity in both malignant growth and typical cell lines (Yoon et al. 2007; Aziz et al. 2019), adjusted cell morphology, decreased cell suitability, and caused oxidative worry in lung fibroblast and glioblastoma cells (Asharani et al. 2009), human and rodent liver cells (Hussain et al. 2005; Kim et al. 2010), HeLa cells (Sonoda et al. 1998) and THP-1 monocytes (Foldbjerg et al. 2009). AgNP-initiated cytotoxicity decreases cell feasibility in different cell lines by causing apoptosis through the mitochondrial pathway (Hsin et al. 2008) and creates oxidant species (Hess et al. 2008), which are notable for causing lipid peroxidation of organic films and harm to basic proteins and DNA (Li and Osborne 2008).

The likely poisonousness of the combined Hym-AgNPs against HeLa carcinoma cells has been analyzed. The HeLa cells were treated with various fixations (50–200%) of nanoparticles for 48 h and afterward MTT examine was utilized to gauge. As the decrease in cell suitability by $\sim$50% (inhibitory concentration = IC50) in correlation with the control was accomplished at a portion of 100 μg/ml of Hym-AgNPs, while the best number of cells were slaughtered at 200 μg/ml of Hym-AgNPs. In any case, the cytotoxic impacts might be in part because of direct activity of Ag^+ particles discharged from AgNPs. Singh and Ramarao (2012) detailed that AgNPs were taken-up by macrophages through receptor-intervened phagocytosis and Ag^+ particles were discharged from AgNPs. The free Ag^+ particles thus may meddle with a few cytoplasmic structures and pathways, including mitochondrial capacities actuating pressure pathways and apoptosis. Along these lines, further examinations are expected to light up the AgNP actuated apoptosis and altered form from the viewpoint. In the current examination, to manufacture AgNPs of 50–100 nm in hydrodynamic distance across, which is an extensively littler size for testing cyto-harmfulness. As recently revealed by Sohaebuddin et al.

(2010), nano sized AgNPs are more harmful than bigger and micro sized particles. Also, the AgNPs are synthesized in this examination are very cytotoxic against HeLa cells. At present malignant growth kills around 70,00,000 individuals worldwide on yearly premise. Henceforth, as of late, the quest for the malignant growth therapeutics from normal items expanded step by step. Bioactive mixes in marine life forms have been accounted for against different malignant growth cell lines. Taking everything into account, anticancer movement of silver nanoparticles incorporated by *Hypnea musciformis* appeared impressive cytotoxic impact against human malignant growth cell lines. In this way, the incorporated silver nanoparticles could be considered as a successful anticancer specialist. Be that as it may, an examination in regards to cooperation of microbial orchestrated AgNPs with the disease cell lines should be determined before the broad utilizing clinical applications.

5.5 Effect of Green Synthesized Nanoparticles (*Rhp*-AgNPs, *Grc*-AgNPs, *Hym*-AgNPs, *Srt*-AgNPs, *Stm*-AgNPs and *Dtd*-AgNPs) on Soil Microbial Isolates

Soil quality is characterized as the limit of mud to work, inside normal or oversaw biological system limits, to continue plant and creature efficiency, keep up and upgrade water and air quality, and bolster human wellbeing and home (Karlen et al. 1997). Among the elements impacting soil quality, natural pointers revealed as basically significant (Doran and Zeiss 2000) on the grounds that dirt creatures reasonably impact soil biological system process, particularly the deterioration of soil natural issue and the cycling of supplements (Kennedy and Smith 1995). In this way, security of soil microbial biomass and assorted variety is one of the significant difficulties for economical asset use due to more prominent degrees of microbial biomass and decent variety mean more noteworthy supplement turnover and illness suppressiveness of the dirt (Janvier et al. 2007). The inverse being valid for wiped out soil with low supplement and carbon saves and more prominent degree of contaminants brought about by the nearness of xenobiotics synthetic substances or different variations in the dirt condition. Among the xenobiotics amazing quantities of new nanoparticles designed for modern and natural application or framed by results of human movement, which are as of now finding their way into soils (Maurice and Hochella 2008). While the centralization of most built nanoparticles in the situations despite everything stay obscure, introduction displaying recommend that is soil could be a significant sing of nanoparticles discharged in to the earth and that focus in soil are higher than in water or air (Gottschalle et al. 2009). Builted/ synthesized nanoparticles can be made of single components like carbon (C) or Silver (Ag) or a blend of components. The expanding section of these nanoparticles unavoidably leads to their amassing in soil, which have raised worries about their aggregation on soil microbial action and decent variety. Right now almost no data is accessible on how these NPs influence the filth microbial network. Henceforth, the impact of Rhp-AgNPs, Grc-AgNPs, Hym-AgNPs, Srt-AgNPs, Stm-AgNPs and Dtd-AgNPs on the microbial disconnected from garden soil were examined. The detachment of microorganism was completed utilizing sequential weakening strategy. Aliquots of 100 μl of various weakening of nursery soil were spread onto plates of supplement agar mode for microorganisms and potato dextrose agar for growths. The plates were brooded at 28 °C for 5 days under oxygen consuming conditions. Created provinces were picked and disengaged dependent on morphological models and the separated microbes were sub-refined as unadulterated culture. Unadulterated refined organisms were identified as *Bacillus spp, Bacillus subtilis, Staphylococcus epidermidis, Serratia spp, Pseudomonas spp, Pseudomonas fluorescens* andisolated parasites were *Aspergillus* disinfects, *Aspergillus flavus, Alternaria alternata, Cladosporium spp*. The microscopic organisms detached from the nursery soils are soil N-cycle, nitrifying microorganisms (Mishra and Kumar 2009). To consider the impact of silver nanoparticles on soil organisms to sort of in vitro measure were completed they are calorimetric stock test and agar well dispersion examine.

5.6 Calorimetric Chowder Assay

Antimicrobial action of *Rhp* AgNPs, *Grc*-AgNPs, *Hym*-AgNPs, *Srt*-AgNPs, *Stm*-AgNPs and *Dtd*-AgNPs were concentrated against *Bacillus spp*., *Bacillus subtilis* Gram positive soil nitrifying microbes *Staphylococcus epidermidis* Gram positive denitrifying microscopic organisms, *Serratia spp* is Gram negative soil bio control operator, *Pseudomonas fluorescens, Pseudomonas spp* Gram negative, plant development enhancer and biocontrol specialist. *Aspergillus fumicatus, Aspergillus flavus, Alternaria alternata, and Cladosporium spp*, soil borne growths, recycler of carbon and nitrogen from perished living being.

Test of 3 ml of microbial culture were put into test cylinders and 1, 1.5 and 2 μl of proper weakening of *Rhp*-AgNPs, *Grc*-AgNPs, *Hym*-AgNPs, *Srt*-AgNPs, *Stm*-AgNPs and *Dtd*-AgNPs were included. Following 24 hours brooding, absorbance perusing at 520 nm frequency for every growth were estimated post hatching at 37 °C for 12 hours. Bacterial cell practicality and least inhibitory focus values were controlled by watching the turbidity and the absorbance perusing of the suspension post brooding. The most minimal centralization of blended nanoparticles with clear suspensions was considered as the esteems. The suspensions of disconnected microbial (*Bacillus spp, Bcillus subtilis, Serratia spp, Pseudomonas fluorescens, Pseudomonas spp, Staphylococcus epidermidis, Aspergillus disinfects, Aspergillus flavus, Alternaria alternata* and *Cladosporium spp*) inoculums with every extraordinary focus (1 μl, 1.5 μl, 2 μl) of *Rhp*-AgNPs, *Grc*-AgNPs, *Hym*-AgNPs, *Srt*-AgNPs, *Stm*-AgNPs and *Dtd*-AgNPs in stock examination technique were exceptionally overcast that stayed all through the hatching time frame. This watched the visual suspension for deciding the MIC as the turbidity because of bacterial and contagious development.

The calorimetric stock microbial development examination gave MIC esteem is 1 μl for all the tried 6 microbes and 4 parasites. The lower MIC values got for microorganisms was *Pseudomonas fluorescens* and *Serratia spp* against *Rhp*-AgNPs, *Grc*-AgNPs, *Dtd*-AgNPs of which *Dtd*-AgNPs indicated most extreme development inhibitory action against *Pseudomonas fluorescens*. The consequences of calorimetric stock microbial development examine of silver nanoparticles orchestrated from ocean growth indicated moderate inhibitory impact against a large portion of the dirt confined parasites which differed in the range. The MIC esteems were seen in *Rhp*-AgNPs and *Grc*-AgNPs. The secluded fungal micro-organisms *Aspergillus fumicatus*, *Aspergillus flavus*, *Alternaria alternata* and *Cladosporium spp* had supplementary reticence (less turbid mixture) on *Rhp*-AgNPs and *Grc*-AgNPs. The results of our revision showed momentous antimicrobial behavior in calorimetric broth assay.

5.7 Agar Well Diffusion Assay

The impact of silver nanoparticles blended by eatable ocean growth on the micro-organisms disconnected from the nursery soil were concentrated by calorimetric stock test and further affirmed by Agar well dispersion examination. The consequence of the antimicrobial screening trial of *Rhp*-AgNPs, *Grc*-AgNPs, *Hym*-AgNPs, *Srt*-AgNPs, *Stm*-AgNPs and *Dtd*-AgNPs were tried against soil microbial segregates (*Bacillus spp, Bcillus substilis, Serratia spp, Pseudomonas fluorescens, Pseudomonas spp, Staphylococcusepidermidis, Aspergillus fumicatus, Aspergillus flavus, Alternaria alternata*, and *Cladosporium spp*) utilizing agar well dispersion procedure. The blended AgNPs were seen as increasingly compelling against tried microbes. *Grc*-AgNPs, *Hym*-AgNPs, *Srt*-AgNPs, *Stm*-AgNPs and *Dtd*-AgNPs were seen as insufficient or demonstrated helpless restraint on bacterial and contagious development. The bigger zone of hindrance was watched for *Rhp*-AgNPs against *Serratia spp* (8 mm) and *Cladosporium spp* (9 mm).

Despite the fact that base hindrance on soil microorganisms by orchestrated silver nano particles have been seen in our examination. The possibility that microorganisms are safe, resilent, and practically repetitive is invasion in biology (Allison and Martiny 2008). Serious extent of metabolic adaptability, physiological resilience to changing ecological conditions (Mayer et al. 2004), high bounties broad dispersal, and potential for quick development rates have likewise lead to the proposal that microbial networks will impervious to change (Fenchel and Finlay 2004). Furthermore, fast developmental adjustment through flat quality exchange could permit touchy microorganisms to adjust to new natural conditions and rapidly return the network to its unique piece (Allison and Martiny 2008). Obviously, these examinations show that filth microbial networks frequently are very strong to bothers. Despite the fact that far reaching comprehension of the associations between metal oxide NPs and microorganisms particularly microbes is still at an early age (Han and Gu 2010), Findings propose that microscopic organisms with a capacity to bear a poisonous specialist may show up with time (Baath 1992) and the antimicrobial movement of incorporated NPs might be decreased by bacterial self security component; for example, *Bacillus subtilis* (Gram positive) reacted to nC60 by modifying layer lipid arrangement, stage progress temperature, and film smoothness (Fang et al. 2007). Earlier, Thill et al. (2006) contemplated the effect of water scattering of CeO_2-NPs (7 nm) on Gram negative bacterium *E. coli* and discovered emphatically charged CeO_2 at nonpartisan pH showed a solid electrostatic fascination towards bacterial external layers and Ce (1 V) was diminished to Ce (111) at the outside of the microorganisms. Microscopic organisms additionally have other defensive reactions (Wu et al. 2010), they demonstrated that bacterial cells have extra cell protein to kill limited quantities of poisonous particles to microorganisms and such defensive components might turn out to be less compelling. Sudheer Khan et al. (2011) found that microscopic organisms emitted exopolysaccharides that topped AgNPs in this manner decreasing its harmfulness to *E.coli, S. aerues* and *Micrococcus luteus* contrasted with the uncapped ones. The AgNP-harmfulness to nitrification microscopic organisms has been accounted for to be exceptionally reliant on their size, where AgNPs with under 5 nm distance across were accounted for to fundamentally restrain the nitrification microbes (Cha and Hu 2008). Our outcome indicated that the normal molecule size in 20 to 100 nm. AgNPs communications with microorganisms have been seen as reliant on the size and states of the NPs. AgNPs have round (7 and 29 nm) and Pseudo spherical shape (89 nm) with a limited size appropriation. Among these, Martinez-Castanon et al. (2008) found that the 7 nm AgNPs introduced best movement against *E. coli* and *S. aureus*. As a result of their size, 7 nm AgNPs can without much of a stretch arrive at the atomic contact of microscopic organisms and they present the best surface region; in this manner the contact with microbes is the best (Lok et al. 2006). Essentially, the littler size they are, the more noteworthy their surface zone to volume proportion and higher their microbial reaching proficiency (Wong et al. 2010).

5.8 Effect of Silver Nanoparticles on Soil Parameters and Earthworms (Morphology and Histology)

Nanoparticles are utilized in science and medication in a wide assortment of ways, including direct application to patients (Salata 2004). Toward the finish of their item life, NPs are probably going to wind up in the earth, especially in soil and water bodies, and consequently influence life forms in those media. NPs have extraordinary diffusivity and along these lines may cause progressively visit contacts with the surfaces of permeable media contrasted with bigger measured particles (Wiesner et al. 2006). Some metal NPs bear novel and exceptional organic properties that permit them to connect explicitly with chosen proteins and restrain their exercises (Bhattacharya and Mukherjee 2008). NPs can likewise infiltrate numerous sorts of cells and tissues, and travel through the body framework causing tissue harm. Silver

(Ag), a respectable metal, has been utilized over numerous years as unadulterated silver (Ag), silver nitrate ($AgNO_3$), and silver sulfadiazine (Ag SD) for the treatment of consumes, wounds, and a few bacterial contaminations. In any case, because of the rise of progressively powerful anti-infection agents, the utilization of these Ag mixes has declined especially. As of late, nanotechnology has increased huge driving force because of its capacity to deliver metals at nano dimensions, which radically changes the concoction, physical, and optical properties of metals (Bhattacharya and Mukherjee 2008; Srivastava et al. 2021) and the current utilization of such items is expanding. In this manner, concentrates on the eco-toxicology of Ag NPs-one of the most utilized NPs-have gotten increasingly significant.

Night crawlers are key life forms in earthly biological systems. *Lumbricus rubellus* is the most well-known worm species found in farming biological systems (Perez-Losada et al. 2009); consequently, this species is a fascinating contender for use as an animal model to screen soil contamination. So analyzes were completed on *Lumbricus rubellus* (night crawlers), are gathered from Aniyaparanallur, Srivaikundam, Thoothukudi soil. Night crawler of grown-up, weighing around 200–300 mg and having very much evolved clitellum were utilized for all tests and they were developed in earthen pots containing local soil to keep up physical and synthetic boundaries of the dirt. A set with no application were kept up as control. The reference control likewise kept with the utilization of SWC (*Gracilaria corticata* and *Hypnea musciformis*). Another set kept with the utilization of *Grc*-AgNPs and *Hym*-AgNPs are considered as treatment.

Soil tests were drawn when *Gracilaria corticata, Hypnea musciformis*, *Grc*-AgNPs and *Hym*-AgNPs application to get to physical (dampness substance and mass thickness and synthetic (pH and natural issue) boundaries of the muck. Information shows that filth dampness content was diminished in the *Grc*-AgNPs regarded soil as days advanced. There is no adjustment in soil dampness with control and soil getting SWC (*Grc* and *Hym*). A pattern of reduction of mass thickness was seen with the expansion of SWC and *Grc*-AgNPs and *Hym*-AgNPs were recorded and were expanded with long periods of medicines.

The materials properties of soil, for example, pH and natural issue were contemplated and organized, natural issue content is for the most part viewed as one of the key pointers of soil quality (Schjonning et al. 2009). Numerous highlights of good soil structure, for example, security, friability and dampness maintenance, might be influenced by soil natural issue. The dirt natural issue expanded continuously in the control and reference control as days continued. However, utilization of SWC expanded the natural issue in all the dirt examples and the addition was fast with expanding development period. On twentieth day, use of *Hypnea musciformis* upgraded the natural issue more effectively than *Gracilaria corticata*. In any case, the degree of soil natural issue was appeared to be decline in the NPs rewarded soil. Diminished dampness content, Bulk thickness and natural issue actuated by silver nanoparticles decreased via growth in the current examination is reliable with prior reports (Cornelis et al. 2012; Benoit et al. 2013; wang et al. 2013).

Soil with fine surface was accounted for to show higher surface territories which additionally encourage Ag-sorption (Jacobson et al. 2005). Ag sorption and versatility are likewise constrained by soil natural matter (Jones and Peterson 1986). Soils with high natural issue fixations absorb Ag more firmly than to mineral soils. To acquire understanding into the impact of AgNPs in the earthbound condition, *Lumbricus rubellus* worm sort, presented to Ag-NPs, and morphology and histology were seen on each twentieth and 40th days. One lot of reference control were kept with use of SWC.

The introduction of *Grc*-AgNPs and *Hym*-AgNPs had no impact on worm endurance over the multi day time frame. There was 100% endurance at both the nanoparticles introductions. Our outcomes are in concurrence with the discoveries of Roh J-Y et al. 2009. Be that as it may, following 20 days of presentation of AgNPs the earthy colored pigmentation of worm's skin step by step changed into dim shading. The pigmentation is bit by bit diminished with expanded period in all the NPs rewarded night crawler.

Impact of silver nanoparticles introduction on the life structures of the night crawler likewise examined and the transverse segment of fragments from the clitellum area of the control, SWC rewarded and *Grc*-Ag nanoparticles uncovered worms were taken fingernail skin and epidermis of the all the tried night crawler were unblemished. The roundabout and longitudinal muscles are unaffected. In any case, shrinkage of digestive tract divider is seen in the *Grc*-AgNPs uncovered worms.

The histology concentrates on worm gave the end that the green incorporated nanoparticles indicated high harm in the intestinal divider. No different variations from the norm were seen in charge and reference control classification. The AgNPs saw as having negative impact on the night crawler.

5.9 Fluorescence—FL

Atoms have different states alluded to as vitality levels. Fluorescence spectroscopy is principally worried about electronic and vibrational states. For the most part, the species being analyzed has a ground electronic express (a low vitality condition) of intrigue, and an energized electronic condition of higher vitality. Inside every one of these electronic states there are different vibrational states. In fluorescence, the species is first energized, by engrossing a photon, from its ground electronic state to one of the different vibrational states in the energized electronic state. Impacts with different particles cause the energized atom to lose vibrational vitality until it arrives at the most reduced vibrational state from the energized electronic state. The particle at that point drops down to one of the different vibrational degrees of the ground electronic state once more, radiating a photon in the process. As atoms may drop down into any of a few vibrational levels in the ground express, the produced photons will have various energies, and along these lines frequencies. Consequently, by breaking down the various frequencies of light discharged in fluorescent spectroscopy, alongside their relative forces, the structure of the diverse vibrational levels can be resolved. An emanation map is estimated by recording the outflow spectra

Table 5.1 FL data of six types of silver NPs

Nano samples	FL emission (nm)	Band gap value (eV)	Colour
(a) *Rhp*-AgNPs	449	2.7616	Violet
(b) *Grc*-AgNPs	459	2.7015	Blue
(c) *Hym*-AgNPs	479	2.5887	Blue
(d) *Srt*-AgNPs,	490	2.5306	Bluish green
(e) *Stm*-AgNPs	493	2.5152	Bluish green
(f) *Dtd*-AgNPs	497	2.4949	Bluish green

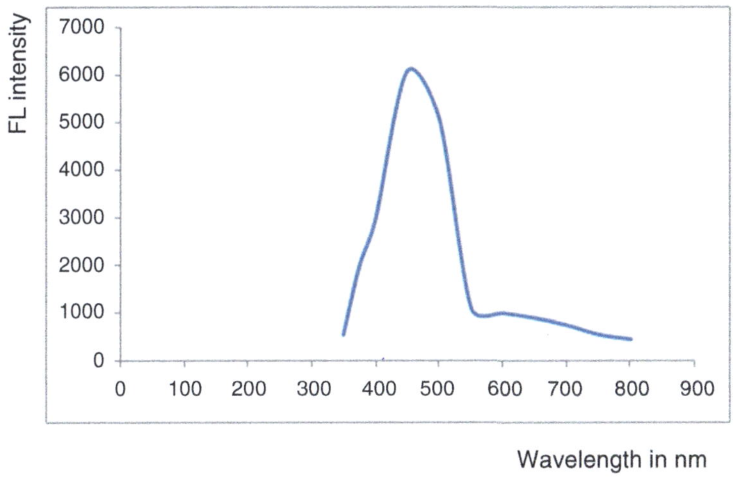

Fig. 5.4 FL spectra of the synthesised silver nanoparticles from the aqueous extracts of seaweeds (**a**) *Rhp*-AgNPs at 449 nm

coming about because of a scope of excitation frequencies and joining them all together. This is a three dimensional surface informational index: outflow power as an element of excitation and discharge frequencies, and is normally portrayed as a form map. The FL behaviour of the six samples of silver nano particles have the FL emission wavelengths of 449, 459, 479, 490, 493 and 497 nm respectively for (a) *Rhp*-AgNPs (b) *Grc*-AgNPs, (c) *Hym*-AgNPs (d) *Srt*-AgNPs, (e) *Stm*-AgNPs (f) *Dtd*-AgNPs and they have the band gap of 2.7616, 2.7015, 2.5887, 2.5306, 2.5152 and 2.4949 eV and shows the violet FL for 449 nm and bluish FL for 459, 479 nm and 490, 493 and 497 shows the bluish green emission which factors the emission for the excitation values and also shows the nano sample variance of transmittance and FL value for the colour change observed in different silver NPs and is mentioned in Table 5.1 and Figs. 5.4, 5.5, 5.6, 5.7, 5.8, 5.9.

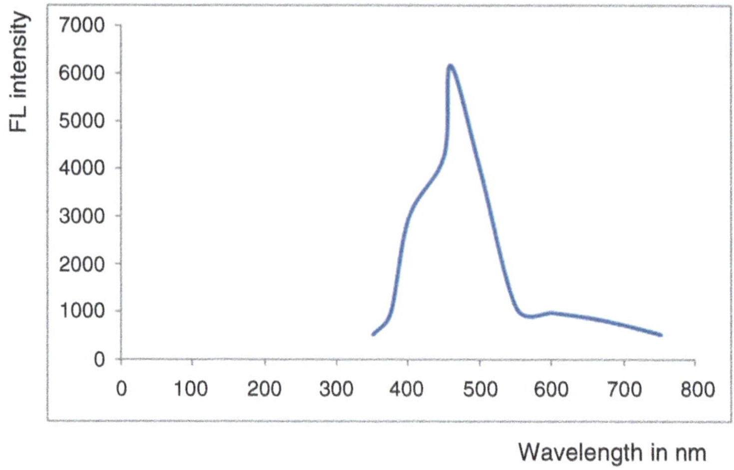

Fig. 5.5 FL spectra of the synthesised silver nanoparticles from the aqueous extracts of seaweeds (**b**) *Grc*-AgNPs at 459 nm

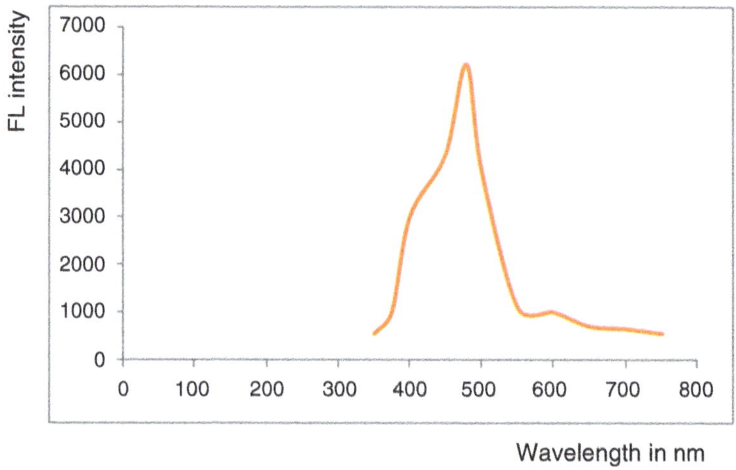

Fig. 5.6 FL spectra of the synthesised silver nanoparticles from the aqueous extract of seaweeds (**c**) *Hym*-AgNPs at 479 nm

5.10 Conclusion

Seaweed extracts have been used for the synthesize of silver nanoparticles (Ag-NPs) as they offer many completion in versatile fields. The use of biological materials in nanoparticles synthesis is preferred as they are eco-friendly and compatible for pharmaceutical and other biomedical applications, here antioxidant and anti-cancerous effect is discussed. Silver metal has toxicity against a wide range of microorganisms especially silver nanoparticles which has promising antimicrobial

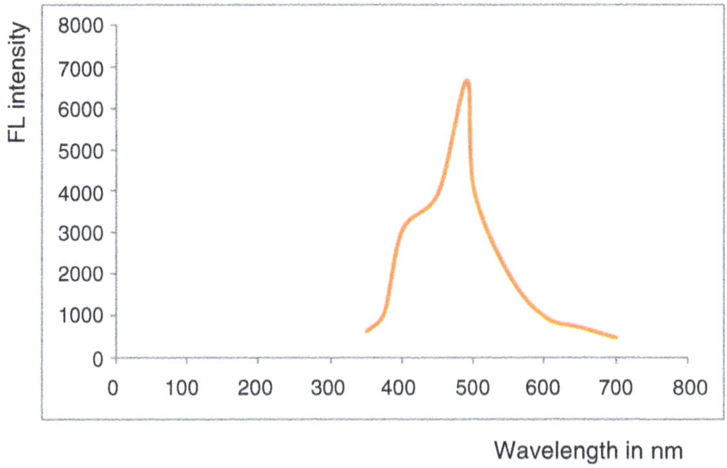

Fig. 5.7 FL spectra of the synthesised silver nanoparticles from the aqueous extract of seaweeds (**d**) *Srt*-AgNPs at 490 nm

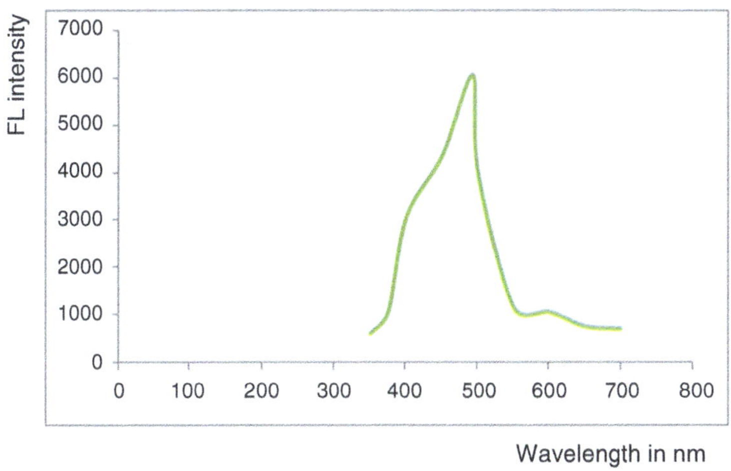

Fig. 5.8 FL spectra of the synthesised silver nanoparticles from the aqueous extracts of seaweeds (**e**) *Stm*-AgNPs at 493 nm

properties. Silver nanoparticles are effective as anti-inflammatory, anti-angiognesis, antiviral and anti-platelet activity against cancer cells. The Ag-NPs synthesized were from the aqueous extracts of seaweeds and were characterized by using spectral analysis. The effect of the synthesized silver nanoparticles on the isolated soil microbes were assessed by different assays such as Calorimetric chowder assay and Agar diffusion assay. The effect of synthesized Ag-NPs on earthworm was also studied by examining its morphology and histology. The toxicity of the combined seaweed-AgNPs against HeLa carcinoma cells has been analyzed and FL data

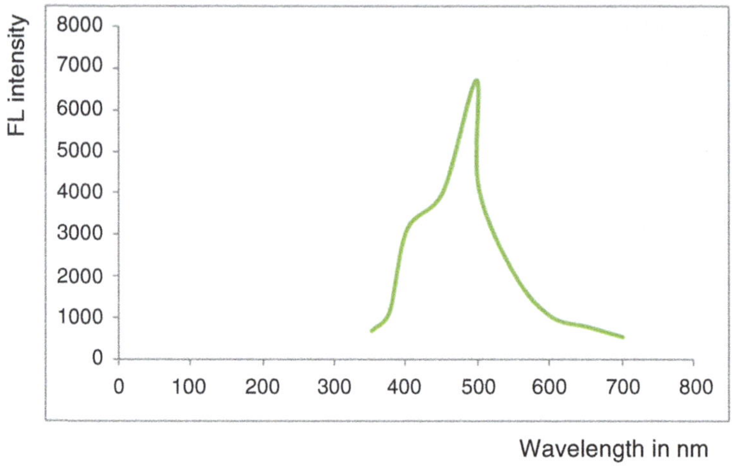

Fig. 5.9 FL spectra of the synthesised silver nanoparticles from the aqueous extracts of seaweeds **(f)** *Dtd*-AgNPs at 497 nm

analyze the band gap and color of FL for each specimen and shows that the synthesized Ag-NPs are free from flaws.

Acknowledgement The corresponding author would like to thank all co-authors and the managements of EGSP group and St.Mary's College for writing the Springer book content and also would like to thank the scholar V Murugammal for her dedicated and untiring effort for this work to get completed.

References

Ahmed M, Karns M, Goodson M, Rowe J, Hussain S, Schlager J, Hong Y (2008) DNA damage response to different surface chemistry of silver nanoparticles in mammalian cells. Toxicol Appl Pharmacol 233:404–410

Allison SD, Martiny JBH (2008) Resistance, resilience, and redundancy in microbial communities. PNAS 105:11512–11519

Asharani PV, Low Kah Mun G, Hande MP, Valiyaveettil S (2009) Cytotoxicity and genotoxicity of silver nanoparticles in human cells. ACS Nano 3:279–290

Ayman MA, Hamad AA, Abdelrahman OE (2014) Synthesis of silver nanoparticles by green method stabilized to synthetic human stomach fluid. Molecules 19:6737–6753

Aziz N, Pandey R, Barman I, Prasad R (2016) Leveraging the attributes of *Mucor hiemalis*-derived silver nanoparticles for a synergistic broad-spectrum antimicrobial platform. Front Microbiol 7:1984. https://doi.org/10.3389/fmicb.2016.01984

Aziz N, Faraz M, Sherwani MA, Fatma T, Prasad R (2019) Illuminating the anticancerous efficacy of a new fungal chassis for silver nanoparticle synthesis. Front Chem 7:65. https://doi.org/10.3389/fchem.2019.00065

Baath E (1992) Measurement of heavy metal tolerance of soil bacteria using thymidine incorporation in to bacteria extracted after homogenization centrifugation. Soil Biol Biochem 24:1167–1172

Baia L, Simon S (2007) UV-vis and TEM assessment of morphological features of silver nanoparticles from phosphate glass matrices. Mod Res Educ Top Microsc 210:577–583

Benoit ME, Hernandez MX, Dinh ML, Benavente F, Vasquez O, Tenner AJ (2013) C1q-induced LRP1B and GPR6 proteins expressed early in Alzheimer disease mouse models, are essential for the C1q-mediated protection against amyloid-beta neurotoxicity. J Biol Chem 288:654–665

Bhattacharya R, Mukherjee P (2008) Biological properties of "naked" metal nanoparticles. Adv Drug Deliv Rev 60:1289–1306

Cao XL, Cheng C, Ma YL, Zhao CS (2010) Preparation of silver nanoparticles with antimicrobial activities and the researches of their biocompatibilities. J Mater Sci Mater 21:2861–2868

Cha O, Hu Z (2008) Size dependent and reactive oxygen species related to nanosilver toxicity to nitrifying bacteria. Environ Sci Technol 42:4583–4588

Chaloupka K, Malam Y, Seifalian AM (2010) Nanosilver as a new generation of nanoproduct in biomedical applications. Trends Biotechnol 28:580–588

Chen X, Schluesener HJ (2008) Nanosilver: a nanoproduct in medical application. Toxicol Lett 176:1–12

Cho K, Park J, Osaka T, Park S (2005) The study of antimicrobial activity and preservative effects of nanosilver ingredients. Electrochim Acta 51:956–960

Cornelis G, Thomas DEM, Laughlin MJMC, Kirby JK, Beale DG, Chitleborough D (2012) Retention and dissolution of engineered silver nanopartilcles in natural soils. Soil Sci Soc Am J 76:891–902

Ding LJ, Su JQ, Xu HJ, Jia ZJ, Zhu YG (2015) Long-term nitrogen fertilization of paddy soil shifts iron-reducing microbial community revealed by RNA-^{13}C-acetate probing coupled with pyro sequencing. ISME J 9:721–734. https://doi.org/10.1038/ismej.2014.159

Doran JW, Zeiss MR (2000) Soil health and sustainability: managing the biotic component of soil quality. Appl Soil Ecol 15:3–11

Fang J, Lyon DY, Weisner MR, Dong J, Alvarez PJJ (2007) Effect of fullerence water suspension on bacterial phospholipids and membrane phase behavior. Environ Sci Technol 41:2636–2642

Fenchel T, Finlay BJ (2004) The ubiquity of small species: patterns of local and global diversity. Bioscience 54:777–784

Feng X, Qi X, Li J, Yang LW, Qiu MC, Yin JJ, Lu F, Zhong JX (2011) Preparation structure and photo-catalytic performances of hybrid Bi2SiO5 modified Si nanowire arrays. Appl Surf Sci 257:5571–5575

Foldbjerg R, Olesen P, Hougaard M, Dang DA, Hoffmann HJ, Autrup H (2009) PVP-coated silver nanoparticles and silver ions induce reactive oxygen species, apoptosis, and necrosis in THP-1 monocytes. Toxicol Lett 190:156–162

Gottschalle K, Sonderer T, Scholz RW, Nowak B (2009) Model environmental concentrations of engineered nanomaterials (TiO2, ZnO, Ag, CNT, fullerenes) for different regions. Environ Sci Technol 41:2022–2027

Hackenberg S, Scherzed A, Kessler M, Froelich K, Ginzkey C, Koehler C, Burghartz M, Hagen R, Kleinsasser N (2010) Zinc oxide nanoparticles induce photocatalytic cell death in human head and neck squamous cell carcinoma cell lines in vitro. Int J Oncol 37:1583–1590

Han X, Gu JD (2010) Sorption and transformation of toxic materials by microorganisms. In: Mitchel R, Gu J-D (eds) Environmental microbiology, 2nd edn. John Wiley & Sons, Inc., pp 153–176

Hayward RC, Saville DA, Aksay IA (2000) Electrophoretic assembly of colloidal crystals with optically tunable micropatterns. Nature 404:56–59

Henglein A (1993) Physicochemical properties of small metal particles in solution: "microelectrode" reactions, chemisorption, composite metal particles, and the atom-to-metal transition. J Phys Chem 97:5457–5471

Hess R, Jones L, Schlager JJ (2008) Unique cellular interaction of silver nanoparticles: size-dependent generation of reactive oxygen species. J Phys Chem B 112:13608–13619

Hsin Y, Chen C, Huang S, Shih T, Lai P, Chueh P (2008) The apoptotic effect of nanosilver is mediated by ROS- and JNK-dependent mechanism involving the mitochondrial pathway in NIH3T3 cells. Toxicol Lett 179:130–139

Hussain SM, Hess KL, Gearhart JM, Geiss KT, Schlager JJ (2005) In vitro toxicity of nanoparticles in BRL 3A rat liver cells. Toxicol In Vitro 1:975–983

Jacobson AR, Bride MC, Baveye P, Steenhuis TS (2005) Environmental factors determining the trace-level sorption o silver and thallium to soils. Sci Total Environ 345:191–205

Janvier C, Villenuve F, Alabouvette C, Edel-Hremann V, Mateille T, Steinberg C (2007) Soil health through soil disease suppression: which strategy from descriptors to indicators? Soil Biol Biochem 39:1–23

Johnston HJ, Hutchison G, Christensen FM, Peters S, Hankin S, Stone V (2010) A review of the in vivo and in vitro toxicity of silver and gold particulates: particle attributes and biological mechanisms responsible for the observed toxicity. Crit Rev Toxicol 40:328–346

Jones KC, Peterson PS (1986) Silver in welsh soils-physical and chemical distribution studies. Geoclerma 37(2):157–174

Joshi N, Jain N, Pathak A, Singh J, Prasad R, Upadhyaya CP (2018) Biosynthesis of silver nanoparticles using *Carissa carandas* berries and its potential antibacterial activities. J Sol-Gel Sci Techn 86(3):682–689. https://doi.org/10.1007/s10971-018-4666-2

Karlen DL, Maubach MJ, Doran JW, Cline RG, Harris RF, Schumann GF (1997) Soil quality: a concept, definition and framework for evaluation. Soil Sci Soc Am J 61:4–10

Kennedy AC, Smith KL (1995) Soil microbial diversity and sustainability of agricultural soils. Plant Soil 170:75–86

Kim J, Kuk E, Yuk N, Kim JH, Park SJ (2006) Mode of antibacterial action of silver nanoparticles. J Proteome Res 5:916–924

Kim YS, Song MY, Park JD, Song KS, Ryu HR, Chung YH, Chang HK, Lee JH, Oh KH, Kelman BJ (2010) Subchronic oral toxicity of silver nanoparticles. Part Fibre Toxicol 7:20–31

Kreuter J, Gelperina S (2008) Use of nanoparticles for cerebral cancer. Tumori 94:271–277

Li GY, Osborne NN (2008) Oxidative-induced apoptosis to an immortalized ganglion cell line is caspase independent but involves the activation of poly(ADP-ribose)polymerase and apoptosis-inducing factor. Brain Res 188:35–43

Lohse SE, Murphy CJ (2012) Applications of colloidal inorganic nanoparticles: from medicine to energy. J Am Chem Soc 134:15607–15620

Lok CN, Ho CM, Chen R, He Q-Y, Yu W-Y, Sun H, Tam PK-H, Chiu J-F, Che C-M (2006) Proteomic analysis of the mode antibacterial action of silver nanoparticles. J Proteomic Res 5:916–924

Martinez-Castanon GA, Nino-Martinez N, Martinez-Gutierrez F, Martinez-Mendoza JR, Ruiz F (2008) Synthesis of antibacterial activity of silver nanoparticles with different sizes. J Nanopart Res 10:123–153

Maurice PA, Hochella MF (2008) Nasoscale particles and process: a new dimension in soil science. Adv Agron 100:123–153

Mayer AF, Lipson DA, Martin AP, Schadt CW, Schmidt SK (2004) Molecular and metabolic characterization and cold tolerant alpine soil *Pseudomonas sensu stricto*. Appl Environ Microbiol 70:483–489

Mishra VK, Kumar A (2009) Impactof metal nanoparticles on plant growth promoting rhizobacteria. Dig J Nanomater Biostruct 4:587–592

Mohan YM, Lee KJ, Premkumar T, Geckeler KE (2007) Hydrogel networks as nanoreactors: a novel approach to silver nanoparticles for antibacterial applications. Polymer 48:158–164

Moon J, Shibamoto T (2009) Antioxidant assays for plant and food components. J Agric Food Chem 57:1655–1666

Mukherjee P, Senapati S, Mandal D, Ahmad A, Khan MI, Kumar R, Sastry M (2002) ChemBioChem 3:461

Oberdorster G, Oberdorster E, Oberdorster J (2005) Nanotoxicology: an emerging discipline evolving from studies of ultrafine particles. Environ Health Perspect 113:823–839

Perez-Losada M, Ricoy M, Marshall JC, Dominquez J (2009) Phylogenetic assessment of the earthworm *Aporrectodea caliginosa* species complex (oligochaeta, Lumbricidae) based on mitochondrial and nuclear DNA sequences. Mol Phylogenet Evol 52:293–302

Prasad R (2014) Synthesis of silver nanoparticles in photosynthetic plants. J Nanoparticles, Article ID 963961. https://doi.org/10.1155/2014/963961

Prasad R, Swamy VS (2013) Antibacterial activity of silver nanoparticles synthesized by bark extract of *Syzygium cumini*. J Nanoparticles. https://doi.org/10.1155/2013/431218

Prasad KS, Pathak D, Patel A, Dalwadi P, Prasad R, Patel P, Kaliaperumal SK (2011) Biogenic synthesis of silver nanoparticles using *Nicotiana tobaccum* leaf extract and study of their antibacterial effect. Afr J Biotechnol 9(54):8122–8130

Prasad R, Swamy VS, Varma A (2012) Biogenic synthesis of silver nanoparticles from the leaf extract of *Syzygium cumini* (L.) and its antibacterial activity. Int J Pharma Bio Sci 3(4):745–752

Ravindra BK, Rajasab AH (2014) A comparative study on biosynthesis of silver nanoparticles using four different fungal species. Int J Pharm Pharm Sci 6:372–376

Roh J-Y, Sim SJ, Yi J, Park K, Chung KH, Ryu D-Y, Choi J (2009) Ecotoxicity of silver nanoparticles on the soil nematode Caenorhabditis elegans using functional ecotoxicogenomics. Environ Sci Technol 43:3933–3940

Salata OV (2004) Application of nanoparticles in biology and medicine. *Journal of Nanobiotechnology* 2:1–12

Schjonning P, Heckrath G, Christenson BT (2009) Threats to soil quality in Denmark; a review of existing knowledge in the content of the EU soil thematic strategy, DJF report, plant science. Department of Agroecology. Faculty of Agricultural Science, Denmark, vol 3, p 1–121

Singh RP, Ramarao P (2012) Cellular uptake, intracellular trafficking, and cytotoxicity of silver nanoparticles. Toxicol Lett 213:249–259

Sohaebuddin SK, Thevenot PT, Baker D, Eaton JW, Tang L (2010) Nanomaterial cytotoxicity is composition, size, and cell type dependent. Part Fibre Toxicol 7:22–44

Sonoda E, Sasaki MS, Buerstedde JM, Bezzubova O, Shinohara A, Ogawa H, Takata M, Yamaguchi-Iwai Y, Takeda S (1998) Rad51-deficient vertebrate cells accumulate chromosomalbreaks prior to cell death. EMBO J 15:598–608

Srivastava S, Usmani Z, Atanasov AG, Singh VK, Singh NP, Abdel-Azeem AM, Prasad R, Gupta G, Sharma M, Bhargava A (2021) Biological nanofactories: using living forms for metal nanoparticle synthesis. Mini-Rev Med Chem 21(2):245–265

Su J, Zhang J, Liu L, Huang Y, Mason RP (2008) Exploring feasibility of multicolored CdTe quantum dots for in vitro and in vivo fluorescent imaging. J Nanosci Nanotechnol 8:1174–1177

Sudheer Khan S, Bharath Kumar E, Mukherjee A, Chandrasekaran N (2011) Bacterial tolerance to silver nonoparticle (SNPs): *Aeromonas punctata* isolated from sewage environment. J Basic microbiol 51:183–190

Tan WB, Jiang S, Zhang Y (2007) Quantum-dot based nanoparticles for targeted silencing of HER2/neu gene via RNA interference. Biomaterials 28:1565–1571

Theivasanthi T, Alagar M (2010) Xray diffraction studies of copper nano powder. Appl Phy Res 1:112–117

Thill A, Zeyons O, Spella O, Chauvat F, Rose J, Auffan M, Flank AM (2006) Cyto-toxicity of CeO_2 nanoparticles for *Escherichia coli*: physic-chemical insight of the cyto-toxicity mechanism. Environ Sci Technol 40:6151–6156

Travan A, Pelillo C, Donati I, Marsich E, Benincasa M, Scarpa T, Semeraro S, Turco G, Gennaro R, Paoletti S (2009) Non-cytotoxic silver nanopartícl-polysaccharide nanocomposites with anti-microbial activity. Biomacromolecules 10:1429–1435

Wang S, Jiang XM, Wang Q (2013) Experiment and grey relational analysis of seaweed particle combustion in a fluidized bed. Energy Convers Manag 66:115–126

Wiesner MR, Lowery GV, Alvarez P, Dionysiou D, Biswas P (2006) Assessing the risks of manufactured nanomaterials. Environ Sci Technol 40:4336–4345

Wong MS, Son DS, Chen HH (2010) Bacterial performance of visible-light responsive titania photo catalyst with silver nanostructures. PLoS One 5:e1094

Wu B, Wang Y, Lee Y-H, Horst A, Wang Z, Chen D-R, Radhakrishna S, Tang YJ (2010) Comparative eco-toxicity of nano ZnO particles under aquatic and aerosol exposure modes. Environ Sci Technol 44:1484–1489

Yoon K, Hoon B, Park JH, Hwang J (2007) Susceptibility constants of *Escherichia coli* and *Bacillus subtilis* to silver and copper nanoparticles. Sci Total Environ 373:572–575

You C, Han C, Wang X, Zheng Y, Li Q, Hu X, Sun H (2012) The progress of silver nanoparticles in the antibacterial mechanism, clinical application and cytotoxicity. Mol Biol Rep 39:9193–9201

Zanette C, Pelin M, Crosera M, Adami G, Bovenzi M, Larese FF, Florio C (2011) Silver nanoparticles exert a long-lasting antiproliferative effect on human keratinocyte HaCaT cell line. Toxicol In Vitro 5:1053–1060

Zargar M, Hamid AA, Bakar FB, Shamsudin MN, Shameli K, Jahanshiri F, Farahani F (2011) Green synthesis and antibacterial effect of silver nanoparticles using *Vitex Negundo* L. Molecules 16:6667–6676

Chapter 6
Epidemiology of COVID-19: Special Emphasis on Nanoscience and Its Implications

Ranjit Pabbati, Afreen Fathima, Jaime Humberto Flores Garcia, and Venkateswar Reddy Kondakindi

Abstract Coronavirus disease 2019 (COVID-19) is a highly contagious pathogenic viral infection caused by SARS-Cov-2. Coronavirus seems to have taken a popular role in the twenty-first century. The first instance of COVID-19 which was reported in Hubei province, Wuhan, China, has now spread to the entire world by human-to-human transmission. The World Health Organization (WHO) declared this infectious disease as a pandemic. Currently this pandemic has created a global health crisis. In this chapter, we analysed the epidemiological characteristics, i.e. occurrence, distribution and transmission of disease in different countries, laboratory diagnosis, prevention, control and treatment of COVID-19. The main objective of this chapter is to provide the latest insights over nanotechnology and its implications in the diagnosis, treatment, prevention and control of COVID-19. In this direction, several emerging issues such as optical biosensor nanotechnology, respiratory masks, Nanofibers Membrane Technology, etc. We are in opinion that this chapter will provide useful insights towards understanding the role of nanotechniques to combat COVID-19.

Keywords COVID-19 · Occurrence · Transmission · Prevention · Control and treatment · Nanotechnology

R. Pabbati · A. Fathima · V. R. Kondakindi (✉)
Centre for Biotechnology, Institute of Science & Technology, Jawaharlal Nehru Technological University Hyderabad, Hyderabad, Telangana, India

J. H. Flores Garcia
Departamento de ciencias de la Enfermería, Universidad técnica de Manabí, Portoviejo, Ecuador

N. R. Maddela et al. (eds.), *Nanotechnology for Advances in Medical Microbiology*, Environmental and Microbial Biotechnology,
https://doi.org/10.1007/978-981-15-9916-3_6

6.1 Introduction

Coronavirus disease is a pathogenic viral infection and is highly contagious caused by SARS-CoV-2 (Wikipedia 2020). It is an RNA virus that causes disease in animals and mammals. This coronavirus disease is also called as COVID-19 pandemic. These viruses are responsible for respiratory tract infections which may range from mild to lethal. These viruses have a positive-sense SS RNA as their genome with an outer envelope and nucleocapsid in the form of helical symmetry. This is enclosed by an icosahedral protein shell. The size of the genome of coronavirus compasses roughly 26 to 32 kilobases, one of the biggest RNA virus among the other RNA viruses (Fig. 6.1). Coronaviruses are enormous, for the most part round, once in a while pleomorphic (variable fit as a fiddle), particles with spherical surface projections (Goldsmith et al. 2004). The normal distance of the infectious particle is around 125 nm. The virus envelope in electron micrographs seems to be a distinct pair of electron-dense shells. Envelope comprises of a lipid bilayer, in which the film (M), envelope (E) and spike (S) auxiliary proteins are tied down (Lai and Cavanagh 1997). Nucleocapsid contains numerous duplicates of N (Nucleocapsid) protein, are bound to the positive sense single-stranded RNA genome in a constant dots on-a-string-type testimony. The lipid bilayer envelope, film proteins, and nucleocapsid ensure the infection when it is exterior the host cell (Neuman et al. 2011).

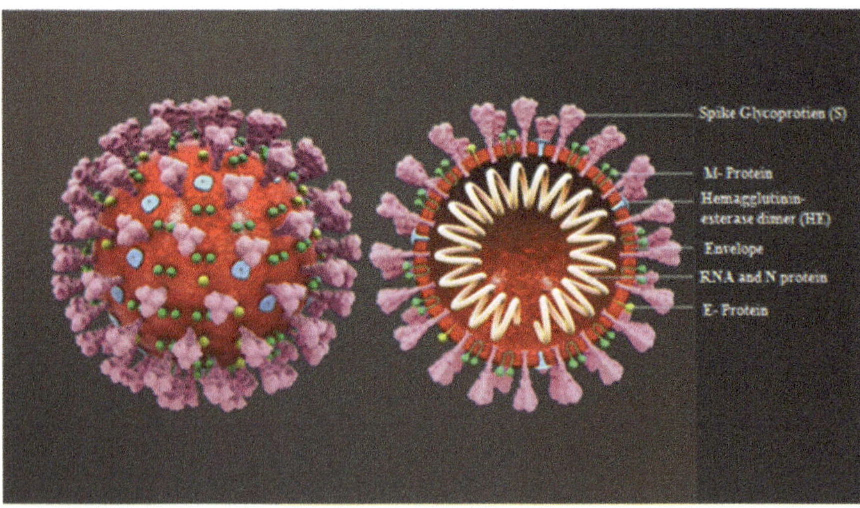

Fig. 6.1 The structure of coronavirus (Wikimedia 2020)

6.2 Epidemiology

Epidemiology is defined as the study of disease spread and examination of the diffusion and determinants of wellbeing-related states or occasions in indicated population and utilization of investigation to the control of medical issues. The word epidemiology is derived from a Greek language in which *epi* means upon, *demos* indicates the people and *logos* mean study of (Last 2001). Epidemiology has its underlying foundations in the examination of what comes upon a populace. The following are the important steps in the epidemiology:

(i) *Disease occurrence*: While there are numerous proportions of disease frequency, epidemiologists frequently gauge the disease occurrence in a population as incidence and prevalence of the infection. The key contrast between these two measures is the hour of ailment beginning.

(ii) *Incidence*: Incidence can be evaluated utilizing information from a disease registry data or a partner preliminary. There is a certain suspicion of a timeframe, for example, new cases within a month.

(iii) *Prevalence*: Prevalence measures the new and existing cases of a disease or outcome. Since prevalence tallies both new and existing cases, the term of the sickness influences the commonness.

Infections with a long span will be more pervasive than those with shorter term. Ceaseless, nonlethal conditions are more predominant than conditions with high mortality. Prevalence of a disease is legitimately identified with the span of the infection (CDC 2020).

6.3 Disease Occurrence and Distribution of COVID-19

In 2019, December the primary instance of coronavirus illness 2019 was accounted for in Wuhan, Hubei Province, China, during an outbreak of viral pneumonia. Since December 2019, patients with unexplained pneumonia have been found in Wuhan, Hubei Province, China. On 7th January 2020, Chinese specialists affirmed that the reason was a novel coronavirus that had not been recently distinguished, unique in relation to different coronaviruses, for example, Middle East respiratory syndrome coronavirus (MERS-CoV) in 2012 to 2015 and severe acute respiratory syndrome coronavirus (SARS-CoV) in 2003 (He et al. 2020). A large number of underlying cases were related to direct exposure to live markets, while ensuing cases were most certainly not. This further notified that infection spreads to human-to-human transmission through contacts, respiratory droplets, various discharges, etc.; on 21st January 2020, the National Health Commission of the People's Republic of China declared 2019-nCoV pneumonia as a class B irresistible infection, by considering the category A infectious disease as a reference (Wang et al. 2020a). In China, 11,791 cases were affirmed, and 17,988 cases were associated in 34 regions similarly as with 24:00, 31st January 2020. Gradually the number of cases attained peak stage in the month of February, and downfall of cases was observed in late March; as of 12th May 2020, the total number of cases recorded were 82,919, and the number of deaths were 4,633. The irresistible respiratory infection COVID-19 has spread quickly inside China and to neighbouring nations and throughout the world in a very short period of time. The first affirmed coronavirus cases outside China happened on 20th January 2020.

On 21st January 2020, the very first case was reported in Washington State of the United States (Holshue 2020). The continuous COVID-19 pandemic was affirmed at the United States in January 2020. On 25th February, the Centers for Disease Control and Prevention (CDC) cautioned the Americans to get ready for a nearby outbreak (Taylor 2020). A national crisis was proclaimed by President Trump on 13th March (Liptak 2020). The alarm in other countries of the world was not gigantic, but rather the number of the contaminated individuals in different nations of the world was additionally expanded. Among these nations South Korea, Japan, Iran, Saudi Arabia, Afghanistan, Ireland, Brazil, Pakistan, India, Russia, Kuwait, Qatar, the United Arab Emirates and France were top nations of the world.

The infection was affirmed to have spread to Brazil on 25th February 2020 when a man from São Paulo tested positive for the infection (Brasil confirma 2020). Starting on 2nd June 2020, 555,383 cases have been affirmed in the nation, causing 31,199 deaths. As of June 2020, Brazil has the second-most noteworthy number of affirmed COVID-19 cases on the planet behind the United States (Wikipedia 2020). The infection spread to Russia on 31st January 2020, when two Chinese residents in Tyumen, Siberia, and Chita, Russian Far East tested positive for the infection, with the two cases being contained. Russia was generally late in enduring a serious outbreak; however it now has become the nation with the third-most cases on the planet, after the United States and Brazil (Wikipedia 2020). As per official figures starting on 3rd June, Russia has 432,277 affirmed cases; 195,957 recuperations; 5,215 passings; and over 11.4 million tests performed. The city of Moscow is as of now the most influenced government subject. The contamination spread from Italy on 2nd March, prompting extra estimates, for example, dropping occasions; shutting schools, theatres and galleries; closing the fringe; and announcing a non-working period which kept going up to 11th May, has been expanded twice. The continuous COVID-19 pandemic was first declared and spread to Spain on 31st January 2020, when a German visitor tested positive for SARS-CoV-2 in La Gomera, Canary Islands (Sanidad confirma 2020). As of 2nd June 2020, there have been 239,932 declared cases and 27,127 deaths (Actualización n° 165 2020). The real number of cases was viewed as a lot higher, and the same number of individuals with just mellow or no side effects was probably not going to have been tested (Lau et al. 2020; Linde et al. 2020). The progressing COVID-19 pandemic spread to the United Kingdom in late January 2020. Starting on 3rd June 2020, there have been 279,856 declared cases and 39,728 deaths (Wikipedia 2020) shown in Fig. 6.2a and b.

The COVID-19 pandemic in India is a part of the overall pandemic of coronavirus ailment 2019 (COVID-19) brought about by SARS-CoV-2. The first instance of COVID-19 in quite a while, which began from China, was accounted on 30th January 2020. Starting on 8th June 2020, the MoHFW has affirmed a sum of

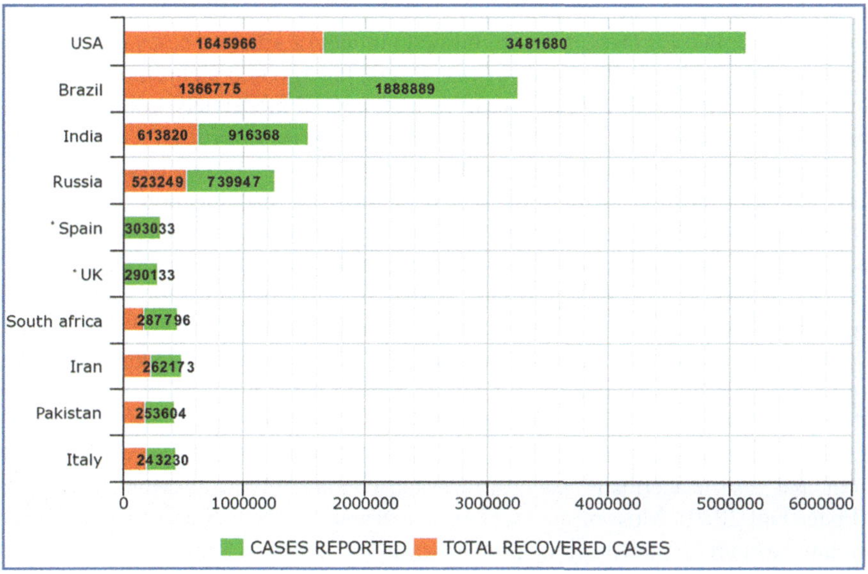

A

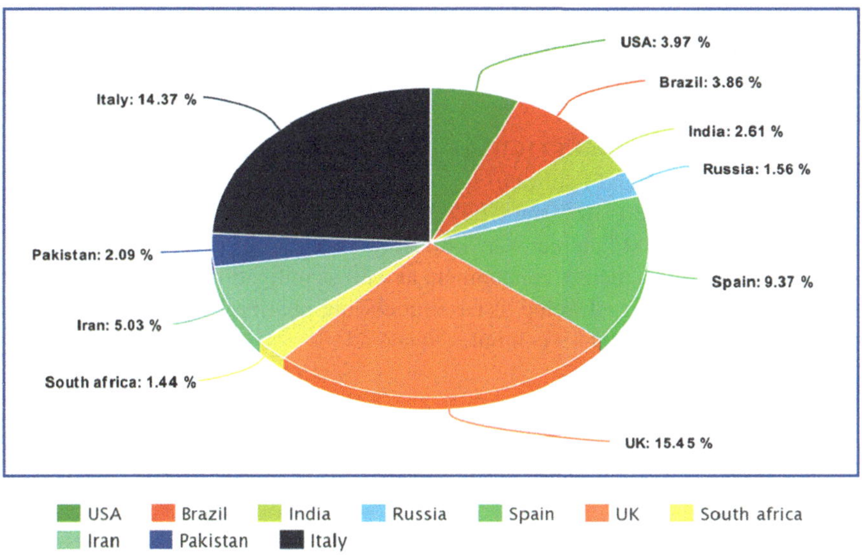

B

Fig. 6.2 Disease transmission details of coronavirus in various countries (Worldometer 2020 as updated till 16th July 2020). (**a**) Information about recovery and reported cases in various countries *(Spain and UK total recovered data is not available). (**b**) Percentage of deaths in various countries

256,611 cases; 1,24,430 recuperations (counting 1 movement); and 7,200 passings in the nation. India at present has the biggest number of affirmed cases in Asia, with the quantity of all-out affirmed cases penetrating the 100,000 blemish on 19th May and 200,000 on 3rd June. India's case casualty rate is moderately lower at 2.80%, against the worldwide 6.13%, starting on 3rd June (Wikipedia 2020).

The infection was first declared and spread to Italy on 31st January 2020, when two Chinese vacationers in Rome tested positive for the infection (Covid-19—Situazione in Italia 2020). Starting on 7th June 2020, Italy has 35,262 dynamic cases, one of the most noteworthy in the world. Overall there have been 234,998 declared cases and 33,899 passings (a pace of 561 passings for each million population, while there have been 165,837 recuperations or dismissals (Wikipedia 2020). By 7th June, Italy had tested around 2,627,000 people. Due to the set number of tests played out, the genuine number of contaminated individuals in Italy, as in different nations, is assessed to be greater than the official check (Aggiornamento 2020; Flaxman et al. 2020; Lau et al. 2020; CNBC 2020). On 27th January 2020, the principal case in Germany was affirmed and contained close to Munich, Bavaria (Erster Fall 2020). Most of the cases in January and early February started from a similar car parts producer as the primary case. On 25th and 26th February, various cases identified with the Italian flare-up were distinguished in Baden-Württemberg. A huge group connected to a festival occasion was framed in Heinsberg, North Rhine-Westphalia, with the primary demise covered on 9th March 2020 (Kreis Heinsberg 2020; Hamburg, Hamburger Abendblatt 2020; Robert Koch Institute 2020). Starting on 8th June 2020, the RKI has formally detailed 184,193 cases, 8,674 passings and around 1,69,600 recoveries (Table 6.1) (Robert Koch Institute Covid-19 dash board 2020).

In the same manner, the COVID-19 pandemic affected lakh of people in Iran, Saudi Arabia, France and Pakistan.

As of 8th June 2020 (10:00am CET), more than 6.9 million cases were reported globally, with 401,000 fatalities. On 14th May to 8th June, over 1.5 million cases were diagnosed. Even though containment as well as mitigation measures have been intensified and efforts to develop disease-modifying pharmacologic compounds are being carried out, COVID-19 continues to spread.

6.4 Laboratory Diagnosis of COVID-19

The COVID-19 pandemic has majorly affected clinical microbiology research facilities in the previous a while. The laboratory diagnosis of COVID-19 is presently being performed using two tests: viral tests and antibody tests. The viral test determines whether you have a current disease, and the antibody test determines if you have an infection in the past. An antibody test may not indicate that you have an ongoing infection, because it can take 1 to 3 weeks' time after infection to produce anticorps. Having antibodies to the virus that triggers COVID-19 may provide defence against the virus being infected again. Rapid and reliable identification of

Table 6.1 Disease transmission details of coronavirus in various countries

Country	Cases reported	Total deaths reported	Total recovered cases	Transmission classification	Total population
USA	3,481,680	138,291	1,645,966	Community transmission	331,081,677
Brazil	1,888,889	72,950	1,366,775	Community transmission	212,620,008
India	916,368	23,952	613,820	Clusters of cases	1,380,530,110
Russia	739,947	11,614	523,249	Clustered cases	145,937,175
*Spain	303,033	28,406	N/A	Community transmission	46,775,564
*UK	290,133	44,830	N/A	Community transmission	67,900,637
South Africa	287,796	4,172	160,693	Community transmission	5,933,662
Iran	262,173	13,211	227,561	Community transmission	84,033,003
Pakistan	253,604	5,320	178,737	Cluster of cases	221,037,609
Italy	243,230	34,967	196,016	Community transmission	60,457,891
Saudi Arabia	237,803	2,283	183,048	Cluster of cases	34,833,307
Germany	200,451	9,139	186,000	Cluster of cases	83,795,199
Bangladesh	190,057	2,424	105,523	Community transmission	164,752,852
France	172,377	30,029	78,820	Community transmission	65,279,649
Canada	108,155	8,790	72,485	Community transmission	37,755,152
China	83,650	4,634	78,719	Cluster of cases	1,439,323,776
Iraq	81,575	3,345	52,621	Community transmission	40,251,640
Indonesia	78,572	3,710	39,050	Community transmission	273,634,614
Philippines	57,545	1,603	20,796	Community transmission	109,635,041
Kuwait	56,174	396	46,897	Cluster of cases	4,272,860
UAE	55,198	396	46,418	Community transmission	9,894,897
Singapore	46,630	26	42,988	Cluster of cases	5,852,170
Israel	41,325	368	19,989	Pending	9,197,590
Afghanistan	34,740	1,048	22,456	Cluster of cases	38,956,620
Switzerland	33,016	1,968	29,800	Community transmission	8,657,154
Japan	21,686	982	18,545	Cluster of cases	126,459029
Australia	10,250	108	8,035	Cluster of cases	25,511,086
Malaysia	8,729	122	8,526	Cluster of cases	32,381,446

Worldometer (2020) as updated till 16th July 2020
*Spain and UK total recovered data is not available—N/A

COVID-19 is critical for community and hospital monitoring of outbreaks (CDC 2020). Current coronavirus diagnostic tests include reverse transcription polymerase chain reaction (RT-PCR), real-time RT-PCR (rRT-PCR) and reverse transcription loop-mediated isothermal amplification (RT-LAMP) tests; RT-LAMP has equal sensitivity to rRT-PCR, which is very precise used for MERS-CoV detection (Bhadra et al. 2015; Chan et al. 2015; Huang et al. 2018).

6.4.1 RT-PCR

RT-PCR is a diagnostic test which uses specimens of nasal swab, tracheal aspirate or bronchoalveolar lavage (BAL). Collection of upper respiratory samples through nasopharyngeal and oropharyngeal swabs is the primary and preferred method for diagnosis. The use of bronchoscopy as a diagnostic method for COVID-19 is not recommended since the generated aerosol poses a significant risk to both patients and healthcare staff. Bronchoscopy would only be used for incubated patients if the upper respiratory tests are negative and other diagnostic methods will alter the clinical management considerably. Bronchoscopy can, however, be suggested when clinical and safety conditions are met and for unclear diagnosis (Wang et al. 2020b). Alternatively, tracheal aspiration and non-bronchoscopic BAL may be used in incubated patients to collect respiratory specimens (WHO 2020). SARS-CoV-2 RNA was extracted from upper and lower respiratory tract specimens, and the virus was isolated from upper respiratory tract secretions and BAL specimens in a cell culture; however, limited RNA data is available. In a set of cases by Zou et al., SARS-CoV-2 RNA levels were found to be higher in upper respiratory tract samples (as shown by lower nose period threshold values), and the first 3 days after symptom onset and elevated SARS-CoV-2 RNA levels were also observed in asymptomatic patient samples obtained from upper respiratory tract samples (Zou et al. 2020). Several studies have shown that SARS-CoV-2 RNA in the blood and stool specimens can also be detected. It is possible that viral RNA will be detectable for weeks, as seen in some instances of SARS-CoV or MERS-CoV infection. The specificity of the RT-PCR test appears to be very high although, particularly in asymptomatic patients, false-positive results may occur due to swab contamination (Memish et al. 2014; Rodriguez-Morales et al. 2020; Zumla et al. 2015; Chan et al. 2004; Cheng et al. 2004; Hung et al. 2004; Peiris et al. 2003). The sensitivity levels are not explicit, but are estimated at around 66–80%.The validity of tests in asymptomatic persons in close contact with symptomatic persons is much less clear; the rate of positivity could exceed 50% without symptoms or confirmed infection (Ai et al. 2020). A single negative test does not preclude SARS-CoV-2 infection, especially in highly exposed individuals where the test is performed using a nasopharyngeal swab specimen and at the onset of the infection (Zhuang et al. 2020).

Infection severity can be estimated through a standard real-time RT-PCR set-up that usually runs through 35 cycles, meaning that around 35 billion new copies of the viral DNA sections are created from each virus strand present in the sample by the end of the process.

6.4.2 TrueNat Testing

Current PCR tests can distinguish a wide variety of pathogens in real time, with clear results, while the chain reaction is in progress. TrueNat is a chip-based, portable RT-PCR computer which was originally developed by Goa-based startup Molbio Diagnostics as a portable device for tuberculosis. The platform is on its way to becoming one of COVID-19's principal diagnostic tools as India seeks to increase its testing capacity. The new versions of the TrueNat system will detect a coronavirus SARS-CoV-2 enzyme (called RdRp) located in the RNA (First post 2020).

6.4.3 Antibody Tests

Antibody testing (or serological testing) is a bloodstream protein test to see whether a person was affected with COVID-19. A diseased person would have unique antibodies to certain pathogens to which they were exposed. As part of a bigger cycle, the immune system develops antibodies to protect itself against an infection. In comparison to a nasal or throat swab test that searches for genetic signatures of the body's virus, an antibody test looks for signs of the body's reaction to the virus. Antibodies are found in the blood, and either a finger prick or a blood sample taken with a needle produces a blood sample. In an antibody test, two different forms of an antibody are sought: IgM antibodies against SARS-CoV-2, which evolve in an infection early on IgG antibodies to SARS-CoV-2, often discovered after someone recovered from the infection. With respect to COVID-19 diagnosis, RT-PCR and TrueNat antibody tests are varied in their strengths and limitations. The advantage of antibody tests is to provide clarification as to their COVID-19 status for authorities and individuals. This test also gives the information related to infection rate, prevention, control and treatment of COVID-19. As part of the vaccine-enabled response, antibodies are produced faster and more abundantly. Antibody testing is crucial for determining who had the virus, particularly when a number of infected people do not appear to have any symptoms. An antibody test will yield results in less than an hour. The test can be used 7 to 10 days after someone has been contaminated, although it does have a greater percentage of error than swab tests. Antibody tests that look for IgM antibodies are typically quick finger prick tests that can deliver results in less than 20 minutes. But IgG testing requires a blood sample to be sent to a laboratory—a process that could take a week to produce results. IgG tests are more accurate than quick IgM tests, but don't provide as much detail about anyone being infected with COVID-19.

False negatives and false positives: The various test kits currently in use are a necessity—but they are also products of fast-track acceptance and production of COVID-19 science, where demand and pace for public health are the focus (First post 2020).

6.4.4 Antigen Test

Antigen tests look for different proteins present only in the virus, which is regarded as 'foreign' by the body's immune response. Most COVID-19 antigen tests target the 'spike protein' that studs the surface of the coronavirus. In this examination, a swab from the nose is obtained, where there is a high likelihood that virus particles are present. The swab is first dipped into a virus-inactivating solution and then moved to a test sheet. The test strip houses antibodies binding and holding coronavirus proteins in place as the fluid spreads. Checking for antigens has some primary benefits. One of the major benefits of using antigen test is that it decreases the pressure of relying on only RT-PCR tests to classify patients with COVID-19. These tests are also inexpensive compared to RT-PCR, costing around Rs 450 per test. The following are the boundaries of antigen selection. Antigen test can only disclose whether a person is currently experiencing SARS-CoV-2 infection. Antigens shall not be present before or after the infection has passed. Because antigen detection does not require any amplification processes of the virus or its genetic material, a swab sample can have too little antigen to detect. This may result in a false-negative outcome. To confirm a true negative for COVID-19, a negative test should be followed up by the more reliable RT-PCR test as a precaution. Accuracy is the single biggest issue with antigen checks, which as a diagnostic method are much less sensitive than RT-PCR (First post 2020).

6.4.5 Nanotechnology Approaches for Diagnosis of COVID-19

Nowadays nano-based products are being developed and install for the containment, diagnosis and treatment of COVID-19 (Future markets 2020). Nanosensors are now a reality, showing great ability to detect extremely low concentrations of bacteria and viruses and thus alert clinicians even before symptoms have shown or on patients with very low viral loads. As several nations around the world struggle with the rising number of cases involving coronavirus, the monitoring of suspected carriers is also being increased. A broad range of detection kits are in the production line, among which the ground-breaking rapid nanogold-based test will ease the burden on healthcare systems caused by the COVID-19 pandemic (Lu et al. 2020).

A recent review paper (*ACS Nano*, 'Diagnosing COVID-19: Disease and Detection Tools') discusses the existing nanotechnology-based diagnostic methods, e.g. nucleic acid and computed tomography testing) and potential new techniques (e.g. protein and point-of-care testing).

It allows researchers to move beyond architecture to advance their technologies. Although tremendously helpful for the current pandemic, it would be useful to build plug-and-play diagnostics to handle the outbreak of SARS-CoV-2 to avoid potential epidemics too (Nanowerk 2020).

When nanoparticles enter a biological system, such as human blood, they come into contact with different biomolecules, such as proteins, instantly; such biomolecules form a coating layer on the surface of the nanoparticles—the so-called biomolecular corona—thus giving the nanoparticles a special biological identity, which may be very different from the surface of the pure nanoparticles. Researchers have also shown that disease-specific corona protein can be used in conjunction with advanced classifiers for early detection and screening of cancers (Nanowerk 2020).

6.4.5.1 Magnetic Levitation (MagLev)

The MagLev approach may furnish useful insights into measuring protein density in solution for a better awareness of the protein's physicochemical properties. As long as various diseases produce substantial difference in the plasma proteome, the levitation development and patterns of plasma proteins may hold some information on an individual's health conditions. More precisely, MagLev's optical images of levitated proteins subjected to machine learning examination provide useful information on the health status of the individual (Nanowerk 2020).

6.4.5.2 Optical Biosensor Nanotechnology

A new device based on optical biosensor nanotechnology would allow the coronavirus to be extracted directly from patient samples within 30 minutes approximately without the need for centralized laboratory tests. The latest technique could quickly establish if a patient is infected by coronavirus or influenza virus. The initiative should be used for more than the present pandemic and for the treatment of humans. The new biosensor tool will also be used to examine various forms of coronavirus present in reservoir animals, such as fleas, to detect and track the possible evolution of these viruses and to prevent future human outbreaks (Hu et al. 2020).

Scientists from the University of Maryland School of Medicine (UMSOM) have developed an experimental COVID-19 diagnostic test that can visually detect the virus' presence in 10 minutes. This uses a simple assay that includes plasmonic nanoparticles in gold to detect a shift in colour when the virus is present.

Based on our preliminary findings, we conclude that as early as the first day of infection this exciting new test can detect RNA material from the virus. Nevertheless, more studies are required to confirm whether this is actually the case, said study leader Dipanjan Pan, PhD, UMSOM Professor of Diagnostic Radiology and Nuclear Medicine and Pediatrics (Sciencedaily 2020).

The test does not require the use of any specialized laboratory methods for research, such as those widely used to amplify DNA. The authors published their research last week in nanotechnology journal *ACS Nano* of the American Chemical Society. When a patient obtains a nasal swab or saliva sample, the RNA is extracted from the sample through a simple process which takes about 10 minutes. To detect a particular protein, the test uses a highly specific molecule that is bound to the gold nanoparticles. This protein is part of the genetic code the novel coronavirus is related to. Once the biosensor binds to the gene sequence of the virus, by turning the liquid reagent from purple to blue, the gold nanoparticles react. This RNA-based research looks very promising in terms of virus detection. The ground-breaking solution yields results without the need for an extensive laboratory facility. While further clinical trials are needed, the manufacturing and processing of this test will be much less costly than a typical COVID-19 laboratory study; it does not require laboratory equipment or specialized staff to conduct the test and interpret the results. If this new test meets the FDA standards, it may potentially be used as a monitoring tool to track any outbreak of infections in day-care centres, nursing homes, college campuses and work places (Tectales 2020).

6.5 Prevention and Control of COVID-19

Because presently no vaccine is available to prevent coronavirus disease (COVID-19), preventing exposure to this virus is the only way to prevent illness. Because the primary transmission mode is droplet transmission, the preventive measures revolve around avoiding droplet transfer.

6.5.1 General Preventive and Control Measures of COVID-19

- *Wash hands frequently*: Wash hands with a hand sanitizer, soap and running water or disposable alcoholic sanitizer. This kills viruses present on hands. In daily life, the general public will often wash their hands, and therefore gloves are not required. However, health professionals, close contacts, nurses and staff working in crowded environments will have to wear gloves to reduce the possibility of transmitting through communication. Wearing gloves, however, is no substitute for washing your face (Web MD 2020).
- *Restrict travelling*: In order to contain the spread of COVID-19 pandemic, international travel of passengers has been prohibited under MHA's (Ministry of Home Affairs) orders related to lockdown measures (Web MD 2020).

- *Practise social distancing*: Stay home as much as possible, as one can get and spread the infection without realizing it. Six-feet distance must be kept from everyone if one decides to go out (Web MD 2020).
- *Cover your mouth and nose when in public*: Wearing a mask protects one from COVID-19 especially when you encounter people, go to public places, enter crowded or enclosed spaces, take public transportation, etc. There is no need of wearing a mask when alone or at home in an open environment. If one has COVID-19, even if they do not feel sick, they will spread it. Wearing a mask of fabric can shield others. This is not a substitution for distancing from society (Web MD 2020).
- *Cover your mouth and nose with elbows or tissues*: Never place hands over your mouth and nose while coughing or sneezing. If your nose and mouth are covered with your elbows, the viruses will remain in clothing without contaminating other objects' surfaces (Web MD 2020).
- *Don't touch your face*: Coronaviruses may live several hours on surfaces that we touch. They can get into the body if they come onto the hands and then into the eyes, nose or mouth when touched with those hands (Web MD 2020).
- *Clean and disinfect*: Using soap and water, first clean but also disinfect objects that are in constant reach, such as chairs, doorknobs, light switches, toilets, faucets and sinks using a combination of household bleach and water (1/3 cup of bleach per gallon of water or 4 teaspoons of bleach per quarter of water) or a home cleaner certified to treat SARS-CoV-2. Wear gloves while cleaning/ disinfecting and throw them away when you're done. There is no evidence that herbal therapies and tea can prevent infection (Web MD 2020).
- *Home quarantine*: The aim of home quarantine is to prevent transmission of virus from person-to-person contacts, thereby preventing the development of cases of second and third generation. When there are a large number of asymptomatic near contacts or potential patients, home quarantine should be an effective choice for solving the problems which medical institutions cannot solve. If there are any suspicious symptoms, seek guidance from the medical personnel. Home quarantine must be suggested for the following people: people who have travelled or lived in countries or regions where local cases begin to rise and individuals with close contact with suspects and confirmed cases (Wenhong Zhang 2020).
- *Isolation*: Keeping people who are ill away from healthy people like, where possible, using a separate 'ill' bedroom and bathroom. A report published by the WHO on 9th July also stated that COVID-19 airborne transmission cannot be ruled out entirely. Rare situations like surgical procedures and noisy, enclosed spaces raise the possibility of airborne transmission. They also mentioned that when it comes to researching the airborne transmission of the virus, 'more studies are desperately needed', now that transmission by contact and respiratory droplets have been concluded to their best understanding. The latest protection measures would now have to provide adequate and efficient ventilation in public buildings, office settings, schools, hospitals and nursing homes with these new possibilities. Ventilated spaces must now be supplemented with methods for managing airborne pathogens, such as local exhausts, highly efficient air filters and ultraviolet

germicidal lights mounted far away to minimize exposure to the skin and eyes (Wenhong Zhang 2020).

- *Overcrowding, as previously observed, must be avoided at all costs*: Safe physical distance maintenance, while not the ultimate method for prohibiting transmission, may lead to lower chances of infection. This must be followed with strict measures in public transport and buildings particularly. These steps are easy to execute and are also relatively inexpensive. Some, like distance maintenance, may not entail money but a strict rule-and-punishment program; though it might sound extreme, it's the only way to control the virus spread and stop the pandemic as we know it (Wenhong Zhang 2020).

6.5.2 Scope of Nanotechnology in Prevention and Control of COVID-19

The outbreak of COVID-19 puts a global pressure on modern societies and particularly the infrastructure related to health care. Nanotechnology offers fresh opportunities for developing inexpensive and scalable methods of detection, secure equipment for personal safety and more successful medical solutions of COVID-19 (Chan 2020). Recently a nano-filter was developed which is believed to retain filtering efficiency through the use of nanofibers, even after hand washing. A reusable, nano-filtered face mask could help ease the pressures of face mask supply shortages. For years now, researchers have developed the potential nanoparticles to treat bacterial and viral infections (Statnano 2020). For example, gold nanoparticles are designed to specific viruses such as Ebola and influenza. These nanoparticles can break the structure of the viruses by heating the particles with infrared wavelength of light. Nanoparticles can also be used for drug delivery.

Nanotechnology approaches should reduce acute and chronic effects of the COVID-19 pandemic from the perspective of identification, safety and treatment. Nanotechnology contributions may include but are not limited to the following topics and their application to address the challenges of COVID-19.

Different applications have been found in medical field, such as Ag NPs, which can be used for biosensors, as they have a wide antiviral spectrum; the suppression of viruses can be expected. The silver nanoparticles are having a strong and broad-spectrum microbicidal activity (Prasad 2014). Ag NPs produce reactive oxygen species resulting in oxidative stress and generation of free silver ions; therefore, healthcare workers can be protected from the risk of acquiring infection by the use of these nanoparticles (Aziz et al. 2014, 2015, 2016; Joshi et al. 2018). Usually healthcare workers use primary protection equipment like clothing to avoid contact with infected individuals. Research is being carried out to develop protective clothing using Ag NPs which are absorbed on a polymer sheet with a nanoscale fibre like structure (Nanoshel 2020).

There are numerous nanotechnology products available for equipping people to battle COVID-19. While enhanced respiratory masks and gloves are used to reduce the infection and people are exposed to external environment, soaps, sanitizers, disinfectants, shampoos and detergents, made up of antiviral and antibacterial non-materials, are used internally to counter the disease. Graphene, nano-diamond, polymer nanofibers (e.g. polyacrylonitrile) and nanoparticles such as platinum, titanium dioxide and copper oxide are generally included in these product groups which add to their expertise (Statnano 2020).

Below are the latest updates on COVID-19 nano-products:

6.5.3 *Nano-based Vaccines*

(i). RNA- Based Vaccine

A type of candidate vaccine: LNP-encapsulated mRNA

(ii). Viral Vector-Based Vaccine

A type of candidate vaccine: Adenovirus type 5 vector
Nanotechnology part: Designing a nanoscale viral vector to deliver vaccine agent

(iii). Protein Subunit-Based Vaccine

A type of candidate vaccine: Recombinant protein nanoparticle vaccine
Nanotechnology part: Designing the recombinant F-proteins to self-assemble into nanoparticle constructs that approximate the size of the RSV virus

(iv). DNA-Based Vaccine

A type of candidate vaccine: Proteo-lipid vehicle (PLV)
Nanotechnology part: Using a neutral lipid formulation (liposome-based) with the high efficacy of the fusogenic protein-mediated delivery technology (Statnano 2020)

6.5.4 *Respiratory Masks*

(i) Nanofibers Membrane Technology

Queensland University of Technology (QUT) researchers have constructed and examined a highly breathable nanofiber-based cellulose material which is able to remove virus-size nanoparticles. This nanoparticle-removing new material is developed to be used as a disposable filter cartridge in biodegradable, anti-pollution masks. This is an important factor for people who have to wear masks for long periods or those with existing respiratory conditions. The higher the breathability, the greater the comfort and reduction in fatigue (Statnano 2020).

(ii) Air Filtration Systems

Clean air filtration systems are designed by Mack Antonoff HVAC to counter COVID-19 (coronavirus), mould, mildew and pet dander. Installed in your existing ductwork, the five HEPA, PECO, and nano-filters with UV light are important for families with allergies, asthma, bronchitis and other health problems. Turn-Key Environmental Consultants sells HealthPro® Compact Air Purifiers which capture up to 0.003 microns of 99.5% of particles (e.g. viruses and bacteria). The exclusive IQAir HyperHEPA® filtration technology lies at the core of this revolutionary system. A thick nanofiber network effectively traps particles of all sizes (Statnano 2020).

(iii) Photoelectrochemical Oxidation Technology

In March 2020, a newly designed air purification device called Molekule is set to be tested against a virus that serves as a substitute for coronavirus, which effectively killed air pollutants such as bacteria, mould spores and viruses, according to the University of South Florida. The device uses photoelectrochemical oxidation (PECO), a process that uses UV-A light to trigger a catalyst in the filter covered by nanoparticles of Molekule to produce free radicals which oxidize air pollutants (Statnano 2020).

6.6 Treatment of COVID-19

To date, there is still no specific approved treatment for COVID-19 and no cure for an infection though treatments and vaccines are under investigation and will be tested through clinical trials. Instead, treatment focuses on symptom management as the virus continues its course (Cascella et al. 2020).

Types of treatments practised for COVID-19 include:

- Antiviral and/or retroviral drugs
- Breathing helps, including mechanical ventilation
- Steroids to reduce inflammation in the lungs
- Transfusion of plasma into the blood

The FDA has not licensed any medications or biologics for the prevention or treatment of COVID-19. On 1st May 2020, remdesivir obtained an FDA Emergency Use Authorization (EUA), based on preliminary data showing a quicker recovery period for hospitalized patients with serious illness (Cascella et al. 2020).

(i) O_2 Fast Challenge

For a patient with $SpO_2 < 93–94\%$ ($< 88–90\%$ if COPD) or respiratory rate $>$ 28–30 min or dyspnoea, 40% of Venturi mask administration of oxygen is needed.

After a reassessment of 5 to 10 minutes, if the clinical and instrumental image has improved, the patient will continue the treatment and undergo a re-evaluation within 6 hours. In case of failure improvement or new worsening, the patient undergoes a noninvasive treatment, if not contraindicated (Cascella et al. 2020).

(ii) HFNO and Noninvasive Ventilation

As for HFNO or NIV, the panel of experts point out that such approaches carried out by devices with good interface fitting do not produce widespread dispersion of exhaled air, and their use can be regarded at low risk of airborne transmission (Cascella et al. 2020).

6.6.1 Other Therapies

(i) Corticosteroids

Among other therapeutic approaches, while systemic corticosteroids were not prescribed for treating viral pneumonia or acute respiratory distress syndrome (ARDS), these medications are typically used in extreme CARDS (e.g. methylprednisolone 1 mg/Kg/day). Of note, a recent large-size RCT (the RECOVERY trial) has shown that dexamethasone decreases deaths in critically ill COVID-19 patients by one-third. In the intervention group, 2,100 patients received dexamethasone (6 mg/day for 10 days), while patients in the control group ($n = 4{,}300$) received the disease standard treatment (Ledford 2020).

A medication named dexamethasone is the talk of the hour despite recent reports that in patients with extreme types of COVID-19, it may reduce mortality by one-third. The allegations emerged in a press release from the University of Oxford released by scientists participating in the RECOVERY (short for 'COVID-19 Therapy Randomized Evaluation') trial. They said the test data suggest that dexamethasone in patients with ventilator support cuts the risk of death by one-third and one-fifth in patients requiring oxygen support. Dexamethasone is not something of a novel drug. It is a steroid often used with potent anti-inflammatory properties. Dexamethasone falls within a broader drug class called corticosteroids for COVID-19 patients; when it is too aggressive to control by any means, dexamethasone helps to dampen the body's immune response (Healthworld 2020).

(ii) Antiviral Agents

Although no antiviral therapies have been approved, several approaches such as lopinavir/ritonavir have been suggested (orally 400/100 mg per 12 hours) (Bimonte et al. 2020). Preclinical studies have indicated that remdesivir (GS5734)—an RNA polymerase inhibitor with in vitro activity against multiple RNA viruses, including Ebola—may be useful for both prophylaxis and HCoV infection therapy (Gordon et al. 2020). This drug has been tested positively in a rhesus macaque model of MERS-CoV infection and recently in SARS-CoV-2-infected macaques. Alpha interferon (e.g. 5 million units by aerosol inhalation twice per day) was also used (De Wit et al. 2020; Williamson et al. 2020).

Several anti-flu drugs such as oseltamivir were used to treat COVID-19 patients (Chen et al. 2020). Another anti-flu medication, favipiravir, has shown some in vitro effect against SARS-CoV-2. A retrospective analysis found once again that the broad-spectrum antiviral arbidol would increase the rate of discharge and decrease the rate of COVID-19 patient's mortality (Wang et al. 2020c).

(iii) Antiviral/Immunomodulatory Drugs

As immunomodulatory therapy, chloroquine (500 mg every 12 hours) and hydroxychloroquine (200 mg every 12 hours) were suggested. Of note, Gautret et al. (2020) found in a non-randomized trial that hydroxychloroquine was significantly correlated with viral load reduction before viral disappearance, and this effect was strengthened by the azithromycin macrolides (Gautret et al. 2020). Indeed, in vitro and in vivo studies have shown that macrolides can mitigate inflammation and modulate the immune system. In particular, these drugs can induce the downregulation of cell surface adhesion molecules, decrease the development of proinflammatory cytokines, stimulate phagocytosis by alveolar macrophages and inhibit neutrophil activation and mobilization (Zarogoulidis et al. 2012). Further studies are required to support the use of azithromycin, alone or combined with other medications, such as hydroxychloroquine, outside of any bacterial overlaps. Also, attention must be given to the concomitant usage of hydroxychloroquine with azithromycin as the interaction may result in a higher risk of prolongation of the QT period and cardiac arrhythmias. Chloroquine can also cause QT prolongation (Mercuro et al. 2020).

(iv) Serotherapy

Antibodies extracted from the healed individual's blood constitute a therapeutic choice currently under review. It is determined that the dosage of antibodies required to treat a single patient with SARS-CoV-2 demands that at least three patients recovered from the SARS-CoV-2 infection eliminate anticorps. A clinical trial for investigating an antibody cocktail for the prevention and treatment of COVID-19 has been initiated on 11th June 2020 (Cascella et al. 2020).

(v) Anticoagulant

Since COVID-19 patients have a higher risk of venous thromboembolism and the correlation of anticoagulant therapy with decreased ICU mortality, it is recommended that thromboprophylaxis should be offered to patients. In addition, complete therapeutic-intensity anticoagulation (e.g. enoxaparin 1 mg/kg twice daily) is indicated in the case of suspected thrombophilia or thrombosis (Kollias et al. 2020).

(vi) Inflammation Inhibitors

Throughout Italy, apart from traditional treatments, a major study led by the Istituto Nazionale Tumori, Fondazione Pascale di Napoli, focuses on the use of tocilizumab. This is a humanized IgG1 monoclonal antibody, directed against the IL-6 receptor and widely used to treat rheumatoid arthritis, juvenile arthritis, giant cell arthritis and Castleman's syndrome and to mitigate toxicity due to inhibitors of the immune control level (Buonaguro et al. 2020). In addition, a randomized, double-blind, placebo-controlled, phase 2/3 research on sarilumab, which is another anti-IL-6 receptor antibody, is underway in the United States. Many specific approaches were put to the test. Anakinra is a recombinant antagonist of the IL-1 receptor used to treat auto-inflammatory disorders such as adult-onset still's disease, systemic-onset juvenile idiopathic arthritis and Mediterranean fever. The authors of a retrospective study found that the use of anakinra in patients with mild to extreme ARDS and hyper-inflammation (C-reactive protein in the range of 100 mg/L, ferritin in the range of 900 ng/mL or both) led to clinical improvement in 72% of patients (Cavalli et al. 2020). Acalabrutinib is a selective inhibitor of Bruton tyrosine kinase which regulates the signalling and activation of macrophages. Roschewski et al. (2020) tested this agent in a prospective off-label clinical trial on 19 patients hospitalized with severe COVID-19. They proved the medication increased oxygenation in most patients, enhancing inflammation indicators such as C-reactive protein and IL-6 (Roschewski et al. 2020).

Eli Lilly and Co may have a drug specifically developed to treat COVID-19 that is approved for use as early as September if all goes well with one of the two antibody therapies it is testing. The drugs belong to a class of synthetic medicines widely used to treat cancer, rheumatoid arthritis and many other disorders, called monoclonal antibodies. A monoclonal antibody drug developed against COVID-19 is expected to be more successful than currently being tested on repurposed drugs against the virus (Carl O'Donnell and Michael Erman 2020).

6.6.2 Nanotechnology in Treatment of COVID-19

Nanotechnology and nano-medicine have an excellent promise in solving a variety of specific health concerns, including viruses, which are considered a significant medical problem. Nano-biotechnology applications could represent a new avenue for virus treatment or disinfection. The possibilities of using non-materials effectively in this area as vaccines and nanosensors are also highlighted. Interesting and surprising properties of chemical compounds, especially nano-drugs, can contribute significantly not only to medicine and pharmaceuticals, but promising solutions to stop the deadly COVID-19 outbreak worldwide can also emerge.

Theranostics is a newly emerging drug that involves the identification and neutralization of viruses using nano-drugs and nanomedicine, with an emphasis on diagnosis and treatment. Accordingly, there are studies of nanoparticles being used to combat pathogens that cause influenza and tuberculosis. Thanks to the potential surface modification and functionalization, nanoparticles have the ability to detect the pathogens and viruses with a huge amount of reports in the literature. Nanoparticles may be modified or functionalized to dissolve the virus' lipid membrane or even bind to the spike proteins at S1 and/or penetrate into the envelope, encapsulating nucleocapsid and RNA. Nanoparticles can be modified/functionalized to target a specific virus, bacteria and other pathogens or a rage. Given their size, modified nanoparticles in the bloodstream can travel through the body without causing problems or affecting other functions, particularly those that participate in the human immune system and can remain in the body much longer to detect viruses (Nanografi 2020).

(i) Pharmaceuticals

(a) Novochizol™

Nanotechnology part: Chitosan-based nanoparticles aerosol formulation

Novochizol™ is a nanoparticle based on chitosan that is completely biocompatible and firmly adheres to the lung epithelial tissues and ensures continuous release without systemic dissemination. Extensive preclinical research, performed by academic collaborators at Bioavanta-Bosti, suggests that Novochizol™ is a safe and efficient technology for drug delivery (Statnano 2020).

(b) Peptide nanostructures against coronavirus' spike proteins

In the preclinical evaluation stage, researchers are engineering a new nanostructured therapy which could potentially disable the virus and prevent human cells from infecting it. The MIT team discovered a peptide molecule that binds directly and tightly to the spike protein of the coronavirus. The Simpson Querrey Institute (SQI) in Northwestern has been working on 'gluing' millions of peptides into a nanostructure that is the carrier of the precious drugs. The drug and carrier's identical chemistry helps scientists to develop nanostructures that shield the peptide medication as it circulates throughout the body until the disease's perpetrator, the novel coronavirus, is encountered. The nanostructures of the SQI carrier have water-filled channels which could keep and defend the antiviral therapies against destructive enzymes. The SQI team tested the idea using a possible Alzheimer's disease drug, and it was found that the general approach is highly successful in in vitro experiments (Statnano 2020).

6.7 Conclusions

The following are the main conclusions of this topic:

(i) Coronavirus disease is a highly contagious pathogenic viral infectious disease which causes acute and mild infection of the upper respiratory tract.
(ii) This epidemic resulted in more than 13 million cases worldwide and around 6 lakhs of deaths causing a global health crisis.
(iii) The primary transmission is by direct person-to-person contact. The infection also spreads through close contact via aerosol droplets. The rate of transmission of the virus has not yet been known as it differs based on different environmental factors.

(iv) Community transmission of infection resulted in rapid spike in the cases and deaths making it a global health emergency.
(v) Various methods have been developed for the diagnosis of COVID-19. Diagnosis is mainly done by chest CT examination. RT-PCR is routinely used in detection of acute respiratory infection. Apart from RT-PCR, antibody testing, antigen testing is also being used.
(vi) Different types of diagnostic detection kits are being developed using nanotechnology for rapid and error-free testing. Magnetic levitation techniques and optical biosensor nanotechnology may provide useful insights for the detection of viruses.
(vii) the As there is no treatment available currently, preventing exposure is the only way to prevent the illness. Physical distancing, maintaining personal hygiene and self-isolation are the main important preventive measures to be followed.
(viii) Development of vaccines is a major preventive measure and nanotechnology-based protective equipment for efficient control of the infection.
(ix) The most serious part of this disease is that there is no definite treatment to cure the infection. Various drugs and vaccines are currently being studied as there is a severe necessity to develop drugs and vaccines to reduce the effect of COVID-19.
(x) This chapter also underlined the development of more efficient treatment methods by utilizing innovative techniques like nanotechnology by taking into consideration the cost of producing the clinical drugs.
(xi) As this pandemic is still in the spreading phase and not over yet, we are learning the new data about COVID-19 every day. There is a need to follow the updates in order to monitor the risk factors and also the therapy modalities to suppress the virus and the disease.

References

Actualización n° 165 (2020) Enfermedad por el coronavirus (COVID-19) (PDF). https://doi.org/10.1016/j.jmii.2020.03.013. PMC 7102572. PMID 32205091

Aggiornamento (2020) (PDF). 27 April 2020. Accessed 28 Mar 2020. http://www.salute.gov.it/imgs/C_17_notizie_4623_0_file.pdf

Ai T, Yang Z, Hou H et al (2020) Correlation of chest CT and RT-PCR testing in coronavirus disease 2019 (COVID-19) in China: a report of 1014 cases. Radiology 296:E32–E40. https://doi.org/10.1148/radiol.2020200642

Aziz N, Fatma T, Varma A, Prasad R (2014) Biogenic synthesis of silver nanoparticles using *Scenedesmus abundans* and evaluation of their antibacterial activity. J Nanoparticles, Article ID 689419. https://doi.org/10.1155/2014/689419

Aziz N, Faraz M, Pandey R, Sakir M, Fatma T, Varma A, Barman I, Prasad R (2015) Facile algae-derived route to biogenic silver nanoparticles: synthesis, antibacterial and photocatalytic properties. Langmuir 31:11605–11612. https://doi.org/10.1021/acs.langmuir.5b03081

Aziz N, Pandey R, Barman I, Prasad R (2016) Leveraging the attributes of *Mucor hiemalis*-derived silver nanoparticles for a synergistic broad-spectrum antimicrobial platform. Front Microbiol 7:1984. https://doi.org/10.3389/fmicb.2016.01984

Bhadra S, Jiang YS, Kumar MR, Johnson RF, Hensley LE, Ellington AD (2015) Real-time sequence-validated loop-mediated isothermal amplification assays for detection of Middle East respiratory syndrome coronavirus (MERS-CoV). PLoS One 10(4):e0123126

Bimonte S, Crispo A, Amore A, Celentano E, Cuomo A, Cascella M (2020) Potential antiviral drugs for SARS-Cov-2 treatment: preclinical findings and ongoing clinical research. In Vivo 34 (3):1597–1602

Brasil confirma primeiro caso da doença (2020) Ministerio da Saude. Archived from the original on 6 March 2020. Accessed 4 Mar 2020. https://www.gov.br/saude/pt-br

Buonaguro FM, Puzanov I, Ascierto PA (2020) Anti-IL6R role in treatment of COVID-19-related ARDS. J Transl Med 18(1):165

Cascella M, Rajnik M, Cuomo A, Dulebohn SC, Di Napoli R (2020) Features, evaluation and treatment coronavirus (COVID-19). https://www.ncbi.nlm.nih.gov/books/NBK554776/. Accessed 4 Jul 2020

Cavalli G, De Luca G, Campochiaro C, Della-Torre E, Ripa M, Canetti D, Oltolini C, Castiglioni B, Tassan Din C, Boffini N, Tomelleri A, Farina N, Ruggeri A, Rovere-Querini P, Di Lucca G, Martinenghi S, Scotti R, Tresoldi M, Ciceri F, Landoni G, Zangrillo A, Scarpellini P, Dagna L (2020) Interleukin-1 blockade with high-dose anakinra in patients with COVID-19, acute respiratory distress syndrome, and hyperinflammation: a retrospective cohort study. Lancet Rheumatol 2(6):e325–e331

Centre for disease control and prevention -CDC (2020). https://www.cdc.gov/csels/dsepd/ss1978/lesson3/section2.html. Accessed 18 May 2020.

Chan WCW (2020) Nano research for COVID-19. ACS Nano 14:3719–3720. https://pubmed.ncbi.nlm.nih.gov/3222, https://statnano.com/technology-against-covid-19-nano-insights7916/

Chan KH, Poon LLLM, Cheng VCC et al (2004) Detection of SARS coronavirus in patients with suspected SARS. Emerg Infect Dis 10:294–299

Chan JF, Choi GK, Tsang AK, Tee KM, Lam HY, Yip C (2015) Development and evaluation of novel real-time reverse transcription-PCR Assays with locked nucleic acid probes targeting leader sequences of human-pathogenic coronaviruses. J Clin Microbiol 53:2722–2726

Chen N, Zhou M, Dong X, Qu J, Gong F, Han Y, Qiu Y, Wang J, Liu Y, Wei Y, Xia J, Yu T, Zhang X, Zhang L (2020) Epidemiological and clinical characteristics of 99 cases of 2019 novel coronavirus pneumonia in Wuhan, China: a descriptive study. Lancet 395(10223):507–513

Cheng PK, Wong DA, Tong LK et al (2004) Viral shedding patterns of coronavirus in patients with probable severe acute respiratory syndrome. Lancet 363:1699–1700

CNBC (2020) The total number of Italian coronavirus cases could be '10 times higher' than known tally, according to one official. CNBC. March 24. https://www.cnbc.com/2020/03/24/italian-coronavirus-cases-seen-10-times-higher-than-official-tally.html. Accessed 31 Mar 2020.

COVID 19 pandemic (2020). https://en.wikipedia.org/wiki/COVID-19_pandemic. Accessed 25 Jul 2020. https://doi.org/10.1016/j.jsb.2010.11.021. PMC 4486061. PMID 21130884

COVID 19 pandemic in Brazil (2020) Wikipedia. https://en.wikipedia.org/wiki/COVID-19_pandemic_in_Brazil. Accessed 22 Jul 2020.

COVID 19 pandemic in India (2020) Wikipedia. https://en.wikipedia.org/wiki/COVID-19_pandemic_in_India. Accessed 25 Jul 2020.

COVID 19 pandemic in Russia (2020) Wikipedia. https://en.wikipedia.org/wiki/COVID-19_pandemic_in_Russia. Accessed 26 Jul 2020.

COVID 19 pandemic in United Kingdom (2020) Wikipedia. https://en.wikipedia.org/wiki/COVID19_pandemic_in_the_United_Kingdom#cite_note-nb02-4. Accessed 26 Jul 2020.

Covid-19—Situazione in Italia (2020). salute.gov.it (in Italian). Ministero della Salute. http://www.salute.gov.it/portale/nuovocoronavirus/homeNuovoCoronavirus.jsp?lingua=english. Accessed 4 May 2020

De Wit E, Feldmann F, Cronin J, Jordan R, Okumura A, Thomas T, Scott D, Cihlar T, Feldmann H (2020) Prophylactic and therapeutic remdesivir (GS-5734) treatment in the rhesus macaque model of MERS-CoV infection. Proc Natl Acad Sci USA 117(12):6771–6776

Erster Fall des Coronavirus in Deutschland bestätigt (in German) (2020) January 28. Accessed 28 Jan 2020. https://www.spiegel.de/consent

First post (2020) COVID-19 testing: how antibody, antigen, RT-PCR, TrueNat tests differ, their strengths and limitations. https://www.firstpost.com/health/covid-19-testing-how-antibody-antigen-rt-pcr-truenat-tests-differ-their-strengths-and-limitations-8548691.html. Accessed 9 Jul 2020

Flaxman S, Swapnil M, Gandy A et al. (2020) Report 13: estimating the number of infections and the impact of non-pharmaceutical interventions on COVID-19 in 11 European countries. Imperial College London: 1–35. https://doi.org/10.25561/77731

Future markets (2020) Diagnostic testing, antiviral and antimicrobial coatings and surfaces, air-borne filtration, facemasks, PPE, drug delivery and therapeutics. https://www.futuremarketsinc.com/nanotechnology-and-nanomaterials-solutions-for-covid-19-diagnostic-testing-antiviral-and-antimicrobial-coatings-and-surfaces-air-borne-filtration-facemasks-ppe-drug-delivery-and-therapeutics/. Accessed 25 May 2020

Gautret P, Lagier JC, Parola P, Hoang VT, Meddeb L, Mailhe M, Doudier B, Courjon J, Giordanengo V, Vieira VE, Dupont HT, Honoré S, Colson P, Chabrière E, La Scola B, Rolain JM, Brouqui P, Raoult D (2020) Hydroxychloroquine and azithromycin as a treatment of COVID-19: results of an open-label non-randomized clinical trial. Int J Antimicrob Agents 56:105949

Goldsmith CS, Tatti KM, Ksiazek TG, Rollin PE, Comer JA, Lee WW et al (2004) Ultrastructural characterization of SARS coronavirus. Emerg Infect Dis 10(2):320–326. https://doi.org/10.3201/eid1002.030913

Gordon CJ, Tchesnokov EP, Feng JY, Porter DP, Götte M (2020) The antiviral compound remdesivir potently inhibits RNA-dependent RNA polymerase from Middle East respiratory syndrome coronavirus. J Biol Chem 295(15):4773–4779

Hamburg, Hamburger Abendblatt (2020) Coronavirus: Zwei Tote in Deutschland—Italien sperrt das ganze Land. abendblatt.de. Accessed 10 Mar 2020. https://www.abendblatt.de/vermischtes/article228637475/Coronavirus-Corona-News-Live-Ticker-Covid-19-Zahl-Infektionen-Deutschland-631-Tote-Italien.html

He F, Yu D, Li W (2020) Coronavirus disease 2019: what we know? J Med Virol 92(7):719–725

Healthworld (2020) Dexamethasone: game-changer in the COVID-19 battle? (2020). https://health.economictimes.indiatimes.com/news/pharma/dexamethasone-game-changer-in-the-covid-19-battle/76817741. Accessed 24 Jul 2020

Holshue ML (2020) First case of 2019 novel coronavirus in the United States. N Engl J Med 382:9296–9936

Hu X, Wang X, Han L, Chen T, Wang L et al. (2020) Reverse transcription loop-mediated isothermal amplification combined with nanoparticles-based biosensor for diagnosis of COVID-19. medRxiv. https://doi.org/10.1101/2020.03.17.20037796

Huang P, Wang H, Cao Z, Jin H, Chi H, Zhao J (2018) A rapid and specific assay for the detection of MERS-CoV. Front Microbiol 9:1101

Hung IF, Cheng VC, Wu AK et al (2004) Viral loads in clinical specimens and SARS manifestations. Emerg Infect Dis 10:1550–1557

Joshi N, Jain N, Pathak A, Singh J, Prasad R, Upadhyaya CP (2018) Biosynthesis of silver nanoparticles using *Carissa carandas* berries and its potential antibacterial activities. J Sol-Gel Sci Techn 86(3):682–689

Kollias A, Kyriakoulis KG, Dimakakos E, Poulakou G, Stergiou GS, Syrigos K (2020) Thromboembolic risk and anticoagulant therapy in COVID-19 patients: emerging evidence and call for action. Br J Haematol 189(5):846–847

Kreis Heinsberg (2020) Kreis Heinsberg. Accessed 29 Feb 2020. https://www.kreis-heinsberg.de/aktuelles/aktuelles/?pid=5138

Lai MM, Cavanagh D (1997) The molecular biology of coronaviruses. Adv Virus Res 48:1–100. https://doi.org/10.1016/S0065-3527(08)60286-9

Last JM (2001) Dictionary of epidemiology, 4th edn. Oxford University Press, New York, p 61

Lau H, Khosrawipour V, Kocbach P, Mikolajczyk A, Ichii H, Schubert J, Bania J, Khosrawipour T (2020) Internationally lost COVID-19 cases. J Microbiol Immunol Infect. https://doi.org/10.1016/j.jmii.2020.03.013. ISSN 1684-1182. PMC 7102572. PMID 32205091

Ledford H (2020) Coronavirus breakthrough: dexamethasone is first drug shown to save lives. Nature 582(7813):469

Linde P, Oriol G, Elena S (2020) Lack of testing hampering Spain's efforts to slow coronavirus outbreak. EL PAÍS.

Liptak K (2020) Trump declares national emergency—and denies responsibility for coronavirus testing failures. CNN. Accessed 18 Apr 2020. https://edition.cnn.com/2020/03/13/politics/donald-trump-emergency/index.html

Lu R, Zhao X, Li J, Niu P, Yang B et al (2020) Genomic characterisation and epidemiology of 2019 novel coronavirus: implications for virus origins and receptor binding. Lancet 395:565–574

Memish ZA, Assiri AM, Al-Tawfiq JA (2014) Middle East respiratory syndrome coronavirus (MERS-CoV) viral shedding in the respiratory tract: an observational analysis with infection control implications. Int J Infect Dis 29:307–328

Mercuro NJ, Yen CF, Shim DJ, Maher TR, McCoy CM, Zimetbaum PJ, Gold HS (2020) Risk of QT interval prolongation associated with use of hydroxychloroquine with or without concomitant azithromycin among hospitalized patients testing positive for coronavirus disease 2019 (COVID-19). JAMA Cardiol 5:1036–1041

Nanografi (2020). https://nanografi.com/blog/covid19-and-nanotechnology/. Accessed 06 Apr 2020

Nanoshel (2020) Nanotechnology for health workers: COVID 19 control and prevention. https://www.nanoshel.com/Nanotechnology-for-Health-workers-COVID-19-Control-.

Nanowerk (2020) Nanotechnology-based approaches to testing for COVID-19 infections in high-risk individuals https://www.nanowerk.com/spotlight/spotid=55130.php. Accessed 12 May 2020

Neuman BW, Kiss G, Kunding AH, Bhella D, Baksh MF, Connelly S et al (April 2011) A structural analysis of M protein in coronavirus assembly and morphology. J Struct Biol 174(1):11–22

O'Donnell C, Erman M (2020) Lilly COVID-19 treatment could be authorized for use as soon as September—chief scientist. https://in.reuters.com/article/health-coronavirus-lilly-exclusive/exclusive-lilly-covid-19-treatment-could-be-authorized-for-use-as-soon-as-september-chief-scientist-idINKBN23I0A6. Accessed 11 Jun 2020

Peiris JS, Chu CM, Cheng VC et al (2003) Clinical progression and viral load in a community outbreak of coronavirus-associated SARS pneumonia: a prospective study. Lancet 361:1767–1772

Prasad R (2014) Synthesis of silver nanoparticles in photosynthetic plants. J Nanoparticles, Article ID 963961. https://doi.org/10.1155/2014/963961

Robert Koch Institute (2020) SARS-CoV-2: Fallzahlen in Deutschland, China und weltweit. Robert Koch Institute (in German) https://www.rki.de/DE/Content/InfAZ/N/Neuartiges_Coronavirus/Fallzahlen.html?nn=13490888. Archived from the original on 3 March 2020. Accessed 2 Mar 2020

Robert Koch-Institute: COVID-19-Dashboard (2020) Robert Koch Institute in German. Accessed 27 May 2020. https://www.rki.de/DE/Content/InfAZ/N/Neuartiges_Coronavirus/Fallzahlen.html?nn=13490888

Rodriguez-Morales AJ, Cardona-Ospina JA, Gutiérrez-Ocampo E, Clinical, laboratory and imaging features of COVID-19 et al (2020) A systematic review and meta-analysis. Travel Med Infect Dis 34:101623

Roschewski M, Lionakis MS, Sharman JP, Roswarski J, Goy A, Monticelli MA, Roshon M, Wrzesinski SH, Desai JV, Zarakas MA, Collen J (2020) Inhibition of Bruton tyrosine kinase in patients with severe COVID-19. Sci Immunol 5(48):eabd0110

Sanidad confirma en La Gomera el primer caso de coronavirus en España (2020) El Pais (in Spanish. Archived from the original on 2020-01-31. Accessed 31 Jan 2020. https://elpais.com/sociedad/2020/01/31/actualidad/1580509404_469734.html

Sciencedaily (2020) Researchers develop experimental rapid COVID-19 test using nanoparticle technique. https://www.sciencedaily.com/releases/2020/05/200529150643.htm. Accessed 29 May 2020

Statnano (2020) Technology against covid-19: nano insights into prevention, diagnosis, and treatment. https://statnano.com/technology-against-covid-19-nano-insights

Taylor M (2020) Exclusive: U.S. axed CDC expert job in China months before virus outbreak. Reuters. Retrieved March 24, 2020. https://www.reuters.com/article/us-health-coronavirus-china-cdc-exclusiv/exclusive-u-s-axed-cdc-expert-job-in-china-months-before-virus-outbreak-idUSKBN21910S

Tectales (2020) Nanotechnology provides rapid visual detection of COVID-19 (2020). https://tectales.com/bio-tech-it/nanotechnology-provides-rapid-visual-detection-of-covid-19.html.

Wang W, Xu Y, Gao R et al (2020a) Detection of SARS-CoV-2 in different types of clinical specimens. JAMA 323:1843–1844

Wang Z, Yang B, Li Q, Wen L, Zhang R (2020b) Clinical features of 69 cases with coronavirus disease 2019 in Wuhan, China. Clin Infect Dis 71(15):769–777. https://doi.org/10.1093/cid/ciaa272

Wang C, Peter WH, Frederick GH, George FG (2020c) A novel coronavirus outbreak of global health concern. Lancet 395(10223):470–473

Web MD (2020) Coronavirus and COVID-19: what you should know. https://www.webmd.com/lung/coronavirus. Accessed 24 Jul 2020

WHO (2020) Laboratory testing for 2019 novel coronavirus (2019-nCoV) in suspected human cases. Interim Guid. Geneva, Switzerland, World Health Organization site, 2020.

Wikimedia commons (2020) file; 3D medical animation coronavirus structure.jpg. https://commons.wikimedia.org/wiki/File:3D_medical_animation_coronavirus_structure.jpg. Accessed 25 Jul 2020

Williamson BN, Feldmann F, Schwarz B, Meade-White K, Porter DP, Schulz J, van Doremalen N, Leighton I, Yinda CK, Pérez-Pérez L, Okumura A, Lovaglio J, Hanley PW, Saturday G, Bosio CM, Anzick S, Barbian K, Cihlar T, Martens C, Scott DP, Munster VJ, de Wit E (2020) Clinical benefit of remdesivir in rhesus macaques infected with SARS-CoV-2. Nature 583:273–276

Worldometer (2020) COVID 19 coronavirus pandemic. https://www.worldometers.info/coronavirus/. Accessed 24 Jul 2020

Zarogoulidis P, Papanas N, Kioumis I, Chatzaki E, Maltezos E, Zarogoulidis K (2012) Macrolides: from in vitro anti-inflammatory and immunomodulatory properties to clinical practice in respiratory diseases. Eur J Clin Pharmacol 68(5):479–503

Zhang W (2020) Prevention and control of COVID-19. World Century Publishing Corp. https://www.worldscientific.com/doi/pdf/10.1142/11834

Zhuang GH, Shen MW, Zeng LX et al (2020) Potential false-positive rate among the 'asymptomatic infected individuals' in close contacts of COVID-19 patients. Zhonghua Liu Xing Bing Xue Za Zhi 41(4):485–488

Zou L, Ruan F, Huang M et al (2020) SARS-CoV-2 viral load in upper respiratory specimens of infected patients. N Engl J Med 382:1177–1179

Zumla A, Hui DS, Perlman S (2015) Middle East respiratory syndrome. Lancet 386:995–1007

Chapter 7
Surface-Modified Noble Metal Nanoparticles as Antimicrobial Agents: Biochemical, Molecular and Therapeutic Perspectives

Nabarun Chandra Das, Bishnupada Roy, Ritwik Patra, Abhigyan Choudhury, Madhureema Ghosh, and Suprabhat Mukherjee

Abstract Despite the progress of the development of antimicrobial therapeutics, the whole world is still under pressure of several microbial diseases. Antimicrobial drug development is therefore considered as one of the most practicable research works at present time. Even most of the pharmaceutical companies are investing to develop better therapeutic solutions against the life-threatening infectious diseases caused by *Mycobacterium tuberculosis*, *Helicobacter pylori*, *Vibrio cholerae*, *Entamoeba histolytica*, *Plasmodium falciparum* and many others. These microbial pathogens are not only a curse for human health but also a result in huge economic losses by affecting the health of economically important animals like poultry, cattle and other livestock. Considering the urgency, several effective antibiotics have been developed to combat microbial diseases and are available in the market. However, emergence of resistance against these drugs due to the maluses has created an alarming situation. In this scenario, the use of bioactive noble metal nanoparticles (silver, gold and platinum nanoparticles) has shown better therapeutic efficiency in terms of low treatment dose, less toxicity and absence of microbial resistance. Moreover, the use of several surface modifiers, coating and stabilizing agents resulted in enhancement of the bioactivity, rapid delivery and controlled drug release, improvement of biocompatibility and cytotoxicity. In this chapter, we have presented a comprehensive overview on the antimicrobial efficacy of noble metal nanoparticles along with the mechanistic insights behind their activity at the cellular and molecular level.

Nabarun Chandra Das, Bishnupada Roy, and Ritwik Patra have been contributed equally to this chapter.

N. C. Das · R. Patra · A. Choudhury · M. Ghosh · S. Mukherjee (✉)
Department of Animal Science, Kazi Nazrul University, Asansol, West Bengal, India
e-mail: suprabhat.mukherjee@knu.ac.in

B. Roy
Department of Chemistry, Visva-Bharati University, Santiniketan, West Bengal, India

N. R. Maddela et al. (eds.), *Nanotechnology for Advances in Medical Microbiology*, Environmental and Microbial Biotechnology,
https://doi.org/10.1007/978-981-15-9916-3_7

Keywords Noble metal nanoparticles · Antimicrobial activity · Drug delivery · Biocompatibility · Cytotoxicity

7.1 Introduction

The human healthcare industries throughout the world are facing tremendous challenges due to the increasing emergence of resistance in pathogenic bacteria, fungi and parasites against the conventional antimicrobial drugs. In particular, maluses of antimicrobial agents including inadequate dose and duration of treatment, long-term use as well as increasing environmental pollution are the major reasons (Sánchez-López et al. 2020). Moreover, an increasing number of deaths due to the infection of antibiotic-resistant pathogens are being reported almost every corner of the globe. Reports on new classes of resistance mechanism in life-threatening infectious pathogens, viz. *Mycobacterium tuberculosis*, *Helicobacter pylori*, *Vibrio cholerae*, *Entamoeba histolytica*, *Plasmodium falciparum* and many others, are continuously threatening mankind.

In this alarming situation, the exploration of noble metal nanoparticles has eventually come out as a boon for all. Bioactive nanoparticle itself or in combination with any antimicrobial drug can provide better intervention, especially against the drug-resistant pathogens (Dey et al. 2016; Chowdhury et al. 2018, 2020). Amongst the metal nanoparticles, silver (Ag), gold (Au) and platinum (Pt) are majorly explored for synthesizing therapeutic nanoparticles. These metal nanoparticles possess excellent antimicrobial activity and good penetration efficacy and exert a little or no toxic side effects on nontargeted cells and tissues (Dey et al. 2016; Chowdhury et al. 2018, 2020; Aziz et al. 2019; Inamuddin et al. 2021).

Another interesting feature in designing bioactive nanoparticles is appropriate surface modification. These metal nanoparticles (mainly Ag, Au) are usually functionalized with a variety of functional groups, such as polysaccharides, peptides, antibodies, RNA and DNA to promote their biomedical applications (Lee et al. 2020). Surface modification on metal nanoparticles provides several significant advantages. First, the modification provides an opportunity to stabilize nanoparticles against agglomeration. Second, it helps empower their self-organization, and, third, it creates interest to offer compatibility with others (Viswanathan et al. 2019). The clinical advantages achieved after surface modification are mentioned as good antimicrobial effect, high bioactivity, good cell growth and increased fatigue power (Izman et al. 2012). The organic ligands such as polysaccharides, peptides, amino acids, proteins, etc. are also considered as one of the good methods of surface modification to achieve the better outcomes. The organic groups are adequate to keep nanoparticles against accumulation; functional groups on nanoparticles surface may permit careful interaction of molecules with metal nanoparticles. The detailed working mechanisms of all these methods are explained previously in literature by many research groups such as Kango et al. (2013), Roy et al. (2014), Asri et al.

(2017), Chowdhury et al. (2018), Qi et al. (2017), Mozetič (2019), Oun et al. (2020) and Liu et al. (2020). Hitherto, surface-modified AgNPs and AuNPs have been successfully explored as efficient antimicrobial and anticancer agents, drug and gene delivery vehicles, radiotherapy enhancement agents, important component in diagnostic assays and imaging, and many other healthcare sectors (Zhang et al. 2016; Lee and Jun 2019; Prasad et al. 2016; Aziz et al. 2019). The surface properties of newly synthesized nanoparticles remain insufficient many times in terms of low biocompatibility, toxicity and weak adhesion properties. Therefore, proper fabrication approach and selection of appropriate surface-modifying/surface-capping agent is considered as key prior to aim any kind of practical applications.

There are many reported metallic (noble and non-noble) and nonmetallic antimicrobial agents available, but many of them are found toxic to most of the living organisms. Therefore, to overcome this problem, different inorganic and metal-based antibacterial agents with sustainability, enhanced stability and biocompatibility are synthesized under strict processing conditions (Rajawat and Qureshi 2012; Hossain et al. 2015; Vijayakumari et al. 2019). Currently, Ag and Au are the major metallic-based nanoparticles utilized as antibacterial agents because of their long-term stability and excellent biocompatibility. Studies have proved that metal-based nanoparticles show biocidal activity against Gram-negative and Gram-positive bacteria (Roy et al. 2014; Franci et al. 2015; Chiriac et al. 2016; Rajeshkumar et al. 2016; Wang et al. 2017; Ovais et al. 2019; Chowdhury et al. 2020). The antimicrobial effects of metal nanoparticles have been attributed to their nano-size and high surface-area-to-volume ratio, which permits them to penetrate the bacterial membranes (Prasad and Swamy 2013; Aziz et al. 2014, 2015, 2016). The mechanisms of antibacterial effect of metallic nanoparticles are metal ion release, oxidative stress and non-oxidative-based stress existing instantaneously. These nanoparticles can serve only when nanoparticles interact with microbe's cell walls; several approaches for the contact of microbes with the nanoparticles were used such as van der Waals forces, electrostatic attraction, receptor/ligand and hydrophobic interactions. After successful contact, metallic nanoparticles can pass through inner membranes, interact with metabolic paths and induce variations in membrane morphology. Once nanoparticles interact with microbes inside cellular machinery, it acts to prevent enzyme functions, disable proteins and electrolyte imbalance, induce oxidative stress and change gene expression scale (Vijayakumari et al. 2019). However, excessive quantity of microbes produces barrier that resists antimicrobial mediators and microbes avoiding the resistant system by forming superantigens. The extracellular polymeric secretion also produces everlasting attachment of microbes. In this chapter, we have presented a comprehensive overview on the therapeutic efficacy of noble metal nanoparticles against pathogenic microbes with a special emphasis on the mechanistic insights of their action at the biochemical, cellular and molecular level.

7.2 Chemistry of Noble Metal Nanoparticles

7.2.1 Design, Synthesis and Characterization

The work efficiency of noble metal nanoparticles mostly depends on the size of the nanoparticle and the stabilizing or capping or surface-modifying agents associated with it. The surface-modified nanomaterials should have biocompatibility to be useful for clinical purpose because without biocompatibility it cannot access the living cells. In addition to that, the synthesis procedure should be green in keeping the concern regarding environment in mind. There are two procedures for the synthesis of nanomaterial, i.e., top-down and bottom-up. Top-down is the scaling down of bulk material to the nano one by some mechanical process, and the bottom-up approach follows the reverse path. The process used in laboratories is the bottom-up approach. In this process, suitable bulk material containing the noble metal is reduced by a reducing agent and then scaled up to the nano one. A suitable capping agent is required here (Fig. 7.1). The efficiency of the capping material is very much important as it determines the size of the nanoparticles by stopping agglomeration to bulk. Chowdhury et al. (2018) presented the change in size distribution of AuNPs by using different polymeric (e.g., chitosan) and non-polymeric (e.g., tyrosine) substances as capping agents. The reduction procedure converting Ag^+ to Ag^0 also involves some amount of heat. Depending on the medium, reducing and capping

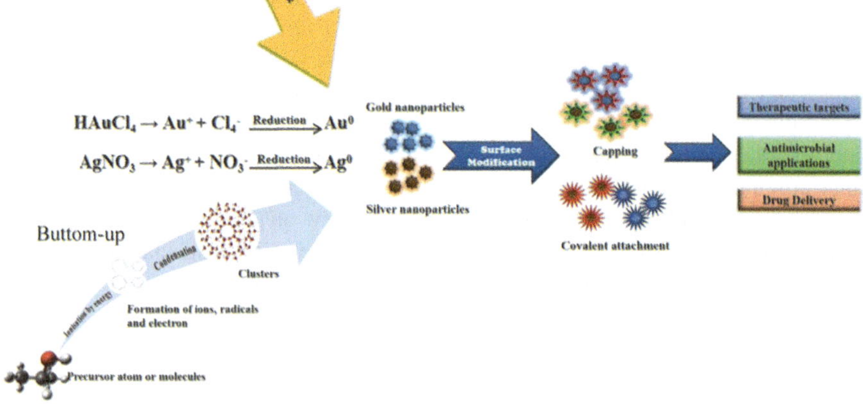

Fig. 7.1 Biosynthesis of noble metal nanoparticles, surface modification and its applications

agent, requirement of heat varies. Roy et al. (2014) showed the varying yield of nanoparticles depending on the source of heat.

Various inorganic and organic reducing and capping or surface-modifying agents are available globally. Some commonly used reducing agent are sodium borohydride, ß-D-glucose, starch, negatively charged heparin, saccharides, polyoxometalates, bamboo hemicelluloses, sodium citrate, ascorbic acid, potassium bitartrate, maltose, etc. On the other side, surface-modifying agents are organic thiol compounds, surfactants, long-chain amine, carboxylates, starch, gum Arabic, gelatin, carboxymethyl cellulose, hydrogels, etc. (Roy et al. 2014; Dey et al. 2015, 2016).

But keeping the environmental concern in mind, the protocol for synthesizing surface-modified noble metal nanoparticles should be so designed that the solution medium, reducing agent and heating procedure, remains environment-friendly, i.e., green synthesis procedure should be adapted complying with 12 fundamental principles of green chemistry which focuses on minimization or total elimination of generated hazardous waste and maximization of the efficiency of chemical processes without compromising the safety concern of the products (Roy et al. 2014).

In recent times, studies conducted by Roy et al. (2014) and Chowdhury et al. (2018, 2020) showed that surface-modified AgNPs and AuNPs synthesized through green route are very much efficient in exerting lethal action on microbes including Gram-positive and Gram-negative bacteria, pathogenic fungus (*Pichia guilliermondii*) as well as parasite (microfilaria of *Setaria cervi*). Noble metal nanoparticles produced from the microbes have also been used for antimicrobial purpose against pathogenic organisms. For example, Ag nanoparticles synthesized from *Bhargavaea indica*, *Brevibacterium frigoritolerans* and *Sporosarcina koreensis* exhibited antimicrobial properties against *Salmonella enterica*, *Vibrio parahaemolyticus*, *B. cereus*, *Bacillus anthracis* and *E. coli* (Singh et al. 2016). Similarly, bowl-shaped AgNPs synthesized by *Bacillus subtilis* have been shown to possess excellent antibacterial, antifungal and antifilarial activities (Dey et al. 2016).

For characterization of nanomaterials, a lot of methods are available. The first-hand information about the formation of nanoparticles can be obtained from the signature peak UV-visible spectroscopy. To obtain the size and structure, transmission electron microscopy (TEM) is used. Scanning electron microscope (SEM) gives the information about the surface topology and shape of the material, while dynamic light scattering (DLS) gives the size distribution of the nanomaterial. The zeta potential study tells us about the stability of the nanoparticles. If the potential is more than +30V or less than −30V, then the nanoparticles is highly stable (Roy et al. 2014; Chowdhury et al. 2018, 2020). X-ray diffraction and Fourier transform infrared spectrometry are used to study the interaction between the nanoparticles and capping/surface-modifying agents. Spectrofluorometer is used to study luminescent nanoparticles. Herein, we have included a representative figure (Fig. 7.2) containing most of the commonly used characterization data for surface (chitosan)-modified AuNP (named as GC).

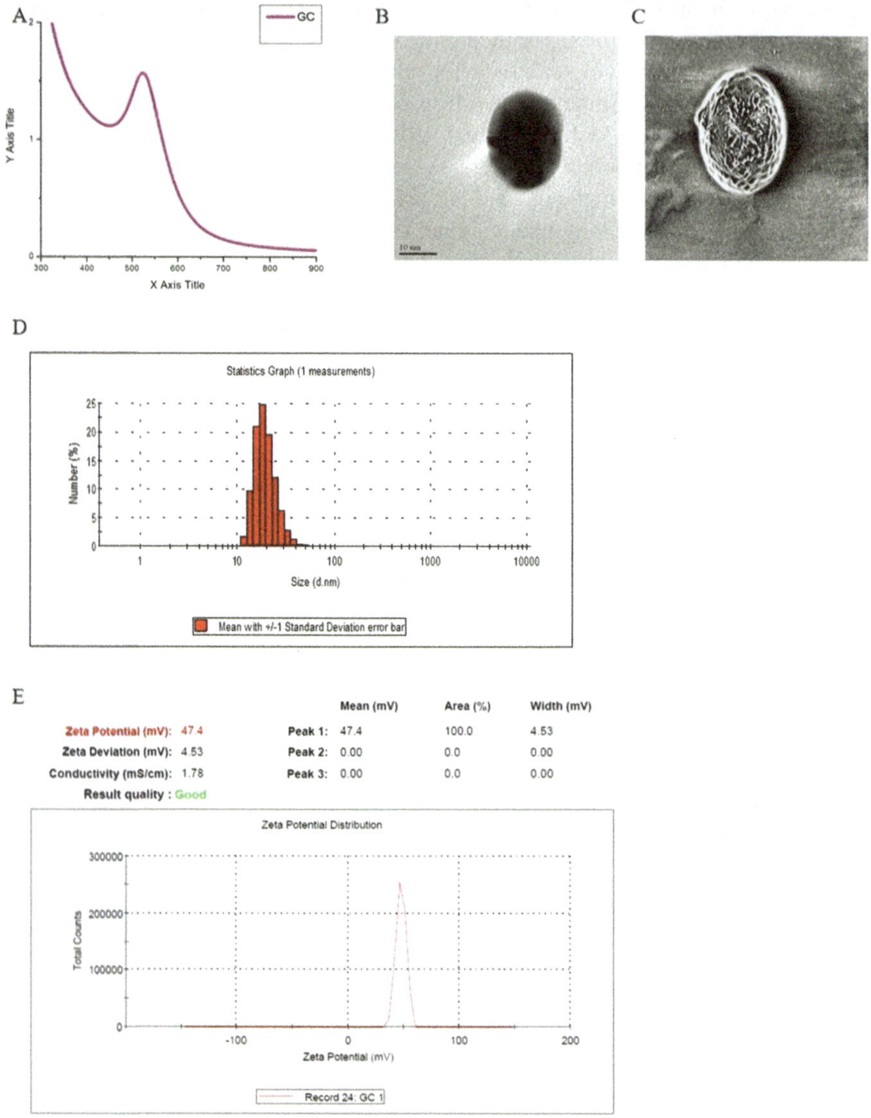

Fig. 7.2 Major characterization techniques for surface-modified noble metal nanoparticles. (**A**) UV-visible spectroscopic analysis. (**B**) TEM micrograph. (**C**) SEM study. (**D**) DLS study. (**E**) Zeta pot study

7.3 Facets of Nanoparticles-Microbe Interactions

Since 1928, the discovery of antibiotics in medical science has brought an epoch, but the use of metals from the historic age has been helping us keep our body healthy. Metals have been found to be used as water disinfectant, food preserver, surgical wound healer as well as in treatment of leprosy, tuberculosis, gonorrhea and syphilis; metals have several satisfactory aspects (Lemire et al. 2013). Metals have numerous beneficial roles in maintaining our health but have limited use as antimicrobial therapeutics. Regarding the reports of drug resistance as well as lack of new antibiotics, researchers have found to turn their works towards metals to develop antimicrobials. There are several antimicrobial agents like ammonium compound, *N*-halamine siloxanes and heterocyclic compounds. However, these molecules possess low efficiency along with toxic effects on the living organism as well as environment, and most importantly nontargeted actions made those agents as out-listed (Shahid-ul-Islam et al. 2016). In this context, the use of nanoparticles of silver, gold and platinum are found to reflect potent antimicrobial activity along with low toxicity to human cells, long-term durability and improved biocompatibility. The effectiveness of these nanoparticles is accredited to its composition and surface properties (Ugru et al. 2018). In this section, we have discussed about different facets of nanoparticle-microbe interaction with a special emphasis on AgNPs and AuNPs.

7.3.1 Interaction of AgNPs with Pathogenic Microbes and Their Antimicrobial Effect

Beside various successful histories in textile, food storing and environmental programmes (Wijnhoven et al. 2009), AgNPs with its broad-spectrum antimicrobial activity against bacteria, virus, fungi and parasites prove their ‘oligodynamic property’ (Gaiser et al. 2009). One of the fascinating features of AgNP is its affinity and capacity of attachment with the microbial membrane/surface. AgNPs can be easily attached with the bacterial cell wall due to the presence of carboxyl, phosphate and amino groups (Abbaszadegan et al. 2015). In this regard, it has been evidenced that Gram-positive bacteria are less susceptible than that of Gram-negative (Malanovic and Lohner 2016). Phagocytosis and passive diffusion are the two main ways for the AgNPs to enter inside the bacterial cell (AshaRani et al. 2009; Carlson et al. 2008). In addition, it has been reported that AgNP also exploits the copper transport system (CTR) to enter the bacterial cell (Ghandour et al. 1988). The cytotoxic effects of the AgNPs are majorly imparted through the damage of membrane structure, leakage of cellular components (specially cytoplasm), DNA damage, inhibition of respiratory chain and collapsing protein motive force (Bragg and Rainnie 1974; Sondi and Salopek-Sondi 2004; Morones et al. 2005; Prasad et al. 2011, 2012; Swamy and Prasad 2012). Size is an important criterion in the bioactivity and biocompatibility of

AgNP. The study of Morones et al. (2005) revealed that AgNPs of 1–10 nm can efficiently bind to the surface of *E. coli*, *V. cholera*, *S. typhus* and *P. aeruginosa* and impart lethal action on these bacteria. Beside size, the shape of AgNPs is also known to play a crucial role in exerting microbicidal activity. Truncated triangular shape and sharp-edge triangular shape show better result over spherical and rod shaped (Pal et al. 2007; Dong et al. 2012). On the other hand, hydrogel-capped hexagonal and bowl-shaped AgNPs also have been documented for higher bioactivity against both Gram-positive and Gram-negative bacteria at a very low dose (Dey et al. 2015, 2016).

Interaction with cell membrane following induction of lipid peroxidation (LPO) is considered as a principal attribute in bioactive AgNPs. AgNP-induced LPO is usually diagnosed by estimating malondialdehyde, the end product of LPO (Chowdhury et al. 2020). Free radicals generated from LPO further enters into the chain reaction to generate more reactive oxygen species (ROS) that collectively induce oxidative stress leading to death of the microbial cells (Hwang et al. 2008; Kora and Arunachalam 2011; Saha et al. 2016). This postulation has been experimentally shown in AgNP-induced growth inhibition of *E. coli* (Hwang et al. 2008) and *P. aeruginosa* (Kora and Arunachalam 2011). Sulphur affinity is also an important characteristic of AgNPs which also facilitates its binding with the membrane proteins (Roy et al. 2019).

Beside antibacterial activities, AgNPs also found to inhibit viral growth using varieties of ways. PVP-coated AgNPs of 1–10 nm size have a potent antiviral effect in inducing interaction with envelope glycoprotein gp120 against HIV-1 (Lara et al. 2010a), while HSV-1 was found to be inhibited by MES (mercaptoethane sulfonate)-coated silver NPs (Baram-Pinto et al. 2009). Besides these, poxvirus, hepatitis B virus and influenza viruses were also found to be repressed by AgNPs (Rogers et al. 2008; Papp et al. 2010; Lu et al. 2008).

Several groups of protozoa have also been demonstrated as the targets for AgNPs. *Plasmodium falciparum*, the causative protozoan for malaria also found to be controlled by several silver nanoformulations with a significant efficacy in culture condition (Rai et al. 2017). In addition, surface-modified AgNPs are also capable of killing larvae of mosquitoes in terms of combat malaria (Saha et al. 2016).

Interestingly, AgNPs also interact with the various extracellular and cellular components of microfilaria (microscopic larval form of filarial parasites). In recent past, surface-modified green AgNPs have been found as extremely potent antifilarial agents. For example, AgNPs capped with chitosan, polyvinyl alcohol and hydrogel have been documented to interact with the cell membrane, induce ROS generation and lead to death of the microfilaria (Saha et al. 2014, 2016). In addition, lipid-coated AgNPs have been shown to possess anti-Wolbachian activity to reduce microfilarial growth (Ali et al. 2013). Wolbachia is an endosymbiotic bacterium that controls many important physiological functions of the filarial parasite (Mukherjee et al. 2018).

Nevertheless, AgNPs with exposed coating show a significant oligodynamic effect, but graphene oxide-silver NP composite (Ag-GO)-coated nanoplatform exert a more stronger effect as antimicrobial along with low side effects (Ghosh et al. 2019). The study of Wierzbicki et al. (2019) disclosed a higher effectivity of Ag-GO over *E. coli*, *S. aureus* and *Staphylococcus epidermidis* than AgNPs. GO sheets are decorated on AgNPs via thiol groups which provide effective results against both Gram-negative and Gram-positive bacteria. While chitosan-capped AgNPs were also found as both antibacterial and antifungal when treated against *E. coli* and *P. guilliermondii* (Roy et al. 2014), there are also several synergistic applications of AgNPs and antibiotics with satisfactory outcomes. Surface modification of AgNPs with amoxicillin (Kirthi et al. 2019), cephradine (Masri et al. 2018), streptomycin (Kora and Rastogi 2013), ampicillin (Tippayawat et al. 2017), vancomycin and amikacin (Kaur and Kumar 2019) has reflected better efficacy than separate use of antibiotics or AgNPs against bacterial pathogens, viz. *S. aureus*, *C. albicans*, *Acinetobacter baumannii*, *Enterococcus faecalis*, *Mycobacterium tuberculosis* and *E. coli*, respectively. Recently conjugation of AgNPs with various antibiotics and anthelmintics has been reported to provide better therapeutic efficiency (Dey et al. 2016). Therapeutic potential of AgNPs including their interactions with different microbes as well as their molecular targets have been presented in Table 7.1.

7.3.2 AuNP-Microbe Interaction and Antimicrobial Effect of AuNPs

The use of gold nanoparticles (AuNPs) are highly advantageous in diagnosis (e.g., microscopy) and treatment of human diseases including microbial infection due to its ability to scatter light in visible light regions and relative nontoxic nature (Khan et al. 2014). Besides that, potency to detoxify pollutants along with uses in formulation of biosensors and disease markers makes it more precious (Dykman and Khlebtsov 2011; Lopez et al. 2004). Antimicrobial activity of the gold complexes is a century-old discovery (Glišić and Djuran 2014). Various studies have pointed out that nanoparticles or formulation formed of gold produce different outcomes that depend on the characteristics of the formulation. First of all, the size that majorly influences the bioactivity (Brayner et al. 2006): the smaller one resembles greater toxicity than the larger one (Lin et al. 2013) and also achieves better permeation through the microbial membrane (Lopez-Chaves et al. 2018). A study by Ahmad et al. (2013) distinctly envisioned the differences between 7 nm and 15 nm AuNPs in inhibiting *Candida* sp., whereas other studies also found to indicate same story between 15 and 35 nm AuNPs (AshaRani et al. 2011; Chen et al. 2013a, b). Regarding the shape of AuNPs, diverse opinions have been reported in the context of the antimicrobial effects of AuNPs. Few studies have demonstrated spherical AuNPs as more toxic than rod-shaped, while non-spherical AuNPs have also demonstrated better toxicity (Liu et al. 2018). A study by Sultana et al. (2015) showed that the toxicity of flower-shaped AuNPs is more efficacious than the spherical ones. However, the toxicity of AuNPs not only depends on size and

Table 7.1 Therapeutic applications of AgNPs as an antimicrobial agent

Reducing agent	Surface modification	Size and shape	Therapeutic applications	References
Peel extract of *Carica papaya*	–	10–35 nm and spherical	Against Gram-negative bacteria: *Escherichia coli* and *Klebsiella pneumonia* Against Gram-positive bacteria: *Staphylococcus aureus* and *Bacillus subtilis*	Kokila et al. (2016)
Tuber extract of *Curcuma longa*	–	18 ± 0.56 nm and spherical	Against Gram-negative bacteria: *E. coli* O157:H7 Against Gram-positive bacteria: *Listeria monocytogenes*	Alsammarraie et al. (2018)
Fruit extract of *Tamarind indica*	–	10 nm and spherical	Against Gram-negative bacteria: *K. pneumonia*, *Salmonella typhi*, *E. coli* and *Pseudomonas aeruginosa* Against Gram-positive bacteria: *B. subtilis*, *S. aureus*, *Micrococcus luteus* and *Bacillus cereus*	Jayaprakash et al. (2017)
Fruit extract of *Carambola* sp.	–	10–40 nm and spherical	Against Gram-negative bacteria: *E. coli* and *P. aeruginosa*	Gavade et al. (2015)
Leaf extract of *Azadirachta indica*	–	5–20 nm, spherical	Against Gram-negative bacteria: *E. coli* Against Gram-positive bacteria: *S. aureus*	Ahmed et al. (2016a)
Leaf extract of *Ocimum sanctum*	–	12–16 nm and spherical	Against Gram-negative bacteria	Jain and Mehata (2017)
Leaf extract of *Eriobotrya japonica*	–	20 nm and spherical	Against Gram-negative bacteria: *E. coli* Against Gram-positive bacteria: *S. aureus*	Rao and Tang (2017)
Leaf extract of *Lantana camara*	–	410–450 nm and spherical	Against Gram-negative bacteria: *E. coli* and *P. aeruginosa* Against Gram-positive bacteria: *S. aureus*	Shriniwas and Subhash (2017)
Extract of *Caulerpa racemosa* marine algae	–	5–25 nm, spherical and triangular	Against Gram-negative bacteria: *Proteus mirabilis* Against Gram-positive bacteria: *S. aureus*	Kathiraven et al. (2015)

Fungal biomass of *Penicillium polonicum* (ARA 10)	–	10–15 nm, spherical and oval	Against Gram-negative bacteria: *Salmonella enterica* serovar Typhimurium	Neethu et al. (2018)
Fungal biomass of *Phanerochaete chrysosporium* (MTCC-787)	–	34–90 nm, spherical and oval	Against Gram-negative bacteria: *P. aeruginosa* and *K. pneumonia* Against Gram-positive bacteria: *S. aureus* and *Staphylococcus epidermidis*	Saravanan et al. (2018)
Fungal biomass of *Trichoderma longibrachiatum*	–	Variable size and spherical	Against fungi: *Fusarium verticillioides*, *F. moniliforme*, *Penicillium brevicompactum*, *Helminthosporium oryzae* and *Pyricularia grisea*	Elamawi et al. (2018)
Mycelial cell filtrate of *Aspergillus brasiliensis*	–	6–21 nm and spherical	Against Gram-negative bacteria: *E. coli*, and *P. aeruginosa* Against Gram-positive bacteria: *B. subtilis* and *S. aureus* Against fungi: *Candida albicans*	Omran et al. (2018)
Cell-free supernatant of *Pseudomonas aeruginosa* (ATCC27853)	–	25–45 nm and spherical	Against Gram-negative bacteria: *E. coli* and *Acinetobacter baumannii* Against Gram-positive bacteria: *S. aureus*	Quinteros et al. (2016)
Mycelial cell filtrate of *Trichoderma atroviride* (KNUP001)	Amino acid capping	15–25 nm and variable shape	Against Gram-negative bacteria: *E. coli* and *P. aeruginosa* Against Gram-positive bacteria: *S. aureus* Against fungi: *C. albicans*	Kumar et al. (2018)
Tyrosine	Chitosan capped	13–22 nm and spherical	Against Gram-negative bacteria: *E. coli* Against fungi *Pichia guilliermondii* Against Microfilaria of *Setaria cervi*	Roy et al. (2014)
Tyrosine	Polyvinyl alcohol capped	13–15 nm	Against microfilaria of *S. cervi*	Saha et al. (2014)
–	Sodium 2-mercaptoethanesulfonate (MES)-capped	4 nm	Against *Herpes simplex virus type 1 (HSV-1)*	Baram-Pinto et al. (2009)
–	poly-N-vinyl-2-pyrrolidone (PVP)	1–10 nm	Against *Human immunodeficiency virus type 1 (HIV-1)*	Lara et al. (2010a)

(continued)

Table 7.1 (continued)

Reducing agent	Surface modification	Size and shape	Therapeutic applications	References
–	DNA-hydrogel capped and SHGel capped	20 nm and bowl shaped	Against Gram-negative bacteria, *E. coli* Against Gram-positive bacteria, *B. subtilis* Against fungi *P. guilliermondii* Against microfilaria of *S. cervi*	Dey et al. (2016)
–	GOSHGel capped	5 nm	Against Gram-negative bacteria, *E. coli* Against Gram-positive bacteria, *B. subtilis* Against fungi *P. guilliermondii*	Ghosh et al. (2019)
–	Ag-GO composite coated	Spherical	Against Gram-negative bacteria, *Salmonella enteritidis*	Wierzbicki et al. (2019)
Sodium borohydride ($NaBH_4$)	Citrate, SDS and PVP capped	21–70 nm and spherical	Increase activity of streptomycin, ampicillin and tetracycline in killing bacteria	Kora and Rastogi (2013)
Sodium borohydride (NaBH4)	Cephradine-conjugated silver nanoparticles (Ceph-AgNPs) and vildagliptin-conjugated silver nanoparticles (Vgt-AgNPs)	30–80 nm and spherical	Against Gram-negative bacteria, *E. coli* K1, *P. aeruginosa*, *K. pneumonia*	Masri et al. (2018)
Sodium borohydride ($NaBH_4$)	PVP, vancomycin and amikacin capped	5–35 nm and spherical	Against Gram-negative bacteria, *E. coli* Against Gram-positive bacteria, *S. aureus*	Kaur and Kumar 2019
Ethylene glycol	PVP coated	25 ± 4 nm and spherical	Against Gram-negative bacteria, *E. coli* Against Gram positive bacteria, *S. aureus*	Wang et al. (2016)
D-Maltose	D-Maltose coated	86.81 ± 13.39 nm	Against Gram-negative bacteria, *E. coli* Against Gram-positive bacteria, *S. aureus*	Tippayawat et al. (2017)
Sodium hydroxide	Amoxicillin coated	35.50 nm, spherical and oval	Against Gram-negative bacteria, *E. coli*	Kirthi et al. (2019)

shape but also on the surface chemistry, especially coating types and properties of the particles. According to Freese et al. (2012), AuNP coated with ethanediamine showed a better result in internalization of the particles inside the target cells. AuNP conjugated with different drugs have found to confer better delivery and treatment outcome. For example, AuNP-kanamycin complex applied against *S. epidermidis* and *Enterobacter aerogenes* revealed a very significant efficacy (Payne et al. 2016). On the other side, improved efficiency of levofloxacin against *S. aureus*, *P. aeruginosa* and *E. coli* has been reported after conjugating with AuNPs (Bagga et al. 2017). Similarly, gallic acid-AuNP conjugate has been found effective against pathogenic bacteria like *Shigella flexneri* and *Plesiomonas shigelloides* (Rattanata et al. 2016). AuNPs of 4 nm size with sodium 2-mercaptoethanesulfonate (MES) was reported to be lethal against virus HSV-1 (Baram-Pinto et al. 2010), while AuNPs coated with amphiphilic sulphate-ended ligand can inhibit growth of HIV-1 (Di Gianvincenzo et al. 2010).

AuNPs are also effective against protozoan and helminth parasites. In this connection, antileishmanial activity of AuNP with 30 nm size found to cause around 75% inhibition of the parasite count which indicates towards the potency of this nanoformulation in pharmaceutical industries (Ahmad et al. 2016). AuNPs in single or in conjugated with indolicidin (a short 13-residue antimicrobial and cytolytic peptide) have several successful applications against fungal growth (Ahmad et al. 2013; Rahimi et al. 2019). Alike AgNPs, surface modification also plays a critical role in regulating the bioactivities and biocompatibility of AuNPs. Recent studies by Chowdhury et al. (2018, 2020) revealed chitosan-coated AuNPs as excellent antimicrobial agent displaying lethal effects on bacteria, fungus and microfilaria at a relatively lower dose than that of uncapped AuNPs. Moreover, surface modification with chitosan also found to enhance the stability and biocompatibility of the AuNPs (Chowdhury et al. 2018, 2020).

AuNPs have been found to execute better performance in delivering drug with hampering the physiological homeostasis. CHrPfs25 (codon-harmonized recombinant Pfs25, a *Plasmodium falciparum* protein used as vaccine antigen) when delivered in conjugation with AuNPs induces the expression of malaria transmission-blocking antibodies (Kumar et al. 2015). As a delivery vehicle, AuNP is not only restricted within the delivery of drugs, but also it has been found suitable to deliver proteins, genes and vaccines (Kong et al. 2017). In order to develop antimicrobial activity, internalization, i.e., cellular uptake of the AuNPs, is the primary criterion, and it is usually achieved by the attachment to the cell surface following clathrin-mediated endocytosis, non-specific endocytosis and phagocytosis (Mironava et al. 2010). Interaction of AuNPs with microbial cell membrane resulting in distortion of the membrane architecture (Huo et al. 2016; Rattanata et al. 2016) that finally leads to leakage of the cellular components (Payne et al. 2016). Inhibition of microbial growth also found to be mediated by blocking the H^+-ATPase proton pumping (Wani et al. 2013) along with enhancing yield of ROS (Roy et al. 2018; Chowdhury et al. 2018). The study of Ahmad et al. (2015) showed how AuNP helps in elevating the number of ROS and in due time destroying cellular components to finally kill *Leishmania*. Moreover, AuNPs also can destroy transmembrane

Table 7.2 Therapeutic applications of AuNPs as an antimicrobial agent

Reducing agent	Surface modification	Size and shape	Therapeutic applications	References
Fruit extract of *Punica granatum*	–	5–17 nm, spherical and triangular	Against Gram-negative bacteria like *Salmonella typhi*, *Vibrio cholerae* and *Pseudomonas aeruginosa* Against Gram-positive bacteria, *Staphylococcus aureus* Against fungi *Aspergillus flavus* and *Candida albicans*	Lokina et al. (2014)
Fruit extract of *Solanum lycopersicum*	–	14 nm, diverse	Against Gram-negative bacteria, *P. aeruginosa* Against Gram-positive bacteria, *S. aureus*	Bindhu and Umadevi (2014)
Nuts extract of *Areca catechu*	–	13.7 nm and spherical	Against Gram-negative bacteria, *Escherichia coli*, *Klebsiella pneumonia*, *Enterobacter* and *P. aeruginosa* Against Gram-positive bacteria, *S. aureus*	Rajan et al. (2015)
Flowers of *Plumeria alba*	–	28 ± 5.6–15.6 ± 3.4 and spherical	Against Gram-negative bacteria, *E. coli*	Mata et al. (2016)
Root extract of *Trianthema decandra*	–	33–99 nm and triangular	Against Gram-negative bacteria, *E. coli*, *Proteus vulgaris*, *Yersinia enterocolitica* and *P. aeruginosa* Against Gram-positive bacteria, *B. subtilis* Against fungi, *C. albicans*	Geethalakshmi and Sarada (2013)
Root extract of *Mammea suriga*	–	22–50 nm and square	Against Gram-negative bacteria, *E. coli* and *P. aeruginosa* Against Gram-positive bacteria, *B. subtilis* and *S. aureus*	Poojary et al. (2016)
Leaf extract of *Euphorbia hirta* L.	–	6–71 nm and spherical	Against Gram-negative bacteria, *E. coli*, *K. pneumonia* and *P. aeruginosa*	Annamalai et al. (2013)
Leaf extract of *Solanum nigrum*	–	32 ± 6 nm and spherical	Against Gram-negative bacteria, *E. coli* and *P. aeruginosa*	Muthuvel et al. (2014)

			Against Gram-positive bacteria, *B. subtilis* and *Staphylococcus saprophyticus*	
Shell extract of *Chenopodium formosanum*	–	8 ± 6 nm and spherical	Against Gram-negative bacteria, *E. coli* Against Gram-positive bacteria, *S. aureus*	Chen et al. (2019)
Aerial parts of *Rivea hypocrateriformis*	–	10–50 nm and spherical	Against Gram-negative bacteria, *E. coli*, *P. aeruginosa* and *K. pneumonia* Against Gram-positive bacteria, *S. aureus* and *B. subtilis* Against *Chrysosporium indicum* and *Trichophyton rubrum*	Godipurge et al. (2016)
Rhizome extract of *Acorus calamus*	–	Lesser than 100 nm and spherical	Against Gram-negative bacteria, *E. coli* Against Gram-positive bacteria, *S. aureus*	Ganesan and Gurumallesh Prabu (2019)
Stem extract of *Hibiscus cannabinus*	–	13 nm and spherical	Against Gram-negative bacteria, *P. aeruginosa* Against Gram-positive bacteria, *S. aureus*	Bindhu et al. (2014)
Stem extract of *May-tenus royleanus*	–	30 nm and hexagonal	Against parasite *Leishmania* sp.	Ahmad et al. (2015)
Sclerotial extract of *Lignosus rhinocerotis*	–	10–25 nm, variable shape	Against Gram-negative bacteria, *P. aeruginosa* Against Gram-positive bacteria: *S. aureus*	Katas et al. (2019)
Sodium citrate	Kanamycin capped	20 ± 5 nm and spherical	Gram-negative bacteria, *Enterobacter aerogenes* Gram-positive bacteria, *Staphylococcus epidermidis*	Payne et al. (2016)
–	Bromelain capped and levofloxacin conjugated	38.11 ± 2 nm	Against Gram-negative bacteria, *E. coli* Against Gram-positive bacteria, *S. aureus*	Bagga et al. (2016)
–	Human serum albumin and levofloxacin conjugated	27.2 ± 1 nm	Against Gram-negative bacteria, *P. aeruginosa* and *E. coli* Against Gram-positive bacteria, *S. aureus*	Bagga et al. (2017)
–	Indolicidin conjugated	30 nm and spherical	Against fungi, *C. albicans*	Rahimi et al. (2019)

(continued)

Table 7.2 (continued)

Reducing agent	Surface modification	Size and shape	Therapeutic applications	References
Sodium borohydride ($NaBH_4$)	Sulphate-ended ligand coated	variable	Against human immunodeficiency virus (HIV)	Di Gianvincenzo et al. (2010)
Sodium borohydride ($NaBH_4$)	Amoxicillin coated	79 ± 43 nm	Against Gram-positive bacteria, *S. aureus*	Silvero et al. (2018)
–	Sodium 2-mercaptoethanesulfonate (MES) capped	4 nm	Against *Herpes simplex virus type 1* (HSV-1)	Baram-Pinto et al. (2010)
–	Galic acid conjugated	17 nm	Against Gram-negative bacteria, *Shigella flexneri* and *Plesiomonas shigelloides*	Rattanata et al. (2016)
Essential oil of *Nigella sativa* (NsEO)	NsEO coated	15.6 to 28.4 nm and spherical	Against Gram-positive bacteria, *S. aureus*	Manju et al. (2016)
Dextran	Dextran coated	22 ± 3 nm	Against Gram-negative bacteria, *E. coli*	Nath et al. (2008)
Piper nigrum extract	Chitosan capped	1–10 nm	Against microfilaria of *Setaria cervi*	Saha et al. (2017)
Terminalia chebula extract	Chitosan capped	10 nm	Against microfilaria of *Wuchereria bancrofti* and *S. cervi*	Roy et al. (2018)
Chitosan	Chitosan functionalized	10–50 nm	Against microfilaria of *S. cervi*	Chowdhury et al. (2018)

electrostatic efflux (Li et al. 2010) to execute their antimicrobial activities. Herein, Table 7.2 describes the composition and antimicrobial effects of the AuNPs available till date.

7.3.3 *Antimicrobial Effect of PtNPs*

The novel nano-tool PtNP is now in top of interest. Platinum is now being used extensively in automotive and chemical industries to develop catalytic convertor and new chemical compounds (Shi et al. 2015). For instance, platinum is also found to use in generating eco-friendly energy sources (Madsen et al. 2011). Beside of all these, several reports have also highlighted that platinum is also beneficial for pharmaceutical industries as platinum has activity in producing bio-imaging, detecting biological molecules and modulating nanomedicine (Tanaka et al. 2011; Moglianetti et al. 2016; Rao et al. 2016). Several physical and physio-chemical properties like, size, shape, surface structure and capping agent along with the dispersion state and stability help optimize and finally lead to formulate the desired PtNPs with specific target-based activity. However antibacterial activity of the platinum first came in public in 1965 (Pedone et al. 2017), but antimicrobial activity of the nanostructured platinum has not been explored yet. PVP-conjugated PtNPs with 1 to 3 nm size have been reported as effective against the bacterium, *P. aeruginosa* (Gopal et al. 2013). In another study, pectin-capped PtNPs showed an efficacy against both Gram-positive and Gram-negative bacteria (Pedone et al. 2017). Green synthesis of PtNPs using *Garcinia mangostana* fruit extract also found as successful in inhibiting the growth of various bacteria like *P. aeruginosa*, *K. pneumonia*, *B. subtilis* and *S. aureus* (Nishanthi et al. 2019). There are limited a number of reports published yet, but effective radical-scavenging property will make PtNP as the most success nanomedicine in the upcoming future. However, the use of PtNP as biomedicine is still in contradictory stage. Antimicrobial activities of PtNPs alongside their interactions of with different microbes based on the available reports are listed in Table 7.3.

7.4 Intracellular and Intercellular Targeted Delivery of Nanoparticles

The intracellular and intercellular delivery of nanoparticles is the most important aspect in the development of therapeutic nanoparticles for various applications like its use as antimicrobial and immunomodulatory drugs, anticancer agents, cellular modulators and nanodevices for studying cell organelles (Paulo et al. 2011; Prasad et al. 2017). The development of smart noble nanoparticles requires appropriate surface modifications which facilitate the use of the nanoparticles directly as

Table 7.3 Therapeutic applications of PtNPs as an antimicrobial agent

Reducing agent	Surface modification	Size and shape	Therapeutic applications	References
Polyaniline and Ag-Pt bimetallic colloidal	–	2–3 nm	Against Gram-positive bacteria: *S. aureus* and *Streptococcus* sp.	Boomi et al. (2013)
Extract of seaweed *Padina gymnospora*	–	25 nm, truncated, octahedral, tetrahedral and spherical	Against Gram-negative bacteria: *Escherichia coli*, *Salmonella typhi* and *Klebsiella pneumonia* Against Gram-positive bacteria: *Staphylococcus aureus*, *Streptococcus pneumoniae*, *Lactococcus lactis* and *Streptococcus mutants*	Ramkumar et al. (2017)
Fruit extract of *Garcinia mangostana*	–	20–25 nm and spherical	Against Gram-negative bacteria: *P. aeruginosa* and *K. pneumonia* Against Gram-positive bacteria: *B. subtilis and S. aureus*	Nishanthi et al. (2019)
Hexachloroplatinate and apigenin	–	1–2 nm and spherical	Against Gram-negative bacteria: *P. aeruginosa* Against Gram positive bacteria: *S. aureus*	Gurunathan et al. (2019)
Sodium hydroxide	PVP	1–3 nm, sphere, cuboids and flower shaped	Against Gram-negative bacteria: *P. aeruginosa*	Gopal et al. (2013)
Sodium borohydride ($NaBH_4$)	Pectin capped	2–5 nm	Against Gram-negative bacteria: *Escherichia coli* and *P. aeruginosa* Against Gram-positive bacteria: *B. subtilis* and *S. aureus*	Ahmed et al. (2016b)
Plant extract of *Taraxacum laevigatum*	Phytochemicals capped	2–7 nm, spherical	Against Gram-negative bacteria: *Pseudomonas aeruginosa* Against Gram-	Tahir et al. (2017)

(continued)

Table 7.3 (continued)

Reducing agent	Surface modification	Size and shape	Therapeutic applications	References
			positive bacteria: *Bacillus subtilis*	
Sodium borohydride ($NaBH_4$)	Jacalin capped	3.1 ± 1.6 nm	Against Gram-negative bacteria: *Aeromonas hydrophila*	Ahmed et al. (2018)
Sodium hydroxide	Curcumin stabilized	3.8 nm	Against Gram-negative bacteria: *E. coli* Against Gram-positive bacteria: *S. aureus*	Yu et al. (2019)
Doxycycline	Doxycycline capped	10–20 nm	Against Gram-negative bacteria: *E. coli, Salmonella typhimurium* Against Gram-positive bacteria: *Streptococcus pyogenes* and *S. aureus*	Safdar et al. (2020)

therapeutics and/or cargo for the drugs (Karimi et al. 2016). In order to deliver the therapeutics to different intracellular cellular targets, internalization of AgNPs/AuNPs involves binding with the membrane receptors leading to receptor-mediated endocytosis (Panzarini et al. 2018). In addition, indirect incorporation through hydrophobic and electrostatic interaction with phospholipid bilayer has also been reported (Chou et al. 2011). Upon entering the cytoplasm, the mobility of the ingested nanoparticles depends on the size and biological interactions with the various organic and inorganic constituents of cytoplasm and targeted organelles. In this context, peptide conjugation with the nanoparticles enhances its distribution across the cell by recognizing the nuclear localization signal (NLS), mitochondrial localization signal and trafficking to endoplasmic reticulum (ER) (Paulo et al. 2011). Taking clue from AgNPs and AuNPs, superparamagnetic iron oxide NPs (SPIONs) are nowadays used along with the mitochondrial targeting peptide (MTP) to differentiate the intracellular proteins and play an important function in regulating the cellular trafficking across the endocytotic pathway, localization of the protein within the plasma membrane and for the cellular uptake of basic amino acids (Salaklang et al. 2008).

In recent times, AuNPs, AgNPs and silica nanoparticles have been utilized/attempted for targeting various biomarkers associated with cancer, autoimmune and infectious microbial and parasitic diseases. Owing to its surface properties and affinity towards cell surface molecules (as discussed in the earlier section), AgNPs can easily cross the cell membrane and can deliver a drug of choice. However, this phenomenon is dependent on the size of the AgNPs. Previously, Dey et al. (2016) demonstrated the uptake of hydrogel-capped AgNPs inside by mouse macrophages, and the presence of AgNPs within the cell was confirmed by HR-TEM following EDX. These AgNPs were also reported to load streptomycin, diethylcarbamazine and albendazole (Dey et al. 2016). On the other side, nontoxic nature of AuNPs enables smooth delivery of drugs which can control the protein expression within the cell and used as an intracellular sensor in many diseases (Rosi et al. 2006). AuNPs formed by poly(γ-glutamic acid) conjugated with L-phenylalanine (40–200 nm) are assigned for the protein delivery through absorption and release within the cytoplasm and can be used as an agent for the vaccine development and signalling pathway modulators (Akagi et al. 2011). The polymeric NPs consist of the self-assembly structure and dendrimers loaded with hydrophilic and hydrophobic drugs and conjugated to targeting moieties to serve as an active cellular target for human diseases, mainly cancer (Nag and Delehanty 2019). Thus, the application of both AgNPs and AuNPs for the intracellular and intercellular targeting for specific cellular and subcellular components enhances the potential of these nanomaterials as molecular fluorescent agents, detection probes and the therapeutic factors for drug development to counteract various human diseases.

Several developmental research and modifications are currently going on to improve the efficacy of the nanoparticles, especially the cell-penetrating property. Quantum dots (QDs) are the best examples of the modern version of nanotherapeutics. QDs are semiconductor crystals that are made up of the group II–VI, III–V or IV–VI atoms from the periodic table and perform an extensive function of size-tunable fluorescent emission and broad excitation spectra, leading to beneficial for single as well as multiple molecule tracking (Ruedas-Rama et al. 2012). The quantum dot coated with poly(ethylene glycol) (PEG) and NLS is microinjected showing active transport and accumulation across the nucleus and when coated with fluorescent protein shows association with cell line across the cytoplasm (Derfus et al. 2004; Medintz et al. 2008). The microinjection of phospholipid-coated quantum dots to the early embryo of *Xenopus* to monitor the physiological changes along the development leads to developmental abnormalities (Dubertret et al. 2002). The main characteristic features of liposomes include high biocompatibility, simple surface modifications and flexibility across switching between both hydrophobic and hydrophilic drugs. Liposomes can easily incorporate the functional phospholipids across various target moieties and increase its efficiency for cellular targeting and therapeutic efficacy (Gabizon et al. 2006; Nag and Delehanty 2019). Poly(lactic-co-glycolic acid) (PLGA) serves as both an intercellular and intracellular target and is being used for co-staining the actin and mitochondria using antibody-conjugated quantum dots and MitoTracker Red, respectively (Chou et al. 2011). It is distributed across the early endosomes, Golgi apparatus and endoplasmic reticulum based on the cellular internalization and types of cells to enhance the specificity of drug delivery and therapeutic development (Cartiera et al. 2009). In a recent report by Mondal et al. (2019), luminous benign QDs have been reported for excellent intracellular imaging and delivery of antimicrobial drug (streptomycin). These streptomycin-loaded QDs were found to cure

Table 7.4 Intracellular and intercellular targets of noble metal nanoparticles and their mechanism of antimicrobial action

Types of nanoparticles	Mechanism of action and damage	References
Silver NP	Internalized via scavenger receptor-mediated phagocytosis, mitochondrial damage, induces apoptosis and cell death	Singh and Ramarao (2012)
Gold NP	Passive delivery, controls protein expression in cell, nontoxic	Rosi et al. (2006)
Silica NP	Inhibits kinase activity by delivering antibody against phospho-Akt intracellularly, translation inhibitors ribosome-inactivating proteins (RIPs)	Bale et al. (2010)
Quantum dot	Microinjected quantum dot coated with florescent protein, phospholipid coat, penetrates peptide, vesicle fusion, reminiscent actin/kinesin-mediated active transport	Ruan et al. (2007)
Liposomes	Targeting cancer cells, incorporate functional phospholipids and enhance cellular targeting	Patil et al. (2016)
PLGA	Dispersed across endosome, Golgi apparatus, ER, causes cell internalization, greater intracellular drug accumulation	Cartiera et al. (2009)
Superparamagnetic iron oxide NPs (SPIONs)	Mitochondrial targeting, regulate cellular trafficking, endocytotic pathway	Salaklang et al. (2008)
NPs formed by poly(γ-glutamic acid) conjugated with L-phenylalanine (40–200 nm)	Protein delivery: absorb protein and release in cytoplasm, targeted for vaccine development, signalling pathway interference	Akagi et al. (2011)
Polymeric NPs	Active cellular targeting for cancer, loaded with hydrophilic and hydrophobic drugs and conjugated to targeting moieties	Nag and Delehanty (2019)
Poly(propyleneimine) (PPI) dendrimers	siRNA-mediated cancer cell targeting, accumulation of siRNA in the cytoplasm of cancer cells and gene silencing	Taratula et al. (2009)

peritonitis in mice model, and the efficacy was higher than free streptomycin (Mondal et al. 2019). The various types of nanoparticles associated with the targeted delivery across the intracellular and intercellular components of cells are listed in Table 7.4.

7.5 Cellular and Molecular Mechanism of Antimicrobial Action of Noble Nanoparticles

Considering the broad-spectrum antimicrobial activity of the noble metal nanoparticles, the mechanisms of action of different nanoparticles have been studied at cellular and molecular levels employing both in vitro and in vivo experimental setup. Till date a number of approaches have been exploited to synthesize noble metal nanoparticles, and intriguingly the surface-modified nanoparticles have been found to possess more efficient antimicrobial activity (Roy et al. 2014; Chowdhury et al. 2018; Prasad et al. 2020). Various studies conducted on exploring the antimicrobial activity of noble metal nanoparticles revealed AgNPs as more reactive than AuNPs, while AuNPs are more benign than that of the AgNPs (Roy et al. 2014; Chowdhury et al. 2018). Interestingly surface modification of AgNPs and AuNPs by polysaccharide (like chitosan) was found to improve the bioactivities of both nanoparticles (Roy et al. 2014; Chowdhury et al. 2018, 2020). Firstly, such enhancement in bioactivity is not solely contributed by the surface-modifying agent, rather the capping agent helps in the movement of the encapsulated nanoparticles in it the biological medium; secondly, it gets easily attached at the binding sites of the cell membrane of the targeted cell; and, thirdly, it remains attached for a long time favouring the release of the particles into the cell. All of these activities of the surface-modifying agent helping the targeted drug delivery in turn help the nanoparticles reach the requisite site and execute the task.

Once the size of particles approaches the nanoscale (1–100 nm), various a typical classical and quantum mechanical phenomena appear to be extant which are usually absent in matter beyond the upper and lower limits of this size range. This lays down the foundation of the antimicrobial properties of the noble nanoparticles. It is well observed that these physiochemical properties owe to and are regulated by their physical characteristics like size, shape, overall crystal structure, surface-area-to-mass ratio, as well as the ζ-potential at their slipping planes. However, mechanisms of action of the nanoparticles can be generalized under three varied models such as reactive oxygen species (ROS) induction, non-ROS mechanisms and metal ion release mechanisms as we have depicted in Fig. 7.3. However, it is to be well noted that a nanoparticle may function through one or more mechanisms simultaneously.

In general, noble metal nanoparticles first interact with the membrane of the target microbial cells. This interaction results in the induction of lipid peroxidation (LPO) of membrane lipids. LPO generates several free radicals that initiates chain reaction to generate more radicals by damaging the cellular biomacromolecules. In fact, free radical species have a very special property of generating furthermore free radicals by cleaving existing covalent bonds in the surrounding molecules, and the resulting free radicals produce more such species in a chain reaction fashion; this occurs at an exponential rate. It wreaks havoc in any biochemical system by disrupting the existing bonding interactions holding the system together, and this capacity is known as oxidative stress.

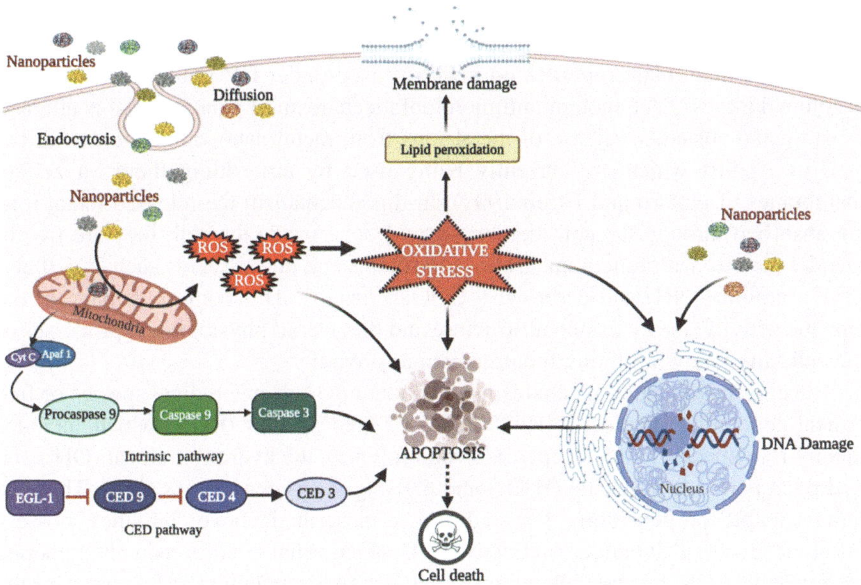

Fig. 7.3 Mechanistic insights of the function of surface-modified noble metal nanoparticles at a cellular and molecular level

Additionally, quantitative real-time polymerase chain reaction (RT-PCR) studies show that ROS also increases the expression levels of a general stress response gene (Dna K) and two oxidative stress genes (Kat A and Ahp C) of oxidative proteins, which thereby causes extensive damage to the intracellular components (Gurunathan et al. 2012). ROS production can be induced by noble nanoparticles (Gurunathan et al. 2012) via different mechanisms—one such mechanism is illustrated in the photocatalytic model; it states that when nanoparticles absorb photons of energy equal to or greater than the band gap, their electrons get promoted to the conduction band from the valence band leaving behind a hole in the valence band—which are definite theoretical antivalents of the physical electrons and thus carry a positive charge and flow against the direction of electron movement. These holes are present on the surface of the metal oxides and when they react with the surrounding H_2O or (OH^-) or O_2 species, they produce hydroxyl ($OH^{\cdot}$) and superoxide radicals (O_2^-). Furthermore, it was also observed that ultrasonic activation is yet another mechanism for inducing ROS formation by which nanoparticles split the surrounding H_2O into H^+ and can react with dissolved O_2 to generate H_2O_2. Studies show that the negatively charged superoxide and hydroxyl radicals are unable to penetrate the cell membrane and function while being attached to the cell surface; H_2O_2 on the other hand is able to penetrate the cell membrane and cause extensive cellular damage. As already mentioned, ROS axes are not the only mechanisms by which the antimicrobial properties of the nanoparticles are mediated. Observations against several studies using Fourier transform infrared (FTIR) analysis, electron spin resonance, liquid chromatography-mass spectrometry, transmission electron microscopy (TEM), proteomics tools and flat cultivation show that various nanoparticles have efficient antimicrobial properties no matter if used under UV light, natural light or absolute darkness. Yet another antimicrobial mechanism of nanomaterial is attained by slow and sustained release of metal ions from metal nanocomposites, as in the case of AgNPs which are currently being used by embedding them on zeolite membranes (Tavolaro and Drioli 1999). In this mechanism the released metal ions are absorbed through the cell membrane, and once inside the cell, they are free to interact directly with functional groups of proteins and nucleic acids such as sulfanyl (–SH), amino (–NH) and carboxyl (–COOH) groups. This confers damage to the enzyme activity, the cytoskeletal structure and the overall physiological processes of the cell, ultimately inhibiting the organism as a whole.

The cells have natural mechanisms of producing such free radical species or free radical-generating species, called reactive oxygen species (ROS) which includes mainly four species like the superoxide anion (O_2^-), the hydroxyl radical ($OH^·$), the hydrogen peroxide molecule (H_2O_2) and the singlet oxygen species ($^1O_2$). They are quite capable of generating free radicals as described above, but they possess different levels of dynamics and activity. Under normal conditions, cells generate cytoplasmic ROS, but this alteration in cellular redox potential is balanced by the generation of antioxidant species which counters the exponential increase of ROS, keeping the species concentrations within redox equilibrium limits. This equilibrium is useful in maintaining regular cellular homeostasis and appears to play an important role in cellular signalling and disease pathophysiology. However, upsetting this equilibrium produces oxidative stress which results in a change in permeability and integrity of the cell membrane mostly by causing peroxidation of the membrane lipids (Cheloni et al. 2016); it also deals damage to nucleic acids and various proteins. NPs as filaricidal diminished the activities of enzymes SOD, catalase and GPx that are most vital in antioxidant defence mechanism (Jeeva et al. 2015). On the other side, GSH is a sulphur-containing protein that functions as antioxidant by carrying the reactive electrons from the peroxide and its level is usually depleted after filaricidal induction (Mukherjee et al. 2016). But, an inclined level of glutathione-S-transferase (GST) maintains a high ROS level in mitochondria, endoplasmic reticulum and peroxisome which is known to signal apoptosis (Mukherjee et al. 2016).

ROS-mediated apoptosis has a close relation with DNA damage, while noble nanoparticles also display DNA-binding property. Previous studies in this direction revealed that chitosan-capped and supramolecular hydrogel-capped AgNPs preferentially bind at the minor groove of bacterial as well as parasitic DNA (Roy et al. 2014; Dey et al. 2016). Therefore, AgNP-induced ROS and direct binding of AgNP to microbial DNA result in p53 activation that most likely signals activation of apoptotic pathways. In case of microfilaria, chitosan-/hydrogel-capped AgNPs activate cell death abnormal (CED) pathway, and caspase mediates pathways to cause cell death (Roy et al. 2014; Dey et al. 2016). Egg-laying defective (EGL)-1, CED-3, CED-4 and CED-9 are four essential proteins essential for CED pathway in microfilaria. During apoptosis CED-9 is negatively regulated by EGL-1 and fails to show dominancy over CED-4 and CED-3 (Shaham and Horvitz 1996). In normal-living filarial cell, CED-4 dimers are sequestered with CED-9 on the outer surface of the mitochondria and inhibit apoptosis (Lettre and Hengartner 2006). In time, stimulation after an apoptotic induction increased the level of EGL-1 and makes bonding with CED-9 by BH3 domain that in turn disrupt CED-4–CED-9 complex (Yan et al. 2004, 2005). After dissociation of CED-9, two asymmetric CED-4 dimers oligomerize to make a tetrameric apoptosome, and this tetrameric structure then recruits proCED-3 molecules (Huang et al. 2013). Next to that, CED-3 (a cystine protease) becomes activated and executes apoptosis. A study by Mukherjee et al. (2016) revealed a rich level of cystine protease family protein, caspases, which direct a new path suitable for microfilarial apoptosis. Caspase proteins, viz. cas-9, cas-8 and cas-3, cytochrome c and poly (ADP-ribose) polymerase (PARP) are the main mediators that are responsible for intrinsic and extrinsic apoptosis. The generalized model of the mechanism of action of noble metal nanoparticles is demonstrated in Fig. 7.4.

7.6 Therapeutic Promises of Nanoparticles as Antimicrobial Agents, Prospects and Challenges

Nanotechnologies most specifically nanoparticles are now in good demand in pharmaceutical industries. This new therapeutic strategy can deliver drug most accurately and exert the desired function at a very low dose. As we all know, treatment of HIV needs antiretroviral drugs, ritonavir, lopinavir and efavirenz, but they have very low sustainability in physiological conditions. In this context, the use of nanoparticles formulated using poly(lactic-co-glycolic acid) (PLGA) can enhance the sustained release of the anti-HIV drugs from 48 h to 4 weeks (Rizvi and Saleh 2018). Similar kind of phenomena also found as evidence when poly(lactic-co-glycolic) (PLG) AgNPs encapsulated with rifampin, isoniazid and pyrazinamide drugs were used against tuberculosis. Detection of rifampicin for 4 days in blood and 9 days in tissues along with 9 to 11 days of retention for isoniazid and pyrazinamide in blood and tissue strongly indicates higher potency of nanoformulation over unbound drugs (Gelperina et al. 2005). Moreover, esculentin-1a-capsulated nanostructures are found as 17 times more effective than free esculentin-1a as anti-*P. aeruginosa* therapeutic (Yeh et al. 2020). Another study on chitosan-coated AgNP-conjugated form with ciprofloxacin displayed a better MIC than the free drug against *E. coli* (Kumar et al. 2016), while the encapsulated daptomycin form can enhance the release time up to 4 h (Silva et al. 2015). Besides that, phosphate- and polyphosphate-conjugated PEG nanoparticles also can sustain the release of phosphate to 100 h (Yin et al. 2017). According to the findings of Fan et al. (2019), polyethylene glycol-functionalized AuNP when applied in conjugation with ampicillin almost 18% lower MIC was found than ampicillin alone against *S. aureus*. Several surface-modified nanoparticles have also been reported for affecting

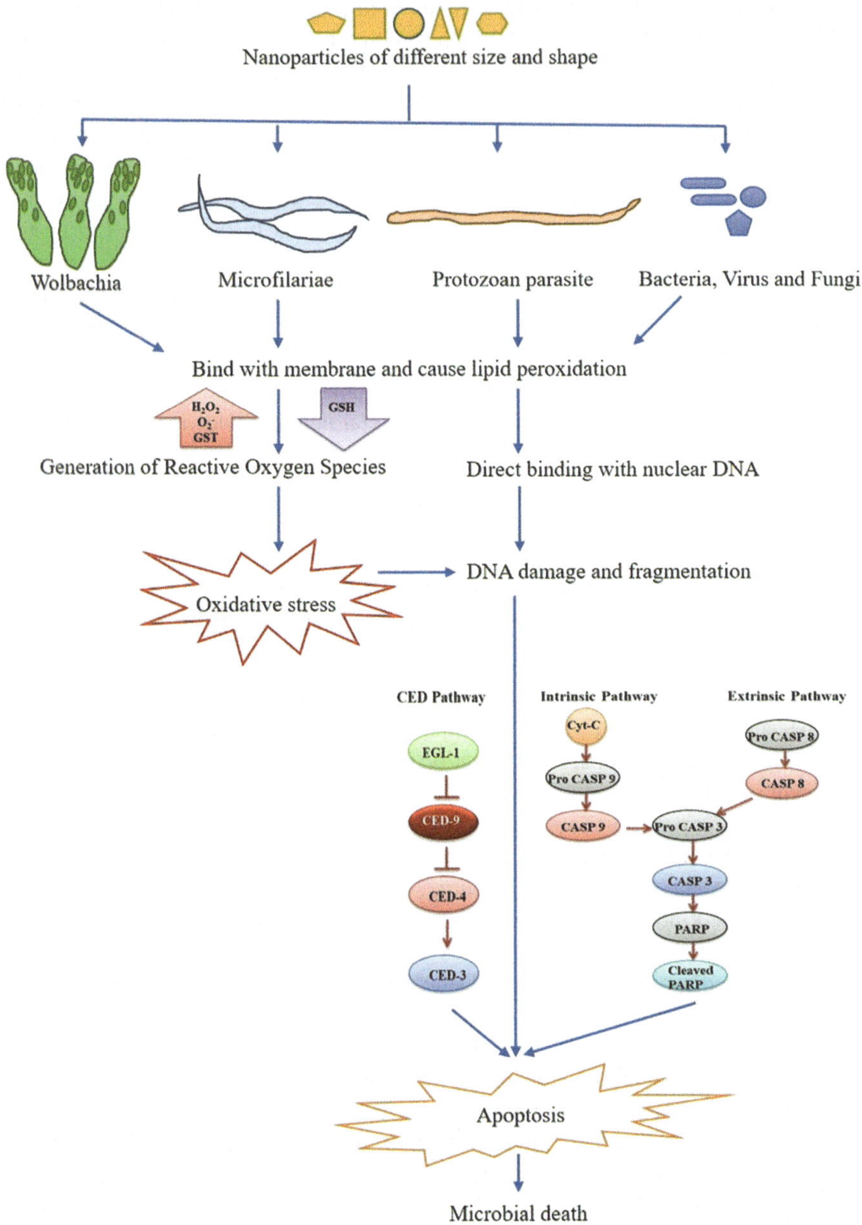

Fig. 7.4 Insights of ROS-mediated apoptosis induced by noble metal nanoparticles

bacterial biofilm. PLGA/chitosan-coated nanoparticles when conjugated with colistin anti-*P. aeruginosa* biofilm effect found to last for 72 h that is much an improved report than free colistin (d'Angelo et al. 2015). Interestingly, 0.0156 μg/ml of ciprofloxacin encapsulated with PLGA was found sufficient to eradicate *P. aeruginosa* biofilm within 3 days only (Baelo et al. 2015).

Bacterial synthesis of nanoparticles is now considered as a new trend as bacteria are able to hydrolyse metal compounds and reduce metal ions to form nano in a green way. While the study of Prakash et al. (2011) showed the level of potentiality of AgNP synthesized by *Bacillus cereus* in diminishing the growth of *E. coli* and *Streptococcus*. Another work summarized that only 5 μg/ml of concentrated AgNP of *Bacillus* is sufficient to generate antimicrobial activity against *E. coli*, *Pseudomonas aeruginosa*, *Proteus mirabilis*, *Serratia marcescens* and *Klebsiella* sp. (Yokesh Babu et al. 2013). A report by Dey et al. (2016) demonstrated a synthesis of flower-shaped novel AgNPs by *B. subtilis*, and these AgNPs were found to be active even at a dose of 1.25 μg/ml against bacteria, fungus and parasite. In biological methods of nanoparticles, formulation with bacteria is not only a single option, but the use of fungus also has several successes in works. A recent study on sclerotial extract of *Lignosus rhinocerotis*-mediated synthesis of AuNPs explained very high antimicrobial activity against Gram-negative *P. aeruginosa* and *E. coli* and Gram-positive bacteria *S. aureus* and *Bacillus* sp. (Katas et al. 2019).

Nanoparticle exploitation is not only limited to bacterial, viral and fungal inhibition, but it is also in evidence that they have much potency in eradicating microparasites. The study of Saini et al. (2016) and Saha et al. (2017) found that AgNP as well as AuNP both have the ability to eliminate microfilariae of *Setaria cervi*. Similarly, the study of Roy et al. (2018) showed the efficacy of *Terminalia chebula* extract. AuNP fully depends on degradation of nuclear DNA. This data was further supported by the work of Yadav et al. (2020) that described AgNPs synthesized using *Andrographis paniculata* leaf extract elevate the ROS level and generate oxidative stress that finally leads to filarial death via induction of apoptosis. In addition, *Wolbachia* depletion using polyanhydrated nanoparticle as delivery medium of doxycycline can also reduce microfilarial abundance at a very low dose. Besides these, it has also been reported that the use of transferrin-conjugated solid lipid-coated AgNPs is a more effective antimalarial than unconjugated form (Gupta et al. 2007), while a little improvement was documented when AgNPs were used with violacein as anti-plasmodium (Rahman et al. 2019).

All of the reports or findings descripted above are showing the improvement of therapeutics, and all the success is due to development of nanoformulation (Prasad et al. 2019). Effective and specific targeting, electrostatic interactions, stabilizing and reducing capabilities and drug delivery role along with toxic nature to microbes with low side effects to humans play a crucial role to achieve the therapeutic potentiality. Even the interaction property with intracellular components helps enhance the therapeutic ability and that finally differs it from other treatments. The DNA-binding ability of the nanoparticles creates a milestone in medicinal science. Polymer-stabilized and surface-modified (using chitosan, polyethylene glycol, polyvinyl alcohol and styrene) AgNPs can bind to *E. coli* DNA molecule and disrupt that to inhibits bacteria replication (Roy et al. 2014). In support another study of Li et al. (2013) clearly visualized with AFM topography that AgNPs and citrate-modified

AuNPs are capable of binding to microbial DNA molecule to cease the DNA replication.

The use of nanoparticles is advantageous, but the synthesis procedure should follow biological eco-friendly as well as benign green synthesis approach rather than heftiest physical and chemical processes. Moreover, surface modification using modifying agents has an opportunistic effect to ameliorate toxicity of the nanoparticles. For example, chitosan-functionalized AgNPs/AuNPs suggest that the use of chitosan minimizes the cytotoxic level of the noble metal nanoparticle along with an increase of the bioactivity activity inducing ROS generation (Roy et al. 2014; Chowdhury et al. 2018, 2020). In addition, two interesting studies conducted by Dey et al. (2015, 2016) revealed capping of AgNPs using supramolecular hydrogel (SHGel) and DNA hydrogel increases the bioactivity and decreases the toxic effects. Other coating agents like PVP and citrate also have been reported to minimize the cytotoxicity of noble nanoparticles (Akter et al. 2017). All of these surface modifications also enable the drug-loading capacity of the noble nanoparticles of metal and improvement of the efficacy of loaded drug. The antimicrobial activity of amoxicillin, penicillin G, clindamycin, vancomycin and erythromycin was found to be improved owing to the conjugation of polymer-coated AgNPs (Rai et al. 2009). Though noble nanoparticles possess an excellent ability to inhibit microbial growth in vivo and in vitro, still there are some contradictions against their use as therapeutics regarding the side effects and bioavailability. In this connection, so modern approaches of nanotechnologies like tunable nanoparticles, quantum dots, nanogels, etc. are evolving rapidly to meet the current need.

Microbial diseases are the continuous threat to all living creatures from the prehistoric times. Discovery of antibiotics was a challenge to ride above those diseases. But accumulation of resistance against antibiotics has created a difficult situation nowadays. While in several experiments, the use of noble metal nanoparticles provides a hope that they can be effective over microbes. In addition, metal and metal oxide-formulated nanoparticles have been found as the potent killer of multidrug-resistant bacteria as well. The study of Franci et al. (2015) pointed the ability of AgNPs against methicillin-resistant *S. aureus* (MRSA) and methicillin-resistant *S. epidermidis* (MRSE), while efficacious report against ampicillin-resistant *E. coli*, multidrug-resistant *Pseudomonas aeruginosa* and erythromycin-resistant *Streptococcus pyogenes* highlighted AgNPs as a potent antibiofilm and antibacterial formulation (Lara et al. 2010b). In this regard AuNPs are also proved as highly active as several studies showed effectivity in inhibition of growth of multidrug-resistant bacterial strain of *E. coli*, *S. aureus* and *Salmonella typhimurium* (Bresee et al. 2011; Dasari et al. 2015). Despite these successes, more and more researches on developing new smart noble metal nanoparticles as well as new approaches for tuning the size and physico-biochemical properties of the nanoparticles also are in progress to counteract the emerging microbial disease.

7.7 Conclusion and Future Directions

In the mid-2020, the whole world is still under pressure of several microbial diseases. Antimicrobial drug development is therefore considered as one of the most practicable research works at present time. Even most of the pharma companies are investing to develop better therapeutic solutions against the life-threatening infectious diseases caused by *Mycobacterium tuberculosis*, *Helicobacter pylori*, *Vibrio cholerae*, *Entamoeba histolytica*, *Plasmodium falciparum*, and many others. These microbial pathogens are not only a curse for human health but also a result in huge economic losses by affecting the health of economically important animals like poultry, cattle and other livestock. Considering the urgency, several effective antibiotics have been developed to combat microbial diseases and are available in the market. However, emergence of resistance against these drugs due to the maluses has created an alarming situation. In this scenario, the use of bioactive noble metal nanoparticles has shown better therapeutic efficiency in terms of low treatment dose, less toxicity and absence of microbial resistance. Moreover, the use of several surface modifiers, coating and stabilizing agents, resulted in enhancement of the bioactivity, rapid delivery and controlled drug release, improvement of biocompatibility and cytotoxicity. In this chapter, we have presented a comprehensive overview on the antimicrobial efficacy of noble metal nanoparticles along with the mechanistic insights behind their activity at a cellular and molecular level.

Making or formulation of nanoparticles for their commercialization prospect, more specifically for modulating nanomedicine and various biological applications, is now the most rapidly growing field in nanotechnology. Several companies are investing to develop various pharmaceutical applications on the basis of drug delivery, imaging and bio-diagnostic properties of the noble metal nanoparticles. Nanoparticles with drug delivery ability have been found as the most successful over other modes of treatment in the last few years. Even the exploitation is not only limited as antimicrobial, but NP is also found to be advantageous for cancer treatment.

Considering the COVID-19 pandemic, noble metal nanoparticles could be utilized as therapeutics as well as drug delivery vehicles. Some interesting findings on the efficacy of bioactive AgNPs and AuNPs against the antigenic proteins of SARS and MERS (Kim et al. 2018; Lin et al. 2019; Sekimukai et al. 2020) indicated that these nanoparticles could be the useful options to treat COVID-19. Particularly, rapid delivery of anti-SARS-CoV-2 antibody and/or vaccine could be aimed for COVID-19 treatment shortly.

Taken together, it is clearly evident that surface-modified noble metal nanoparticles, especially AgNPs and AuNPs, are advantageous for therapeutic uses due to their benign nature, broad-spectrum bioactivities, ability of conjugating/immobilizing several drugs/enzymes, tunable delivery and release of drugs, imaging/luminating potential and several other physio-biochemical attributes. Future research is therefore needing more emphasis in applying these nanoparticles as in situ nanotrackers or nanobiosensors for diagnosing as well as treating life-threatening diseases of humans.

Acknowledgement BR and SM acknowledge all the collaborators with whom they have previously worked. RP thanks the Department of Higher Education, Govt. of West Bengal, for awarding Swami Vivekananda Merit Cum Means Fellowship.

References

Abbaszadegan A, Ghahramani Y, Gholami A et al (2015) The effect of charge at the surface of silver nanoparticles on antimicrobial activity against gram-positive and gram-negative bacteria: a preliminary study. J Nanomater 2015:720654. https://doi.org/10.1155/2015/720654

Ahmad T, Wani IA, Lone IH et al (2013) Antifungal activity of gold nanoparticles prepared by solvothermal method. Mater Res Bull 48:12–20. https://doi.org/101016/jmaterresbull201209069

Ahmad A, Syed F, Shah A et al (2015) Silver and gold nanoparticles from *Sargentodoxa cuneata*: synthesis, characterization and antileishmanial activity. RSC Adv 5:73793–73806. https://doi.org/10.1039/C5RA13206A

Ahmad A, Syed F, Imran M et al (2016) Phytosynthesis and antileishmanial activity of gold nanoparticles by *Maytenus royleanus*. J Food Biochem 40:420–427. https://doi.org/10.1111/jfbc.12232

Ahmed S, Saifullah, Ahmad M et al (2016a) Green synthesis of silver nanoparticles using *Azadirachta indica* aqueous leaf extract. J Radiat Res Appl Sci 9:1–7. https://doi.org/10.1016/j.jrras.2015.06.006

Ahmed KBA, Raman T, Anbazhagan V (2016b) Platinum nanoparticles inhibit bacteria proliferation and rescue zebrafish from bacterial infection. RSC Adv 6:44415–44424. https://doi.org/10.1039/C6RA03732A

Ahmed KBA, Raman T, Veerappan A (2018) Jacalin capped platinum nanoparticles confer persistent immunity against multiple *Aeromonas* infection in zebrafish. Sci Rep 8:2200. https://doi.org/10.1038/s41598-018-20627-3

Akagi T, Shima F, Akashi M (2011) Intracellular degradation and distribution of protein-encapsulated amphiphilic poly(amino acid) nanoparticles. Biomaterials 32:4959–4967. https://doi.org/10.1016/j.biomaterials.2011.03.049

Akter M, Sikder MT, Rahman MM et al (2017) A systematic review on silver nanoparticles-induced cytotoxicity: physicochemical properties and perspectives. J Adv Res 9:1–16. https://doi.org/10.1016/j.jare.2017.10.008

Ali M, Afzal M, Bhattacharya SM et al (2013) Nanopharmaceuticals to target antifilarials: a comprehensive review. Expert Opin Drug Deliv 10:665–678. https://doi.org/10.1517/17425247.2013.771630

Alsammarraie FK, Wang W, Zhou P et al (2018) Green synthesis of silver nanoparticles using turmeric extracts and investigation of their antibacterial activities. Colloids Surf B Biointerfaces 171:398–405. https://doi.org/10.1016/j.colsurfb.2018.07.059

Annamalai A, Christina VLP, Sudha D et al (2013) Green synthesis, characterization and antimicrobial activity of Au NPs using Euphorbia hirta L. leaf extract. Colloids Surf B Biointerfaces 108:60–65. https://doi.org/10.1016/j.colsurfb.2013.02.012

AshaRani PV, Low Kah Mun G, Hande MP, Valiyaveettil S (2009) Cytotoxicity and genotoxicity of silver nanoparticles in human cells. ACS Nano 3:279–290. https://doi.org/10.1021/nn800596w

AshaRani PV, Lianwu Y, Gong Z, Valiyaveettil S (2011) Comparison of the toxicity of silver, gold and platinum nanoparticles in developing zebrafish embryos. Nanotoxicology 5:43–54. https://doi.org/10.3109/17435390.2010.489207
Asri RIM, Harun WSW, Samykano M et al (2017) Corrosion and surface modification on biocompatible metals: a review. Korean J Couns Psychother 77:1261–1274. https://doi.org/10.1016/j.msec.2017.04.102
Aziz N, Fatma T, Varma A, Prasad R (2014) Biogenic synthesis of silver nanoparticles using *Scenedesmus abundans* and evaluation of their antibacterial activity. J Nanoparticles, Article ID 689419. https://doi.org/10.1155/2014/689419
Aziz N, Faraz M, Pandey R, Sakir M, Fatma T, Varma A, Barman I, Prasad R (2015) Facile algae-derived route to biogenic silver nanoparticles: synthesis, antibacterial and photocatalytic properties. Langmuir 31:11605–11612. https://doi.org/10.1021/acs.langmuir.5b03081
Aziz N, Pandey R, Barman I, Prasad R (2016) Leveraging the attributes of *Mucor hiemalis*-derived silver nanoparticles for a synergistic broad-spectrum antimicrobial platform. Front Microbiol 7:1984. https://doi.org/10.3389/fmicb.2016.01984
Aziz N, Faraz M, Sherwani MA, Fatma T, Prasad R (2019) Illuminating the anticancerous efficacy of a new fungal chassis for silver nanoparticle synthesis. Front Chem 7:65. https://doi.org/10.3389/fchem.2019.00065
Baelo A, Levato R, Julián E et al (2015) Disassembling bacterial extracellular matrix with DNase-coated nanoparticles to enhance antibiotic delivery in biofilm infections. J Control Release 209:150–158. https://doi.org/10.1016/j.jconrel.2015.04.028
Bagga P, Ansari TM, Siddiqui HH et al (2016) Bromelain capped gold nanoparticles as the novel drug delivery carriers to aggrandize effect of the antibiotic levofloxacin. EXCLI J 15:772–780. https://doi.org/10.17179/excli2016-710
Bagga P, Siddiqui HH, Akhtar J et al (2017) Gold nanoparticles conjugated levofloxacin: for improved antibacterial activity over levofloxacin alone. Curr Drug Deliv 14:1114–1119. https://doi.org/10.2174/1567201814666170316113432
Bale SS, Kwon SJ, Shah DA et al (2010) Nanoparticle-mediated cytoplasmic delivery of proteins to target cellular machinery. ACS Nano 4:1493–1500. https://doi.org/10.1021/nn901586e
Baram-Pinto D, Shukla S, Perkas N et al (2009) Inhibition of herpes simplex virus type 1 infection by silver nanoparticles capped with mercaptoethane sulfonate. Bioconjug Chem 20:1497–1502. https://doi.org/10.1021/bc900215b
Baram-Pinto D, Shukla S, Gedanken A, Sarid R (2010) Inhibition of HSV-1 attachment, entry, and cell-to-cell spread by functionalized multivalent gold nanoparticles. Small 6:1044–1050. https://doi.org/10.1002/smll.200902384
Bindhu MR, Umadevi M (2014) Antibacterial activities of green synthesized gold nanoparticles. Mater Lett 120:122–125. https://doi.org/10.1016/j.matlet.2014.01.108
Bindhu MR, Vijaya Rekha P, Umamaheswari T, Umadevi M (2014) Antibacterial activities of *Hibiscus cannabinus* stem-assisted silver and gold nanoparticles. Mater Lett 131:194–197. https://doi.org/10.1016/j.matlet.2014.05.172
Boomi P, Prabu HG, Mathiyarasu J (2013) Synthesis and characterization of polyaniline/Ag–Pt nanocomposite for improved antibacterial activity. Colloids Surf B Biointerfaces 103:9–14. https://doi.org/10.1016/j.colsurfb.2012.10.044
Bragg PD, Rainnie DJ (1974) The effect of silver ions on the respiratory chain of *Escherichia coli*. Can J Microbiol 20:883–889. https://doi.org/10.1139/m74-135
Brayner R, Ferrari-Iliou R, Brivois N et al (2006) Toxicological impact studies based on Escherichia coli bacteria in ultrafine ZnO nanoparticles colloidal medium. Nano Lett 6:866–870. https://doi.org/10.1021/nl052326h
Bresee J, Maier KE, Boncella AE et al (2011) Growth inhibition of *Staphylococcus aureus* by mixed monolayer gold nanoparticles. Small 7:2027–2031. https://doi.org/10.1002/smll.201100420

Carlson C, Hussain SM, Schrand AM et al (2008) Unique cellular interaction of silver nanoparticles: size-dependent generation of reactive oxygen species. J Phys Chem B 112:13608–13619. https://doi.org/10.1021/jp712087m

Cartiera MS, Johnson KM, Rajendran V et al (2009) The uptake and intracellular fate of PLGA nanoparticles in epithelial cells. Biomaterials 30:2790–2798. https://doi.org/10.1016/j.biomaterials.2009.01.057

Cheloni G, Marti E, Slaveykova VI (2016) Interactive effects of copper oxide nanoparticles and light to green alga *Chlamydomonas reinhardtii*. Aquat Toxicol 170:120–128. https://doi.org/10.1016/j.aquatox.2015.11.018

Chen H, Dorrigan A, Saad S et al (2013a) In vivo study of spherical gold nanoparticles: inflammatory effects and distribution in mice. PLoS One 8:e58208. https://doi.org/10.1371/journal.pone.0058208

Chen J, Wang H, Long W et al (2013b) Sex differences in the toxicity of polyethylene glycol-coated gold nanoparticles in mice. Int J Nanomedicine 8:2409–2419. https://doi.org/10.2147/IJN.S46376

Chen M-N, Chan C-F, Huang S-L, Lin Y-S (2019) Green biosynthesis of gold nanoparticles using *Chenopodium formosanum* shell extract and analysis of the particles' antibacterial properties. J Sci Food Agric 99:3693–3702. https://doi.org/10.1002/jsfa.9600

Chiriac V, Stratulat DN, Calin G et al (2016) Antimicrobial property of zinc-based nanoparticles. IOP Conf Ser Mater Sci Eng 133:12055. https://doi.org/10.1088/1757-899x/133/1/012055

Chou LYT, Ming K, Chan WCW (2011) Strategies for the intracellular delivery of nanoparticles. Chem Soc Rev 40:233–245. https://doi.org/10.1039/C0CS00003E

Chowdhury P, Roy B, Mukherjee N et al (2018) Chitosan biopolymer functionalized gold nanoparticles with controlled cytotoxicity and improved antifilarial efficacy. Adv Compos Hybrid Mater 1:577–590. https://doi.org/10.1007/s42114-018-0040-7

Chowdhury P, Roy B, Mukherjee S et al (2020) Polymer anchored gold nanoparticles: synthesis, characterization and antimicrobial activities. Nanosci Nanotechnol – Asia 10:1–13. https://doi.org/10.2174/2210681210666200128155244

d'Angelo I, Casciaro B, Miro A et al (2015) Overcoming barriers in *Pseudomonas aeruginosa* lung infections: engineered nanoparticles for local delivery of a cationic antimicrobial peptide. Colloids Surf B Biointerfaces 135:717–725. https://doi.org/10.1016/j.colsurfb.2015.08.027

Dasari TPS, Zhang Y, Yu H (2015) Antibacterial activity and cytotoxicity of gold (I) and (III) ions and gold nanoparticles. Biochem Pharmacol open access 4:199. https://doi.org/10.4172/2167-0501.1000199

Derfus AM, Chan WCW, Bhatia SN (2004) Intracellular delivery of quantum dots for live cell labeling and organelle tracking. Adv Mater 16:961–966. https://doi.org/10.1002/adma.200306111

Dey B, Mondal RK, Mukherjee S et al (2015) A supramolecular hydrogel for generation of a benign DNA-hydrogel. RSC Adv 5:105961–105968. https://doi.org/10.1039/C5RA19172F

Dey B, Mukherjee S, Mukherjee N et al (2016) Green silver nanoparticles for drug transport, bioactivities and a bacterium (*Bacillus subtilis*)-mediated comparative nano-patterning feature. RSC Adv 6:46573–46581. https://doi.org/10.1039/C5RA27886D

Di Gianvincenzo P, Marradi M, Martínez-Avila OM et al (2010) Gold nanoparticles capped with sulfate-ended ligands as anti-HIV agents. Bioorg Med Chem Lett 20:2718–2721. https://doi.org/10.1016/j.bmcl.2010.03.079

Dong PV, Ha CH, Binh LT, Kasbohm J (2012) Chemical synthesis and antibacterial activity of novel-shaped silver nanoparticles. Int Nano Lett 2:9. https://doi.org/10.1186/2228-5326-2-9

Dubertret B, Skourides P, Norris DJ et al (2002) In vivo imaging of quantum dots encapsulated in phospholipid micelles. Science 298:1759–1762. https://doi.org/10.1126/science.1077194

Dykman LA, Khlebtsov NG (2011) Gold nanoparticles in biology and medicine: recent advances and prospects. Acta Naturae 3:34–55. https://doi.org/10.32607/20758251-2011-3-2-34-55

Elamawi RM, Al-Harbi RE, Hendi AA (2018) Biosynthesis and characterization of silver nanoparticles using *Trichoderma longibrachiatum* and their effect on phytopathogenic fungi. Egypt J Biol Pest Control 28:28. https://doi.org/10.1186/s41938-018-0028-1

Fan Y, Pauer AC, Gonzales AA, Fenniri H (2019) Enhanced antibiotic activity of ampicillin conjugated to gold nanoparticles on PEGylated rosette nanotubes. Int J Nanomedicine 14:7281–7289. https://doi.org/10.2147/IJN.S209756

Franci G, Falanga A, Galdiero S et al (2015) Silver nanoparticles as potential antibacterial agents. Molecules 20:8856–8874. https://doi.org/10.3390/molecules20058856

Freese C, Gibson MI, Klok H-A et al (2012) Size- and coating-dependent uptake of polymer-coated gold nanoparticles in primary human dermal microvascular endothelial cells. Biomacromolecules 13:1533–1543. https://doi.org/10.1021/bm300248u

Gabizon AA, Shmeeda H, Zalipsky S (2006) Pros and cons of the liposome platform in cancer drug targeting. J Liposome Res 16:175–183. https://doi.org/10.1080/08982100600848769

Gaiser BK, Fernandes TF, Jepson M et al (2009) Assessing exposure, uptake and toxicity of silver and cerium dioxide nanoparticles from contaminated environments. Environ Health 8:S2. https://doi.org/10.1186/1476-069X-8-S1-S2

Ganesan RM, Gurumallesh Prabu H (2019) Synthesis of gold nanoparticles using herbal *Acorus calamus* rhizome extract and coating on cotton fabric for antibacterial and UV blocking applications. Arab J Chem 12:2166–2174. https://doi.org/10.1016/j.arabjc.2014.12.017

Gavade SJM, Nikam GH, Dhabbe RS et al (2015) Green synthesis of silver nanoparticles by using carambola fruit extract and their antibacterial activity. Adv Nat Sci Nanosci Nanotechnol 6:45015. https://doi.org/10.1088/2043-6262/6/4/045015

Geethalakshmi R, Sarada DVL (2013) Characterization and antimicrobial activity of gold and silver nanoparticles synthesized using saponin isolated from *Trianthema decandra* L. Ind Crop Prod 51:107–115. https://doi.org/10.1016/j.indcrop.2013.08.055

Gelperina S, Kisich K, Iseman MD, Heifets L (2005) The potential advantages of nanoparticle drug delivery systems in chemotherapy of tuberculosis. Am J Respir Crit Care Med 172:1487–1490. https://doi.org/10.1164/rccm.200504-613PP

Ghandour W, Hubbard JA, Deistung J, Hughes MN, Poole RK (1988) The uptake of silver ions by *Escherichia coli* K12: toxic effects and interaction with copper ions. Appl Microbiol Biotechnol 28:559–565. https://doi.org/10.1007/BF00250412

Ghosh D, Dhibar S, Dey A et al (2019) Graphene oxide dispersed supramolecular hydrogel capped benign green silver nanoparticles for anticancer, antimicrobial, cell attachment and intracellular imaging applications. J Mol Liq 282:1–12. https://doi.org/10.1016/j.molliq.2019.03.010

Glišić BĐ, Djuran MI (2014) Gold complexes as antimicrobial agents: an overview of different biological activities in relation to the oxidation state of the gold ion and the ligand structure. Dalton Trans 43:5950–5969. https://doi.org/10.1039/C4DT00022F

Godipurge SS, Yallappa S, Biradar NJ et al (2016) A facile and green strategy for the synthesis of Au, Ag and Au–Ag alloy nanoparticles using aerial parts of R. hypocrateriformis extract and their biological evaluation. Enzyme Microb Technol 95:174–184. https://doi.org/10.1016/j.enzmictec.2016.08.006

Gopal J, Hasan N, Manikandan M, Wu H-F (2013) Bacterial toxicity/compatibility of platinum nanospheres, nanocuboids and nanoflowers. Sci Rep 3:1260. https://doi.org/10.1038/srep01260

Gupta Y, Jain A, Jain SK (2007) Transferrin-conjugated solid lipid nanoparticles for enhanced delivery of quinine dihydrochloride to the brain. J Pharm Pharmacol 59:935–940. https://doi.org/10.1211/jpp.59.7.0004

Gurunathan S, Han JW, Dayem AA et al (2012) Oxidative stress-mediated antibacterial activity of graphene oxide and reduced graphene oxide in *Pseudomonas aeruginosa*. Int J Nanomedicine 7:5901–5914. https://doi.org/10.2147/IJN.S37397

Gurunathan S, Jeyaraj M, Kang M-H, Kim J-H (2019) Graphene oxide-platinum nanoparticle nanocomposites: a suitable biocompatible therapeutic agent for prostate cancer. Polymers (Basel) 11:733. https://doi.org/10.3390/polym11040733

Hossain KMZ, Patel U, Ahmed I (2015) Development of microspheres for biomedical applications: a review. Prog Biomater 4:1–19. https://doi.org/10.1007/s40204-014-0033-8

Huang W, Jiang T, Choi W et al (2013) Mechanistic insights into CED-4-mediated activation of CED-3. Genes Dev 27:2039–2048. https://doi.org/10.1101/gad.224428.113

Huo S, Jiang Y, Gupta A et al (2016) Fully zwitterionic nanoparticle antimicrobial agents through tuning of core size and ligand structure. ACS Nano 10:8732–8737. https://doi.org/10.1021/acsnano.6b04207

Hwang ET, Lee JH, Chae YJ et al (2008) Analysis of the toxic mode of action of silver nanoparticles using stress-specific bioluminescent bacteria. Small 4:746–750. https://doi.org/10.1002/smll.200700954

Inamuddin, Ahamed MI, Prasad R (2021) Advanced antimicrobial materials and applications. Springer, Singapore.ISBN: 978-981-15-7098-8. https://www.springer.com/gp/book/9789811570971

Izman S, Abdul-Kadir MR, Anwar M et al (2012) Surface modification techniques for biomedical grade of titanium alloys: oxidation, carburization and ion implantation processes. In: Titanium alloys-towards achieving enhanced properties for diversified applications 42:201–228. https://doi.org/10.5772/36318

Jain S, Mehata MS (2017) Medicinal plant leaf extract and pure flavonoid mediated green synthesis of silver nanoparticles and their enhanced antibacterial property. Sci Rep 7:15867. https://doi.org/10.1038/s41598-017-15724-8

Jayaprakash N, Vijaya JJ, Kaviyarasu K et al (2017) Green synthesis of Ag nanoparticles using tamarind fruit extract for the antibacterial studies. J Photochem Photobiol B Biol 169:178–185. https://doi.org/10.1016/j.jphotobiol.2017.03.013

Jeeva JS, Sunitha J, Ananthalakshmi R et al (2015) Enzymatic antioxidants and its role in oral diseases. J Pharm Bioallied Sci 7:S331–S333. https://doi.org/10.4103/0975-7406.163438

Kango S, Kalia S, Celli A et al (2013) Surface modification of inorganic nanoparticles for development of organic–inorganic nanocomposites—a review. Prog Polym Sci 38:1232–1261. https://doi.org/10.1016/j.progpolymsci.2013.02.003

Karimi M, Ghasemi A, Sahandi Zangabad P et al (2016) Smart micro/nanoparticles in stimulus-responsive drug/gene delivery systems. Chem Soc Rev 45:1457–1501. https://doi.org/10.1039/c5cs00798d

Katas H, Lim CS, Nor Azlan AYH et al (2019) Antibacterial activity of biosynthesized gold nanoparticles using biomolecules from *Lignosus rhinocerotis* and chitosan. Saudi Pharm J SPJ Off Publ Saudi Pharm Soc 27:283–292. https://doi.org/10.1016/j.jsps.2018.11.010

Kathiraven T, Sundaramanickam A, Shanmugam N, Balasubramanian T (2015) Green synthesis of silver nanoparticles using marine algae *Caulerpa racemosa* and their antibacterial activity against some human pathogens. Appl Nanosci 5:499–504. https://doi.org/10.1007/s13204-014-0341-2

Kaur A, Kumar R (2019) Enhanced bactericidal efficacy of polymer stabilized silver nanoparticles in conjugation with different classes of antibiotics. RSC Adv 9:1095–1105. https://doi.org/10.1039/C8RA07980C

Khan AK, Rashid R, Murtaza G et al (2014) Gold nanoparticles: synthesis and applications in drug delivery. Trop J Pharm Res 13:1169–1177. https://doi.org/10.4314/tjpr.v13i7.23

Kim Y-S, Son A, Kim J et al (2018) Chaperna-mediated assembly of ferritin-based middle east respiratory syndrome-coronavirus nanoparticles. Front Immunol 9:1093. https://doi.org/10.3389/fimmu.2018.01093

Kirthi AV, Karthik L, Venkatesan J, Chaskar AC (2019) 'Synergistic-cidal' effect of amoxicillin conjugated silver nanoparticles against *Escherichia coli*. bioRxiv:832568. https://doi.org/10.1101/832568

Kokila T, Ramesh PS, Geetha D (2016) Biosynthesis of AgNPs using *Carica papaya* peel extract and evaluation of its antioxidant and antimicrobial activities. Ecotoxicol Environ Saf 134:467–473. https://doi.org/10.1016/j.ecoenv.2016.03.021

Kong F-Y, Zhang J-W, Li R-F et al (2017) Unique roles of gold nanoparticles in drug delivery, targeting and imaging applications. Molecules 22:1445. https://doi.org/10.3390/molecules22091445

Kora AJ, Arunachalam J (2011) Assessment of antibacterial activity of silver nanoparticles on *Pseudomonas aeruginosa* and its mechanism of action. World J Microbiol Biotechnol 27:1209–1216. https://doi.org/10.1007/s11274-010-0569-2

Kora AJ, Rastogi L (2013) Enhancement of antibacterial activity of capped silver nanoparticles in combination with antibiotics, on model gram-negative and gram-positive bacteria. Bioinorg Chem Appl 2013:871097. https://doi.org/10.1155/2013/871097

Kumar R, Ray PC, Datta D et al (2015) Nanovaccines for malaria using *Plasmodium falciparum* antigen Pfs25 attached gold nanoparticles. Vaccine 33:5064–5071. https://doi.org/10.1016/j.vaccine.2015.08.025

Kumar GV, Su C-H, Velusamy P (2016) Ciprofloxacin loaded genipin cross-linked chitosan/heparin nanoparticles for drug delivery application. Mater Lett 180:119–122. https://doi.org/10.1016/j.matlet.2016.05.108

Kumar M, Bansal K, Gondil VS et al (2018) Synthesis, characterization, mechanistic studies and antimicrobial efficacy of biomolecule capped and pH modulated silver nanoparticles. J Mol Liq 249:1145–1150. https://doi.org/10.1016/j.molliq.2017.11.143

Lara HH, Ixtepan-Turrent L, Garza-Treviño EN, Rodriguez-Padilla C (2010a) PVP-coated silver nanoparticles block the transmission of cell-free and cell-associated HIV-1 in human cervical culture. J Nanobiotechnol 8:15. https://doi.org/10.1186/1477-3155-8-15

Lara HH, Ayala-Núñez NV, Ixtepan Turrent L d C, Rodríguez Padilla C (2010b) Bactericidal effect of silver nanoparticles against multidrug-resistant bacteria. World J Microbiol Biotechnol 26:615–621. https://doi.org/10.1007/s11274-009-0211-3

Lee SH, Jun B-H (2019) Silver nanoparticles: synthesis and application for nanomedicine. Int J Mol Sci 20:865. https://doi.org/10.3390/ijms20040865

Lee KX, Shameli K, Yew YP et al (2020) Recent developments in the facile bio-synthesis of gold nanoparticles (AuNPs) and their biomedical applications. Int J Nanomedicine 15:275–300. https://doi.org/10.2147/IJN.S233789

Lemire JA, Harrison JJ, Turner RJ (2013) Antimicrobial activity of metals: mechanisms, molecular targets and applications. Nat Rev Microbiol 11:371–384. https://doi.org/10.1038/nrmicro3028

Lettre G, Hengartner MO (2006) Developmental apoptosis in *C. elegans*: a complex CEDnario. Nat Rev Mol Cell Biol 7:97–108. https://doi.org/10.1038/nrm1836

Li W-R, Xie X-B, Shi Q-S et al (2010) Antibacterial activity and mechanism of silver nanoparticles on *Escherichia coli*. Appl Microbiol Biotechnol 85:1115–1122. https://doi.org/10.1007/s00253-009-2159-5

Li K, Zhao X, Hammer B et al (2013) Nanoparticles inhibit DNA replication by Binding to DNA: modeling and experimental validation. ACS Nano 7:9664–9674. https://doi.org/10.1021/nn402472k

Lin C, Tao K, Hua D et al (2013) Size effect of gold nanoparticles in catalytic reduction of p-nitrophenol with NaBH4. Molecules 18:12609–12620. https://doi.org/10.3390/molecules181012609

Lin LC-W, Huang C-Y, Yao B-Y et al (2019) Viromimetic STING agonist-loaded hollow polymeric nanoparticles for safe and effective vaccination against middle east respiratory syndrome coronavirus. Adv Funct Mater 29:1807616. https://doi.org/10.1002/adfm.201807616

Liu K, He Z, Byrne HJ et al (2018) Investigating the role of gold nanoparticle shape and size in their toxicities to fungi. Int J Environ Res Public Health 15:998. https://doi.org/10.3390/ijerph15050998

Liu Y, Wu S, Saavedra-Mella F et al (2020) Rhizosphere modifications of iron-rich minerals and forms of heavy metals encapsulated in sulfidic tailings hardpan. J Hazard Mater 384:121444. https://doi.org/10.1016/j.jhazmat.2019.121444

Lokina S, Suresh R, Giribabu K et al (2014) Spectroscopic investigations, antimicrobial, and cytotoxic activity of green synthesized gold nanoparticles. Spectrochim Acta A Mol Biomol Spectrosc 129:484–490. https://doi.org/10.1016/j.saa.2014.03.100

Lopez N, Janssens TVW, Clausen BS et al (2004) On the origin of the catalytic activity of gold nanoparticles for low-temperature CO oxidation. J Catal 223:232–235. https://doi.org/10.1016/j.jcat.2004.01.001

Lopez-Chaves C, Soto-Alvaredo J, Montes-Bayon M et al (2018) Gold nanoparticles: distribution, bioaccumulation and toxicity. In vitro and in vivo studies. Nanomedicine 14:1–12. https://doi.org/10.1016/j.nano.2017.08.011

Lu L, Sun RW-Y, Chen R et al (2008) Silver nanoparticles inhibit hepatitis B virus replication. Antivir Ther 13:253–262

Madsen AT, Ahmed EH, Christensen CH et al (2011) Hydrodeoxygenation of waste fat for diesel production: study on model feed with Pt/alumina catalyst. Fuel 90:3433–3438. https://doi.org/10.1016/j.fuel.2011.06.005

Malanovic N, Lohner K (2016) Gram-positive bacterial cell envelopes: the impact on the activity of antimicrobial peptides. Biochim Biophys Acta – Biomembr 1858:936–946. https://doi.org/10.1016/j.bbamem.2015.11.004

Manju S, Malaikozhundan B, Vijayakumar S et al (2016) Antibacterial, antibiofilm and cytotoxic effects of Nigella sativa essential oil coated gold nanoparticles. Microb Pathog 91:129–135. https://doi.org/10.1016/j.micpath.2015.11.021

Masri A, Anwar A, Ahmed D et al (2018) Silver nanoparticle conjugation-enhanced antibacterial efficacy of clinically approved drugs cephradine and vildagliptin. Antibiot (Basel, Switzerland) 7:100. https://doi.org/10.3390/antibiotics7040100

Mata R, Bhaskaran A, Sadras SR (2016) Green-synthesized gold nanoparticles from *Plumeria alba* flower extract to augment catalytic degradation of organic dyes and inhibit bacterial growth. Particuology 24:78–86. https://doi.org/10.1016/j.partic.2014.12.014

Medintz IL, Pons T, Delehanty JB et al (2008) Intracellular delivery of quantum dot-protein cargos mediated by cell penetrating peptides. Bioconjug Chem 19:1785–1795. https://doi.org/10.1021/bc800089r

Mironava T, Hadjiargyrou M, Simon M et al (2010) Gold nanoparticles cellular toxicity and recovery: effect of size, concentration and exposure time. Nanotoxicology 4:120–137. https://doi.org/10.3109/17435390903471463

Moglianetti M, De Luca E, Pedone D et al (2016) Platinum nanozymes recover cellular ROS homeostasis in an oxidative stress-mediated disease model. Nanoscale 8:3739–3752. https://doi.org/10.1039/C5NR08358C

Mondal MK, Mukherjee S, Joardar N et al (2019) Synthesis of smart graphene quantum dots: a benign biomaterial for prominent intracellular imaging and improvement of drug efficacy. Appl Surf Sci 495:143562. https://doi.org/10.1016/j.apsusc.2019.143562

Morones JR, Elechiguerra JL, Camacho A et al (2005) The bactericidal effect of silver nanoparticles. Nanotechnology 16:2346–2353. https://doi.org/10.1088/0957-4484/16/10/059

Mozetič M (2019) Surface modification to improve properties of materials. Dent Mater 12:441. https://doi.org/10.3390/ma12030441

Mukherjee N, Parida PK, Santra A et al (2016) Oxidative stress plays major role in mediating apoptosis in filarial nematode *Setaria cervi* in the presence of trans-stilbene derivatives. Free Radic Biol Med 93:130–144. https://doi.org/10.1016/j.freeradbiomed.2016.01.027

Mukherjee S, Joardar N, Mondal S et al (2018) Quinolone-fused cyclic sulfonamide as a novel benign antifilarial agent. Sci Rep 8:12073. https://doi.org/10.1038/s41598-018-30610-7

Muthuvel A, Adavallan K, Balamurugan K, Krishnakumar N (2014) Biosynthesis of gold nanoparticles using *Solanum nigrum* leaf extract and screening their free radical scavenging and antibacterial properties. Biomed Prev Nutr 4:325–332. https://doi.org/10.1016/j.bionut.2014.03.004

Nag OK, Delehanty JB (2019) Active cellular and subcellular targeting of nanoparticles for drug delivery. Pharmaceutics 11:543. https://doi.org/10.3390/pharmaceutics11100543

Nath S, Kaittanis C, Tinkham A, Perez JM (2008) Dextran-coated gold nanoparticles for the assessment of antimicrobial susceptibility. Anal Chem 80:1033–1038. https://doi.org/10.1021/ac701969u

Neethu S, Midhun SJ, Radhakrishnan EK, Jyothis M (2018) Green synthesized silver nanoparticles by marine endophytic fungus *Penicillium polonicum* and its antibacterial efficacy against biofilm forming, multidrug-resistant *Acinetobacter baumanii*. Microb Pathog 116:263–272. https://doi.org/10.1016/j.micpath.2018.01.033

Nishanthi R, Malathi S, John PS, Palani P (2019) Green synthesis and characterization of bioinspired silver, gold and platinum nanoparticles and evaluation of their synergistic antibacterial activity after combining with different classes of antibiotics. Korean J Couns Psychother 96:693–707. https://doi.org/10.1016/j.msec.2018.11.050

Omran BA, Nassar HN, Fatthallah NA et al (2018) Characterization and antimicrobial activity of silver nanoparticles mycosynthesized by *Aspergillus brasiliensis*. J Appl Microbiol 125:370–382. https://doi.org/10.1111/jam.13776

Oun AA, Shankar S, Rhim J-W (2020) Multifunctional nanocellulose/metal and metal oxide nanoparticle hybrid nanomaterials. Crit Rev Food Sci Nutr 60:435–460. https://doi.org/10.1080/10408398.2018.1536966

Ovais M, Zia N, Khalil AT et al (2019) Nanoantibiotics: recent developments and future prospects. Front Clin Drug Res Anti Infect 5:158–174. https://doi.org/10.2174/9781681086378119050006

Pal S, Tak YK, Song JM (2007) Does the antibacterial activity of silver nanoparticles depend on the shape of the nanoparticle? A study of the Gram-negative bacterium *Escherichia coli*. Appl Environ Microbiol 73:1712–1720. https://doi.org/10.1128/AEM.02218-06

Panzarini E, Mariano S, Carata E et al (2018) Intracellular transport of silver and gold nanoparticles and biological responses: an update. Int J Mol Sci 19:1305. https://doi.org/10.3390/ijms19051305

Papp I, Sieben C, Ludwig K et al (2010) Inhibition of influenza virus infection by multivalent sialic-acid-functionalized gold nanoparticles. Small 6:2900–2906. https://doi.org/10.1002/smll.201001349

Patil Y, Amitay Y, Ohana P et al (2016) Targeting of pegylated liposomal mitomycin-C prodrug to the folate receptor of cancer cells: intracellular activation and enhanced cytotoxicity. J Control Release 225:87–95. https://doi.org/10.1016/j.jconrel.2016.01.039

Paulo CSO, Pires das Neves R, Ferreira LS (2011) Nanoparticles for intracellular-targeted drug delivery. Nanotechnology 22:494002. https://doi.org/10.1088/0957-4484/22/49/494002

Payne JN, Waghwani HK, Connor MG et al (2016) Novel synthesis of kanamycin conjugated gold nanoparticles with potent antibacterial activity. Front Microbiol 7:607. https://doi.org/10.3389/fmicb.2016.00607

Pedone D, Moglianetti M, De Luca E et al (2017) Platinum nanoparticles in nanobiomedicine. Chem Soc Rev 46:4951–4975. https://doi.org/10.1039/C7CS00152E

Poojary MM, Passamonti P, Adhikari AV (2016) Green synthesis of silver and gold nanoparticles using root bark extract of *Mammea suriga*: characterization, process optimization, and their antibacterial activity. Bionanoscience 6:110–120. https://doi.org/10.1007/s12668-016-0199-8

Prakash A, Sharma S, Ahmad N, Ghosh A, Sinha P (2011) Synthesis of AgNPs by *Bacillus cereus* bacteria and their antimicrobial potential. J Biomater Nanobiotechnol 2:156–162. https://doi.org/10.4236/jbnb.2011.22020

Prasad KS, Pathak D, Patel A, Dalwadi P, Prasad R, Patel P, Kaliaperumal SK (2011) Biogenic synthesis of silver nanoparticles using *Nicotiana tobaccum* leaf extract and study of their antibacterial effect. Afr J Biotechnol 9(54):8122–8130

Prasad R, Swamy VS, Varma A (2012) Biogenic synthesis of silver nanoparticles from the leaf extract of *Syzygium cumini* (L.) and its antibacterial activity. Int J Pharma Bio Sci 3(4):745–752

Prasad R, Swamy VS (2013) Antibacterial activity of silver nanoparticles synthesized by bark extract of *Syzygium cumini*. J Nanoparticles. https://doi.org/10.1155/2013/431218

Prasad R, Pandey R, Barman I (2016) Engineering tailored nanoparticles with microbes: quo vadis. WIREs Nanomed Nanobiotechnol 8:316–330. https://doi.org/10.1002/wnan.1363

Prasad R, Pandey R, Varma A, Barman I (2017) Polymer based nanoparticles for drug delivery systems and cancer therapeutics. In: Kharkwal H, Janaswamy S (eds) Natural polymers for drug delivery. CAB International, London, pp 53–70

Prasad R, Kumar V, Kumar M, Choudhary D (2019) Nanobiotechnology in bioformulations. Springer International Publishing, Cham. ISBN 978-3-030-17061-5. https://www.springer.com/gp/book/9783030170608

Prasad R, Siddhardha B, Dyavaiah M (2020) Nanostructures for antimicrobial and antibiofilm applications. Springer International Publishing, Cham. ISBN 978-3-030-40336-2. https://www.springer.com/gp/book/9783030403362

Qi K, Cheng B, Yu J, Ho W (2017) Review on the improvement of the photocatalytic and antibacterial activities of ZnO. J Alloys Compd 727:792–820. https://doi.org/10.1016/j.jallcom.2017.08.142

Quinteros MA, Aiassa Martínez IM, Dalmasso PR, Páez PL (2016) Silver nanoparticles: biosynthesis using an ATCC reference strain of *Pseudomonas aeruginosa* and activity as broad spectrum clinical antibacterial agents. Int J Biomater 2016:5971047. https://doi.org/10.1155/2016/5971047

Rahimi H, Roudbarmohammadi S, Delavari HH, Roudbary M (2019) Antifungal effects of indolicidin-conjugated gold nanoparticles against fluconazole-resistant strains of *Candida albicans* isolated from patients with burn infection. Int J Nanomedicine 14:5323–5338. https://doi.org/10.2147/IJN.S207527

Rahman K, Khan SU, Fahad S et al (2019) Nano-biotechnology: a new approach to treat and prevent malaria. Int J Nanomedicine 14:1401–1410. https://doi.org/10.2147/IJN.S190692

Rai M, Yadav A, Gade A (2009) Silver nanoparticles as a new generation of antimicrobials. Biotechnol Adv 27:76–83. https://doi.org/10.1016/j.biotechadv.2008.09.002

Rai M, Ingle AP, Paralikar P et al (2017) Recent advances in use of silver nanoparticles as antimalarial agents. Int J Pharm 526:254–270. https://doi.org/10.1016/j.ijpharm.2017.04.042

Rajan A, Vilas V, Philip D (2015) Studies on catalytic, antioxidant, antibacterial and anticancer activities of biogenic gold nanoparticles. J Mol Liq 212:331–339. https://doi.org/10.1016/j.molliq.2015.09.013

Rajawat S, Qureshi MS (2012) Comparative study on bactericidal effect of silver nanoparticles, synthesized using green technology, in combination with antibiotics on *Salmonella typhi*. J Biomater Nanobiotechnol 3:480–497. https://doi.org/10.4236/jbnb.2012.34049

Rajeshkumar S, Malarkodi C, Vanaja M, Annadurai G (2016) Anticancer and enhanced antimicrobial activity of biosynthesizd silver nanoparticles against clinical pathogens. J Mol Struct 1116:165–173. https://doi.org/10.1016/j.molstruc.2016.03.044

Ramkumar VS, Pugazhendhi A, Prakash S et al (2017) Synthesis of platinum nanoparticles using seaweed *Padina gymnospora* and their catalytic activity as PVP/PtNPs nanocomposite towards biological applications. Biomed Pharmacother 92:479–490. https://doi.org/10.1016/j.biopha.2017.05.076

Rao B, Tang R-C (2017) Green synthesis of silver nanoparticles with antibacterial activities using aqueous *Eriobotrya japonica* leaf extract. Adv Nat Sci Nanosci Nanotechnol 8:15014. https://doi.org/10.1088/2043-6254/aa5983

Rao D, Sheng Q, Zheng J (2016) Novel nanocomposite of chitosan-protected platinum nanoparticles immobilized on nickel hydroxide: facile synthesis and application as glucose electrochemical sensor. J Chem Sci 128:1367–1375. https://doi.org/10.1007/s12039-016-1146-5

Rattanata N, Klaynongsruang S, Leelayuwat C et al (2016) Gallic acid conjugated with gold nanoparticles: antibacterial activity and mechanism of action on foodborne pathogens. Int J Nanomedicine 11:3347–3356. https://doi.org/10.2147/IJN.S109795

Rizvi SAA, Saleh AM (2018) Applications of nanoparticle systems in drug delivery technology. Saudi Pharm J SPJ Off Publ Saudi Pharm Soc 26:64–70. https://doi.org/10.1016/j.jsps.2017.10.012

Rogers JV, Parkinson CV, Choi YW et al (2008) A preliminary assessment of silver nanoparticle inhibition of monkeypox virus plaque formation. Nanoscale Res Lett 3:129–133. https://doi.org/10.1007/s11671-008-9128-2

Rosi NL, Giljohann DA, Thaxton CS et al (2006) Oligonucleotide-modified gold nanoparticles for intracellular gene regulation. Science 312:1027–1030. https://doi.org/10.1126/science.1125559

Roy B, Mukherjee S, Mukherjee N et al (2014) Design and green synthesis of polymer inspired nanoparticles for the evaluation of their antimicrobial and antifilarial efficiency. RSC Adv 4:34487–34499. https://doi.org/10.1039/C4RA03732D

Roy P, Saha SK, Gayen P et al (2018) Exploration of antifilarial activity of gold nanoparticle against human and bovine filarial parasites: a nanomedicinal mechanistic approach. Colloids Surf B Biointerfaces 161:236–243. https://doi.org/10.1016/j.colsurfb.2017.10.057

Roy A, Bulut O, Some S et al (2019) Green synthesis of silver nanoparticles: biomolecule-nanoparticle organizations targeting antimicrobial activity. RSC Adv 9:2673–2702. https://doi.org/10.1039/C8RA08982E

Ruan G, Agrawal A, Marcus AI, Nie S (2007) Imaging and tracking of tat peptide-conjugated quantum dots in living cells: new insights into nanoparticle uptake, intracellular transport, and vesicle shedding. J Am Chem Soc 129:14759–14766. https://doi.org/10.1021/ja074936k

Ruedas-Rama MJ, Walters JD, Orte A, Hall EAH (2012) Fluorescent nanoparticles for intracellular sensing: a review. Anal Chim Acta 751:1–23. https://doi.org/10.1016/j.aca.2012.09.025

Safdar M, Ozaslan M, Khailany RA et al (2020) Synthesis, characterization and applications of a novel platinum-based nanoparticles: catalytic, antibacterial and cytotoxic studies. J Inorg Organomet Polym Mater 30:2430–2439. https://doi.org/10.1007/s10904-019-01387-7

Saha SK, Chowdhury P, Saini P, Babu SPS (2014) Ultrasound assisted green synthesis of poly (vinyl alcohol) capped silver nanoparticles for the study of its antifilarial efficacy. Appl Surf Sci 288:625–632. https://doi.org/10.1016/j.apsusc.2013.10.085

Saha SK, Roy P, Saini P et al (2016) Carbohydrate polymer inspired silver nanoparticles for filaricidal and mosquitocidal activities: a comprehensive view. Carbohydr Polym 137:390–401. https://doi.org/10.1016/j.carbpol.2015.11.007

Saha SK, Roy P, Mondal MK et al (2017) Development of chitosan based gold nanomaterial as an efficient antifilarial agent: a mechanistic approach. Carbohydr Polym 157:1666–1676. https://doi.org/10.1016/j.carbpol.2016.11.047

Saini P, Saha SK, Roy P et al (2016) Evidence of reactive oxygen species (ROS) mediated apoptosis in *Setaria cervi* induced by green silver nanoparticles from Acacia auriculiformis at a very low dose. Exp Parasitol 160:39–48. https://doi.org/10.1016/j.exppara.2015.11.004

Salaklang J, Steitz B, Finka A et al (2008) Superparamagnetic nanoparticles as a powerful systems biology characterization tool in the physiological context. Angew Chem Int Ed Engl 47:7857–7860. https://doi.org/10.1002/anie.200800357

Sánchez-López E, Gomes D, Esteruelas G et al (2020) Metal-based nanoparticles as antimicrobial agents: an overview. Nanomater (Basel, Switzerland) 10:292. https://doi.org/10.3390/nano10020292

Saravanan M, Arokiyaraj S, Lakshmi T, Pugazhendhi A (2018) Synthesis of silver nanoparticles from *Phenerochaete chrysosporium* (MTCC-787) and their antibacterial activity against human pathogenic bacteria. Microb Pathog 117:68–72. https://doi.org/10.1016/j.micpath.2018.02.008

Sekimukai H, Iwata-Yoshikawa N, Fukushi S et al (2020) Gold nanoparticle-adjuvanted S protein induces a strong antigen-specific IgG response against severe acute respiratory syndrome-related coronavirus infection, but fails to induce protective antibodies and limit eosinophilic infiltration in lungs. Microbiol Immunol 64:33–51. https://doi.org/10.1111/1348-0421.12754

Shaham S, Horvitz HR (1996) An alternatively spliced *C. elegans* ced-4 RNA encodes a novel cell death inhibitor. Cell 86:201–208. https://doi.org/10.1016/s0092-8674(00)80092-6

Shahid-ul-Islam, Butola BS, Mohammad F (2016) Silver nanomaterials as future colorants and potential antimicrobial agents for natural and synthetic textile materials. RSC Adv 6:44232–44247. https://doi.org/10.1039/C6RA05799C

Shi Y, Lin M, Jiang X, Liang S (2015) Recent advances in FePt nanoparticles for biomedicine. J Nanomater 2015:467873. https://doi.org/10.1155/2015/467873

Shriniwas PP, Subhash TK (2017) Antioxidant, antibacterial and cytotoxic potential of silver nanoparticles synthesized using terpenes rich extract of *Lantana camara* L. leaves. Biochem Biophys Rep 10:76–81. https://doi.org/10.1016/j.bbrep.2017.03.002

Silva NC, Silva S, Sarmento B, Pintado M (2015) Chitosan nanoparticles for daptomycin delivery in ocular treatment of bacterial endophthalmitis. Drug Deliv 22:885–893. https://doi.org/10.3109/10717544.2013.858195

Silvero CMJ, Rocca DM, de la Villarmois EA et al (2018) Selective photoinduced antibacterial activity of amoxicillin-coated gold nanoparticles: from one-step synthesis to in vivo cytocompatibility. ACS Omega 3:1220–1230. https://doi.org/10.1021/acsomega.7b01779

Singh RP, Ramarao P (2012) Cellular uptake, intracellular trafficking and cytotoxicity of silver nanoparticles. Toxicol Lett 213:249–259. https://doi.org/10.1016/j.toxlet.2012.07.009

Singh P, Singh H, Kim YJ et al (2016) Extracellular synthesis of silver and gold nanoparticles by *Sporosarcina koreensis* DC4 and their biological applications. Enzyme Microb Technol 86:75–83. https://doi.org/10.1016/j.enzmictec.2016.02.005

Sondi I, Salopek-Sondi B (2004) Silver nanoparticles as antimicrobial agent: a case study on *E. coli* as a model for Gram-negative bacteria. J Colloid Interface Sci 275:177–182. https://doi.org/10.1016/j.jcis.2004.02.012

Sultana S, Djaker N, Boca-Farcau S et al (2015) Comparative toxicity evaluation of flower-shaped and spherical gold nanoparticles on human endothelial cells. Nanotechnology 26:55101. https://doi.org/10.1088/0957-4484/26/5/055101

Swamy VS, Prasad R (2012) Green synthesis of silver nanoparticles from the leaf extract of *Santalum album* and its antimicrobial activity. J Optoelectron Biomed Mater 4(3):53–59

Tahir K, Nazir S, Ahmad A et al (2017) Facile and green synthesis of phytochemicals capped platinum nanoparticles and in vitro their superior antibacterial activity. J Photochem Photobiol B 166:246–251. https://doi.org/10.1016/j.jphotobiol.2016.12.016

Tanaka S-I, Miyazaki J, Tiwari DK et al (2011) Fluorescent platinum nanoclusters: synthesis, purification, characterization, and application to bioimaging. Angew Chem Int Ed 50:431–435. https://doi.org/10.1002/anie.201004907

Taratula O, Garbuzenko OB, Kirkpatrick P et al (2009) Surface-engineered targeted PPI dendrimer for efficient intracellular and intratumoral siRNA delivery. J Control Release 140:284–293. https://doi.org/10.1016/j.jconrel.2009.06.019

Tavolaro A, Drioli E (1999) Zeolite membranes. Adv Mater 11:975–996. https://doi.org/10.1002/(SICI)1521-4095(199908)11:12<975::AID-ADMA975>3.0.CO;2-0

Tippayawat P, Sapa V, Srijampa S et al (2017) d-Maltose coated silver nanoparticles and their synergistic effect in combination with ampicillin. Monatshefte für Chemie – Chem Mon 148:1197–1203. https://doi.org/10.1007/s00706-017-2004-y

Ugru MM, Sheshadri S, Jain D et al (2018) Insight into the composition and surface corona reliant biological behaviour of quercetin engineered nanoparticles. Colloids Surf A Physicochem Eng Asp 548:1–9. https://doi.org/10.1016/j.colsurfa.2018.03.055

Vijayakumari J, Raj TLS, Selvi AA et al (2019) A comparative study of plant mediated synthesis of silver, copper and zinc nanoparticles from *Tiliacora acuminata* (lam.) Hook. F. and their antibacterial activity studies. Synthesis 18:19–34

Viswanathan A, Rangasamy J, Biswas R (2019) Functionalized antibacterial nanoparticles for controlling biofilm and intracellular infections. In: Pathak YV (ed) Surface modification of nanoparticles for targeted drug delivery. Springer International Publishing, Cham, pp 183–206

Wang Y-W, Tang H, Wu D et al (2016) Enhanced bactericidal toxicity of silver nanoparticles by the antibiotic gentamicin. Environ Sci Nano 3:788–798. https://doi.org/10.1039/C6EN00031B

Wang L, Hu C, Shao L (2017) The antimicrobial activity of nanoparticles: present situation and prospects for the future. Int J Nanomedicine 12:1227–1249. https://doi.org/10.2147/IJN.S121956

Wani IA, Ahmad T, Manzoor N (2013) Size and shape dependant antifungal activity of gold nanoparticles: a case study of *Candida*. Colloids Surf B Biointerfaces 101:162–170. https://doi.org/10.1016/j.colsurfb.2012.06.005

Wierzbicki M, Jaworski S, Sawosz E et al (2019) Graphene oxide in a composite with silver nanoparticles reduces the fibroblast and endothelial cell cytotoxicity of an antibacterial nanoplatform. Nanoscale Res Lett 14:320. https://doi.org/10.1186/s11671-019-3166-9

Wijnhoven SWP, Peijnenburg WJGM, Herberts CA et al (2009) Nano-silver – a review of available data and knowledge gaps in human and environmental risk assessment. Nanotoxicology 3:109–138. https://doi.org/10.1080/17435390902725914

Yadav S, Sharma S, Ahmad F, Rathaur S (2020) Antifilarial efficacy of green silver nanoparticles synthesized using *Andrographis paniculata*. J Drug Deliv Sci Technol 56:101557. https://doi.org/10.1016/j.jddst.2020.101557

Yan N, Gu L, Kokel D et al (2004) Structural, biochemical, and functional analyses of CED-9 recognition by the proapoptotic proteins EGL-1 and CED-4. Mol Cell 15:999–1006. https://doi.org/10.1016/j.molcel.2004.08.022

Yan N, Chai J, Lee ES et al (2005) Structure of the CED-4–CED-9 complex provides insights into programmed cell death in *Caenorhabditis elegans*. Nature 437:831–837. https://doi.org/10.1038/nature04002

Yeh Y-C, Huang T-H, Yang S-C et al (2020) Nano-based drug delivery or targeting to eradicate bacteria for infection mitigation: a review of recent advances. Front Chem 8:286. https://doi.org/10.3389/fchem.2020.00286

Yin Y, Papavasiliou G, Zaborina OY et al (2017) De Novo synthesis and functional analysis of polyphosphate-loaded poly(ethylene) glycol hydrogel nanoparticles targeting pyocyanin and pyoverdin production in *Pseudomonas aeruginosa* as a model intestinal pathogen. Ann Biomed Eng 45:1058–1068. https://doi.org/10.1007/s10439-016-1740-1

Yokesh Babu M, Janaki Devi V, Ramakritinan CM et al (2013) Biosynthesis of silver nanoparticles from seaweed associated marine bacterium and its antimicrobial activity against UTI pathogens. Int J Curr Microbiol App Sci 2:155–168

Yu X, Yuan L, Zhu N et al (2019) Fabrication of antimicrobial curcumin stabilized platinum nanoparticles and their anti-liver fibrosis activity for potential use in nursing care. J Photochem Photobiol B 195:27–32. https://doi.org/10.1016/j.jphotobiol.2019.03.023

Zhang X-F, Liu Z-G, Shen W, Gurunathan S (2016) Silver nanoparticles: synthesis, characterization, properties, applications, and therapeutic approaches. Int J Mol Sci 17:1534. https://doi.org/10.3390/ijms17091534

Chapter 8
RETRACTED CHAPTER: Nanomaterials for Advanced Microbiology

Govindh Boddeti, Venu Reddy, and B. S. Diwakar

The Editors have retracted this chapter because it contains meaningless phrases. The content is therefore unreliable. The authors have not responded to correspondence from the Publisher about this retraction.

N. R. Maddela et al. (eds.), *Nanotechnology for Advances in Medical Microbiology*, Environmental and Microbial Biotechnology,
https://doi.org/10.1007/978-981-15-9916-3_8

RETRACTED CHAPTER

Chapter 9
Nanoparticles for Biofilm Control

Ranjit Pabbati, Moulika Aerupula, Firdoz Shaik, and Venkateswar Reddy Kondakindi

Abstract Microorganisms of planktonic types have emerged into biofilm structures by acquiring the ability to attach to the surfaces. The extracellular polymeric substances may play a vital role in anchoring the biofilm onto the surface and also protect it from various antibiotics making it highly resistant. This type of survival is procured by species of bacteria, fungi and protists. These biofilms pose numerous hazards to the human health and environment. Over 80% of the diseases are caused by the biofilms which include colorectal cancer, oral infections, periodontitis, cystic fibrosis and many others. Different strategies have been developed to eliminate the biofilms like the use of nano-sized particles. These nanoparticles provide antimicrobial properties and can also be used as a delivery system of the antibiotics with high stability and good biocompatibility. The high surface-area-to-volume ratio makes the nanoparticles to have high contact surface to the bacterial cell membrane and hence plays an efficient therapeutic way in treating the biofilms. In this review, we discuss the various diseases caused by the biofilm and different antibiofilm strategies using the nanomaterials. There is no doubt that the upcoming antibiofilm strategy which uses nanoscience can gain an ultimately new hope in advancements of coping with bacterial infections and resistance.

Keywords Biofilms · Antibiofilm · Pathogenesis · Nanoscience · Biocompatibility

R. Pabbati · M. Aerupula · V. R. Kondakindi (✉)
Centre for Biotechnology, Institute of science & Technology, Jawaharlal Nehru Technological University Hyderabad, Hyderabad, Telangana, India

F. Shaik
Schulich Faculty of Chemistry, Technion-Israel Institute of Technology, Haifa, Israel

N. R. Maddela et al. (eds.), *Nanotechnology for Advances in Medical Microbiology*, Environmental and Microbial Biotechnology,
https://doi.org/10.1007/978-981-15-9916-3_9

9.1 Introduction

Biofilm is an organized arrangement of microbial cells which are irreversibly attached to a surface contained in a matrix made of self-produced extracellular polymeric substances. It is responsible for many diseases infecting human health and sometimes has a positive influence like the mutualistic bacteria, namely, *Staphylococcus epidermidis*, which can hold back the formation of colony by pathogenic bacteria through triggering the host immune system and thereby preventing the attachment of microbes (Donlan 2002). Cystic fibrosis (CF), an inherited genetic disorder, is the most common human disease infecting the lungs seen in Western Europe. Patients with cystic fibrosis experience persistent infection caused by *P. aeruginosa*. So when this *P. aeruginosa* bacterium invades the CF lung, it starts to acclimatize with the CF lung and survives for decades. The reason is the excess production of matrix polysaccharide alginate which makes up the mucoid biofilm that resists antibiotics, elements of natural and acquired immunity, and thus resists phagocytosis. This whole process again develops noticeable antibody responses in the form of granulocytes and gives rise to chronic inflammation resulting in harsh lung tissue damage in the CF patients. The other negative effects of biofilms include dental caries or cavities which are caused by overproduction of organic acids mainly when sugary drinks are consumed or while frequent snacking. These acids breakdown the teeth enamel and lead to dental caries. It is approximated that 65 out of 100 infections are related to bacterial biofilms (Lewis 2001). They are the device- and non-device-linked infections. Data estimated for the device-related infections are 2% each for breast implants and joint prostheses; 4% each for mechanical heart valves, defibrillators and pacemakers; 10% for ventricular shunts; and 40% for ventricular-assisted. Non-device-related biofilm infections include periodontitis, osteomyelitis and many to list out. It is important to have oral hygiene since a person is susceptible to suffer from periodontitis which infects the gums and damages the soft tissues (Kokare et al. 2007). Osteomyelitis is a disease in bones caused by bacteria or the fungal cells. When the bacteria enter the bloodstream and infect the growth plate of the bone which is the metaphysic portion of the bone, the white blood cells gather at the site and try to phagocytose or kill the pathogen by releasing enzymes. So, these enzymes may break the bone and form pus spreading throughout the blood vessels. This leads to the dysfunctioning of the affected bone areas (Ziran 2007).

The antibiotics which are available to date have shown low efficacy in treating the biofilms associated with infections because of their high values of minimum inhibitory concentration (MIC) and minimum bacterial concentration (MBC) leading to in vivo toxicity. Sometimes, biofilms prevent the phagocytosis of the invading bacteria by impairing the phagocytes and complement system. Hence, it is important to focus on the MBCs, MICs, mechanism of action and chemical structures of the antibiofilm substances which include chelating agents, peptide antibiotics, etc. Different strategies can be implemented to combat with the biofilms like replacing the infected foreign bodies such as stents, implants with the sterile ones, blocking the quorum sensing pathway or by altering the c-di-GMP. LP 3134, LP 3145, LP 4010

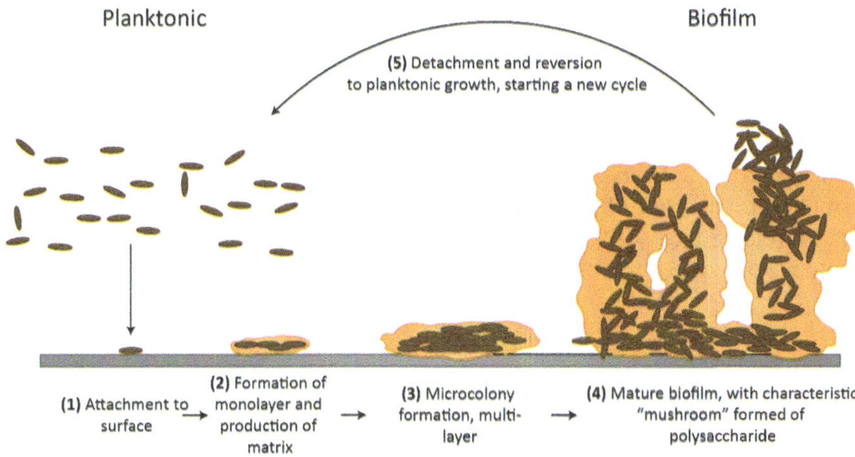

Fig. 9.1 Stages of biofilm formation. (Adapted from Vasudevan 2014)

and LP 1062 can inhibit the formation of biofilm in *P. aeruginosa* and *Acinetobacter baumannii* (Ranita Roy et al. 2018). Nanoscale materials are widely applied in the field of medicine because of their high reactivities with the large surface-area-to-volume ratio. In fact, the physicochemical properties are controllable when it comes to the nanoparticles. The other most important factor why nanotechnology has been more adapted is because of its low toxicity to the host and thus can gain a positive hope over the conventional treatments of biofilms (Ramasamy and Lee 2016). These nanoparticles (NPs) are able to find a way out to prevent the drug resistance mechanisms and other important processes. The combinations of plant-based antimicrobials with the nanoparticles are also studied (Baptista et al. 2018).

In this chapter, we are going to discuss the different stages leading to the formation of biofilms (Fig. 9.1). It is also important to study the types of pathogens involved in biofilm synthesis and the pathogenesis related to the biofilm in humans. As these biofilms give rise to dangerous diseases leading to chronic infections, there is a necessary need to combat the microbes, and also depending on the applications of nanoscience, we will be reviewing various strategies incorporated till now in treating the biofilms.

9.2 Biofilms

In 1947, Antonie van Leeuwenhoek examined the aggregates of "animalcules" scrapped from surfaces of human tooth under the microscope. Almost 100 years later, microbiologists studied the formation of biofilms and also concluded that bacteria grow differently when they attach to a specific surface in immobilized populations. Studies also show that the bacteria eliminated from the native ecosystem grow predominantly as planktonic cells (Costerton 1999). Biofilms are the organized structures comprising of microorganisms where they stick to each other producing extracellular polymers to cohesion. Biofilms are formed due to various parameters like nutritional cues; identification of specific or non-specific binding sites on a surface etiofilms can bind to the surfaces like the tooth, rock or any single species or a diverse group of microbes (Karatan and Watnick 2009; Hoffman et al. 2005). This shift of survival from a planktonic growth to biofilm keeps them safe from toxic factors like antibiotic desiccation and host body's immune system (Tortora et al. 2015).

9.2.1 Bacterial Biofilms

In order to survive, the wild bacterial strains rely on fimbriae which protrude from the thick layer of exopolysaccharides (EPS). Fimbriae facilitate specific adhesion to the surfaces, and non-specific adhesion to inert surfaces is provided by EPS. Fimbriae interaction with the surfaces is not strong as it can be removed easily by simple sonication. Firm adhesion of the bacterial strains needs the elastic polymers of the EPS for effective non-specific interaction in aquatic ecosystem (Costerton 1999). The bacteria release protons and signalling molecules radially diffusing away from the cell. We observe a sharp increase in the concentration of the diffusing molecules finding itself near a surface or interface which would let the cell recognize that it is near the surface because diffusion became limited on that side. Thus, once the bacterial cell has sensed a surface, they start to form colonies in monolayer fashion. and active adhesion starts leading to biofilm formation. The cells aggregate to form microcolonies at a specific location. Now in order to make this reversible attachment to irreversible attachment, the bacteria should synthesize new exopolysaccharide to cement other bacterial cells in developing biofilm. In this process, the attached cells upregulate the genes required for EPS synthesis itself (Costerton 1999).

The basic structural unit of biofilm is the microcolony. It consists of different types of species. Depending on the species, microcolony (mushroom-like shape) may have composed of 10–20% cells and remaining 75–90% EPS matrix (Costerton 1999). Regarding biofilm formation potentials of pure species bacteria, it is interesting that the quantity of biofilms produced is not only different between the genera but also vary among the species of same genus (Maddela and Meng 2020); this could be attributed to multiple factors, such as metabolic properties, quorum sensing properties (Maddela et al. 2019), functional groups of exopolysaccharides (Maddela et al. 2018), etc. In general, cells attach to the surfaces which are rougher and hydrophobic in nature. The presence of fimbriae, flagella and EPS helps an organism when a mixed community is involved (Table 9.1).

Table 9.1 Variables essential in cell adhesion and biofilm synthesis

Properties of the substratum	Properties of the bulk fluid	Properties of the cells
Texture or roughness	Flow velocity	Cell surface hydrophobicity
Hydrophobicity	pH	Fimbriae
	Cations	Extracellular polymeric substance
	Presence of antimicrobial agents	Showing antimicrobial activity

Adapted from Donlan (2002)

Table 9.2 Fungal biofilms resistant to antifungal agents

Type of fungi producing biofilm	Resistance to antifungal agent
Candida albicans and *Candida parapsilosis*	Fluconazole, amphotericin-B, nystatin, voriconazole
Aspergillus fumigates	Itraconazole, caspofungin
Cryptococcal	Fluconazole and voriconazole
Tichosporonasahii	Amphotericin-B, caspofungin, voriconazole and fluconazole
Pneumocystis carinii	Azole and amphotericin-B

Adapted from Fanning and Mitchell (2012)

9.2.2 *Fungal Biofilm*

The important fungal species which produce biofilms are *Aspergillus*, *Candida*, *Cryptococcus*, *Trichosporon*, *Coccidioides* and *Pneumonia* (Table 9.2). Factors involving the fungal biofilm resistivity are structural complexity, existence of extracellular matrix (ECM), metabolic heterogeneity intrinsic to biofilms and upregulation of efflux pump genes (Fanning and Mitchell 2012). The biofilms of *C. albicans* are composed of yeast form and hyphal cells which are important for biofilm formation. Steps involving biofilm formation are attachment to the substrate and multiplication of yeast cells on the surface followed by triggering of hyphal formation (Finkel and Mitchell 2011). As the biofilm matures, the cohesivity appears due to ECM aggregation (Al-Fattani and Douglas 2006). Other species of *Candida* like *C. tropicalis*, *C. glabrata*, though contain ECM, fail to produce true hyphae (Silva et al. 2011).

Biofilms of the cells of *Aspergillus* called conidia bind to the substrate and mycelia forms with biofilm maturation. Hyphae can be differently organized in two forms of *A. fumigatus* biofilm infection. For example, hyphae form into a intertwined ball in *Aspergilloma* and in *aspergillosis* show hyphae in separated form (Loussert et al. 2010). There are species which do not produce hyphae as part of their biofilm. Some of the species are *Cryptococcus neoformans* and *Pneumocystis jirovecii* (Cushion et al. 2009).

9.3 Biofilm-Related Pathogenesis

Back in the 1970s, Nils Høiby perceived the connection between the causes of relentless infection and the clusters of bacteria in cystic fibrosis patients (Høiby 2017). It was then noted that biofilms are involved in clinical infections (Costerton et al. 1999; Hall-Stoodley and Stoodley 2009). Bacteria in biofilm mode of survival protect itself by staying in the dormant state from the immune system, thereby causing local tissue damage. In the later stages, it leads to acute infection (Table 9.3 gives an overview of the biofilm-related diseases in humans) (Vestby et al. 2020).

9.3.1 Native Valve Endocarditis

Native valve endocarditis is caused when the vascular endothelium of the four valves, namely, mitral, aortic, tricuspid and pulmonic valves, interacts with the microbes travelling in the blood stream. The species responsible for NVE are *Pneumococci*, *Candida*, *Aspergillus* and some Gram-negative bacteria. The route of infection of these organisms is in blood stream via the oropharynx, gastrointestinal tract and genitourinary tract. Since the microbes bind very poorly to the

Table 9.3 Biofilm-associated diseases

Body system	Affected organs	Disease
Auditory	Middle ear	Otitis media
Cardiovascular	Cardiac valves	Infective endocarditis
	Arteries	Atherosclerosis
Digestive	Salivary glands	Sialolithiasis (salivary duct stones)
	Gall bladder	Recalcitrant typhoid fever and predisposition to hepatobiliary cancers
	Gastrointestinal tract, especially the small and large intestine	Inflammatory bowel disease and colorectal cancer
Integumentary	Skin and underlying tissue	Wound infections
Reproductive	Vagina	Bacterial vaginosis
	Uterus and fallopian tubes	Chronic endometritis
	Mammary glands (breasts)	Mastitis
Respiratory	Nasal cavity and paranasal sinuses	Chronic rhinosinusitis
	Throat, i.e. pharynx with tonsils and adenoids and larynx with vocal cords	Pharyngitis and laryngitis
	Upper and lower airways	Pertussis (whooping cough) and other border tell infections, cystic fibrosis
Urinary	Prostate gland	Chronic bacterial prostatitis
	Urethra, bladder, ureters, kidneys	Urinary tract infections

Adapted from Vestby et al. (2020)

endothelium, nonbacterial thrombotic endocarditis (NBTE) is established when the endothelium is disrupted. This accumulates platelets, fibrin and red blood cells. Fibronectin is released by the endothelium cells with the result of vascular injury. This fibronectin can adhere to collagen, human cells and also the bacteria which leads to biofilm formation. Multiple medications are followed specific to the species involved like the administration of penicillin for streptococcal endocarditis and fluconazole for *Candida* endocarditis (Donlan and Costerton 2002; Stickler 1996.).

9.3.2 Otitis Media

This is a painful ear infection specific in the middle ear located behind the eardrum. The organisms causing otitis media are *Streptococcus pneumoniae*, *Haemophilus influenzae*, *Moraxella catarrhalis*, *Staphylococcus epidermidis* and *Pseudomonas aeruginosa*. Since only low amounts of antibiotics are penetrated in the middle ear, effective drugs are better used to treat otitis media like amoxicillin, cefaclor and erythromycin (Donlan and Costerton 2002; Wells et al. 1995).

9.3.3 Chronic Bacterial Prostatitis

The infection is seen in the prostate gland which moved from the urethra (Domingue and Hellstrom 1998). Once the bacteria occupy the prostate duct, proliferation occurs very rapidly forming spores of microcolonies and thus forming the biofilms in the dust system. Microbes infecting chronic bacterial prostatitis are *E.coli*, *P. aeruginosa* and species of *Proteus*, *Serratia*, etc. (Donlan and Costerton 2002).

9.3.4 Cystic Fibrosis

This is a type of lower respiratory tract infection. Cystic fibrosis is the absence of cystic fibrosis transmembrane conductance regulatory protein (CFTR). The thickening of the respiratory epithelium is due to the elevated absorption of electrolytes. *Staphylococcus aureus*, *H. influenzae* and *P. aeruginosa* are the microbes involved. The infection can be delayed for several years if treated in the early stages with ciprofloxacin and colistin (Donlan and Costerton 2002).

9.3.5 *Periodontitis*

This disease is the infection in teeth tissues, gums and periodontal tissues. Subgingival crevice is the first site for infection (Govan and Deretic 1996). The bacteria start colonizing the surfaces of the tooth and mucus and also regulate calcium flux releasing the toxins. Once the bacterial plaque becomes significant with minerals of calcium and phosphate ions (calculus or tartar), the defensive mechanism of the saliva can no longer support the enamel of the tooth and thus cause periodontal diseases (Overman 2000). The suggested treatment is through the elimination of biofilms from the subgingival areas along with addition of antimicrobial substances (Donlan and Costerton 2002).

9.4 Antibiofilm Strategies

With respect to the previous studies, there is an easy and effective way to eliminate premature biofilms with the use of antibiotic than the mature biofilms (Ranita Roy et al. 2018). Despite of it, undetectable nature of the premature biofilms in the body leads to development of clinical conditions which involves most of the mature biofilms (Hoiby et al. 2011; Cramton et al. 1999; Götz 2002; McKenney et al. 1998). Implementation of combinational therapy is more appreciable than the antibiotic monotherapy (Aaron et al. 2002). The biofilm-grown bacteria are more resistant to planktonic bacteria. The other strategy to eliminate biofilm is by finding antifouling or antimicrobial surfaces (Brandl et al. 2008). Nanoparticles made of silver and many other metal nanoparticles display antimicrobial properties which seem to be an effective approach to remove the biofilms (Hoyle and Costerton 1991; Moreau-Marquis et al. 2008; Prasad et al. 2020). The antibiofilm molecules block the signalling pathways in almost all types of bacteria; these molecules can be an enzyme, a peptise, an antibiotic, polyphenols, etc. (Parsek and Singh 2003). Membrane technology has been developed for wastewater treatment. But the formation of biofilm on the membrane is deteriorating the life of the membrane and reducing the water flux. This is the critical problem in the membrane technology development (Ding et al. 2019; Xu and Liu 2011).

The metabolic uncouplers are introduced to disrupt oxidative phosphorylation suppressing the microbial attachment and also reducing the extracellular polymeric substance (EPS) secretions (Chen et al. 2002; Jiang and Liu 2012). The study aimed at inducing the uncoupler, 3,3′,4′,5′ tetrachlorosalicylanilide (TCS), to reduce the EPS and aerobic granulation formation. The optimal level of TCS concentration effectively inhibits the cell binding to the membrane. Bacterial motility is also an important factor in determining the initial cell attachment (O'Toole and Kolter 1998). When the concentration of TCS was increased to 100 μg/L, it showed that the reduced motility ultimately resulted in decreased cell attachment (Feng et al. 2020). The effective way to target the destruction of biofilm is quorum sensing

(Sambanthamoorthy et al. 2014; Kareem et al. 2017; Yu et al. 2018). Cells communicate through different signalling molecules but the way by which the expression of virulence genes is controlled by quorum sensing. The molecules or the compounds which interrupt these communications are called quorum quenchers. These quenchers suppress the virulence gene expression which make the proteases, siderophores, toxic compounds and biofilm formation (Antunes et al. 2010; Ali et al. 2020).

Virstatin was employed which cuts the pili binding by *Acinetobacter baumannii* in order to avoid biofilm production (Chabane et al. 2014). Nanoparticles always showed a better path to inhibit the microbes and so put forth (Ansari et al. 2014). It has been experimented that silver nanoparticles inhibited the growth and occupancy of *E. coli* and *Klebsiella pneumoniae* and also eliminated the exopolysaccharides formation. An inhibitor of multidrug resistant of *A. baumannii* biofilm, 5-episinnuleptolide, attenuated the genes expressing the EPS-producing enzymes which completely vanished in exposure to these compounds (Tseng et al. 2016). Sometimes, the binding of bacteria to the surfaces also depends on the physical properties. Therefore, an effective approach to inhibit the biofilm formation is by changing the surface of several nanostructures of multiwalled carbon nanotubes (Malek et al. 2016).

Bacterial biofilms are a huge threat to the humankind when discussing its growth in water distribution pipelines. These pipes are mostly made of iron stainless steel and galvanized steel or copper-based materials, so biofilms growing on these metals corrode the equipments used in the industry and thus deteriorate the quality of water leading to infectious diseases. The bacteria multiplied in number on stainless steel and titanium. The growth is decreased on copper and nickel substrates as a result of oxidative stress and protein dysfunction. The growth of bacteria in the initial stages is much low on copper, Cu and nickel and Ni substrates when compared to stainless steel (SS) and titanium (Ti); it may be because of varied interactions of microbes with various materials. For example, observe a green coloured material on the copper surface; the possibility is that the surface gave copper ions by getting oxidized. This ceases cellular protein or enzyme activity resulting in the prevention of bacterial attachment to Cu substrate (Santo et al. 2011). In his actual work, the bacterial growth restored after a period of incubation; this occurred as the cells experienced extreme membrane damage, though the DNA is not injured as it is protected by the periplasm (Grosse et al. 2014). Studies showed that the *E.coli* was chiefly destroyed through membrane damage, and also they are wide open to toxic portion of cu (II) which upregulates the genes responsible for ROS (reactive oxygen species) elimination (Wang et al. 2020).

In the recent years, "green antimicrobials" derived from green medicinal plants have been a promising replacements to the conventional ones to eradicate biofilms. These include essential oils which are of high essence due to its cheaper cost, biocompatibility and the ability to fight the bacteria without impelling drug resistance. To critical mechanism of its bactericidal effects is that they can separate the lipid layer from the cell membrane, which increases the permeability of the membrane troubling the cell structures (Wang et al. 2019). This is due to the hydrophobic nature leading them to indissoluble and unsteady in aqueous media limiting its application in therapeutics. Studies have displayed that enveloping essential oils into a surface-active colloidal transport channel enhances their stability in aqueous medium and also the antibiofilm activity in the food and beverages (Arfat et al. 2014; Chen and Zhong 2015; Landis et al. 2017). Therefore, according to Zhaojie Wang et al. (2018), prepared regulatable thymol (essential oil made of oxygenated compound, phenol) contains chitosan micelles for treating bacterial biofilms. Here, the chitosan is a well-known polycationic polysaccharide made of dispersed structures of beta-(1-4)-linked D-glucosamine and N=acetyl-D-glucosamine with the best biocompatibility and antimicrobial properties. The micelles were produced through spontaneous assembly by amphipathic copolymer comprising of toluidine blue O (TBO)-implanted chitosan (CHI-TBO) and poly(propylene sulphide) (PPS). And now the chitosan, external region of the micelle, easily sticks to the oppositely charged that are the negatively charged biofilms. The ROS (reactive oxygen species) generator is the TBO which is associated with chitosan acting as a photosensitizer destroying various bacteria. This ROS can alter the hydrophobic sulphide to invariable oxidized hydrophilic sulfoxide which is much needed to kill the bacteria. So, ROS is generated from the TBO from thymol-loaded TBO-CHI-PPS micelles (T-TCP) with a simultaneous release of thymol from it (Wang et al. 2018).

9.5 Nanoscience

Nowadays, the nanoparticles are widely used in the fields of biomedical and physiology. Nanoparticle, the name itself, suggests it to be a tiny particle of nanoscale having the size of 1–100 nm. These nanoparticles also have a specific wavelength which is less than that of light. This property allows them to deploy in cosmetics, packaging and coatings. The physical, optical properties, etc. of nanoparticles make them play a unique role in the daily life when compared to the bulk materials. Sometimes nanoparticles of the desired shape and size can be obtained by controlling the parameters like salt concentration, pH value, temperature, aeration, etc. The most common shapes produced are spherical, triangular and hexagonal. Usually, nanoparticles work best when the size is less than the critical value, i.e. 10–20 nm (Singh et al. 2016). Metal nanoparticles are purely made of the metal precursors. Due to well-known localized surface plasmon resonance (LSPR) characteristics, these nanoparticles possess unique optoelectrical properties. The alkali and noble metal nanoparticles whose absorption band is in the visible region of the electromagnetic solar spectrum are Cu, Ag and Au. The facet-, size- and shape-controlled synthesis of metal nanoparticles is important in present-day cutting-edge materials (Dreaden et al. 2012). Metal nanoparticles with the advanced optical properties find applications in many research fields.

9.5.1 *Classification of Silver Nanoparticles*

Classification of nanoparticles is based on their morphology, size and chemical properties. Some of the well-known classes of NPs are listed below.

9.5.1.1 Carbon-Based NPs

Fullerenes and carbon nanotubes (CNTs) represent two major classes of carbon-based NPs. Fullerenes contain nanomaterial that are made of globular hollow cage such as allotropic forms of carbon. They have created noteworthy commercial interest due to their electrical conductivity, high strength, structure, electron affinity and versatility (Aliana Astefanei 2015). These materials possess arranged pentagonal and hexagonal carbon units, while each carbon is sp^2 hybridized, shows some of the well-known fullerenes consisting of C60 and C70 with the diameter of 7.114 and 7.648 nm, respectively. CNTs are elongated, tubular structures, 1–2 nm in diameter (El-Sherbiny et al. 2013). These are structurally resembling to graphite sheet rolling upon itself. The rolled sheets can be single, double or multiwalled, and therefore they are named as single-walled (SWNTs), double-walled (DWNTs) or multiwalled carbon nanotubes (MWNTs), respectively. Deposition of carbon precursors especially the atomic carbons, vaporized from graphite by laser or by electric arc onto metal particles, is widely required for their synthesis. Lately, they have been synthesized via chemical vapour deposition (CVD) technique (Elliott et al. 2013). Due to their unique physical, chemical and mechanical characteristics, these materials are not only used in pristine form but also in nanocomposites for many commercial applications such as fillers (Saeed and Khan 2014, 2016), as support medium for different inorganic and organic catalysts (Mabena et al. 2011), and for environmental remediation, these can be used as efficient gas adsorbents (Ngoy et al. 2014).

9.5.1.2 Metal NPs

Metal NPs are purely made of metal precursors. Due to well-known localized surface plasmon resonance (LSPR) characteristics, these NPs possess unique optoelectrical properties. NPs of the alkali and noble metals, i.e. the broad absorption band of Cu, Ag and Au, lie in the visible zone of the electromagnetic solar spectrum.

The facet-, size- and shape-controlled synthesis of metal NPs is important in present-day cutting-edge materials (Dreaden et al. 2012). Due to their advanced optical properties, metal NPs find applications in many research areas. Gold NP coating is widely used for the sampling of SEM, to enhance the electronic stream, which helps in obtaining high-quality SEM images.

9.5.1.3 Ceramic NPs

Ceramic NPs are inorganic nonmetallic solids, synthesized via heat and successive cooling. They can exist in the form of amorphous, polycrystalline, dense, porous or hollow structures (Sigmund et al. 2006). With their use in applications such as catalysis, photocatalysis, photodegradation of dyes and imaging applications, nanoparticles are getting great attention of researchers (Thomas et al. 2015).

9.5.1.4 Semiconductor NPs

Semiconductor NPs possess characteristics of metals and nonmetals and therefore found various applications in the literature due to this property (Ali et al. 2017; Khan et al. 2017).

Semiconductor NPs possess wide band gaps and therefore showed significant alteration in their properties with band gap tuning. Therefore, they play a very important role in photocatalysis, photo-optics and electronic devices (Sun et al. 2000). Because of their suitable band gap and band edge positions, a variety of semiconductor NPs are found exceptionally efficient in water-splitting applications (Hisatomi et al. 2014).

9.5.1.5 Polymeric NPs (PNPs)

Polymeric nanoparticles are generally organic based and are mostly found in nanospheres or nanocapsular shaped (Mansha et al. 2017; Prasad et al. 2017). The overall mass of the matrix particles is generally solid, and the other molecules attach to the outer boundary of the spherical surface using the phenomenon called adsorption. In the following case, the solid mass is completely encapsulated within the particle (Rao and Geckeler 2011). The PNPs can readily functionalize and thus find bundles of applications in the literature.

9.5.1.6 Lipid-Based NPs

The lipid nanoparticles contain lipid moieties and have effective applications in biomedicine. Generally, a lipid NP is characteristically spherical with diameter ranging from 10 to 1000 nm. In a similar way to polymeric NPs, lipid NPs also possess a solid core built of lipid and a matrix containing soluble lipophilic molecules. Surfactants or emulsifiers stabilized the external core of these NPs (Rawat et al. 2011). Lipid nanotechnology (Mashaghi et al. 2013) is a special field, focusing on the designing and synthesis of lipid NPs for various applications in drug delivery and as drug carriers (Puri et al. 2009) and RNA release in cancer therapy.

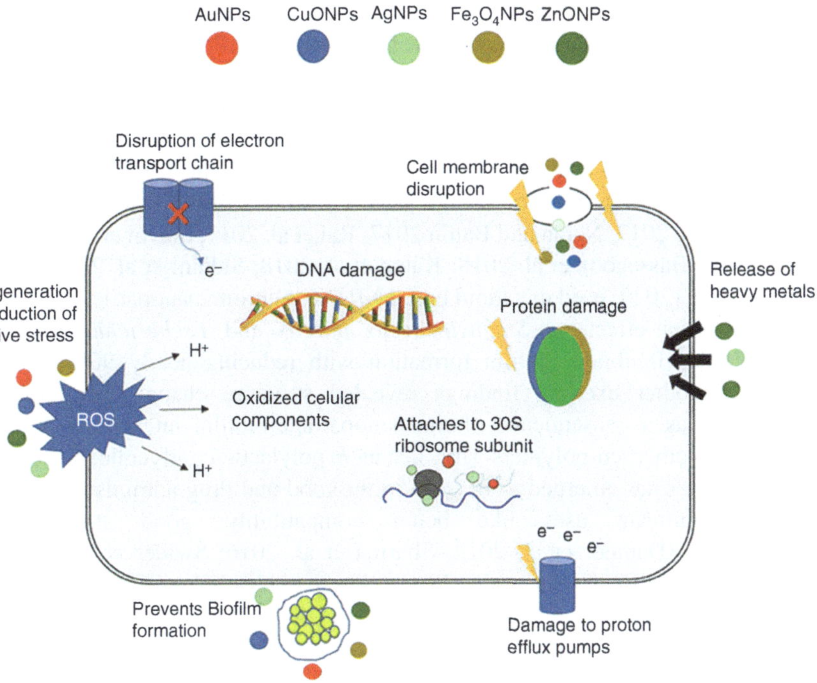

Fig. 9.2 Different mechanisms of action of NPs in bacterial cells. The combination in a single nanomaterial of a multitude of cellular effects may have a tremendous in fighting MDR bacteria. DNA, deoxyribonucleic acid; ROS, reactive oxygen species; AuNPs, gold NPs; CuONPs, copper oxide NPs; Ag NPs, silver NPs; Fe_3O_4 NPs, iron oxide NPs; ZnONPs, zinc oxide NPs. (Adapted from Pedro Bapista et al. 2018)

9.5.2 *Synthesis of Nanoparticles*

Nanoparticles can be synthesized in two following approaches: (1) top-down approach and (2) bottom-up approach. In simple words, the way in which smaller particles assemble into complex particles operated either by physical or chemical forces is called bottom-up approach, while the process in which bulk material is turned into simply smaller particles preferably nano-scaled materials is called top-down approach. Nanoparticles can be synthesized using three different methods. They are:

(1) Physical method: nanoparticles are synthesized by physical means such as by mechanical means and vaporization.

(2) Chemical method: nanoparticles are synthesized using chemicals via sol-gel process, aerosol process, etc. This method is the widely used and accepted method.

(3) Biological method: other name for this method is the green synthesis because it makes use of plant extracts or parts of the leaves, root, flower and fruits and biological species like fungi, yeast, algae and bacteria (Prasad 2014, 2016, 2017; Prasad et al. 2016, 2018a,b; Thangadurai et al. 2020; Srivastava et al. 2021).

There are multiple ways by which nanoparticles exhibit antimicrobial activity; they are (a) interacting directly with the bacterial cell wall, (b) prevention of biofilm formation, (c) evoking the natural and acquired immune responses, (d) production of reactive oxygen species (ROS) and (e) triggering of intracellular effects (Fig. 9.2) (Prasad and Swamy 2013; Joshi et al. 2018; Inamuddin et al. 2021). Since they do not follow similar mechanism of action of standard antibiotic drugs, it can be widely used against resistant bacteria (Singh et al. 2014; Aderibigbe 2017; AlMatar et al. 2017; Hemeg 2017; Natan and Banin 2017; Rai et al. 2017; Slavin et al. 2017; Zaidi et al. 2017; Bassegoda et al. 2018; Katva et al. 2018; Siddiqi et al. 2018). Gómez-Gómez et al. (2020) used metalloid based NP like tellurium nanoparticles (TeNPs) to investigate its effect on *Staphylococcus aureus* and *Escherichia coli*. These nanoparticles inhibited biofilm formation with reducing nearly 90% of biofilm volume; another exciting findings revealed structure change from sphere to rod-shaped as a consequence of the nanoparticle-biofilm interaction. While the increasing, other co-polymers (for instance polylactic-co-glycolic acid, PLGA) other polymer has emerged to be used in the food and drug administration; PLGA shows significant uses like better compatibility, good stability while preparation (Danhier et al. 2012; Sharma et al. 2016; Swider et al. 2018). So, enclosure of antibiotics into these PLGA nanoparticles helps in safe transport and release at the infection site which in general get degraded by enzymes (Huang et al. 2020). But also the relative low drug packing capability and premature or initial burst release limited the use of PLGA-based nanomaterials in in vivo studies. Therefore, quantum dots appeared as an efficient way to combine with polymer-based nanoparticles for added advantage (Huang et al. 2020).

Carbon quantum dots (CQDs) showed an outstanding drug packing capacity because it possesses a large surface area and stable π-π stacking, water repelling and electrostatic interactions or physisorption (Liu et al. 2012; Wang et al. 2017b). The method used for incorporating the CDQs into the PLGA nanoparticles is microvotex-based microfluid which accurately controls the CQD-PLGA hybrid nanoparticle formation with a fine antimicrobial efficacy to fight with *P. aeruginosa* biofilms (Huang et al. 2020). The enclosure efficacy and the packing capacity of the CQDs can be changed by altering the mass ratio of PLGA to CQDs which helps us facilitate a sufficient space for photothermal effect optimization and reducing the toxic effects triggered by the CQDs. The photothermal effect of both CQDs with and without the PLGA nanoparticles was studied by inducing laser of 808 nm with different power densities. The consequences are the increase in temperature from 37 degrees Celsius to 43 degree Celsius which is seen only in the PLGA with CQDs (Huang et al. 2020). Another interesting finding was that the CQD-encapsulated PLGA nanoparticles are able to transform NIR light into thermal energy which lead us to enhance the photothermal effects. In this study, they used azithromycin (AZI) as an antibiotic and loaded into CQD-PLGA hybrid nanoparticles which eliminated more bacteria of *P. aeruginosa* than the ones with the only AZI. This may be due to increased concentration of antibiotic at the reach of the biofilm site (Huang et al. 2020).

9.6 Knowledge Gaps and Future Directions

The purpose of using nanoparticles in preference to antibiotics is because nanoparticles can inhibit microbial drug resistance effectively in specific cases (Wang et al. 2017a). The highly preventable measures of biofilms are accomplished by smaller size with high surface-area-to-mass ratio. The shape of the nanoparticle also has a noticeable effect on biofilm elimination, for example, rod-shaped nanoparticles show a high impact over the spherical-shaped nanoparticles (Slomberg et al. 2013). With high available research studies, developing the effects of nanoparticles is the starting step for any researcher to try his best (Wang et al. 2017a). In spite of it, a research study reported that there was a spread of multiple drug resistance (MDR) not just in the same species of bacteria but also across the genera when it was the positive hope in promoting conjugative transfer of RP4, PK2 and PCF10 plasmids by aluminium nanoparticles (Qiu et al. 2012). The underlying factors may be the range of damage caused to the cell membranes by the aluminium NPs, the amounts of aluminium NPs and the breeding cells, parameters like temperature and pH and selective expression of particular genes (trfAp, trfA and trbB) which is crucial in transferring and replicating RP4 plasmids. The negative effects are also considered to avoid the MDR which may lead to health hazards.

There are limitations regarding the use of NPs in inhibiting the biofilms. As there are various bacterial strains, with different action times, it's difficult to examine the comparative studies of the antibacterial mechanisms. The complexity of the cell membrane structures can be seen as a critical drawback in in vitro studies. Size can become a limitation in transporting all NPs into the bacterial porins which is generally <600 Da. Further research focusing on the intracellular inhibitory mechanisms is left unattended. Huge attention is made towards the NPs which induced oxidative stress, synthesis of proteins and metabolism of bacterial cells. In the view of increasing resistance of the biofilms, nanoparticles are considered to having the greater potential to resolve with low toxic effects (Wang et al. 2017a). However, key factors like NP resistance and surface associations between NP biofilms and hosts need to be sorted out to guarantee fortunate clinical applications (Ramasamy and Lee 2016).

9.7 Conclusions

As of the resistance posed by the biofilms against the conventional antibiotics, there evolved multiple pathogenesis in humans which lead to chronic infections over decades. It has been evident that the therapeutic use of nanoparticles had tremendous antibiofilm effects as far studied. The influence of nanoparticles on bacterial cell membrane permeability, generation of reactive oxygen species and cellular metabolism and reproduction are of high priority. The bacterial attachment and EPS secretions can be reduced by using metabolic uncouplers like 3,3′,4′,5′

tetrachlorosalicylanilide (TCS). Among numerous types of nano-based materials used and studied in antibiofilm strategy, using silver nanoparticles is one of the best possible way in eliminating the microbes. Quorum quenchers can be used to suppress the expression of virulence genes like proteases, siderophore, etc. which is a way to block quorum sensing characteristic. Other compounds like virstatin and 5-episinnuleptolide are also found to be the key inhibitors of some of the bacterial species.

References

Aaron SD, Ferris W, Ramotar K, Vandemheen K, Chan F, Saginur R (2002) Single and combination antibiotic susceptibilities of planktonic, adherent, and biofilm-grown Pseudomonas aeruginosa isolates cultured from sputa of adults with cystic fibrosis. J Clin Microbiol 40:4172–4179

Aderibigbe BA (2017) Metal-based nanoparticles for the treatment of infectious diseases. Molecules 22(8):1370

Al-Fattani MA, Douglas LJ (2006) Biofilm matrix of Candida albicans and Candida tropicalis: chemical composition and role in drug resistance. J Med Microbiol 55:999–1008

Ali S, Khan I, Khan SA, Sohail M, Ahmed R, Atteq UR, Ansari MS, Morsy MA (2017) Electrocatalytic performance of Ni@ Pt core–shell nanoparticles supported on carbon nanotubes for methanol oxidation reaction. J Electroanal Chem 795:17–25

Ali SG, Ansari MA, Alzohairy MA, Alomary MN, Jalal M, AlYahya S, Asiri SMM, Khan HM (2020) Effect of biosynthesized ZnO nanoparticles on multi-drug resistant *Pseudomonas aeruginosa*. Antibiotics 9(5):260

Aliana Astefanei (2015) Analysis and characterization of fullerene nanoparticles. University of Barcelona, julio de

AlMatar M, Makky EA, Var I, Koksal F (2017) The role of nanoparticles in the inhibition of multidrug-resistant bacteria and biofilms. Curr Drug Deliv 15:470–484

Ansari MA, Khan HM, Khan AA, Cameotra SS, Pal R (2014) Antibiofilm efficacy of silver nanoparticles against biofilm of extended spectrum β-lactamase isolates of Escherichia coli and Klebsiella pneumoniae. Appl Nanosci 4:859–868

Antunes LCM, Ferreira RB, Buckner MM, Finlay BB (2010) Quorum sensing in bacterial virulence. Microbiology 156(8):2271–2282

Arfat YA, Benjakul S, Prodpran T, Sumpavapol P, Songtipya P (2014) Properties and antimicrobial activity of fish protein isolate/fish skin gelatin film containing basil leaf essential oil and zinc oxide nanoparticles. Food Hydrocoll 41:265–273

Bapista PV et al (2018) Nano-strategies to fight multidrug resistant bacteria- "A Battle of the titans". Front Microbiol 9:1441

Baptista PV, McCusker MP, Carvalho A, Ferreira DA, Mohan NM, Martins M, Fernandes AR (2018) Nano-strategies to fight multidrug resistant bacteria—"A Battle of the Titans". Front Microbiol 9:1441

Bassegoda A, Ivanova K, Ramon E, Tzanov T (2018) Strategies to prevent the occurrence of resistance against antibiotics by using advanced materials. Appl Microbiol Biotechnol 102:2075–2089

Brandl K, Plitas G, Mihu CN, Ubeda C, Jia T, Fleisher M, Schnabl B, DeMatteo RP, Pamer EG (2008) Vancomycin-resistant enterococci exploit antibiotic-induced innate immune deficits. Nature 455:804–807

Chabane YN, Mlouka MB, Alexandre S, Nicol M, Marti S, Pestel-Caron M et al (2014) Virstatin inhibits biofilm formation and motility of Acinetobacter baumannii. BMC Microbiol 14:62

Chen H, Zhong Q (2015) A novel method of preparing stable zein nanoparticle dispersions for encapsulation of peppermint oil. Food Hydrocoll 43:593–602
Chen G-H et al (2002) Utilization of a metabolic uncoupler, 3,3′,4′,5-tetrachlorosalicylanilide (TCS) to reduce the sludge growth in activated sludge cultures. Water Res 36:2077–2083
Costerton JW (1999) Introduction to biofilm. Int J Antimicrob Agents 11(3–4):217–221
Costerton JW, Stewart PS, Greenberg EP (1999) Bacterial biofilms: a common cause of persistent infections. Science 284:1318–1322
Cramton SE, Gerke C, Schnell NF, Nichols WW, Götz F (1999) The intercellular adhesion (ica) locus is present in Staphylococcus aureus and is required for biofilm formation. Infect Immun 67 (10):5427–5433
Cushion MT, Collins MS, Linke MJ (2009) Biofilm formation by Pneumocystis spp. Eukaryot Cell 8(2):197–206
Danhier F, Ansorena E, Silva JM, Coco R, Le Breton A, Préat V (2012) PLGA-based nanoparticles: an overview of biomedical applications. J Control Release 161(2):505–522
Ding A, Lin D, Zhao Y, Ngo HH, Guo W, Bai L, Liang H (2019) Effect of metabolic uncoupler, 2, 4-dinitrophenol (DNP) on sludge properties and fouling potential in ultrafiltration membrane process. Sci Total Environ 650:1882–1888
Domingue GJ, Hellstrom WJ (1998) Prostatitis. Clin Microbiol Rev 11(4):604–613
Donlan RM (2002) Biofilms: microbial life on surfaces. Emerg Infect Dis 8(9):881
Donlan RM, Costerton JW (2002) Biofilms: survival mechanisms of clinically relevant microorganisms. Clin Microbiol Rev 15(2):167–193
Dreaden EC, Alkilany AM, Huang X, Murphy CJ, El-Sayed MA (2012) The golden age: gold nanoparticles for biomedicine. Chem Soc Rev 41(7):2740–2779
Elliott R, Watson J, Greenberg LS, Timulak L, Freire E (2013) Research on humanistic-experiential psychotherapies. In: Lambert MJ (ed) Bergin & Garfield‘s handbook of psychotherapy and behavior change, 6th edn. Wiley, New York, pp 495–538
El-Sherbiny IM, Salih E, Reicha FM (2013) Green synthesis of densely dispersed and stable silver nanoparticles using myrrh extract and evaluation of their antibacterial activity. J Nanostructure Chem 3(1):8
Fanning S, Mitchell AP (2012) Fungal biofilms. PLoS Pathog 8(4):e1002585
Feng X, Wu Q, Che L, Ren N (2020) Analyzing the inhibitory effect of metabolic uncoupler on bacterial initial attachment and biofilm development and the underlying mechanism. Environ Res 185:109390
Finkel JS, Mitchell AP (2011) Genetic control of Candida albicans biofilm development. Nat Rev Microbiol 9:109–118
Gómez-Gómez B, Sanz-Landaluce J, Pérez-Corona MT, Madrid Y (2020) Fate and effect of in-house synthesized tellurium based nanoparticles on bacterial biofilm biomass and architecture. Challenges for nanoparticles characterization in living systems. Sci Total Environ 719:137501
Götz F (2002) Staphylococcus and biofilms. Mol Microbiol 43(6):1367–1378
Govan JR, Deretic V (1996) Microbial pathogenesis in cystic fibrosis: mucoid Pseudomonas aeruginosa and Burkholderia cepacia. Microbiol Rev 60(3):539–574
Grosse C, Schleuder G, Schmole C, Nies DH (2014) Survival of *Escherichia coli* cells on solid copper surfaces is increased by glutathione. Appl Environ Microbiol 80(22):7071–7078
Hall-Stoodley L, Stoodley P (2009) Evolving concepts in biofilm infections. Cell microbiol 11 (7):1034–1043
Hemeg HA (2017) Nanomaterials for alternative antibacterial therapy. Int J Nanomedicine 12:8211–8225
Hisatomi T, Kubota J, Domen K (2014) Recent advances in semiconductors for photocatalytic and photoelectrochemical water splitting. Chem Soc Rev 43(22):7520–7535
Hoffman LR, D’Argenio DA, MacCoss MJ, Zhang Z, Jones RA, Miller SI (2005) Aminoglycoside antibiotics induce bacterial biofilm formation. Nature 436(7054):1171–1175

Høiby N (2017) A short history of microbial biofilms and biofilm infections. APMIS 125 (4):272–275

Hoiby N, Ciofu O, Johansen HK, Song ZJ, Moser C, Jensen PØ, Bjarnsholt T (2011) The clinical impact of bacterial biofilms. Int J Oral Sci 3(2):55–65

Hoyle BD, Costerton JW (1991) Bacterial resistance to antibiotics: the role of biofilms. Prog Drug Res/Fortschritte der Arzneimittelforschung/Progrès des recherches pharmaceutiques 37:91–105

Huang Z, Zhou T, Yuan Y, Kłodzińska SN, Zheng T, Sternberg C, Wan F (2020) Synthesis of carbon quantum dot-poly lactic-co-glycolic acid hybrid nanoparticles for chemo-photothermal therapy against bacterial biofilms. J Colloid Interface Sci 577:66–74

Inamuddin, Ahamed MI, Prasad R (2021) Advanced antimicrobial materials and applications. Springer, Singapore. ISBN: 978-981-15-7098-8. https://www.springer.com/gp/book/9789811570971

Jiang B, Liu Y (2012) Roles of ATP-dependent N-acylhomoserine lactones (AHLs) and extracellular polymeric substances (EPSs) in aerobic granulation. Chemosphere 88:1058–1064

Joshi N, Jain N, Pathak A, Singh J, Prasad R, Upadhyaya CP (2018) Biosynthesis of silver nanoparticles using *Carissa carandas* berries and its potential antibacterial activities. J Sol-Gel Sci Techn 86(3):682–689. https://doi.org/10.1007/s10971-018-4666-2

Karatan E, Watnick P (2009) Signals, regulatory networks, and materials that build and break bacterial biofilms. Microbiol Mol Biol Rev 73(2):310–347

Kareem SM, Al-Kadmy IM, Al-Kaabi MH, Aziz SN, Ahmad M (2017) Acinetobacter baumannii virulence is enhanced by the combined presence of virulence factors genes phospholipase C (plcN) and elastase (lasB). Microb Pathog 110:568–572

Katva S, Das S, Moti HS, Jyoti A, Kaushik S (2018) Antibacterial synergy of silver nanoparticles with gentamicin and chloramphenicol against Enterococcus faecalis. Pharmacogn Mag 13: S828–S833

Khan I et al (2017) Synthesis of hierarchical WO_3 and Bi_2O_3/WO3 nanocomposites for solar-driven water splitting applications. Int J Hydrog Energy 42:3431–3439

Kokare CR, Kadam SS, Mahadik KR, Chopade BA (2007) Studies on bioemulsifier production from marine Streptomyces sp. S1. IJBT 6:78–84

Landis RF, Gupta A, Lee YW, Wang LS, Golba B, Couillaud B, Rotello VM (2017) Cross-linked polymer-stabilized nanocomposites for the treatment of bacterial biofilms. ACS Nano 11 (1):946–952

Lewis K (2001) Riddle of biofilm resistance. Antimicrob Agents Chemother 45(4):999–1007

Liu C, Zhang P, Zhai X, Tian F, Li W, Yang J, Liu W (2012) Nano-carrier for gene delivery and bioimaging based on carbon dots with PEI-passivation enhanced fluorescence. Biomaterials 33 (13):3604–3613

Loussert C, Schmitt C, Prevost MC, Balloy V, Fadel E et al (2010) In vivo biofilm composition of Aspergillus fumigatus. Cell Microbiol 12:405–410

Mabena LF et al (2011) Nitrogen-doped carbon nanotubes as a metal catalyst support. Appl Nanosci 1(2):67–77

Maddela NR, Meng F (2020) Discrepant roles of a quorum quenching bacterium (Rhodococcus sp. BH4) in growing dual-species biofilms. Sci Total Environ 713:136402

Maddela NR, Zhou Z, Yu Z, Zhao S, Meng F (2018) Functional determinants of extracellular polymeric substances in membrane biofouling: experimental evidence from pure-cultured sludge bacteria. Appl Environ Microbiol 84(15). https://doi.org/10.1128/AEM.00756-18

Maddela NR, Sheng B, Yuan S, Zhou Z, Villamar-Torres R, Meng F (2019) Roles of quorum sensing in biological wastewater treatment: a critical review. Chemosphere 221:616–629

Malek L et al (2016) Vertically aligned multi walled carbon nanotubes prevent biofilm formation of medically relevant bacteria. J Mater Chem B 4:5228–5235

Mansha M, Khan I, Ullah N, Qurashi A (2017) Synthesis, characterization and visible-light-driven photoelectrochemical hydrogen evolution reaction of carbazole-containing conjugated polymers. Int J Hydrog Energy 42(16):10952–10961

Mashaghi S, Jadidi T, Koenderink G, Mashaghi A (2013) Lipid nanotechnology. Int J Mol Sci 14 (2):4242–4282
McKenney D, Hübner J, Muller E, Wang Y, Goldmann DA, Pier GB (1998) The ica locus of Staphylococcus epidermidis encodes production of the capsular polysaccharide/adhesin. Infect Immun 66(10):4711–4720
Moreau-Marquis S, Stanton BA, O'Toole GA (2008) Pseudomonas aeruginosa biofilm formation in the cystic fibrosis airway. Pulm Pharmacol Ther 21:595–599
Natan M, Banin E (2017) From nano to micro: using nanotechnology to combat microorganisms and their multidrug resistance. FEMS Microbiol Rev 41:302–322
Ngoy JM et al (2014) A CO2 Capture technology using multi-walled carbon nanotubes with polyaspartamide surfactant. Energy Procedia 63:2230–2248
O'Toole GA, Kolter R (1998) Flagellar and twitching motility are necessary for Pseudomonas aeruginosa biofilm development. Mol Microbiol 30(2):295–304
Overman PR (2000) Biofilm: a new view of plaque. J Contemp Dent Pract 1(3):18–29
Parsek MR, Singh PK (2003) Bacterial biofilms: an emerging link to disease pathogenesis. Annu Rev Microbiol 57:677–701
Prasad R (2014) Synthesis of silver nanoparticles in photosynthetic plants. J Nanoparticles, Article ID 963961. https://doi.org/10.1155/2014/963961
Prasad R (2016) Advances and applications through fungal nanobiotechnology. Springer International Publishing, Cham. ISBN 978-3-319-42989-2
Prasad R (2017) Fungal nanotechnology: applications in agriculture, industry, and medicine. Springer Nature Singapore Pte Ltd., Singapore. ISBN 978-3-319-68423-9
Prasad R, Swamy VS (2013) Antibacterial activity of silver nanoparticles synthesized by bark extract of *Syzygium cumini*. J Nanoparticles. https://doi.org/10.1155/2013/431218
Prasad R, Pandey R, Barman I (2016) Engineering tailored nanoparticles with microbes: quo vadis. WIREs Nanomed Nanobiotechnol 8:316–330. https://doi.org/10.1002/wnan.1363
Prasad R, Pandey R, Varma A, Barman I (2017) Polymer based nanoparticles for drug delivery systems and cancer therapeutics. In: Kharkwal H, Janaswamy S (eds) Natural polymers for drug delivery. CAB International, London, pp 53–70
Prasad R, Jha A, Prasad K (2018a) Exploring the realms of nature for nanosynthesis. Springer International Publishing, Cham. ISBN 978-3-319-99570-0. https://www.springer.com/978-3-319-99570-0
Prasad R, Kumar V, Kumar M, Wang S (2018b) Fungal nanobionics: principles and applications. Springer Nature Singapore Pte Ltd., Singapore. ISBN 978-981-10-8666-3. https://www.springer.com/gb/book/9789811086656
Prasad R, Siddhardha B, Dyavaiah M (2020) Nanostructures for antimicrobial and antibiofilm applications. Springer International Publishing, Cham. ISBN 978-3-030-40336-2. https://www.springer.com/gp/book/9783030403362
Puri A et al (2009) Lipid-based nanoparticles as pharmaceutical drug carriers: from concepts to clinic. Crit Rev Ther Drug Carrier Syst 26:523–580
Qiu Z, Yu Y, Chen Z, Jin M, Yang D, Zhao Z, Huang A (2012) Nanoalumina promotes the horizontal transfer of multiresistance genes mediated by plasmids across genera. Proc Natl Acad Sci 109(13):4944–4949
Rai M, Ingle AP, Pandit R, Paralikar P, Gupta I, Chaud MV et al (2017) Broadening the spectrum of small-molecule antibacterials by metallic nanoparticles to overcome microbial resistance. Int J Pharm 532:139–148
Ramasamy M, Lee J (2016) Recent nanotechnology approaches for prevention and treatment of biofilm-associated infections on medical devices. Biomed Res Int 2016:1851242
Rao JP, Geckeler KE (2011) Polymer nanoparticles: preparation techniques and size-control parameters. Prog Polym Sci 36:887–913
Rawat MK et al (2011) Studies on binary lipid matrix based solid lipid nanoparticles of repaglinide: in vitro and in vivo evaluation. J Pharm Sci 100:2366–2378

Roy R, Tiwari M, Donelli G, Tiwari V (2018) Strategies for combating bacterial biofilms: a focus on anti-biofilm agents and their mechanisms of action. Virulence 9(1):522–554
Saeed K, Khan I (2014) Preparation and properties of single-walled carbon nanotubes/poly(butylene terephthalate) nanocomposites. Iran Polym J 23:53–58
Saeed K, Khan I (2016) Preparation and characterization of single-walled carbon nanotube/nylon 6,6 nanocomposites. Instrum Sci Technol 44:435–444
Sambanthamoorthy K, Luo C, Pattabiraman N, Feng X, Koestler B, Waters CM, Palys TJ (2014) Identification of small molecules inhibiting diguanylate cyclases to control bacterial biofilm development. Biofouling 30(1):17–28
Santo CE, Lam EW, Elowsky CG, Quaranta D, Domaille DW, Chang CJ, Grass G (2011) Bacterial killing by dry metallic copper surfaces. Appl Environ Microbiol 77(3):794–802
Sharma S, Parmar A, Kori S, Sandhir R (2016) PLGA-based nanoparticles: a new paradigm in biomedical applications. TrAC Trends Anal Chem 80:30–40
Siddiqi KS, Husen A, Rao RA (2018) A review on biosynthesis of silver nanoparticles and their biocidal properties. J Nanobiotechnol 16(1):14
Sigmund W et al (2006) Processing and structure relationships in electrospinning of ceramic fiber systems. J Am Ceram Soc 89:395–407
Silva S, Negri M, Henriques M, Oliveira R, Williams DW et al (2011) Adherence and biofilm formation of non-Candida albicans Candida species. Trends Microbiol 19:241–247
Singh K, Panghal M, Kadyan S, Chaudhary U, Yadav JP (2014) Green silver nanoparticles of Phyllanthus amarus: as an antibacterial agent against multi drug resistant clinical isolates of Pseudomonas aeruginosa. J Nanobiotechnology 12:40
Singh P, Kim YJ, Zhang D, Yang DC (2016) Biological synthesis of nanoparticles from plants and microorganisms. Trends Biotechnol 34(7):588–599
Slavin YN, Asnis J, Hafeli UO, Bach H (2017) Metal nanoparticles: understanding the mechanisms behind antibacterial activity. J Nanobiotechnology 15:65
Slomberg DL, Lu Y, Broadnax AD, Hunter RA, Carpenter AW, Schoenfisch MH (2013) Role of size and shape on biofilm eradication for nitric-oxide-releasing silica nanoparticles. ACS Appl Mater Interfaces 5(19):9322–9329
Srivastava S, Usmani Z, Atanasov AG, Singh VK, Singh NP, Abdel-Azeem AM, Prasad R, Gupta G, Sharma M, Bhargava A (2021) Biological nanofactories: using living forms for metal nanoparticle synthesis. Mini-Rev Med Chem 21(2):245–265
Stickler DJ (1996) Bacterial biofilms and the encrustation of urethral catheters. Biofouling 9 (4):293–305
Sun S, Murray CB, Weller D, Folks L, Moser A (2000) Monodisperse FePt nanoparticles and ferromagnetic FePt nanocrystal superlattices. Science 287(5460):1989–1992
Swider E, Koshkina O, Tel J, Cruz LJ, de Vries IJM, Srinivas M (2018) Customizing poly (lactic-co-glycolic acid) particles for biomedical applications. Acta Biomater 73:38–51
Thangadurai D, Sangeetha J, Prasad R (2020) Functional bionanomaterials. Springer International Publishing, Cham. ISBN 978-3-030-41464-1. https://www.springer.com/gp/book/9783030414634
Thomas S, Kumar Mishra P, Talegaonkar S (2015) Ceramic nanoparticles: fabrication methods and applications in drug delivery. Curr Pharm Des 21(42):6165–6188
Tortora GJ, Funke BR, Case CL (2015) Microbiology: an introduction. Pearson Higher Ed
Tseng SP, Hung WC, Huang CY, Lin YS, Chan MY, Lu PL, Sheu JH (2016) 5-Episinuleptolide decreases the expression of the extracellular matrix in early biofilm Formation of multi-drug resistant Acinetobacter baumannii. Mar Drugs 14(8):143
Vasudevan R (2014) Biofilms: microbial cities of scientific significance. J Microbiol Exp 1 (3):00014
Vestby LK et al (2020) Bacterial biofilm and its role in the pathogenesis of disease. Antibiotics (Basel) 9(2):59
Wang L, Hu C, Shao L (2017a) The antimicrobial activity of nanoparticles: present situation and prospects for the future. Int J Nanomedicine 12:1227–1249

Wang H, Wang K, Mu Q, Stephen ZR, Yu Y, Zhou S, Zhang M (2017b) Mesoporous carbon nanoshells for high hydrophobic drug loading, multimodal optical imaging, controlled drug release, and synergistic therapy. Nanoscale 9(4):1434–1442

Wang H, Qian J, Ding F (2018) Emerging chitosan-based films for food packaging applications. J Agri food Chem 66(2):395–413

Wang Z, Bai H, Lu C, Hou C, Qiu Y, Zhang P, Mu H (2019) Light controllable chitosan micelles with ROS generation and essential oil release for the treatment of bacterial biofilm. Carbohydr Polym 205:533–539

Wang J, Li G, Yin H, An T (2020) Bacterial response mechanism during biofilm growth on different metal material substrates: EPS characteristics, oxidative stress and molecular regulatory network analysis. Environ Res 185:109451

Wells CJ, Leech GJ, Lever AML, Wansbrough-Jones MH (1995) Treatment of native valve Candida endocarditis with fluconazole. J Infect 31(3):233–235

Xu H, Liu Y (2011) Control and cleaning of membrane biofouling by energy uncoupling and cellular communication. Environ Sci Technol 45:595–601

Yu S, Zhu X, Zhou J, Cai Z (2018) Biofilm inhibition and pathogenicity attenuation in bacteria by Proteus mirabilis. R Soc Open Sci 5(4):170702

Zaidi S, Misba L, Khan AU (2017) Nano-therapeutics: a revolution in infection control in post antibiotic era. Nanomed Nanotechnol Biol Med 13:2281–2301

Ziran BH (2007) Osteomyelitis. J Trauma Acute Care Surg 62(6):S59–S60

Chapter 10
Significance of Nanoscience in Food Microbiology: Current Trend and Future Prospects

Ajay Kumar, Amit Pratush, and Surojit Bera

Abstract Nanotechnology is an outstanding discovery for mankind which transform the food industry and conventional food science. Nanomaterial used in food sector could have different properties like solubility, thermodynamic, magnetic, optical, colour, etc. A nanostructure of a material widely differs from its macrostructure in terms of the texture, taste, odour, charge on the surface, etc. Nanosensing, nanostructured ingredients and packaging are the major areas where nanotechnology revolutionized the food industry. As compared to traditional practices, nanotechnology-based active and intelligent methods have greater advantages like increased mechanical strength, nanosensor for pathogen detection and improved gas exchange. The natural nanostructures are mostly proteins such as milk proteins like casein, polysaccharides, lipids, etc., whereas the synthetic nanostructures are polymeric NPs, nanoemulsions and liposomes. Several nanomaterials which control or pause chemical reactions lead to the spoilage used in the food industry for sustainable food characteristic preservation. Nanotechnology also works on a two-way approach to maintain or preserve food characteristics, i.e. modifying the packaging technology or changing the bioactivity of active food components in the packaging environment. However, there are some odds like safety issues creating challenges in modern nanotechnology. In the long run, proper training of the public is needed to ensure the benefits and safety of the application of nanotechnology in the food industry.

Keywords Nanomaterial · Active packaging · Encapsulation · Biosensor · Bioactive molecules

A. Kumar · S. Bera (✉)
Department of Microbiology, School of Bioengineering and Biosciences, Lovely Professional University, Phagwara, Punjab, India

A. Pratush
Arboreal Bio-Innovations Pvt. Ltd., Lucknow, Uttar Pradesh, India

N. R. Maddela et al. (eds.), *Nanotechnology for Advances in Medical Microbiology, Environmental and Microbial Biotechnology*,
https://doi.org/10.1007/978-981-15-9916-3_10

10.1 What is Nanoscience?

Nanoscience or nanotechnology is the functional area of modern era, where unification of science with engineering opens a new door in everyday science. In this area productive molecule size varies between 1 and 100 nanometre at least in one dimension. It is also the foundation and deployment of ingredients, manoeuvre or system, through the regulation of the properties and structure of the matter at the nanometric scale. However according to some researchers, there is a fine line between nanoscience and nanotechnology which distinguishes them between each other. For example, nanoscience is a junction physical science, chemical science, biological science and mathematics which deals with the fabrication of materials at the atomic and molecular level; on the other hand, nanotechnology is the measurement, congregation, control and manufacturing of matters at nanoscale. It is the most truly immerging technology of the twenty-first century. The concept of nanotechnology was first familiarized by a Nobel laureate Richard Feynman in 1959 (Bayda et al. 2020). He was an American physicist who showed the pathway towards this small but powerful field of study. Almost 15 years later, a Japanese scientist Norio Taniguchi first defined the term "nanotechnology" in 1974 as "nanotechnology mainly consists of the processing of separation, consolidation, and deformation of materials by one atom or one molecule". Taniguchi was also considered the father of modern nanotechnology by many scientists. It would be very clear from an example of economic investment in United States (US). In 2015 financial year, US federal R&D investment was calculated approximately $20 billion, more than double that by the private sector. The revenues generated from nanoproducts is continuously increasing with over $200B in financial year 2012 in the United States alone and over $700B worldwide (Bhushan 2016).

Now, the main question comes into mind that why should we go for such small and complicated nanosystem? The answer is embedded in the physical and chemical properties of particles. Matters mostly remain in three states: solid, liquid and gas. Where we cut down the bulk structures of the materials into very small fractions like nanoscale, the regular physical, chemical and biological properties differ. For example, in nanoscale some materials may act stronger, or lighter, and may have magnetic properties, better electricity or heat transfer. Also, they may become more chemically responsive or redirect light better or alter colour as their size or structure is altered. Figure 10.1 shows different aspects of nanoscience in food sectors and their ongoing applications.

10.2 Relation of Food Microbiology with Nanoscience

The topic we are discussing here is related to food microbiology also; and the food microbiology comes under the category of food science. The consumer's preference about food colour, texture and overall quality is continuously changing with a

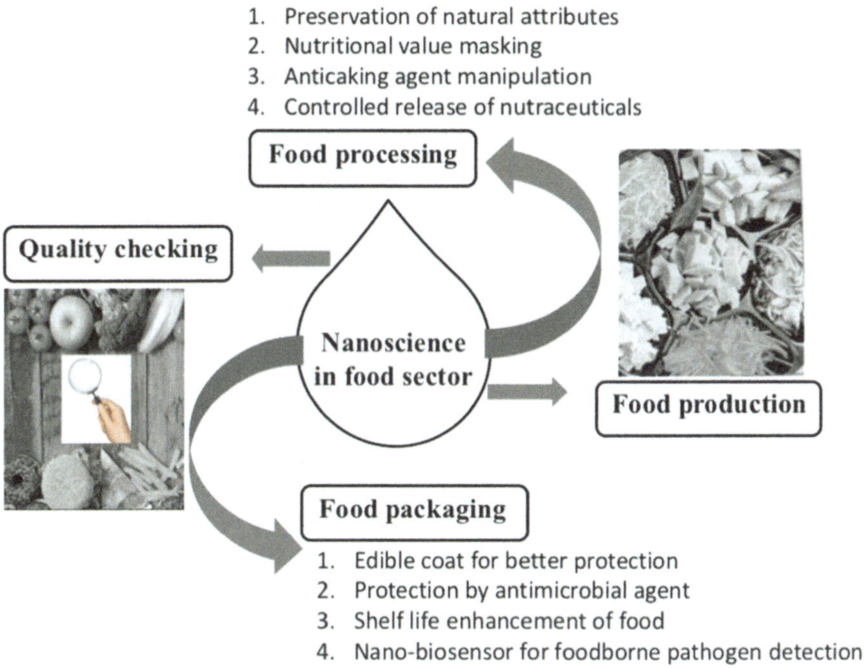

Fig. 10.1 Different aspect-cum-applications of nanoscience in food sector

growing concern over health benefits (Singh 2017). Now, consumers are also aware about the nutritional value of the product; and this nutritional value mostly depends upon various micronutrients and bioactive compounds. However, when microorganisms invaded the intactness of a food, their metabolic activities negatively amend these micronutrients and/or bioactive compounds jeopardizing the health benefits of the foods. As a result, researchers are continuously engaged in searching of new methods to improve the quality of foods without disturbing the basic contents of it.

In the immediate twentieth century, unceasing findings were made to comprehend the connection and significance of microorganisms, particularly pathogenic bacteria in food. Later but before the 1970s, the section food microbiology was considered as a functional science discipline which predominantly involved in the microbial quality control in food. Gradually with time, the technology expended in food manufacture, processing and food consumption patterns has considerably transformed. As a result, complications associated with it have also evolved vividly. These kinds of problems can no longer be solved by only technology application. Hence, research in basic food science and in modern food microbiology is inevitable to understand and effectually resolve the microbiological issues associated with food. The food microbiology discipline not only embraces microbiological aspects of food decomposition, food-borne infections and their efficient control but also fundamental evidence of microbial ecology, physiology, metabolism and genetics. An example of such application gained from food microbiology knowledge is genetically modified bacterial strains to produce fermented foods of better quality. However, emerging nanotechnology shows some prospects in this regard to conglomerate numerous regulatory procedures for effective control of food spoilage and pathogenic microorganisms in food.

Frequent investigation has been conducted to obtain human perception towards nanotechnology in the United States and in Europe. However, the data showed that community knowledge about nanotechnology is extremely narrow (Dingman JS 2008). According to the US Department of Agriculture (USDA), by 2015 the worldwide impression of products in the field of nanotechnology plays a crucial character, and it will be roughly $1 trillion yearly. Food microbiologists are attentive in quality control and quality assurance programs to fabricate innocuous and high-quality food products with zero defects and free of pathogens. In this chapter, we have discussed some key applications of nanotechnology in food microbiology in terms of food safety and human health. The chapter deals with application of nanotechnology in active packaging as antimicrobial agents (e.g. nanocomposite cellulose, etc.); food pathogen detection (e.g. through nanosensors); protection of "good" microbes (e.g. probiotics); food characteristic preservation (e.g. colour, texture, aroma, etc.), i.e. quality attributes; and their future applications.

10.3 General Applications of Nanoscience in Food Technology

Nanotechnology is an excellent technology which revolutionizes the food industry and conventional food science. Nanotechnology proved to be a great tool for the processing and packaging of various foods (Weiss et al. 2006). Nanoparticles used in food could be generated by different technologies which have different properties like solubility, thermodynamic, magnetic, optical, colour, etc. (Feng et al. 2010; Gupta et al. 2016). But till far the acceptability by the public and the agreement on rules for this technology worldwide are not clear (Bieberstein et al. 2013).

Nanotechnology improves the texture and taste, hides bad odour/taste and amends the size of the particle as well as charge on the surface (Powers et al. 2006). Increasing awareness and concern regarding health, nowadays consumers are very much particular about the quality of food and food safety. This necessitates the scientist to look for technology which strengthens the food safety without degrading the nutritional quality of food/product. As nontoxic nanomaterials sometimes contain essential elements and can withstant high pressure and temperature, the demand of these nanoparticles in food industry is very high (Sawai 2003).

The nanoscale development of around 5 nm edible nano-coating from nanoemulsion is used in various foods like bakery goods, vegetables, fruits, meat, fast foods, confectionary, etc. which act as a barrier for moisture and exchange of gases. This edible coating also serves as medium to deliver enzymes, colours, anti-browning compounds, antioxidants and flavour. The shelf life of opened manufactured food is also increased by edible coating (Azeredo et al. 2009; Naoto et al. 2009; Saxena et al. 2017; Nile et al. 2020).

Food nanosensing and food nanostructured ingredients are two major areas where nanotechnology has application in the food industry. Further, food nanostructured ingredients are related to food processing to food packaging, whereas food nanosensing are concerned with food safety and quality. Nanostructures applied in food processing used as antimicrobial agent, nutrient delivery carrier, improve durability and strength of packaging material, food safety, etc. (Ezhilarasi et al. 2013; Prasad et al. 2014, 2017a,b; Thangadurai et al. 2020).

It has been reported that food nanostructured ingredients improve consistency, texture and taste of food along with improvement in shelf life of food (Pradhan et al. 2015; Singh et al. 2017). Nanocarriers are used to deliver food additives without any adverse effect on the morphology of the food. The size of nanocarriers directly related to the absorption efficiency of the material in humans, i.e. if smaller the size, efficiently it is absorbed inside the host (Ezhilarasi et al. 2013; Singh et al. 2017). The following properties should be possessed by the ideal nanocarrier: (i) it must deliver active compound at a specific rate and at a specific time, (ii) it should be able to deliver active compound at a specific target site, and (iii) it must maintain active compound for longer period of time without any modification. The formation of biopolymer matrices, encapsulation, simple solution and emulsion can be done by the application of nanotechnology. The release efficiency and encapsulation properties of nanoparticles are better as compared to other systems. The encapsulation by the nanoparticles has shown excellent results like it has good control over the release of active compounds; it removes/captures odours or taste in food; with other compounds in food, it showed excellent compatibility; and during storage, processing and utilizing, it protects from chemicals, heat, moisture and biological degradation; and finally it also has an outstanding control over the availability of active agent at a target time and at a controlled rate (Weiss et al. 2006; Singh et al. 2017). Besides, the nanoparticles are very small in size so they are efficiently absorbed by the body and deliver the active compound at a target site (Singh et al. 2017).

Nanotechnology has an important role in improving food taste as well as improving the quality of food. The colour from juice of beetroot was removed using nanofilters without deteriorating the flavour. Also, for lactose-intolerant people, nanofilters were used to remove lactose from milk and substitute it with another sugar. Bioencapsulation of tomato with L-lactide improves the shelf life of tomato, and this can be used on other perishable foods to increase the shelf life (Yadav 2017; Nile et al. 2020). Nanofilters are also used to remove microbial contamination from water or milk (Sekhon 2010). The use of nanotechnology improves the photostability and thermal stability of recombinant soybean seed by encapsulating the molecules of cyanidin 3-O-beta-D-glucoside (anthocyanin, a plant pigment) which is present in the inner cavity of the seed (Zhang et al. 2014). Another important application of nanoencapsulation is for flavonoid rutin which has pharmacological activity, but due to less solubility, its use is limited in the food industry. The encapsulated rutin showed more solubility and improved stability (Yang et al.

2015; Singh et al. 2017). Nanomaterials of silicon dioxide and titanium dioxide are used in food as colour agents, and also these are used as flavour/fragrance carrier specially silicon dioxide (Dekkers et al. 2011). Vitamins, proteins, lipids, etc. are important bioactive molecules present in the food which are very sensitive to pH in the human stomach, and also they have less solubility. Encapsulation leads to the increased resistance to the adverse condition in the human body, and also it improves the solubility of these bioactive molecules which help in the increased assimilation. To efficiently deliver various important bioactive molecules using encapsulation method, different techniques are available like nanostructuration, nanoemulsion and nanocomposition. For example, vitamins and flavonoids are well encapsulated using polymeric nanoparticles which help in the efficiently transport and protection of these bioactive molecules (Langer and Peppas 2003). The bioactive molecules in the food get degraded due to the adverse conditions which may be inside the host or outside the host. The encapsulation of these bioactive molecules slows down the degradation rate and provides sufficient time to deliver these bioactive molecules at a target site. This encapsulation also acts as a carrier for the delivery of enzymes, flavour, antioxidant, colours, etc. and also prevents gas exchange or moisture absorption which improves the shelf life of even opened food (Renton 2006; Weiss et al. 2006; Singh et al. 2017). The encapsulation of curcumin which is the bioactive molecule present in turmeric leads to the increased stability of this molecule to ionic strength of various concentrations and also showed stability to pasteurization temperature (Sari et al. 2015; Singh et al. 2017).

10.4 Nanotechnology in Active Packaging of Food

Food packaging is an important part of food processing to improve shelf life. The ideal characteristic of packaging material includes strength, gas permeability and biodegradability (Couch et al. 2016; Singh et al. 2017). As compared to the methods of traditional packaging, nanotechnology-based "active" and "intelligent" methods have greater advantages like increased mechanical strength, nanosensor for pathogen detection and improved gas exchange (Mihindukulasuriya and Lim 2014; Singh et al. 2017). In the food industry, nanotechnology uses nanomaterials like nanoemulsions, nanoparticles and nanoclays to improve the shelf life of food. Nanocomposites are an important part of food packaging system. Many organic compounds (bacteriocins, essential oils, etc.) were tested by research workers as antimicrobial agents in polymeric matrices, but these organic compounds are susceptible to the various food processing steps like high temperature, etc. (Schirmer et al. 2009). So, these difficulties which are encountered during the use of organic compounds can be overcome by the use of nanoparticles (copper, iron, silver, zinc oxide, magnesium oxide, titanium oxide, etc.) which are stable under adverse conditions and even at low concentration showing powerful antimicrobial activities (Singh et al. 2017). Reactive oxygen species generated by titanium dioxide is highly lethal to microbes which make it as a suitable antimicrobial agent. The amendment of polymers with nanoparticles generates cost-effective strong packaging material. Nanocomposites used for coating and packaging purpose and their use improved the mechanical strength and heat resistance, generate low weight material and improve the gas/moisture exchange (Pinto et al. 2013; Mihindukulasuriya and Lim 2014). The amendment of polymeric matrix with active nanoparticles improves the thermal stability, decreases the permeability of gases, makes the matrix resistant to fire and generates light weight material (Duncan 2011). This amendment also improves the shelf life of food by providing scavenging, antimicrobial and antioxidant activity (Sorrentino et al. 2007). In packed food industries, the utilization of nanotechnology influences the aroma and flavour characteristic of food, growth of microbes and moisture regulation (Brody et al. 2008).

In the food industry, active and intelligent nanotechnology is widely used to improve the shelf life of various food products specially the packed one. In active packaging system, the incorporated component either absorbs moisture, oxygen and carbon dioxide or releases antioxidant or antimicrobial molecules either into the food or out of the food or in the vicinity of food environment which helps in the increase of shelf life of food. The quality and shelf life of food product improved by the blend of these active compounds with polymers (Ranjan et al. 2014; Majid et al. 2018). Many inorganic nanoparticles have important application in the food industry. The most common used inorganic nanoparticles are silver, iron, zinc, gold and oxides of metals like silicon oxide, titanium oxide, magnesium oxide, zinc oxide, etc. (Bikiaris and Triantafyllidis 2013). These nanoparticles either slowly migrate and react or directly react with organic molecule of food. These nanoparticles directly kill the microbes by damaging the cell envelop or by interfering the electron transport or indirectly by producing various reactive species or by oxidizing various components of cells (Li et al. 2008). Silver nanoparticles are most commonly used as an antimicrobial agent against most microbial strains. These silver nanoparticles are active not only against bacteria, but also they showed detrimental activity against fungi and viruses (Duncan 2011). These silver nanoparticles show bacteriostatic activity by binding to different proteins, enzymes and DNA of bacteria (Cavaliere et al. 2015; Aziz et al. 2014, 2015, 2016; Joshi et al. 2018).

Many researchers have successfully used silver nanoparticles to inhibit the persistence of common food-borne pathogens. The use of silver nanoparticles decreases the level of pathogenic *Clostridium perfringens* and *E. coli* in the animal feed which helps in the cut-down use of antibiotics in livestocks (Fondevila et al. 2009; Pineda et al. 2012; Elkloub et al. 2015; Adegbeye et al. 2019). Silver nanoparticles are also successfully used in the treatment of water by integrating these particles to filter (Zodrow et al. 2009; Dankovich and Gray 2011). In the food industry, the application of silver nanoparticles is still limited, but attempts are done to substitute the use of sulphur dioxide with these nanoparticles with antimicrobial activity specially in the wine industry (Izquierdo-Canas et al. 2012; Garde-Cerdan et al. 2014; Garcia-Ruiz et al. 2015). Many natural phytochemicals are used as antimicrobial agents in the packaging material for the food to improve the shelf life (Manso et al. 2013; Medina-Jaramillo et al. 2017; Moreno et al. 2019). But silver nanoparticles have greater antimicrobial activity as compared to these phytochemicals. The silver nanoparticles interact with food directly or with the polymer matrix. Stable nature of silver nanoparticles and slow release into the food make these nanoparticles an excellent option in the packaging of food (Duncan 2011). Various research workers used silver nanoparticles in combination with different polymers like laponite and cellulose nanofibrils (Wu et al. 2018; Yu et al. 2019) which showed a detrimental effect on *E. coli*, *S. aureus*, *L. monocytogenes*, *P. citrinum* and *A. niger*. The coating/covering of nanocomposite film (made of polyvinyl pyrrolidone along with silver nanoparticles) around the asparagus improved the shelf life at refrigerated temperature (An et al. 2008). A packaging film was prepared from banana powder, agar and silver nanoparticles that showed antibacterial activity against *L. monocytogenes* and *E. coli* (Orsuwan et al. 2016).

Besides silver, copper nanoparticles also showed promising antimicrobial activity results. These copper nanoparticles showed detrimental activity by degrading the DNA, enzymes, proteins, lipid peroxidation and production of reactive oxygen species (Chatterjee et al. 2014; Yadav et al. 2017). Cioffi et al. (2005) showed antimicrobial activity of copper nanoparticles against *S. aureus*, *E. coli*, *L. monocytogenes* and *S. cerevisiae*. Copper nanoparticles in polyurethane nanofibers showed excellent antibacterial activity against *B. subtilis* and *E. coli* (Sheikh et al. 2011). Besides this, various oxides of nanoparticles [silicon oxide (SiO_2), zinc oxide (ZnO), titanium dioxide (TiO_2) and magnesium oxide (MgO)] showed a promising result in food packaging. Silicon oxide nanoparticles showed improvement in barrier property and mechanical strength of the matrices of polymer. The silicon oxide amendment at the rate of 5% in nanocomposites showed enhancement in physical as well as mechanical properties (Salami-Kalajahi et al. 2012). The good amount of antibacterial activity is shown by zinc oxide nanoparticles. The zinc oxide nanoparticles produce reactive oxygen species which is detrimental to microbial cell, as well as it also produces zinc ions which have antimicrobial activity. Zinc oxide nanoparticles directly interact with the cell wall of the bacteria which leads to the lysis of bacterial cell (Sirelkhatim et al. 2015; Bhuyan et al. 2015). These zinc oxide nanoparticles are activated by visible light (Kim et al. 2020). Zinc ions generated from zinc oxide interact with respiratory enzymes in the bacterial cell and inhibit their activity. The absorption of zinc oxide nanoparticles in the bacterial cell ultimately damages DNA, mitochondria and cell membrane which leads to the formation of free radicals and reactive oxygen species. Because of all these events, oxidative stress is generated which hampers the activity of respiratory enzymes that cause the bacterial cell death (Kim et al. 2020). In a study it was found that different forms of zinc oxide nanoparticles (PVP-capped/coating, film, powder) in egg white and liquid culture media showed excellent antimicrobial activity against *Salmonella enteritidis* and *Listeria monocytogenes* (Jin et al. 2009). Petchwattana et al. (2016) in their study showed antibacterial activity of composite of zinc oxide and polybutylene succinate against *S. aureus* and *E. coli*.

Among all the oxides of nanoparticles, titanium oxide nanoparticles showed favourable and encouraging results (Farhoodi 2016; Sharma et al. 2017). These titanium oxide nanoparticles are activated only in the presence of ultraviolet light (Sharma et al. 2017). The activated titanium oxide nanoparticles interact with DNA, proteins and peptides of microbial cell which leads to the detrimental effect on microbial growth and eventually leads to the death (Brown et al. 2008; Sharma et al. 2017). The oxidative stress is generated by titanium oxide nanoparticles which strike on the cell membrane of bacteria; also hydroxyl radicals generated damage the bacterial DNA and modify enzyme activity which depends upon coenzyme A (Kubacka et al. 2014). In a study it was found that application of composite EVOH-titanium oxide nanoparticles after irradiation of 30 minutes showed 5-log reduction against tested pathogens (Cerrada et al. 2008). Recently, Azizi-Lalabadi et al. (2019) reported that titanium oxide nanoparticles in support into 4A zeolite decline the population of tested pathogenic bacterial species significantly.

10.5 Nanotechnology in Food Pathogen Detection

Food-borne diseases are caused by various biological agents like bacteria, fungi, viruses, etc. The conventional methods (microscopic method, immunological methods and nucleic acid methods) for the detection of pathogens are available, but they are complex, time-consuming, laborious, expensive reagents/equipment, and skilled manpower is required (Kaittanis et al. 2010) which leads to the delay in detection of toxin/pathogen and ultimately slows down the treatment process of infected host (Salyers and Whitt 2002; Manguiat and Fang 2013). Besides microbes, toxins produced by the microbes are also very detrimental to the human which damage the plasma membrane of the cell and may interfere with physiological activity (Salyers and Whitt 2002; Sonawane et al. 2014). The methods used for the detection of toxins also suffer from the same limitations which are encountered in the detection methods of microbes (Valdes et al. 2009; Lopez and Merkoci 2011; Salyers and Whitt 2002; Sonawane et al. 2014).

So, to overcome these problems/limitations, nanotechnology will play an important role with enhanced detection efficiency to detect toxins and microbial pathogens (Valdes et al. 2009; Lopez and Merkoci 2011; Kaittanis et al. 2010; Ali et al. 2011). In nanotechnology devices can be designed which can detect contaminant in shorter time, increase sensitivity and have other multifunctional roles (Jain 2005; Rosi and Mirkin 2005; Nath et al. 2008). Besides normal characteristic features possessed by the nanomaterial, they can be modified such a way that can be used for the detection of ligand which will be specific for each pathogen (Kaittanis et al. 2010; Jain 2005; Rosi and Mirkin 2005). The most common nanomaterial used in the detection of toxins and microbial pathogen includes magnetic nanoparticles, gold nanoparticles, silver nanoparticles, gold nanorods, quantum rods, etc. (Valdes et al. 2009; Lopez and Merkoci 2011).

Nanomaterial can be used to design a sensor that can detect food-borne pathogen/toxins by sensing/detecting the volatile compounds, chemical transduction or specific biological entity (biosensor) (Singh et al. 2020). In biosensors receptor which is biological in nature recognizes the specific toxin/microbial cell which leads to the signal generation by various means like thermal, mass, optical or electrochemical processes. Most of the biosensors are based on changes in electric current or colours.

Heavy metals, microbial toxins and synthetic toxins (like pesticides) pose a great health risk to human population. Fungal toxins are extremely toxic even at a low level which is a challenge for the health sector. Mycotoxins reported generally contaminate around 25% of food grains worldwide. Mycotoxins' intoxication leads to cancer, kidney problem, vomiting and liver problem, and in severe cases, it leads to the death of the patient. *Penicillium verrucosum* and *Aspergillus ochraceus* are reported to produce one of the potent mycotoxin known as ochratoxin A. This toxin is highly toxic to human kidneys and reported to have immunotoxic and teratogenic effects (Petzinger and Ziegler 2000; Cheli et al. 2008; Hayat et al. 2012).

For the detection of ochratoxin A, the biosensor was developed using cerium oxide particles and single-stranded DNA aptamer which is specific for this toxin (Bulbul et al. 2016). The attachment of ochratoxin to the aptamer results into redox property changes in cerium oxide, and this will be quantified by the colour changes of tetramethylbenzidine. This biosensor is very sensitive even 0.15 nM ochratoxin can be detected by this biosensor.

The bacterial heat-stable toxin produced by *S. aureus* is known as staphylococcal enterotoxins which is associated with food-borne illness caused by the contaminated food consumption (Sonawane et al. 2014; Yang et al. 2008; Yang et al. 2009) which leads to vomiting, nausea, diarrhoea and anorexia (Sonawane et al. 2014). The current method of identification of this toxin is based on immunological methods like ELISA which are rapid but have less sensitivity. An optical biosensor was devised in which carbon nanotubes were immobilized with anti-staphylococcal enterotoxin antibodies which is further bounded to secondary antibody with attached horseradish peroxidase (Yang et al. 2008). The sensitivity of this biosensor is six to eight times more as compared to normal immunological sensors. Another important bacterial toxin is shiga-like toxin (stx) produced by the food-borne *E. coli* specially O157:H7 strain. For the detection of shiga-like toxin, a biosensor was prepared by using surface plasmon resonance assay. In this biosensor gold nanoparticles were attached with globotriose antigen which is highly specific to shiga-like toxin (Chien et al. 2008). Another toxin, brevetoxin, which is produced by *Karenia brevis* (marine dinoflagellate) is a possible contaminant in sea food. Brevetoxin is neurotoxic, and consumption leads to diarrhoea, loss of coordination, muscle pain, respiratory problems, paresthesia, etc. (Sonawane et al. 2014). A electrochemical immunosensor was developed to detect brevetoxin using gold nanoparticles attached with poly (amidoamine) which was bounded to conjugate of brevetoxin-bovine serum albumin with attached anti-brevetoxin antibodies with horseradish peroxidase (Tang et al. 2011). This biosensor showed high sensitivity to detect brevetoxin in the range of 0.03 to 8 ng/ml.

Besides toxin, the presence of pathogenic microorganisms in the contaminated food is a big challenge to the food industry. The conventional methods/techniques currently available to detect the presence of these food-borne pathogens are laborious, are time-consuming and required skilled personnel (de Boer and Beumer 1999). One of the common pathogenic strains of *E. coli* reported to be highly pathogenic is *E. coli* O157:H7 which is able to cause disease even if its 100 cells are present

(Sonawane et al. 2014). The infection of *E. coli* O157:H7 leads to symptoms like haemorrhagic colitis, cramps in the stomach, anaemia, haemolytic uremic syndrome, etc. (Sonawane et al. 2014). Through faecal matter contamination of food and water, this organism is able to cause outbreaks. For the detection of *E. coli* O157:H7, electrochemical immune sensor was devised using quantum dots of cadmium sulphide entrapped in the zeolitic imidazolate framework made of metal organic material (Zhong et al. 2019). In another study DNA-based sensor using single-stranded thiolated DNA probe specific to *eaeA* gene present in *E. coli* O157:H7 was used to develop quartz crystal microbalance sensor (Mao et al. 2006). Lin et al. (2008) devised biosensor for the detection of *E. coli* O157:H7 in milk. In this biosensor they used screen-printed carbon electrode on which double antibodies were fabricated using gold nanoparticles. The first antibody interacts with *E. coli* O157:H7, whereas the second antibody is attached to horseradish peroxidase enzyme. Another important food-borne pathogen in the food industry is *Salmonella*. The infection caused by this microbe causes a disease known as salmonellosis. The infection symptoms include diarrhoea, fever, vomiting, abdominal pain, etc. (Lopez and Merkoci 2011, Sonawane et al. 2014). A biosensor was devised using polystyrene on which monoclonal antibodies were immobilized which is specific for *Salmonella* which is further attached to conjugate of polyclonal antibody and gold nanoparticles (Dungchai et al. 2008). In another study for the detection of *Salmonella* in milk, optical nanocrystal probe and magnetic nanoparticles were used to devised biosensor. In this biosensor, antibodies attached to magnetic nanoparticles were used to capture bacteria, and later bacteria were separated from magnetic nanoparticles using magnetic field. Further, bacteria were attached to titanium oxide nanoparticles immobilized with antibody specific for *Salmonella* for UV light absorption. Then complex of magnetic nanoparticle—*Salmonella*—titanium oxide was separated from solution by using a magnetic field, and nanocrystals of unbound titanium oxide were analysed (Joo et al. 2012).

10.6 Nanotechnology in Healthy Food Production

Nanotechnology deals with the nanostructures (size varies between 1 and 100 nm) also known as nanomaterials (Pathakoti et al. 2017). These nanomaterials possess some characteristic features which make them more unique, important and useful than there native forms (Gokularaman et al. 2017). Due to their specific physicochemical properties, the nanomaterials are used in various fields such as food preservation, food safety, etc. (Giner et al. 2020). In food nanotechnology, there are mainly two types of nanostructures that are present, i.e. natural and synthetic. The natural nanostructures are mostly food proteins such as milk proteins like casein, polysaccharides, lipids, etc. (Pathakoti et al. 2017), whereas the synthetic nanostructures are polymeric NPs, nanoemulsions and liposomes (Chang and Chen 2005). Nanotechnology in the food industry plays its role in several forms like food packaging material, farming practices, food processing as well as in food itself (Bajpai et al. 2018). Nanotechnology helps maintain sustainability in the food industry as it offers nanosensors for monitoring the physical, chemical and biological properties of the food production process (Alfadul and Elneshwy 2010). It also provides various sensor technologies for contamination (microbial as well as chemical) detection in food (Lin 2012); besides this these technologies are also helpful for controlling pathogen's growth and ensure food safety and also minimize food wastage (Rodrigues et al. 2012). The use of nanotechnology in the food industry is still in its early stage (Singh et al. 2017), and many nanotechnology-based food applications are still under development (Singh et al. 2017). Nanotechnology also provides an opportunity to improve nutrient content in food, make food nutrients and vitamins easily absorbable to the body as well as help in the masking of the unfavourable taste of some extremely healthy foods (Ravichandran 2010).

10.7 Nanotechnology in Food Characteristic Preservation

Rise in industrialization and fast transportation facilities of the intercontinental food become very much common (Sadiku et al. 2019), but this also raises a new challenge in front of food technologists to maintain or preserve the food characteristics for a long time (Moschopoulou et al. 2019). In other words, we can say that shelf life of the packed food must be increased so that every customer at any corner of the globe got the same native taste of the food (Dobruka and Cierpiszewsk 2014). Besides this nanotechnology also contributes to the manufacturing of healthier and safe foods (Gokularaman et al. 2017). Some food ingredients are present in food materials that work on nanoscale commonly known as nanostructures. These nanostructures have specific properties and are playing a very active role in the improvement of food taste, texture and consistency (Gokularaman et al. 2017). They also enhance the shelf life of food (Pradhan et al. 2015). Studies showed that the major cause of food characteristic deterioration in the chemical reactions took place between the various components of food and its surrounding environment (Bajpai et al. 2018; Dimitrijevic et al. 2015). There are several nanomaterials which control or pause these chemical reactions used in the food industry for sustainable food characteristic preservation (Ghosh et al. 2019). Nanotechnology works on a two-way approach to maintain or preserve food characteristics, i.e. modifying the packaging technology and changing the bioactivity of active food components in the packaging environment (Shafiq et al. 2020).

In the food industry, nanotechnology plays a crucial role in the improvements in food taste and its value. Nanoencapsulation is the globally known as the most common and effective method used for the progressive taste along with maintaining a culinary balance (Nakagawa 2014). This technology is applied to those food materials which are extremely reactive such as plant pigments or photoreactive (Shafiq et al. 2020). For example, polymeric nanoparticles can be used for the encapsulation of bioactive compounds such as vitamins and flavonoids which are actively released in the acidic environment of the stomach without any change in their characteristic (Singh et al. 2017). Another nanotechnique is known as nano-emulsion which is created by using fragile bioactive compounds that are soluble in lipids. This technique is used for the improvement of bioavailability and water dispersion (Kumar 2015; Shafiq et al. 2020; Singh et al. 2017). The above-discussed techniques are used for the safe and secure delivery of vitamins, fragile micronutrients and medicines (Haider and Kang 2015).

10.8 Safety Issues and Future Perspective

There is no doubt that over the last few decades the importance and use of nanotechnology in the food sector are rising up; however some objectionable safety concerns associated with nanomaterial are raising the alarm which cannot be neglected. Few studies showed that nanoparticles may adulterate food materials by migrating into it from packaging materials leading to serious health issues. Researchers claim that physicochemical properties of a substance are entirely different in its nanoform with respect to macrostate. As a result any nanomaterial being used in the food sector must possess the GRAS (generally regarded as safe) status as per authority rules and regulation. Though the small quantity of used materials appears to be safe mostly, bioaccumulation within organs and tissues may pose a distant threat. For example, sometimes silica nanoparticles are used as anti-caking agents, but they can be toxic to human lung cells too depending upon the exposure time and pattern. Need for more useful material preparation in present competitive market, pushing the existing knowledge of nanotechnology to go more smaller and sensitive. Food packaging and food safety are two pillar areas for nanouse. However a variety of factors like particle surface morphology, concentration, surface energy, aggregation, etc. may affect dissolution. Significant benefits are now being provided by nanomaterials for food preservation also like protection from moisture, lipids, gases, off-flavours and odours. They are also showing great results in the field of targeted bioactive compound delivery. However, aforementioned heath safety issues still persist, and those challenges must need to be addressed in order to elevate the use of nanomaterial in food the industry. Consumer concerns must to be kept in mind while designing smart useful nanomaterials, and therefore mandatory testing of nanofoods should be a prerequisite before they are introduced to the global market. To be successful in the long run, systematic and focused education of the public is also needed additionally to utilize the maximum benefits of nanofoods.

Acknowledgement The authors express their thanks to Lovely Professional University, Punjab, India, for their necessary support.

References

Adegbeye MJ, Elghandour MMMY, Barbabosa-Pliego A, Monroy JC, Mellado M, Ravi Kanth Reddy P et al (2019) Nanoparticles in equine nutrition: mechanism of action and application as feed additives. J Equine Vet Sci 78:29–37

Alfadul SM, Elneshwy AA (2010) Use of nanotechnology in food processing, packaging and safety review. Afr J Food Agric Nut Dev 10:21 pages

Ali ME, Hashim U, Mustafa S, Man YB, Yusop MH, Bari MF et al (2011) Nanoparticle sensor for label free detection of swine DNA in mixed biological samples. Nanotechnology 22:195503

An J, Zhang M, Wang S, Tang J (2008) Physical, chemical and microbiological changes in stored green asparagus spears as affected by coating of silver nanoparticles-PVP. LWT Food Sci Technol 41:1100–1107

Azeredo HMC, Henrique CLM, Wood D, Williams TG, Roberto JAB, Tara HM (2009) Nanocomposite edible films from mango puree reinforced with cellulose nanofibers. J Food Sci 74:31–35

Aziz N, Fatma T, Varma A, Prasad R (2014) Biogenic synthesis of silver nanoparticles using Scenedesmus abundans and evaluation of their antibacterial activity. J Nanoparticles, Article ID 689419. https://doi.org/10.1155/2014/689419

Aziz N, Faraz M, Pandey R, Sakir M, Fatma T, Varma A, Barman I, Prasad R (2015) Facile algae-derived route to biogenic silver nanoparticles: synthesis, antibacterial and photocatalytic properties. Langmuir 31:11605–11612. https://doi.org/10.1021/acs.langmuir.5b03081

Aziz N, Pandey R, Barman I, Prasad R (2016) Leveraging the attributes of Mucor hiemalis-derived silver nanoparticles for a synergistic broad-spectrum antimicrobial platform. Front Microbiol 7:1984. https://doi.org/10.3389/fmicb.2016.01984

Azizi-Lalabadi M, Ehsani A, Divband B, Alizadeh-Sani M (2019) Antimicrobial activity of Titanium dioxide and Zinc oxide nanoparticles supported in 4A zeolite and evaluation the morphological characteristic. Sci Rep 9:17439

Bajpai VK, Kamle M, Shukla S, Mahato DK, Chandra P, Hwang SK et al (2018) Prospects of using nanotechnology for food preservation, safety, and security. J Food Drug Anal 26:1201–1214

Bayda S, Adeel M, Tuccinardi T, Cordani M, Rizzolio F (2020 Jan) The history of nanoscience and nanotechnology: from chemical–physical applications to nanomedicine. Molecules 25(1):112

Bhushan B (2016) Introduction to nanotechnology: history, status, and importance of nanoscience and nanotechnology education. In: Global perspectives of nanoscience and engineering education. Springer, Cham, pp 1–31

Bhuyan T, Mishra K, Khanuja M, Prasad R, Varma A (2015) Biosynthesis of zinc oxide nanoparticles from *Azadirachta indica* for antibacterial and photocatalytic applications. Mater Sci Semicond Process 32:55–61

Bieberstein A, Roosen J, Marette S, Blanchemanche S, Vandermoere F (2013) Consumer choices for nano-food and nanopackaging in France and Germany. Eur Rev Agric Econ 40:73–94

Bikiaris DN, Triantafyllidis KS (2013) HDPE/Cu-nanofiber nanocomposites with enhanced antibacterial and oxygen barrier properties appropriate for food packaging applications. Mater Lett 93:1–4

de Boer E, Beumer RR (1999) Methodology for detection and typing of foodborne microorganisms. Int J Food Microbiol 50:119–130

Brody AL, Bugusu B, Han JH, Sand CK, Mchugh TH (2008) Innovative food packaging solutions. J Food Sci 73:107–116

Brown EM, Paunesku T, Wu A, Thurn KT, Haley B, Clark J, Priester T et al (2008) Methods for assessing DNA hybridization of peptide nucleic acid–titanium dioxide nanoconjugates. Anal Biochem 383:226–235

Bulbul G, Hayat A, Andreescu S (2016) ssDNA-functionalized nanoceria: a redox-active apta switch for biomolecular recognition. Adv Healthcare Mat 5:822–828

Cavaliere E, De Cesari S, Landini G, Riccobono E, Pallecchi L et al (2015) Highly bactericidal Ag nanoparticle films obtained by cluster beam deposition. nanomedicine: nanotechnology. Biol Med 11:1417–1423

Cerrada ML, Serrano C, Sanchez-Chaves M, Fernandez-Garcia M, Fernandez-Martin F, de Andres A (2008) Self-sterilized EVOH TiO_2 nanocomposites: effect of TiO_2 content on biocidal properties. Adv Funct Mater 18:1949–1960

Chang YC, Chen DGH (2005) Adsorption kinetics and thermodynamics of acid dyes on a carboxymethylated chitosan-conjugated magnetic nano-adsorbent. Macromol Biosci 5:254–261

Chatterjee AK, Chakraborty R, Basu T (2014) Mechanism of antibacterial activity of copper nanoparticles. Nanotechnology 25:135101

Cheli F, Pinoti L, Campagnoli A, Fusi E, Rebuci R, Baldi A (2008) Mycotoxin analysis, mycotoxin-producing fungi assays and mycotoxin toxicity bioassays in food mycotoxin monitoring and surveillance. Ital J Food Sci 20:447–462

Chien YY, Jan MD, Adak AK, Tzeng HC, Lin YP, Chen YJ et al (2008) Globotriose functionalized gold nanoparticles as multivalent probes for Shiga-like toxin. Chembiochem 9:1100–1109

Cioffi N, Torsi L, Ditaranto N, Tantillo G, Ghibelli L, Sabbatini L et al (2005) Copper nanoparticle/ polymer composites with antifungal and bacteriostatic properties. Chem Mater 17:5255–5262

Couch LM, Wien M, Brown JL, Davidson P (2016) Food nanotechnology: proposed uses, safety concerns and regulations. Agro Food Ind Hitech 27:36–39

Dankovich TA, Gray DG (2011) Bactericidal paper impregnated with silver nanoparticles for point-of-use water treatment. Environ Sci Technol 45:1992–1998

Dekkers S, Krystek P, Peters RJ, Lankveld DX, Bokkers BG, van Hoeven-Arentzen PH et al (2011) Presence and risks of nanosilica in food products. Nanotoxicology 5:393–405

Dimitrijevic M, Karabasil N, Boskovic M, Teodorovic V, Vasilev D, Djordjevic V et al (2015) Safety aspects of nanotechnology applications in food packaging. Procedia Food Sci 5:57–60

Dingman JS (2008) Nanotechnology: its impact on food safety, guest commentary. J Environ Health 70(6):47–50

Dobruka R, Cierpiszewsk R (2014) Active and intelligent packaging food—research and development—a review. Poland J Food Nutra Sci 64:7–15

Duncan TV (2011) Applications of nanotechnology in food packaging and food safety: barrier materials, antimicrobials and sensors. J Colloid Interface Sci 363:1–24

Dungchai W, Siangproh W, Chaicumpa W, Tongtawe P, Chailapakul O (2008) *Salmonella typhi* determination using voltammetric amplification of nanoparticles: a highly sensitive strategy for metalloimmunoassay based on a copper-enhanced gold label. Talanta 77:727–732

Elkloub K, El Moustafa ME, Ghazalah AA, Rehan A (2015) Effect of dietary nanosilver on broiler performance. Int J Poult Sci 14:177–182

Ezhilarasi PN, Karthik P, Chhanwal N, Anandharamakrishnan C (2013) Nanoencapsulation techniques for food bioactive components: a review. Food Bioprocess Technol 6:628–647

Farhoodi M (2016) Nanocomposite materials for food packaging applications: characterization and safety evaluation. Food Eng Rev 8:35–51

Feng T, Xiao Z, Tian H (2010) Recent patents on nano flavour preparation and its application. Recent Pat Food Nutr Agric 2:243–250

Fondevila M, Herrer R, Casallas MC, Abecia L, Ducha JJ (2009) Silver nanoparticles as a potential antimicrobial additive for weaned pigs. Anim Feed Sci Tech 150:259–269

Garcia-Ruiz A, Crespo J, Lopez-de-Luzuriaga JM, Olmos ME, Monge M, Rodriguez-Alfaro MP et al (2015) Novel biocompatible silver nanoparticles for controlling the growth of lactic acid bacteria and acetic acid bacteria in wines. Food Cont 50:613–619

Garde-Cerdan T, Lopez R, Garijo P, Gonzalez-Arenzana L, Gutierrez AR, Lopez-Alfaro I et al (2014) Application of colloidal silver versus sulfur dioxide during vinification and storage of Tempranillo red wines. Aust J Grape Wine Res 20:51–61

Ghosh C, Bera D, Roy L (2019) Role of nanomaterials in food preservation. In: Microbial nanobionics. Springer, Cham, pp 181–211

Giner ST, Prieto C, Lagaron JM (2020) Nanomaterials to enhance food quality, safety, and health impact. Nano 10:941
Gokularaman S, Cruz SA, Pragalyaashree M, Nishadh A (2017) Nanotechnology approach in food packaging-review. J Pharm Sci Res 9:1743–1749
Gupta A, Eral HB, Hatton TA, Doyle PS (2016) Nanoemulsions: formation, properties and applications. Soft Matter 12:2826–2841
Haider A, Kang IK (2015) Preparation of silver nanoparticles and their industrial and biomedical applications: a comprehensive review. Adv Mat Sci Eng 2015:1–16
Hayat A, Paniel N, Rhouati A, Marty JL, Barthelmebs L (2012) Recent advances in ochratoxin A-producing fungi detection based on PCR methods and ochratoxin A analysis in food matrices. Food Cont 26:401–415
Izquierdo-Canas PM, Garcia-Romero E, Huertas-Nebreda B, Gomez-Alonso S (2012) Colloidal silver complex as an alternative to sulphur dioxide in winemaking. Food Cont 23:73–81
Jain KK (2005) Nanotechnology in clinical laboratory diagnostics. Clin Chim Acta 358:37–54
Jin T, Sun D, Su JY, Zhang H, Sue H-J (2009) Antimicrobial efficacy of zinc oxide quantum dots against enteritidis, and O157:H7. J Food Sci 74(1):M46–M52
Joo J, Yim C, Kwon D, Lee J, Shin HH, Cha HJ et al (2012) A facile and sensitive detection of pathogenic bacteria using magnetic nanoparticles and optical nanocrystal probes. Analyst 137:3609–3612
Joshi N, Jain N, Pathak A, Singh J, Prasad R, Upadhyaya CP (2018) Biosynthesis of silver nanoparticles using *Carissa carandas* berries and its potential antibacterial activities. J Sol-Gel Sci Techn 86(3):682–689
Kaittanis C, Santra S, Perez JM (2010) Emerging nanotechnology based strategies for the identification of microbial pathogenesis. Adv Drug Deliv Rev 62:408–423
Kim I, Viswanathan K, Kasi G, Thanakkasaranee S, Sadeghi K, Seo J (2020) ZnO Nanostructures in active antibacterial food packaging: preparation methods, antimicrobial mechanisms, safety issues, future prospects, and challenges. Food Rev Int. https://doi.org/10.1080/87559129.2020.1737709
Kubacka A, Diez MS, Rojo D, Bargiela R, Ciordia S, Zapico I et al (2014) Understanding the antimicrobial mechanism of TiO_2-based nanocomposite films in a pathogenic bacterium. Sci Rep 4:4134
Kumar LY (2015) Role and adverse effects of nanomaterials in food technology. J Toxicol Health 2:1–11
Langer R, Peppas NA (2003) Advances in biomaterials, drug delivery, and bionanotechnology. AICHE J 49:2990–3006
Li Q, Mahendra S, Lyon DY, Brunet L, Liga MV, Li D et al (2008) Antimicrobial nanomaterials for water disinfection and microbial control: potential applications and implications. Water Res 42:4591–4602
Lin M (2012) Nanotechnology-based approaches for rapid detection of chemical and biological contaminants in foods. Nanotechnol Food Bev Nutra Ind 2012:317–334
Lin YH, Chen SH, Chuang YC, Lu YC, Shen TY, Chang CA et al (2008) Disposable amperometric immunosensing strips fabricated by Au nanoparticles-modified screen-printed carbon electrodes for the detection of foodborne pathogen *Escherichia coli* O157:H7. Biosens Bioelectron 23:1832–1837
Lopez BP, Merkoci A (2011) Nanomaterials based biosensors for food analysis applications. Trends Food Sci Technol 2:625–639
Majid I, Nayik GA, Dar MS, Nanda V (2018) Novel food packaging technologies: innovations and future prospective. J Saudi Soc Agric Sci 17:454–462
Manguiat LS, Fang TJ (2013) Evaluation of DADTM kits for the detection of food-borne pathogens in chicken- and meat based street-vended foods. J Food Drug Anal 21:198–205
Manso S, Cacho-Nerin F, Becerril R, Nerin C (2013) Combined analytical and microbiological tools to study the effect on *Aspergillus flavus* of cinnamon essential oil contained in food packaging. Food Cont 30:370–378

Mao X, Yang L, Su X, Yi Y (2006) A nanoparticle amplification based quartz crystal microbalance DNA sensor for detection of *Escherichia coli* O157:H7. Biosens Bioelectron 21:1178–1185

Medina-Jaramillo C, Ochoa-Yepes O, Bernal C, Fama L (2017) Active and smart biodegradable packaging based on starch and natural extracts. Carbohydr Polym 176:187–194

Mihindukulasuriya SDF, Lim LT (2014) Nanotechnology development in food packaging: a review. Trends Food Sci Technol 40:149–167

Moreno MA, Orqueda ME, Gomez-Mascaraque LG, Isla MI, Lopez-Rubio A (2019) Crosslinked electrospun zein-based food packaging coatings containing bioactive chilto fruit extracts. Food Hydrocoll 95:496–505

Moschopoulou E, Moatsou G, Syrokou MK, Paramithiotis S, Drosinos EH (2019) Food quality changes during shelf life. Food Qual Shelf Life 2019:1–31

Nakagawa K (2014) Nano-and microencapsulation of flavor in food systems. In: Nano-and microencapsulation for foods. John Wiley & Sons, Inc., Hoboken, NJ, pp 249–272

Naoto S, Hiroshi O, Mitsutoshi N (2009) Micro and nanotechnology for food processing. (Food safety series) resource: engineering and technology for a sustainable. World Am Soc Agri Eng 16:19

Nath S, Kaittanis C, Tinkham A, Perez JM (2008) Dextran-coated gold nanoparticles for the assessment of antimicrobial susceptibility. Anal Chem 80:1033–1038

Nile SH, Baskar V, Selvaraj D, Nile A, Xia J, Kai G (2020) Nanotechnologies in food science: applications, recent trends, and future perspectives. Nano-Micro Lett 12:45

Orsuwan A, Shankar S, Wang LF, Sothornvit R, Rhim JW (2016) Preparation of antimicrobial agar/banana powder blend films reinforced with silver nanoparticles. Food Hydrocoll 60:476–485

Pathakoti K, Manubolu M, Hwanga HM (2017) Nanostructures: current uses and future applications in food science. J Food Drug Anal 25:245–253

Petchwattana N, Covavisaruch S, Wibooranawong S, Naknaen P (2016) Antimicrobial food packaging prepared from Poly (Butylene Succinate) and Zinc Oxide. Measurement 93:442–448

Petzinger E, Ziegler K (2000) Ochratoxin A from a toxicological perspective. J Vet Pharmacol Ther 23:91–98

Pineda L, Chwalibog A, Sawosz E, Lauridsen C, Engberg R, Elnif J et al (2012) Effect of silver nanoparticles on growth performance, metabolism and microbial profile of broiler chickens. Arch Anim Nutr 66:416–429

Pinto RJB, Daina S, Sadocco P, Neto CP, Trindade T (2013) Antibacterial activity of nanocomposites of copper and cellulose. Biomed Res Int 2013:280512

Powers KW, Brown SC, Krishna VB, Wasdo SC, Moudgil BM, Roberts SM (2006) Research strategies for safety evaluation of nanomaterials. Part VI. Characterization of nanoscale particles for toxicological evaluation. Toxicol Sci 90:296–303

Prasad R, Kumar V, Prasad KS (2014) Nanotechnology in sustainable agriculture: present concerns and future aspects. Afr J Biotechnol 13(6):705–713

Pradhan N, Singh S, Ojha N, Srivastava A, Barla A, Rai V et al (2015) Facets of nanotechnology as seen in food processing, packaging, and preservation industry. Bio Med Res Int 2015:365672

Prasad R, Kumar V, Kumar M (2017a) Nanotechnology: food and environmental paradigm. Springer Nature Singapore Pte Ltd., Singapore. ISBN 978-981-10-4678-0

Prasad R, Bhattacharyya A, Nguyen QD (2017b) Nanotechnology in sustainable agriculture: recent developments, challenges, and perspectives. Front Microbiol 8:1014. https://doi.org/10.3389/fmicb.2017.01014

Ranjan S, Dasgupta N, Chakraborty AR, Samuel SM, Ramalingam C, Shanker R et al (2014) Nanoscience and nanotechnologies in food industries: opportunities and research trends. J Nanopart Res 16:2464

Ravichandran R (2010) Nanotechnology applications in food and food processing: innovative green approaches, opportunities and uncertainties for global market. Int J Green Nanotechnol Phys Chem 1:72–96

Renton A (2006) Welcome to the world of nano foods. http://observer.guardian.co.uk/foodmonthly/futureoffood/story/0,1971266,00.html

Rodrigues SM, Demokritou P, Dokoozlian N, Hendren CO, Karn B, Mauter MS et al (2012) Nanotechnology for sustainable food production: promising opportunities and scientific challenges. Environ Sci Nano 4:767–781
Rosi NL, Mirkin CA (2005) Nanostructures in biodiagnostics. Chem Rev 105:1547–1562
Sadiku MNO, Musa SM, Ashaolu TJ (2019) Food industry: an introduction. Int J Trend Sci Res Dev 3:128–130
Salami-Kalajahi M, Haddadi-Asl V, Roghani-Mamaqani H (2012) Study of kinetics and properties of polystyrene/silica nanocomposites prepared via in situ free radical and reversible addition-fragmentation chain transfer polymerizations. Sci Iran 19:2004–2011
Salyers AA, Whitt DD (2002) Bacterial pathogenesis: a molecular approach, 2nd edn. ASM Press, Washington DC
Sari P, Mann B, Kumar R, Singh RRB, Sharma R, Bhardwaj M, Athira S (2015) Preparation and characterization of nanoemulsion encapsulating curcumin. Food Hydrocol 43:540–546
Sawai J (2003) Quantitative evaluation of antibacterial activities of metallic oxide powders (ZnO, MgO and CaO) by conductimetric assay. J Microbiol Methods 54(2):177–182
Saxena A, Maity T, Paliwal A, Wadhwa S (2017) Technological aspects of nanoemulsions and their applications in the food sector. In: Oprea AE, Grumezescu AM (eds) Nanotechnology applications in food: flavor, stability, nutrition, and safety. Academic Press, pp 129–152
Schirmer BC, Heiberg R, Eie T, Moretro T, Maugesten T, Carlehog M (2009) A novel packaging method with a dissolving CO_2 headspace combined with organic acids prolongs the shelf life of fresh salmon. Int J Food Microbiol 133:154–160
Sekhon BS (2010) Food nanotechnology—an overview. Nanotechnol Sci Appl 3:1–15
Shafiq M, Anjum S, Hano C, Anjum I, Abbasi BH (2020) An overview of the applications of nanomaterials and nano-devices in the food industry. Foods 9:1–27
Sharma C, Dhiman R, Rokana N, Panwar H (2017) Nanotechnology: an untapped resource for food packaging. Front Microbiol 8:1735
Sheikh FA, Kanjawal MA, Saran S, Chung WJ, Kim H (2011) Polyurethane nanofibers containing copper nanoparticles as future materials. Appl Surf Sci 257:3020–3026
Singh T, Shukla S, Kumar P, Wahla V, Bajpai VK, Rather IA (2017) Application of nanotechnology in food science: perception and overview. Front Microbiol 8:1501–1507
Singh S, Kumar V, Dhanjal DS, Datta S, Prasad R, Singh J (2020) Biological biosensors for monitoring and diagnosis. In: Singh J, Vyas A, Wang S, Prasad R (eds) Microbial biotechnology: basic research and applications. Springer Nature Singapore, Singapore, pp 317–336
Sirelkhatim A, Mahmud S, Seeni A, Kaus NHM, Ann LC, Bakhori SKM et al (2015) Review on zinc oxide nanoparticles: antibacterial activity and toxicity mechanism. Nano Micro Lett 7:219–242
Sonawane SK, Arya SS, LeBlanc JG, Jha N (2014) Use of nanomaterials in the detection of food contaminants. Eur J Nutr Food Saf 4:301–317
Sorrentino A, Gorrasi G, Vittoria V (2007) Potential perspectives of bio-nanocomposites for food packaging applications. Trends Food Sci Technol 18:84–95
Tang D, Tang J, Su B, Chen G (2011) Gold nanoparticles-decorated amine-terminated poly (amidoamine) dendrimer for sensitive electrochemical immunoassay of brevetoxins in food samples. Biosens Bioelectron 26:2090–2096
Thangadurai D, Sangeetha J, Prasad R (2020) Nanotechnology for food, agriculture, and environment. Springer International Publishing, Cham. ISBN 978-3-030-31937-3. https://www.springer.com/gp/book/9783030319373
Valdes MG, Gonzalez ACV, Calzon JAG, Dıaz-Garcıa ME (2009) Analytical nanotechnology for food analysis. Microchim Acta 166:1–19
Weiss J, Takhistov P, McClements J (2006) Functional materials in food nanotechnology. J Food Sci 71:R107–R116
Wu Z, Huang X, Li YC, Xiao H, Wang X (2018) Novel chitosan films with laponite immobilized Ag nanoparticles for active food packaging. Carbohydr Polym 199:210–218

Yadav SK (2017) Tissue science & engineering realizing the potential of nanotechnology for agriculture and food technology. J Tissue Sci Eng 8:8–11

Yadav L, Tripathi RM, Prasad R, Pudake RN, Mittal J (2017) Antibacterial activity of Cu nanoparticles against *E. coli*, *Staphylococcus aureus* and *Pseudomonas aeruginosa*. Nano Biomed Eng 9(1):9–14. https://doi.org/10.5101/nbe.v9i1.p9-14

Yang M, Kostov Y, Rasooly A (2008) Carbon nanotubes based optical immunodetection of staphylococcal enterotoxin B (SEB) in food. Int J Food Microbiol 127:78–83

Yang M, Kostov Y, Bruck HA, Rasooly A (2009) Gold nanoparticle based enhanced chemiluminescence immunosensor for detection of staphylococcal enterotoxin B (SEB) in food. Int J Food Microbiol 133:265–271

Yang R, Zhou Z, Sun G, Gao Y, Xu J, Strappe P et al (2015) Synthesis of homogeneous protein-stabilized rutin nanodispersions by reversible assembly of soybean (*Glycine max*) seed ferritin. RSC Adv 5:31533–31540

Yu Z, Wang W, Kong F, Lin M, Mustapha A (2019) Cellulose nanofibril/silver nanoparticle composite as an active food packaging system and its toxicity to human colon cells. Int J Biol Macromol 129:887–894

Zhang Y, Zhu W, Lu Y, Gao Z, Gu J (2014) Nano-scale blocking mechanism of MMT and its effects on the properties of polyisocyanate-modified soybean protein adhesive. Ind Crops Prod 57:35–42

Zhong M, Yang L, Yang H, Cheng C, Deng W, Tan Y et al (2019) An electrochemical immunobiosensor for ultrasensitive detection of *Escherichia coli* O157:H7 using CdS quantum dots-encapsulated metal-organic frameworks as signal-amplifying tags. Biosens Bioelectron 126:493–500

Zodrow K, Brunet L, Mahendra S, Li D, Zhang A, Li Q et al (2009) Polysulfone ultrafiltration membranes impregnated with silver nanoparticles show improved biofouling resistance and virus removal. Water Res 43:715–723

Chapter 11
Nanotechnology in Microbiology

Rajkrishna Mondal

Abstract Pathogenic microorganisms are the immediate cause of several food and waterborne diseases. High-throughput detection of causative agent is the key to preventing rapid and wide spread of these diseases to community. Distinctive physicochemical properties and ability to conjugate with other molecules including biological origin make nanomaterials an outstanding candidate for development of novel diagnostics or improving the sensitivity of pre-existing methods. Nanotechnology-based detection techniques are well-suited for on-site and low-resource settings lacking sophisticated experimental setups. The chapter exploits about the application of nanotechnology in several aspects of microbiology mainly pathogen detection, elimination, and prevention. Nanotechnology has the potential to provide economic access of safe food, water, affordable diagnostic and treatment for a large human population.

Keywords Microbiology · Nanotechnology · Nanoparticles · Biosensor · Nanobiocides · Antimicrobial nanomaterials

11.1 Introduction

"Understanding and control of matter at dimensions between 1 and 100 nm" and "the study of biological organisms that are too small to be seen with the naked eye" are the general definition of nanotechnology and microbiology, respectively. Both are considered modern science branches despite their ancient history. Traditional medical practitioners in India and China had used "Suwarna Bhasma" (gold colloidal) as therapeutics in the fourth and fifth centuries. However, the real interest of nanotechnology was inspired by Richard Phillips Feynman during his legendary talk, "There's Plenty of Room at the Bottom," in the annual meeting of the American

R. Mondal (✉)
Department of Biotechnology, Nagaland University, Dimapur, Nagaland, India
e-mail: rajkrishna@nagalanduniversity.ac.in

N. R. Maddela et al. (eds.), *Nanotechnology for Advances in Medical Microbiology*, Environmental and Microbial Biotechnology,
https://doi.org/10.1007/978-981-15-9916-3_11

Physical Society at the California Institute of Technology (Caltech) in 1959. Modern science proved that human civilization started with the accidental discovery of the brewing process, important microbial process, in Mesopotamia. Both the fields are diverse and integration is required to combat the current problems, e.g. human health, environmental, and ecological damage, innovatively and sustainably. Nanomaterials have tunable optical, magnetic, electrical, thermal, and biological properties. It can be engineered with different shapes, sizes, chemical compositions, and surface functionalities. These properties allow them to be exploited for improving the detection of biological molecules or whole pathogens. Additionally, nanomaterials have much greater surface-area-to-volume ratios than macroscopic materials, which provide a great capacity to functionalize the surface with many molecules. For example, a cube with 1-cm dimensions can be divided into 1021 1-nm cubes, which will increase the surface area by 10 million times. Nanoparticle or its conjugates can be used as a tracer for the detection of whole-cell or the molecule(s) derived from microbes. Nanomaterials conjugated with different polymers are commercially used as packing material to prevent food spoilage. Bioconjugated magnetic nanoparticles (NPs) are successfully employed for the separation of pathogens from a variety of samples, e.g. food, water, and blood. Nanomaterials (NMs) can deliver drugs or vaccines to a target with controlled release. Some nanoparticles are used as an antimicrobial drug to control microbial infection (Inamuddin et al. 2021). Hence, nanotechnology maneuvers several areas of microbiology a few key examples exemplified below (Fig. 11.1).

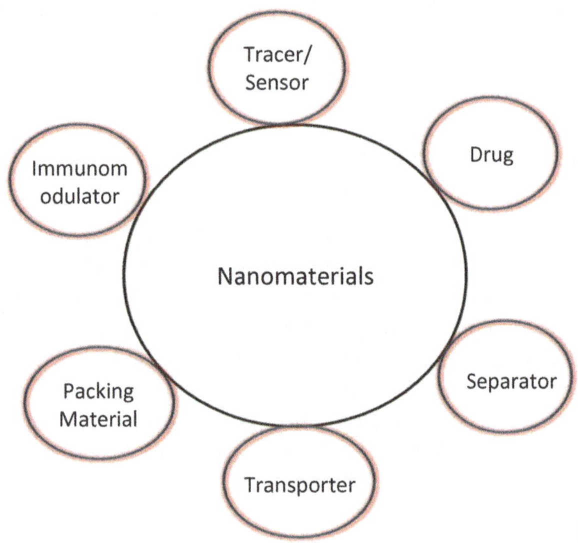

Fig. 11.1 Different aspects of nanomaterials applied in microbiology

11.2 Applications of Nanotechnology in Food Microbiology

Nanotechnology has prospective applications in many aspects of food technology including storage, quality monitoring, food processing, and food packaging (Prasad et al. 2017; Thangadurai et al. 2020). In today's world, people are passionate about ready-to-eat or minimally processed food due to their busy lifestyles. Unfortunately, these kinds of food products are more vulnerable to attack by spoilage and pathogenic microorganisms during processing and distribution (Martı'n-Belloso et al. 2006). Microbial contamination in baby food products may lead to serious pathogenic infection and malnutrition to children. Safety and quality assurance programs to produce safe and high-quality food products that have zero defects and free of pathogens are the topmost priorities of food microbiologists. Wastage and increase of self-life of food are other concerns. One-third of the edible parts of food produced for human consumption get wasted due to spoilage or mishandling. Nanotechnology can deal with multiple aspects of food technology, i.e. detection of the foodborne pathogen, food preservation, and packaging.

11.2.1 Application of Nanotechnology for Detection of Foodborne Pathogen

High-throughput detection of microbes in food materials is the key to preventing several foodborne diseases. Nanotechnology-based techniques developed recently have shown great potential for successful, reliable, and sensible detection of food pathogens. Gold, magnetic, and fluorescent nanoparticles (NPs) are used mostly among all types of nanoparticles. Nanoparticle properties like great permeability, reactivity, and high surface-to-volume ratio can be utilized to make sensitive pathogen detection method. Apart from its potential exploitation, the exposure of NPs could lead to a potential risk to human health.

According to the World Health Organization (WHO) consumption of unsafe food causes 600 million cases of foodborne diseases leading to 420,000 deaths every year worldwide. Children below 5 years of age are more susceptible to foodborne diseases. Unsafe food contaminated with harmful bacteria, viruses, parasites, or chemical substances causes simple food poisoning to life-threatening diseases like diarrhea even cancer. *Listeria monocytogenes*, *Escherichia coli* O157: H7, *Staphylococcus aureus*, *Salmonella enterica*, *Bacillus cereus*, *Vibrio* spp., *Campylobacter jejuni*, *Clostridium perfringens*, and Shiga toxin producing *Escherichia coli* (Oliver et al. 2005; Zhao et al. 2014), *Clostridium botulinum*, *Yersinia enterocolitica*, and *Coxiella burnetii* are common foodborne bacteria. Norovirus and Hepatitis A virus spread typically through raw or undercooked seafood or contaminated raw produce (Iwamoto et al. 2010). Infected food handlers are often the source of food contamination. Some parasites like *Echinococcus spp*, *Ascaris*, *Cryptosporidium*, and *Entamoeba histolytica* also cause foodborne diseases (Pozio 2003). Prion diseases can spread by consuming bovine products containing specified risk material such as brain tissue (Lücker et al. 2001). Naturally occurring toxins (e.g., mycotoxins, marine biotoxins, cyanogenic glycosides, and toxins occurring in poisonous mushrooms), persistent organic pollutants (POPs), and heavy metals such as lead, cadmium, and mercury also cause foodborne diseases(Khin 2015; Fung et al. 2018). The occurrence of foodborne diseases can be reduced through fast, cost-effective, sensitive, and accurate detection of the causative agent. The conventional culture-based methods are time-consuming, low sensitive, and laborious. Biochemical and serological testing (e.g., nucleic acid-based, biosensor-based, and immunological-based methods) provide various advantages over culture-based method but also have limitations. Advancement of nanotechnology-based research has shown high fascinating and promising results for microbial detection in food. Additionally, the amalgamation of nanotechnology in biosensing has improved the sensitivity and detection limits of biological events, microbial toxins, bacterial and viral agents (Doria et al. 2012).

11.2.1.1 Gold Nanoparticle (AuNP)

Gold NPs exhibit the following important properties which make it suitable for a broad spectrum of applications of NP based assays for microbial detection and identification:

(1) Size-dependent physicochemical properties empower the gold NPs an ideal material to be used in the detection of biomolecules at the lowest concentration (Syed and Bokhari 2011). Use of gold NPs depends on their size, for example, small size NPs (2–15 nm in diameter) are used merely in immunohistochemistry, microscopy and large size NPs (80–250 nm diameter) are used in forensics, electronic devices.
(2) Gold NPs are biocompatible with a living cell. Bioconjugation of gold NPs with molecules like DNA, RNA, antibodies, and peptides can be done.
(3) By manipulating their size, shape, optical properties of gold NPs can be easily controlled, which turn out different colors.
(4) Association of other molecules to the metal (gold) surface of a nanoparticle may change the plasmon resonance directly.
(5) Aggregation of gold NPs changes SPR frequency resulting in a change of color from red to violet/blue (Yeh et al. 2012; Ali et al. 2014; Koedrith et al. 2015).
(6) Gold NPs act as a quencher of many fluorophores (Sperling et al. 2008).

Few applications of gold nanoparticles for the detection of foodborne pathogens are presented in Table 11.1.

Table 11.1 Gold NPs or NMs used in foodborne pathogen detection

Food pathogens	Detection Limit	Food material	Biosensors	Reference
Escherichia coli O157: H7	50 cfu/strip in milk	Milk	Amperometric immunosensing Strips	Lin et al. (2008)
Norovirus	60 copies/mL	Lettuce	Electrochemical biosensor	Hong et al. (2015)
L. monocytogenes	2 log cfu/g	Blueberries	Electrode-AuNPs biosensor	Davis et al. (2013)
S. aureus and *S. Typhimurium*	1×10^2–1×10^5 cfu/mL	Pork	AuNP aptasensor via surface-enhanced Raman spectroscopy (SERS)	Zhang et al. (2015)
Pseudomonas aeruginosa	100 ng/mL in the buffer and 200 ng/mL in pasteurized whole milk	Milk	MAB–DSNB–sulfur–AuNPs; SERS-based sandwich immunoassay	Weeks et al. (2003)
Vibrio parahaemolyticus	7.4×10^4 cfu/mL;10^5–10^9 cfu/mL	Shellfish	Agarose-AuNPs; amperometry	Nordin et al. (2017)

11.2.1.2 Fluorescent NPs

Fluorescence is a popular technique to detect different biological events because of its high sensitivity. For many years, different organic dyes and fluorescent proteins have been used for medical and biological purposes with simple, quick, and non-destructive methods. Classical fluorophores have also some limitations, e.g. short lifetime, poor detecting sensitivity. These problems can be solved using fluorescent NPs which have higher signal intensity, stability, and the capability of multiplexing to study the diversity and versatility of biological events (Hotzer et al. 2012; Ruedas-Rama et al. 2012). Some fluorescent NPs have been successfully implemented into the detection of pathogens. Quantum dots (QDs) revolutionize real-time sensitive imaging and sensing. QDs have some unique properties like (1) high brightness, (2) extremely photostable, (3) long excited-state lifetimes, (4) resistant to chemical degradation, (5) high thermal stability, and (6) possible to form bioconjugate with biomolecules such as vitamin, proteins, peptides, antibodies, etc. Above properties make QDs an appropriate tool for various biomedical applications, such as sensor and detecting tool for biomarkers, pathogens, immunolabeling of cells and tissue (Kaittanis et al. 2010; Geszke-Moritz and Moritz 2013).

Streptavidin-conjugated CdSe/ZnS (core/shell) QDs can detect *E. coli* O157: H7 strain two-fold more efficiently compare to normal classical dye (FITC) (Hahn et al. 2005). A combination of biotin-labeled anti-salmonella antibody and streptavidin-coated QDs was used for the detection of *S. typhimurium* in chicken carcass wash water with the detection limit of 10^3 cfu/mL(Yang and Li 2005).

11.2.1.3 Magnetic NPs

There are two types of magnetic NPs. The first one is metallic magnetic NPs (Fe, Ni, Co) which has better magnetic properties but quickly oxidizes in preparation of ferrofluids. The second type is different types of iron oxide (magnetite, maghemite) and other ferrites based NPs. Besides their magnetic properties size, high stability in suspension, infiltration, and interaction on the cellular and molecular level give rise to their applicability in biomedicine science. The application of magnetic NPs in the diagnostic of foodborne pathogens is inadequate. It can be used for separation, purification, and concentration of different biomolecules, including those from microbes or complex matrices (Liu et al. 2006; Huang et al. 2010; Abd-Elsalam et al. 2019). *S aureus* and *E. coli* were detected in apple juice/lettuce by concentrating the bacteria using magnetic zirconia nanoparticles followed by nano LC–MS (Chen et al. 2019). *Escherichia coli* O157: H7 in ground beef samples was detected by immunomagnetic separation with magnetic nanoparticle-antibody conjugates (Varshney et al. 2005).

11.2.1.4 Silica NPs

Silica NPs conjugated with fluorophores have several applications including the detection of pathogens. Silica NPs are robust material, mechanically stable, and transparent. It gives stability and protection once encapsulated with a fluorophore. Multiple fluorophores can be incorporated into the silica matrix resulting in brighter detection compared to a single fluorophore. Rapid detection of *E. coli* O157: H7 was performed with fluorescent dye-doped silica NPs (Tuitemwong et al. 2013) using a fluorescence microscope. The assay was also applicable to water and food samples.

11.2.2 Nano Based Sensor or Assay

In general, nanostructures used in diagnostics must have certain recognition mechanisms specified to the analyte, for example, nucleic acid, antibodies, or enzymes. They must have the capability to generate a distinguishing signal from the analyte by self or by signaling molecules immobilized in the nanostructures (Singh et al. 2020). Some of the nanostructure-based sensor or assay systems are highlighted below:

11.2.2.1 Carbon Nanotube-Based Sensor

Carbon nanotubes (CNTs) are hollow cylindrical tubes with one or more concentric layers of graphite enclosed by fullerenic hemispheres. This unique structure light weight nanomaterial has high thermal conductivity, high thermal and mechanical stability, and high surface-to-volume ratio (Pandit et al. 2016). Single-wall surface carbon nanotubes (SWCNTs) and multi-walled carbon nanotube (MWCNTs) conjugated with ssDNA aptasensor were reported for the detection of *Salmonella* spp. with higher response compared to PCR detection (Weber et al. 2011; Hasan et al. 2018). *E. coli* k-12 strain was detected specifically using anti-*E. coli* antibodies conjugated with SWCNT. This sensor showed a detection limit of 10^2 cfu/mL in <5 min.

11.2.2.2 Lateral Flow Strip Assay (LFSA)

Lateral flow strip assays (LFSAs) are widely being used in diagnostic technology over the past two decades. They are affordable, sensitive, specific, user-friendly, rapid, and robust, equipment-free, and deliverable to end-users. They are normally made in strip type and the analyte is transported utilizing capillary force through a series of the capillary pad, including sample pad, conjugation pad, detection pad, and absorbent pad (Fig. 11.2). The analyte is loaded to a sample pad that migrates to the conjugation pad where analyte interacts with the capture antibody conjugated with NPs. Then analyte-NP conjugated antibody (Ab) moves to the test line in nitrocellulose pad. The test line contains detection antibody specific to pathogen. Analyte-NP conjugated Ab moves father and interacts with the control line containing the second antibody specific to first antibody. The color signal in both the test line and control line indicates the presence of the analyte in the sample. Several commercial

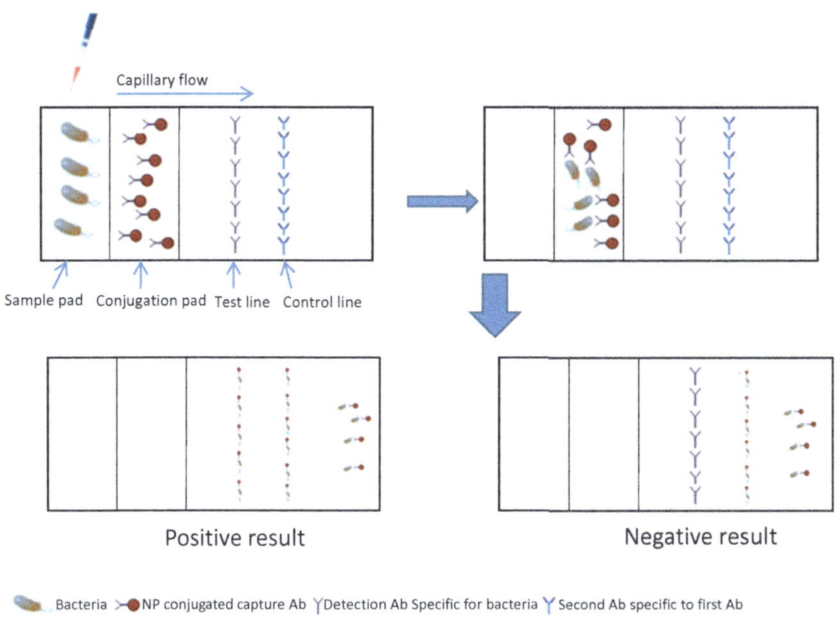

Fig. 11.2 Common structure of Lateral flow strip assay (LFSA)

lateral flow strip assay kits are available for the detection of *E. coli O157; Salmonella; Listeria; Shigella; V. cholerae; V. parahaemolyticus* (Luo et al. 2020).

11.2.2.3 Peptide Nanotubes for Highly Sensitive Pathogen Sensors Chips

de la Rica and Matsui 2010 reported about the development of a label-free sensor chip assembled from peptide nanotubes that enables the electrical detection of food pathogens with an extremely low detection limit. On a 500-μm-thick Pyrex wafer, triple layer of Ti/Ni/Au (500 Å/500 Å/500 Å) deposited followed by deposition of 1-μm-thick silicon oxide passivation coating. The peptide nanotubes prepared by self-assembly from the bolaamphiphilic peptide monomer bis(N-a-amido-glycylglycine)-1,7-heptane dicarboxylate which were used as templates for the immobilization of antibodies (anti-*E. coli* or anti-*S. typhi*). Free adsorption sites on the tubes were blocked by incubation with BSA. Then the nanotubes were allowed to incubate with bacteria sample and the sedimentation rate was measured by impedance analyzer. The sedimentation rate increased when bacteria interact with peptide nanotubes compare to free bacteria.

11.2.2.4 One-Step Detection of Pathogens with Magnetic Beads

A portable NMR biosensor assay system has been reported for detection of foodborne bacteria *E. coli* O157: H7 in food sample. It has detection limit up to 10^1 CFU/mL. In brief, a magnetic nanoparticle was conjugated with anti-*E. coli* O157: H7 monoclonal antibody. Then Ab conjugated MNPs were incubated with sample containing *E. coli* O157: H7. Unbound conjugates were removed by filtration. Bacteria inbound with MNP AB conjugate were separated and concentrated. The MNP induced spatial and temporal disturbance in the homogeneity and strength of the local magnetic field. Water proton's spin-spin relaxation time, T2, has a linear relationship with magnetic particle concentration. Based on T2 signal the presence and concentration of the target pathogen could be measured (Luo and Alocilja 2017).

11.2.2.5 Aptasensors for Foodborne Pathogen Detection

Aptamers are the single-stranded nucleic acid molecules having length 25–99 nt that can be either DNA or RNA. It can recognize various types of ligands like ions, drugs, toxins, and whole pathogen (Dong et al. 2015; Dupont et al. 2015). It shows higher pH, temperature, ionic stability compared to antibodies. Easier synthesis and modification, long self-life, and low cost make aptamer more convenient tool for diagnostic. Several aptamer-based assays and sensors have been reported for foodborne pathogen detection such as *E. coli* (Zou et al. 2018), *S. aureus* (Qiao et al. 2018), *Salmonella* spp. (Shin et al. 2018), *Vibrio parahaemolyticus*,

Campylobacter spp. (Bruno and Sivils 2017), *Listeria* spp. (Liu et al. 2018), and *Shigella flexneri* (Duan et al. 2013). Integration of nanotechnology into aptasensor drastically increases the sensitivity, specificity of pathogen detection techniques. Aptamer conjugated with MNPs could enhance the capture capability and selectivity. Aptasensor may be classified into different groups based on their detection method like visible spectrum, lateral flow, colorimetric, surface plasmon resonance, fluorescence, electrochemical. For example, G-Quadruplex Horseradish Peroxidase-Mimicking DNAzyme aptasensor has been reported for *V. parahaemolyticus* detection based on the colorimetry reaction using catalytic activity of horseradish peroxidase (HRP) (Sun et al. 2019). HRP has some disadvantages, such as difficulty in storage and instable catalytic activity affected by external conditions. Fe_3O_4, CeO_2 NPs, Co_3O_4 NPs, $ZnFe_2O_4$ nanoparticles were reported to exhibit intrinsic peroxidase-like activity as HRP (Wu et al. 2020). New colorimetric aptasensor reported which can detect *S. typhimurium* based on the $ZnFe_2O_4$-reduced graphene oxide ($ZnFe_2O_4$/rGO) nanoparticles having limit of detection 11 cfu/mL (Wu et al. 2017).

11.2.3 Nanoparticle as a Potential Antimicrobial for Food Packaging

More than one-third of all food produced for human consumption is wasted due to food contamination and wrong handling. Seasonal fruits and some vegetables prolonged preservation require special packaging. Nanoparticles have enormous potency to prevent food deterioration and extend the shelf-life of food by inhibiting the growth of food spoiling microbes. Metal and metal oxide nanoparticles neutralize the pathogen by excessive formation of reactive oxygen species leading to oxidative stress and subsequent cell damage (Fu et al. 2014; Wu et al. 2014; Prasad and Swamy 2013; Aziz et al. 2014, 2015, 2016). Metal nanoparticle can also enter into the cell and alter the function of biomolecules which lead to break down cell functions. Nanoparticles are not used directly into food system rather used in coating or packaging materials to get the antimicrobial effect without sacrificing the quality of the packaged food. Several parameters like film thickness, presence of additives in the polymer matrix such as fillers, antifogging and antistatic agents, lubricants, stabilizers, and plasticizers can affect the activity of antimicrobial agents. Nanomaterial based antimicrobial packaging materials could be self-sterilizing or sanitizing that reduce the potential for recontamination of processed food products (Shankar et al. 2016). Various types of organic and inorganic nanofillers are used for food packaging applications. They can be classified depending upon their size, e.g., nanoparticle, nanofribrils, nanoplates, nanorods, and nanotubes.

11.2.3.1 Organic Nanofillers

Chitin nanofibrils exhibited strong antibacterial against *L. monocytogenes* (Shankar et al. 2015). Many biopolymer-based nanocomposite films were prepared with grapefruit seed extract (GSE) used for food packaging applications (Wang and Rhim 2016). Cationic nanostructures manufactured from a combination of the cationic lipid dioctadecyl dimethyl ammonium bromide (DODAB), the antimicrobial peptide gramicidin D (Gr), the antimicrobial cationic polymer poly (diallyldimethylammonium chloride) (PDDA), and the biocompatible polymer poly(methyl methacrylate)(PMMA) have shown great potential against *Escherichia coli, Salmonella Typhimurium, Staphylococcus aureus,* and *Listeria monocytogenes* (Carrasco et al. 2016).

11.2.3.2 Inorganic Nanofillers

Tested antimicrobial activity of AgNPs against a wide range of food pathogens (*Bacillus cereus*, *Listeria monocytogenes*, *Staphylococcus aureus*, *Escherichia coli*, *Streptococcus Typhimurium*, *Candida albicans*, *Candida glabrata*, Candida geochares, and *Candida saitoana*) made them widely accepted in food industry. For example, AgNPs absorbed in simple cotton fibers could eliminate *E. coli*, *S. aureus*, and *S. Typhimurium* (Bhardwaj et al. 2017). Several silver containing zeolites are approved by FDA.

Low cost, low toxicity, and ultraviolet barrier properties of ZnO NPs are the reason for its preference over silver nanoparticle. ZnO NPs-coated polyvinyl chloride films exerted antibacterial properties against *S. aureus* than *E. coli* (Li et al. 2009). UV-blocking, gas barrier, and mechanical properties of ZnO nanocomposite along with antimicrobial property make it more convenient as packaging material.

Copper oxide NPs conjugated with LDPE were used for cheese packaging resulting increase of self-life while refrigerated.

Decay of strawberries, cheese, and shrimp have been delayed by covering TiO_2-incorporated nanocomposite films (Gumiero et al. 2013; Luo et al. 2015).

11.3 Application of Nanotechnology in Water Microbiology

Waterborne diseases claim millions of lives every year worldwide. Presence of pathogens must be monitored to ensure safe drinking water. Viability of the pathogen present in drinking water determines potential infectivity of the pathogen. There are many conventional methods available to monitor the pathogen load in drinking water. Major problems with the existing methods are their inability to determine pathogen species, viability, low sensitivity, poor recovery rate, more time-consuming, and require intensive labor (Aw and Rose 2012). Advancement of nanotechnology empowers rapid and sensitive sampling and detection of pathogens. An adequate supply of safe drinking water is one of the necessities for a healthy life. An important challenge is therefore the rapid, specific, and sensitive detection of waterborne pathogens. Presently, innovations in nanotechnologies and nanosciences are having a significant impact in detection, control, and elimination of waterborne pathogens. Several nanoparticle-based assays and nanodevices are commercially available.

11.3.1 Pathogen Detection

Escherichia coli (Enterohemorrhagic), *Shigella, Vibrio cholerae* like bacteria; norovirus, rotavirus, and protozoa (e.g., *Cryptosporidium* and *Giardia*) are the major waterborne pathogens. They contaminate water resources when infected people or animals shed microbes in faces. Water resources can also be contaminated by mixing with undertreated/untreated sewage. Some zoonotic diseases can transmit from water animals to human also. Nanotechnology has been reported to enhance the detection of waterborne pathogens using the following detection methods.

11.3.1.1 Optical Detection

The optical properties of nanomaterials, e.g., surface plasmon resonance, fluorescence, were exploited to ease the detection of pathogens in water.

Gold (Au) and Silver (Ag) NMs

The colorimetric contrast of gold nanomaterials caused by surface plasmon resonance has shown tremendous potential for the detection of waterborne microbes. Simple colorimetric assays are suitable for low resources on-site testing settings. Gold nanoparticles reduced and stabilized by sialic acid showed change in optical density in presence of influenza B/Victoria and influenza B/Yamagata. Virus (Lee et al. 2013), bacterial antigens and toxins of *Salmonella, E. coli*, and *Vibrio cholerae* were detected by AuNMs (Wang et al. 2010; Jyoti et al. 2010; Schofield et al. 2007). Rotavirus was detected with graphene oxide (GO) surface and a secondary gold nanomaterial using FRET method (Jung et al. 2010).

Highly intense black color contrast of silver nanoparticle was utilized for the detection of *Salmonella* and *Giardia* at very low concentration.

Fluorescent NMs

Cryptosporidium and *Giardia lamblia* in water were detected using chemical fluorophore previously but a higher sensitivity was achieved using semiconductor quantum dot-conjugated antibodies (Zhu et al. 2004). *E. coli* and *hepatitis* were detected in <1 h and with 50 times greater sensitivity than existing methods using multiplexed microfluidic system. Attachment of pathogen to QDs leads to blue shift in their emission spectra that were used for the detection of bacterial spores and *E. coli* aggregation with a Limit of detection (LOD) of 10^4 bacteria/mL (Fig. 11.3) (Dwarakanath et al. 2004).

11.3.1.2 Biosensors

Biosensor consists of a biological target recognition element coupled with a mechanism of signal transduction capable of providing selective quantitative or semi-quantitative analytical information in the form of optical, electrical, mass sensitive, etc.

Optical Biosensors

The optical properties of NMs can be exploited in the development of biosensors, for pathogen detection. For example, photonic crystals are nanoarrays of dielectric scatterers. Rotavirus has been detected using antibody-based photonic crystal biosensors at a LOD of 36 virus focus forming units in 30 mins assay (Pineda et al. 2009). Silicon microring resonators are another example that has been employed as

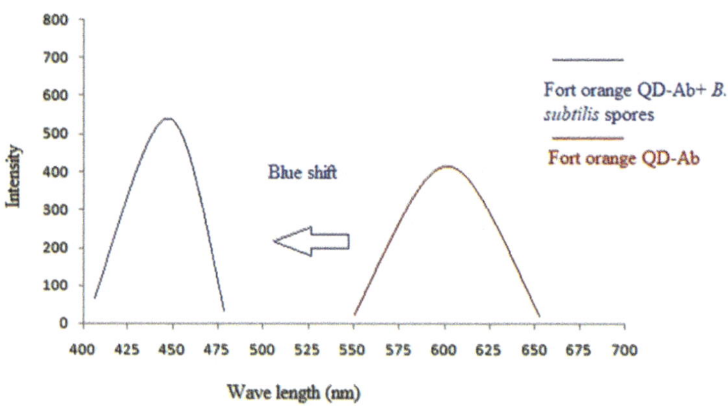

Fig. 11.3 Fluorescence emission spectra of Fort Orange QDs conjugated Antibody (Ab) in free (*Red*) and bound (*Blue*) form with *B. subtilis* spore

sensors for detecting norovirus using carbohydrate functionalized silicon microrings with a LOD of 250 ng/mL.

Electrical Biosensors

Nanotechnology offers a solution to the problems of electrochemical biosensors when detecting waterborne pathogens. For example, cholera toxin with a LOD of 10^{-16} g detected electrochemically using liposome coated carbon nanotubes. In another study, carbon nanotubes coated in Fe_3O_4 NMs have used the rapid detection of coliforms with the help of an amperometric biosensor. *E. coli O157: H7* and other bacteria were detected to the range of 10 cfu/mL using gold NMs by electrochemical detection.

Mass Sensitive Biosensors

Mass sensitive biosensors for water quality monitoring have been mainly used for detection of waterborne bacteria. RNA and toxin of pathogens were detected by surface plasmon resonance (SPR) of Au NMs and hybrid Au − Ag NMs. *E. coli* O157: H7 was detected at 50 cfu/ml using two-photon Rayleigh scattering properties of gold nanorods. Using silicon nitride cantilevers *Salmonella enterica* was detected by monitoring the cantilever's surface bending which was directly related to the number of bacteria associating with the cantilever (Weeks et al. 2003).

11.3.2 Pathogen Elimination

Two billion people worldwide do not have access to safe water according to the World Health Organization (WHO). Reliability on currently employed water purification systems is based on chemical and membrane-based under scanner. Emergence of resistant pathogens and toxic by-products from chemical disinfectants are also major concerns for water purification. Filter membrane with anti-biofouling and antiviral properties is therefore in urgent demand. Nanofiber with low basis weight, high permeability, and small pore size can remove the smaller particle <0.3 μm (Barhate and Ramakrishna 2007).

11.3.2.1 Nanobiocides

Nanobiocides are the nanomaterials with antimicrobial potency. Electrospinning is a novel technology for the fabrication of nanofiber conjugated with different types of nanobiocides such as metal nanoparticles (Table 11.2), antibiotics.

Table 11.2 Nanobiocides used in water disinfection

Nanobiocides	Mode of action	Reported Target	Reference/ Commercial use
TiO_2	Generates hydroxyl free radicals and peroxide formed under UV-A irradiation via oxidative and reductive pathways	Removes micron and submicron particles, polish iron, manganese, silica, H_2O_2 and contaminants such as *Giardia*, *Cryptosporidium*, and oocysts.	Purifics (www.purifics.com)
Silver	Nanosilver releases silver ions in water binding to –SH groups in vital enzymes and damaging them. Also, silver ions interfere with DNA replication and induce structural changes in the cell envelope.	*E. coli*, *Legionella pneumophila* (LP), *Pseudomonas mendocina KR1*, MS2 bacteriophage	Lv et al. (2009)
ZnO	Photocatalytic generation of H_2O_2 may be responsible for antimicrobial action of ZnO. ZnO nanoparticles penetrate the bacterial cell envelope and disorganize the cell membrane	*S. aureus*	Vicentini et al. (2009)
CuO	Bacteriostatic action of CuO nanoparticles mainly originates from the direct interaction of Cu^{2+} species with the cell components	Methicillin-resistant *S. aureus* (MRSA) and *E. coli*, *Chlamydomonas* sp. strain PCC 6803 (freshwater), *Synechococcus sp.* strain (marine) and *Phaeodactylum tricornutum*	Botes and Cloete (2010)
Carbon nanotubes	Prevention of biofilm formation in water filtration membranes. Irrecoverable damages to the outer membrane, releasing the intracellular content	*E. coli*, *S. epidermidis*, *S. typhimurium*, *B. subtilis*, *S. aureus*	Al-Jumaili et al. (2017)
Fullerenes	Increase in membrane hydrophobicity might play a role in antimicrobial activity	*E. coli* K12	Chae et al. (2009)
Polyethylene mine (PEI)	Disruption of bacterial cell membranes	*Streptococcus mutants. Staphylococcus aureus*, *Staphylococcus epidermidis*, *Acinetobacter baumannii, and Candida albicans*	Azevedo et al. (2014)

11.3.2.2 Antimicrobial Nano Polymers/Nanofibers

Polymeric nanoparticles neutralize microorganisms either by releasing antibiotics, antimicrobial peptides, and antimicrobial agents or by contact-killing cationic surfaces, such as quaternary ammonium compounds, alkyl pyridiniums, or quaternary phosphonium. Nanopolymeric antimicrobial materials show long-term antimicrobial activity. They are nonvolatile and chemically stable. They can bind to the surface of interest and hardly permeate through biological membranes.

Chitosan-Based Nanopolymer

Chitosan has antimicrobial activity towards bacteria, especially Gram-positive bacteria, and highly effective against fungi and viruses. Increasing membrane permeability and disruption lead to leakage of intracellular components that are considered as the main antimicrobial mechanisms of chitosan (Qi et al. 2004). The degree of antimicrobial activity depends on the organism, pH, molecular weight, degree of polymerization, and the presence of lipids and proteins. The antibacterial action of improved chitosan derivatives containing quaternary ammonium groups, such as *N*, *N*,*N*-trimethyl chitosan, *N*-propyl-*N*,*N*-dimethyl chitosan, and *N*-furfuryl-*N*,*N*-dimethyl chitosan has also been described. Chitosan-based nanomaterials have been reported to kill 80% *S. aureus* (CCRC10779) and 30% *E. coli* (CCRC 10324) (Don et al. 2005).

N-Bromo-Hydantoin Grafted Polystyrene Beads

The release of active bromine from the grafted polystyrene beads into the running water exhibited excellent antimicrobial activity against *E. coli* and MS2 phage. Moreover, besides the effect of the *N*-haloamine structure on lasting disinfection activities, bead's nanomicro characteristics were found critical for oxidative halogen release control: rate stabilization and modulation, extension and consequently influence antimicrobial activity. These materials can have great potential due to the ability to decontaminate large volumes of contaminated water, low costs, and bromine rechargeability (Farah et al. 2015).

Polyurethane (PU) Nanofibers

Polyurethane (PU) nanofibers when treated with argon plasma produce surface oxide and peroxide groups. Immersion of these PU fibers in 4-vinylpyridine monomer solution with exposure to UV irradiation produces poly(4-vinylpyridine) grafted PU fibers. Antimicrobial activity was obtained when the grafted pyridine groups were functionalized through quaternization with hexylbromide. 99.9% reduction in the viability was observed for Gram-positive *Staphylococcus aureus* and Gram-negative *Escherichia coli* after 4 h contact with the PU fibers (Yao et al. 2008).

Polycarbonate (PC)/Chloroform

Polycarbonate (PC) solution with quaternary ammonium salt benzyl triethylammoniumchloride (BTEAC) is a potential antimicrobial nanofibrous membrane for ultrafiltration. The growth of *S. aureus* ATCC6538, *E. coli* ATCC 25922, and *Klebsiella pneumoniae* ATCC 4352 was totally inhibited with the addition of small amounts of BTEAC to the PC solution. Also, it showed 99.97% filtration efficiency of 0.3 μm size particles in comparison to a polypropylene HEPA (high-efficiency particulate air) filter (Botes and Cloete 2010).

11.4 Application of Nanotechnology in Clinical Microbiology

Infectious diseases caused by pathogens such as viruses, bacteria, fungi, and parasite are the major cause of death worldwide. Infectious diseases dominate health threats all over the world according to a report by the WHO in 2019. Of the many infectious diseases influenza, pneumonia, AIDS, tuberculosis, malaria are the top-ranked and cause of millions of death every year. Despite the pharmaceutical advancement to produce drugs at low cost emergence and dissemination of drug resistance infectious agents is a global concern. The quest for new strategies, pharmaceuticals, devices to diagnose and treat diseases accurately, easily, and efficiently have intensified. Recently various nanotechnology-based materials have been reported with the aim of early detection, effective control, and prevention of infectious diseases (Fig. 11.4).

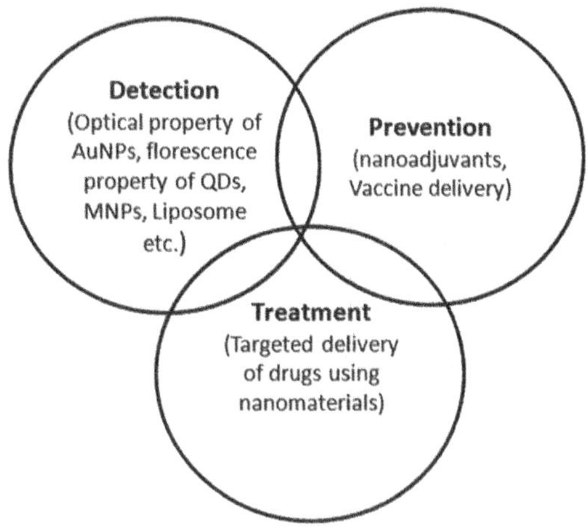

Fig. 11.4 A brief overview of the applications of nanotechnology in clinical microbiology

11.4.1 Detection

NPs can interact with biomolecules with appropriate surface modification. Their unique physicochemical properties allow accurate, rapid, sensitive, and cost-effective diagnostics. Many of the current diagnostics are lacking at least one of the critical parameters mentioned above. NPs which have been successfully utilized for the diagnosis of infectious diseases are particularly fluorescent nanoparticles, metallic nanoparticles, and magnetic nanoparticles.

11.4.1.1 Metallic Nanoparticles

The gold nanoparticles are the first nanomaterials used as nanodiagnostics for the detection of DNA in 1996.The unique color changing capability of AuNPs due to shifting in their SPR is utilized for the detection of malaria pathogen. The aggregation of AuNPs through the interaction with benzalkonium chloride, surfactant, is controlled by an aptamer specific to malaria biomarker PfLDH (*P. falciparum* lactate dehydrogenase). Aptamer binds PfLDH biomarker present in positive sample by leaving the surfactant free which allows aggregation of AuNPs. In absence of biomarker AuNPs remain dispersed in solution. Red color of stable AuNPs changes to blue upon aggregation due to presence of biomarker in the sample (Jain et al. 2016). A simple paper-based assay employing unmodified gold nanoparticles has been utilized to detect *Mycobacterium tuberculosis* (Tsai et al. 2013). Figure 11.5 highlighted the method which can be implemented using a smartphone app for image processing as claimed by the authors. Study also demonstrated the applicability of dextrin-capped AuNP as label for lateral flow assay for early detection of dengue infection in high-risk resource-limited areas (Yrad et al. 2019). Gold nanowire arrays (GNWA) linked with specific antibodies against *E. coli* have been developed to detect urinary tract infection (Basu et al. 2004).

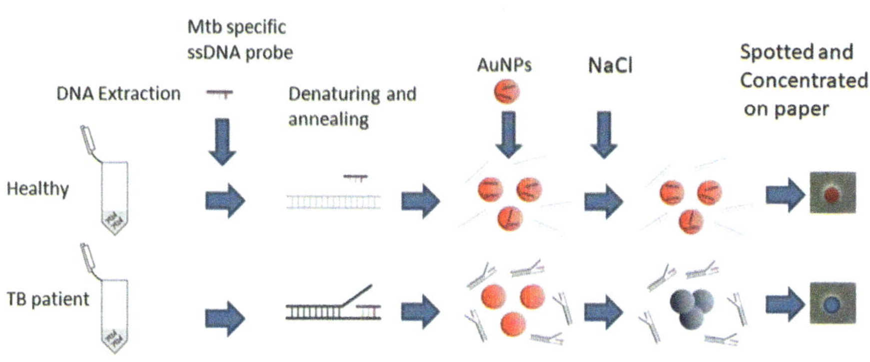

Fig. 11.5 The schematic illustration of the TB diagnostic method using AuNP and Smartphone by paper-based assay

11.4.1.2 Fluorescent Nanoparticles

Metallic fluorescent dyes (Tris(2,2 bipyridyl) osmium bis (hexafluorophosphate) (OsBpy) and Tris (bipyridine) ruthenium(II) dichloride (RuBpy)) encapsulated in silica nanoparticle have been used to detect MTB from sputum sample with high sensitivity and within 4 h (Qin et al. 2007). Fluorescence property of SiO2:Eu improves the sensitivity of traditional methods such as the enzyme-linked immunosorbent assay (ELISA).This silica nanoparticle immunoassay was employed for the detection of HIV(Chunduri et al. 2017). FNPs also improved the detection of TB. Biomarker for TB, lipoarabinomannan (LAM) in urine can be detected at pg/mL utilizing copper-containing bait dye, Reactive Blue 221, to sequester LAM in a hydrogel nanocage (Paris et al. 2017). Similarly, anthrax protective agent (PA) has been detected using fluorescence ELISA that incorporated Eu(III) polymeric NPs which has 100-fold higher sensitivity than conventional ELISA. Many quantum dots (Qdots), nano-crystal semiconductors made of metals such as Si or Ge, have been developed to detect pathogens with enhanced sensitivity and specificities over conventional organic fluorophores. A dual-core Qdots, composed of two fluorescent NPs green color and red, labeled secondary Ab can recognize antibodies against respiratory syncytial virus (RSV) (Agrawal et al. 2005). Barcode made from Qdots can be used for multiplex identification of toxins in the blood. Such a barcode is employed for the detection of HCV and HIV from human serum (Gao et al. 2006).

11.4.1.3 Magnetic Nanoparticles

Foodborne pathogens such as *E. coli* O157:H7 and *Salmonella* can be detected using magnetic nanoparticle (Fe_3O_4 or Fe_2O_3) conjugated antibodies by immunogenic separation (Qasim et al. 2014). Vancomycin-conjugated FePt MNPs were developed to capture and identify vancomycin-resistant enterococci (VRE) and other Gram-positive bacteria present at very low concentrations in samples. A combination of vancomycin-conjugated FePt MNPs with fluorophore increases sensitivity of pathogen detection in blood (Gao et al. 2006). HSV-1 and adenoviruses can be detected at one digit level/μL of serum samples using biocompatible superparamagnetic iron oxide (SPIO) nano-biosensors (Josephson et al. 2001).

11.4.1.4 Liposome

Cholera toxin (CT) has been detected using ganglioside-incorporated liposomes. CT was detected as a colored band on the nitrocellulose membrane strip, where CT bound to GM1-liposomes can be captured by immobilized antibodies in a sandwich immunoassay. Engineered liposome nanoparticle conjugated with fluorescent rhodamine dye was farther applied for detection of CT (Singh et al. 2000). Bacterial toxins of tetanus, botulinum were also detected by similar strategy.

11.4.2 Treatment

Rapid evolution of resistance is making the treatment of infectious diseases more challenging. Nanotechnology has potential of targeted delivery of the drugs to disease sites. For example, liposome, functionalized with half antibodies to recognize proteins that are uniquely found on the surface of plasmodium-infected RBCs, can selectively deliver chloroquine to the infected RBC over non-infected cells in 90 min (Urbán et al. 2011). Above delivery system requires fewer drugs and also reduces side effects of treatment. Treatment of many diseases has the requirement for frequent administration of drugs and sometimes high drug load is used for long-duration treatment. For example, in TB treatment patient needs to take multiple drugs for a period of 6–9 months. Nanotechnology has the prospective to reduce the required frequency of administration by combining these drugs into a single dose (Pandey et al. 2003). Gallium(III) meso-tetraphenylporphyrin nanoparticles were reported to provide sustained release of Ga(III) which inhibit the growth of *M. tuberculosis* and HIV in macrophages. Gallium(III) replaces the iron and inactivates their enzyme important for growth(Choi et al. 2017). Some NPs have potential to be used as drugs. Metallic NPs and their oxides upon UV exposure produce reactive oxygen species with antimicrobial activity. Nitric oxide-releasing NPs (NONPs) have been shown to inhibit the growth of antibiotic-resistant strains of *P. aeruginosa*, *E. faecalis*, *K. pneumoniae*, and *E. coli* and MRSA (Hajipour et al. 2012).

11.4.3 Prevention

Vaccine provides acquired immunity against a particular antigen and can rapidly respond upon subsequent infection. Vaccination is considered as the cheapest and most effective strategy for combating diseases. Millions of lives are saved because of vaccine. For example, smallpox has been successfully eradicated through widespread vaccination programs. Target specific vaccine delivery can be achieved using nanotechnology application. Controlled release of vaccine encapsulated in nanoparticle vessel can boost the immune response. Nanotechnology can also mimic the natural target. Engineered heparin–polymersome complex has both drug- and vaccine-like activity against malaria infection as it prevents invasion of the parasite (drug function) (Najer et al. 2014). Fullerenol nanoparticles were used as an adjuvant in the development of HIV-1 vaccine (Xu et al. 2013). Attenuated MTB delivered using aerosol system shown enhanced immune response against MTB (Garcia-Contreras et al. 2008). Self-life of the NP formulation can be increased by lyophilizing it.

11.5 Conclusion

Safe drinking water, healthy food, affordable diagnostic, and treatment are primary necessities for a healthy life. Microbial contamination in water, food, and pathological specimens can be detected by nanotechnology-based techniques more rapidly and sensitively at low cost compared to conventional testing methods. One major advantage of nanotechnology-based techniques is on-site detection that solves the requirement of sophisticated instruments at resource-limited situation. Again microbiology can also enrich nanotechnology. Nanomicrobiology deals with the synthesis of diverse types of nanoparticles using microbes. Perniciousness of nanomaterials on the environment and human health remains unexplored. More research needs to be carried out to address all the negative impacts of nanomaterials. Interdisciplinary approach integrating nanotechnology and microbiology can be employed for human well-being and environmental sustainability.

References

Abd-Elsalam K, Mohamed AA, Prasad R (2019) Magnetic nanostructures: environmental and agricultural applications. Springer International Publishing, Cham. ISBN 978-3-030-16438-6. https://www.springer.com/gp/book/9783030164386

Agrawal A, Tripp RA, Anderson LJ, Nie S (2005) Real-time detection of virus particles and viral protein expression with two-color nanoparticle probes. J Virol 79(13):8625–8628

Ali M, Mustafa S, Che Man Y, Adam K, Humayun K (2014) Nanobiosensor for the detection and quantification of pork adulteration in meatball formulation. J Exp Nanosci 9(2):152–160

Al-Jumaili A, Alancherry S, Bazaka K, Jacob MV (2017) Review on the antimicrobial properties of carbon nanostructures. Materials (Basel) 10(9):1066

Aw TG, Rose JB (2012) Detection of pathogens in water: from phylochips to qPCR to pyrosequencing. Curr Opin Biotechnol 23(3):422–430

Azevedo MM, Ramalho P, Silva AP, Teixeira-Santos R, Pina-Vaz C, Rodrigues AG (2014) Polyethyleneimine and polyethyleneimine-based nanoparticles: novel bacterial and yeast biofilm inhibitors. J Med Microbiol 63(Pt 9):1167–1173

Aziz N, Fatma T, Varma A, Prasad R (2014) Biogenic synthesis of silver nanoparticles using *Scenedesmus abundans* and evaluation of their antibacterial activity. J Nanoparticles, Article ID 689419. https://doi.org/10.1155/2014/689419

Aziz N, Faraz M, Pandey R, Sakir M, Fatma T, Varma A, Barman I, Prasad R (2015) Facile algae-derived route to biogenic silver nanoparticles: synthesis, antibacterial and photocatalytic properties. Langmuir 31:11605–11612. https://doi.org/10.1021/acs.langmuir.5b03081

Aziz N, Pandey R, Barman I, Prasad R (2016) Leveraging the attributes of *Mucor hiemalis*-derived silver nanoparticles for a synergistic broad-spectrum antimicrobial platform. Front Microbiol 7:1984. https://doi.org/10.3389/fmicb.2016.01984

Barhate RS, Ramakrishna S (2007) Nanofibrous filtering media, filtration problems and solutions from tiny materials. J Membr Sci 296:1–8

Basu M, Seggerson S, Henshaw J, Jiang J, del A Cordona R, Lefave C, Boyle PJ, Miller A, Pugia M, Basu S (2004) Nano-biosensor development for bacterial detection during human kidney infection: use of glycoconjugate-specific antibody-bound gold NanoWire arrays (GNWA). Glycoconj J 21(8–9):487–496

Bhardwaj AK, Shukla A, Mishra RK, Singh SC, Mishra V, Uttam KN et al (2017) Power and time dependent microwave assisted fabrication of silver nanoparticles decorated cotton (SNDC) fibers for bacterial decontamination. Front Microbiol 8:330. https://doi.org/10.3389/fmicb.2017.00330

Botes M, Cloete TE (2010) The potential of nanofibers and nanobiocides in water purification. Crit Rev Microbiol 36(1):68–81

Bruno JG, Sivils JC (2017) Further characterization and independent validation of a DNA aptamer-quantum dot-based magnetic sandwich assay for *Campylobacter*. Folia Microbiol 62 (6):485–490. https://doi.org/10.1007/s12223-017-0520-0

Carrasco LD, Bertolucci R Jr, Ribeiro RT, Sampaio JLM, Carmona-Ribeiro,- A. M. (2016) Cationic nanostructures against foodborne pathogens. Front Microbiol 7:1804. https://doi.org/10.3389/fmicb.2016.01804

Chae S-R, Wang S, Hendren ZD, Wiesner MR, Watanabe Y, Gunsch CK (2009) Effects of fullerene nanoparticles on Escherichia coli K12 respiratory activity in aqueous suspension and potential use for membrane biofouling control. J Membr Sci 329:68–74

Chen Y, Vu J, Thompson MG, Sharpless WA, Chan LJG, Gin JW, Keasling JD, Adams PD, Petzold CJ (2019) A rapid methods development workflow for high-throughput quantitative proteomic applications. PLoS One 14(2):e0211582

Choi S-R, Britigan BE, Narayanasamy P (2017) Ga(III) nanoparticles inhibit growth of both mycobacterium tuberculosis and HIV and release of interleukin-6 (IL-6) and IL-8 in coinfected macrophages. Antimicrob Agents Chemother 61:e02505–e02516

Chunduri LAA, Kurdekar A, Haleyurgirisetty MK, Bulagonda EP, Kamisetti V, Hewlett IK (2017) Femtogram level sensitivity achieved by surface engineered silica nanoparticles in the early detection of HIV infection. Sci Rep 7:7149–7158

Davis D, Guo X, Musavi L, Lin CS, Chen SH, Wu VCH (2013) Gold nanoparticle-modified carbon electrode biosensor for the detection of *Listeria monocytogenes*. Ind Biotechnol 9:31–36

Don T-M, Chen C-C, Lee C-K, Cheng W-Y, Cheng L-P (2005) Preparation and antibacterial test of chitosan/PAA/PEGDA bi-layer composite membranes. *J Biomater Sci Polym Ed* 16:1503–1519

Dong LL, Tan QW, Ye W, Liu DL, Chen HF, Hu HW, Wen D, Liu Y, Cao Y, Kang JW (2015) Screening and identifying a novel ssDNA aptamer against alpha-fetoprotein using CE-SELEX. Sci Rep UK 5:15552–15562

Doria G, Conde J, Veigas B, Giestas L, Almeida C, Assunção M, Rosa J, Baptista PV (2012) Noble metal nanoparticles for biosensing applications. Sensors (Basel) 12(2):1657–1687

Duan N, Ding X, Wu S, Xia Y, Ma X, Wang Z, Chen J (2013) In vitro selection of a DNA aptamer targeted against *Shigella dysenteriae*. J Microbiol Methods 94(3):170–174. https://doi.org/10.1016/j.mimet.2013.06.016

Dupont DM, Larsen N, Jensen JK, Andreasen PA, Kjems J (2015) Characterisation of aptamer-target interactions by branched selection and high-throughput sequencing of SELEX pools. Nucleic Acids Res 43:e139–e139

Dwarakanath S, Bruno JG, Shastry A, Philipps T, John A, Kumar A, Stephenson LD (2004) Quantum dot-antibody conjugates shift fluorescence upon binding bacteria. Biochem Biophys Res Commun 325:739–743

Farah S, Aviv O, Daif M, Kunduru KR, Laout N, Ratner S, Beyth N, Domb AJ (2015) *N*-bromohydantoin grafted polystyrene beads: synthesis and nano-micro beads characteristics for achieving controlled release of active oxidative bromine and extended microbial inactivation efficiency. *J Polym Sci A* 54:596–610

Fu PP, Xia Q, Hwang HM, Ray PC, Yu H (2014) Mechanisms of nanotoxicity:generation of reactive oxygen species. J Food Drug Anal 22:64–75

Fung F, Wang HS, Menon S (2018) Food safety in the 21st century. Biom J 41(2):88–95

Gao J, Li L, Ho P-L, Mak GC, Gu H, Xu B (2006) Combining fluorescent probes and biofunctional magnetic nanoparticles for rapid detection of Bacteria in human blood. *J Am Chem Soc* 125:15702

Garcia-Contreras L, Wong YL, Muttil P, Padilla D, Sadoff J, Derousse J, Germishuizen WA, Goonesekera S, Elbert K, Bloom BR, Miller R, Fourie PB, Hickey A, Edwards D (2008) Immunization by a bacterial aerosol. Proc Natl Acad Sci U S A 105(12):4656–4660

Geszke-Moritz M, Moritz M (2013) Quantum dots as versatile probes in medical sciences: synthesis, modification and properties. Mater Sci Eng C Mater Biol Appl 33:1008–1021

Gumiero M, Peressini D, Pizzariello A, Sensidoni A, Iacumin L, Comi G (2013) Effect of TiO2 photocatalytic activity in a HDPE-based food packaging on the structural and microbiological stability of a short-ripened cheese. Food Chem 138:1633–1640

Hahn MA, Tabb JS, Krauss TD (2005) Detection of single bacterial pathogens with semiconductor quantum dots. Anal Chem 77:4861–4869

Hajipour MJ, Fromm KM, Ashkarran AA, Jimenez de Aberasturi D, de Larramendi IR, Rojo T, Serpooshan V, Parak WJ, Mahmoudi M (2012) Antibacterial properties of nanoparticles. Trends Biotechnol 30(10):499–511

Hasan MR, Pulingam T, Appaturi JN, Zifruddin AN, Teh SJ, Lim TW, Ibrahim F, Leo BF, Thong KL (2018) Carbon nanotube-based aptasensor for sensitive electrochemical detection of whole-cell *Salmonella*. Anal Biochem 554:34–43. https://doi.org/10.1016/j.ab.2018.06.001

Hong SA, Kwon J, Kim D, Yang S (2015) A rapid, sensitive and selective electrochemical biosensor with concanavalin A for the preemptive detection of norovirus. Biosens Bioelectron 64:338–344

Hotzer B, Medintz IL, Hildebrandt N (2012) Fluorescence in nanobiotechnology: sophisticated fluorophores for novel applications. Small 8:2297–2326

Huang YF, Wang YF, Yan XP (2010) Amine-functionalized magnetic nanoparticles for rapid capture and removal of bacterial pathogens. Environ Sci Technol 44:7908–7913

Inamuddin, Ahamed MI, Prasad R (2021) Advanced antimicrobial materials and applications. Springer, Singapore. ISBN 978-981-15-7098-8. https://www.springer.com/gp/book/9789811570971

Iwamoto M, Ayers T, Mahon BE, Swerdlow DL (2010) Epidemiology of seafood-associated infections in the United States. Clin Microbiol Rev 23(2):399–411

Jain P, Chakma B, Singh NK, Patra S, Goswami P (2016) Aromatic surfactant as aggregating agent for Aptamer-gold nanoparticle-based detection of Plasmodium lactate dehydrogenase. Mol Biotechnol 58(7):497–508

Josephson L, Perez JM, Weissleder R (2001) Magnetic nanosensors for the detection of oligonucleotide sequences. Angew Chem Int Ed Engl 40(17):3204–3206

Jung JH, Cheon DS, Liu F, Lee KB, Seo TS (2010) A Graphene oxide based Immuno-biosensor for pathogen detection. Angew Chem Int Ed 49:5708–5711

Jyoti A, Pandey P, Singh SP, Jain SK, Shanker R (2010) Colorimetric detection of nucleic acid signature of Shiga toxin producing *E. coli* using au NPs. *J Nanosci Nanotechnol* 7:4154–4158

Kaittanis C, Santra S, Perez JM (2010) Emerging nanotechnology based strategies for the identification of microbial pathogenesis. Adv Drug Deliv Rev 62:408–423

Khin PT (2015) Safer food for a safer world. JMAJ 58(4):129

Koedrith P, Thasiphu T, Weon J-I, Boonprasert R, Tuitemwong K, Tuitemwong P (2015) Recent trends in rapid environmental monitoring of pathogens and toxicants: potential of nanoparticle-based biosensor and applications. Sci World J 2015:510982

Lee C, Gaston MA, Weiss AA, Zhang P (2013) Colorimetric viral detection based on sialic acid stabilized gold nanoparticles. Biosens Bioelectron 42:236–241

Li X, Xing Y, Jiang Y, Ding Y, Li W (2009) Antimicrobial activities of ZnO powder coated PVC film to inactivate food pathogens. Int J Food Sci Technol 44:2161–2168

Lin YH, Chen SH, Chuang YC, Lu YC, Shen TY, Chang CA, Lin C (2008) Disposable amperometric immunosensing strips fabricated by au nanoparticles-modified screen-printed carbon electrodes for the detection of foodborne pathogen *Escherichia coli* O157:H7. Biosens Bioelectron 23(12):1832–1837

Liu R, Munro S, Nguyen T, Siuda T, Suciu D, Bizak M, Slota M, Hs F, Danley D, Mcshea A (2006) Integrated microfluidic CustomArray device for bacterial genotyping and identification. J Assoc LabAutom 11:360–367

Liu Y, Wang J, Song X, Xu K, Chen H, Zhao C, Li J (2018) Colorimetric immunoassay for *Listeria monocytogenes* by using core gold nanoparticles, silver nanoclusters as oxidase mimetics, and aptamer-conjugated magnetic nanoparticles. MikrochimActa 185:360–367

Lücker E, Horlacher S, Eigenbrodt E (2001) Brain in human nutrition and variant Creutzfeldt-Jakob disease risk (vCJD): detection of brain in retail liver sausages using cholesterol and neuron specific enolase (NSE) as markers. Br J Nutr 86(Suppl 1):S115–S119

Luo Y, Alocilja EC (2017) Portable nuclear magnetic resonance biosensor and assay for a highly sensitive and rapid detection of foodborne bacteria in complex matrices. J Biol Eng 11:14

Luo Z, Qin Y, Ye Q (2015) Effect of nano-TiO2-LDPE packaging on microbiological and physicochemical quality of Pacific white shrimp during chilled storage. Int J Food Sci Technol 50:1567–1573

Luo K, Kim H, Oh M, Kim Y (2020) Paper-based lateral flow strip assay for the detection of foodborne pathogens: principles, applications, technological challenges and opportunities. Crit Rev Food Sci Nutr 60(1):157–170

Lv Y, Liu H, Wang Z, Liu S, Hao L, Sang Y, Liu D, Wang J, Boughton RI (2009) Silver nanoparticle–decorated porous ceramic composite for water treatment. J Membr Sci 331:50–56

Martı'n-Belloso O, Soliva-Fortuny R, Oms-Oliu G (2006) Fresh-cut fruits. In: Hui YH (ed) Handbook of fruits and fruit processing. Blackwell Publishing, Ames, IA, pp 129–144

Najer A, Wu D, Bieri A, Brand F, Palivan CG, Beck H, Meier W (2014) Nanomimics of host cell membranes block invasive malaria parasites. ACS Nano 8:12560–12571

Nordin N, Yusof NA, Abdullah J, Radu S, Hushiarian R (2017) A simple, portable, electrochemical biosensor to screen shellfish for *Vibrio parahaemolyticus*. AMB Express 7:41

Oliver SP, Jayarao BM, Almeida RA (2005) Foodborne pathogens in milk and the dairy farm environment: food safety and public health implications. Foodborne Pathog Dis 2:115–129

Pandey R, Zahoor A, Sharma S, Khuller GK (2003) Nanoparticle encapsulated antitubercular drugs as a potential oral drug delivery system against murine tuberculosis. Tuberculosis 83:373–378

Pandit S, Dasgupta D, Dewan N, Ahmed P (2016) Nanotechnology based biosensors and its application. PharamInnov J 5:18–25

Paris L, Magni R, Zaidi F, Araujo R, Saini N, Harpole M, Coronel J, Kirwan DE, Steinberg H, Gilman RH, Petricoin EF, Nisini R, Luchini A, Liotta L (2017) Urine lipoarabinomannan glycan in HIV-negative patients with pulmonary tuberculosis correlates with disease severity. Sci Transl Med 9:eaal2807

Pineda MF, Leo Li-Ying C, Kuhlenschmidt T, Choi CJ, Kuhlenschmidt M, Cunningham BT (2009) Rapid specific and label-free detection of porcine rotavirus using photonic crystal biosensors. IEEE Sensors J 9(4):470–477

Pozio E (2003) Foodborne and waterborne parasites. Acta Microbiol Pol 52(Suppl):83–96

Prasad R, Swamy VS (2013) Antibacterial activity of silver nanoparticles synthesized by bark extract of *Syzygium cumini*. J Nanoparticles. https://doi.org/10.1155/2013/431218

Prasad R, Kumar V, Kumar M (2017) Nanotechnology: food and environmental paradigm. Springer Nature Singapore Pte Ltd., Singapore. ISBN 978-981-10-4678-0

Qasim M, Lim DJ, Park H, Na D (2014) Nanotechnology for diagnosis and treatment of infectious diseases. J Nanosci Nanotechnol 14(10):7374–7387

Qi L, Xu Z, Jiang X, Hu C, Zou X (2004) Preparation and antibacterial activity of chitosan nanoparticles. Carbohydr Res 339(16):2693–700

Qiao J, Meng X, Sun Y, Li Q, Zhao R, Zhang Y, Wang J, Yi Z (2018) Aptamer-based fluorometric assay for direct identification of methicillin-resistant *Staphylococcus aureus* from clinical samples.*J. Microbiol.* Methods 153:92–98. https://doi.org/10.1016/j.mimet.2018.09.011

Qin D, He X, Wang K, Zhao XJ, Tan W, Chen J (2007) Fluorescent nanoparticle-based indirect immunofluorescence microscopy for detection of *Mycobacterium tuberculosis*. J Biomed Biotechnol 2007(7):89364 de la Rica R, Matsui H (2010) Applications of peptide and protein-based materials in bionanotechnology. Chem Soc Rev 39(9):3499–3509

Ruedas-Rama MJ, Walters JD, Orte A, Hall EAH (2012) Fluorescent nanoparticles for intracellular sensing: *a review*. Anal Chim Acta 751:1–23

Schofield CL, Field RA, Russell DA (2007) Glyconanoparticles for the colorimetric detection of cholera toxin. Anal Chem 79:1356–1361

Shankar S, Reddy JP, Rhim JW, Kim HY (2015) Preparation, characterization, and antimicrobial activity of chitin nanofibrils reinforced carrageenan nanocomposite films. Carbohydr Polym 117:468–475

Shankar S, Jaiswal L, Rhim JW (2016) Gelatin-based nanocomposite films: potential use in antimicrobial active packaging. In: Antimicrobial food packaging. Elsevier, Amsterdam, pp 339–348

Shin WR, Sekhon SS, Kim SG, Rhee SJ, Yang GN, Won K, Rhee SK, Ryu H, Kim K, Min J et al (2018) Aptamer-based pathogen monitoring for Salmonella enteric ser. Typhimurium. J Biomed Nanotechnol 14(11):1992–2002. https://doi.org/10.1166/jbn.2018.2634

Singh AK, Harrison SH, Schoeniger JS (2000) Gangliosides as receptors for biological toxins: development of sensitive fluoroimmunoassays using ganglioside-bearing liposomes. Anal Chem 72(24):6019–6024

Singh S, Kumar V, Dhanjal DS, Datta S, Prasad R, Singh J (2020) Biological biosensors for monitoring and diagnosis. In: Singh J, Vyas A, Wang S, Prasad R (eds) Microbial biotechnology: basic research and applications. Springer Nature Singapore, Singapore, pp 317–336

Sperling RA, Rivera Gil P, Zhang F, Zanella M, Parak WJ (2008) Biological applications of gold nanoparticles. Chem Soc Rev 37(9):1896–1908

Syed MA, Bokhari SH (2011) Gold nanoparticle based microbial detection and identification. J Biomed Nanotechnol 7(2):229–237

Thangadurai D, Sangeetha J, Prasad R (2020) Nanotechnology for food, agriculture, and environment. Springer International Publishing, Cham. ISBN 978-3-030-31937-3. https://www.springer.com/gp/book/9783030319373

Tsai TT, Shen SW, Cheng CM, Chen CF (2013) Paper-based tuberculosis diagnostic devices with colorimetric gold nanoparticles. Sci Technol Adv Mater 14(4):044404

Tuitemwong P, Songvorawit N, Tuitemwong K (2013) Facile and sensitive Epifluorescent silica nanoparticles for the rapid screening of EHEC. J Nanomater 2013:1–8

Urbán P, Estelrich J, Cortés A, Fernàndez-Busquets X (2011) A nanovector with complete discrimination for targeted delivery to Plasmodium falciparum-infected versus non-infected red blood cells in vitro. *J*. Controlled Release 151:202–211

Varshney M, Yang L, Su XL, Li Y (2005) Magnetic nanoparticle-antibody conjugates for the separation of Escherichia coli O157:H7 in ground beef. J Food Prot 68(9):1804–1811

Vicentini DS, Smania A Jr, Laranjeira MCM (2009) Chitosan/poly. (vinyl alcohol). Films containing ZnO nanoparticles and plasticizers. Mat Sci Eng C 30(4):503–508

Wang LF, Rhim JW (2016) Grapefruit seed extract incorporated antimicrobial LDPE and PLA films: effect of type of polymer matrix. LWT—Food Sci Technol 74:338–345

Wang S, Singh AK, Senapati D, Neely A, Yu H, Ray PC (2010) Rapid colorimetric identification and targeted photothermal lysis of *Salmonella* bacteria by using bioconjugated oval-shaped au NPs. Chem—Eur J 16:5600–5606

Weber JE, Pillai S, Rama MK, Kumar A, Singh SR (2011) Electrochemical impedance-based DNA sensor using a modified single walled carbon nanotube electrode. Mater Sci Eng C 31:821–825. https://doi.org/10.1016/j.msec.2010.12.009

Weeks BL, Camarero J, Noy A, Miller AE, Stanker L, De Yoreo JJ (2003) A microcantilever-based pathogen detector. Scanning 25(6):297–299

Wu H, Yin JJ, Wamer WG, Zeng M, Lo YM (2014) Reactive oxygen species related activities of nano-iron metal and nano-ironoxides. J Food Drug Anal 22:86–94

Wu S, Duan N, Qiu Y, Li J, Wang Z (2017) Colorimetric aptasensor for the detection of *Salmonella enterica* serovar typhimurium using ZnFe2O4-reduced graphene oxide nanostructures as an effective peroxidase mimetics. Int J Food Microbiol 261:42–48. https://doi.org/10.1016/j.ijfoodmicro.2017.09.002

Wu W, Yu C, Wang Q, Zhao F, He H, Liu C, Yang Q (2020) Research advances of DNA aptasensors for foodborne pathogen detection. Crit Rev Food Sci Nutr 60(14):2353–2368

Xu L, Liu Y, Chen Z, Li W, Liu Y, Wang L, Ma L, Shao Y, Zhao Y, Chen C (2013) Morphologically virus-like fullerenol nanoparticles act as the dual-functional nanoadjuvant for HIV-1 vaccine. Adv Mater 25:5928–5936

Yang L, Li Y (2005) Quantum dots as fluorescent labels for quantitative detection of Salmonella typhimurium in chicken carcass wash water. J Food Prot 68(6):1241–1245

Yao C, Li X, Neoh KG, Shi Z, Kang ET (2008) Surface modification and antibacterial activity of electrospun polyurethane fibrous membranes with quaternary ammonium moieties. J Membr Sci 320:259–267

Yeh YC, Creran B, Rotello VM (2012) Gold nanoparticles: preparation, properties, and applications in bionanotechnology. Nanoscale 4(6):1871–1880

Yrad FM, Castañares JM, Alocilja EC (2019) Visual detection of Dengue-1 RNA using gold nanoparticle-based lateral flow biosensor. Diagnostics (Basel) 9(3):74

Zhang H, Ma X, Liu Y, Duan N, Wu S, Wang Z, Xu B (2015) Gold nanoparticles enhanced SERS aptasensor for the simultaneous detection of *Salmonella typhimurium* and *Staphylococcus aureus*. Biosens Bioelectron 15:872–877

Zhao X, Lin CW, Wang J, Oh DH (2014) Advances in rapid detection methods for foodborne pathogens. J Microbiol Biotechnol 24:297–312

Zhu L, Ang S, Liu WT (2004) Quantum dots as a novel immunofluorescent detection system for *Cryptosporidium parvum* and *Giardia lamblia*. Appl Environ Microbiol 70(1):597–598

Zou Y, Duan N, Wu S, Shen M, Wang Z (2018) Selection, identification, and binding mechanism studies of an ssDNA aptamer targeted to different stages of *E. coli* O157:H7. J Agric Food Chem 66(22):5677–5682

Chapter 12
Holistic Approaches for Enhanced Production of Prodigiosin—a Natural Biocolour

Subhasish Dutta

Abstract Nowadays, pigments from microbial sources have gained much importance due to its substantial medical and industrial welfares. In this regard, the present book chapter discusses some impactful methodologies regarding the improved production of prodigiosin drug—a promising supplement of synthetic colour. Prodigiosin is a vivid red colour pigment mainly synthesized by bacteria *Serratia marcescens*. Other than its significant contribution towards the cosmetic and food industry, it also poses antifungal, antibacterial, immunosuppressive and antiproliferative activities. Optimization of media components, scale-up study and biokinetic analysis from wild type and recombinant sources have been thoroughly discussed in this chapter. Particular emphasis was put forward on its nanotechnological synthesis and drug delivery for various diseases. Efficient extraction practice, estimation by different analytical instruments and purification of the final product were also highlighted in a best possible way.

Keywords Prodigiosin · Bioengineering · Optimization · *Serratia marcescens* · Enhance

12.1 Introduction

Supplementation of natural ingredients in the food industry is a need of the hour due to excessive toxicity of synthetic pigments and organic dyes. The biocolour which are readily accessible from plants have numerous shortcomings due to light sensitivity or unsteadiness, temperature or adverse pH, insoluble or minimal water solubility and most importantly unavailability all over the year (Gulani et al. 2012). Less water solubility is sometimes responsible for application limitation in

S. Dutta (✉)
Department of Biotechnology, Haldia Institute of Technology (HIT), Haldia, West Bengal, India
e-mail: subhasish.bt@hithaldia.in

N. R. Maddela et al. (eds.), *Nanotechnology for Advances in Medical Microbiology*, Environmental and Microbial Biotechnology,
https://doi.org/10.1007/978-981-15-9916-3_12

the textile dye industry (Shen and Yang 2013). The microbial pigments being idiolites having a differentiated composition according to their assembly and perhaps accredited to the environmentally harmless compounds in the chemical industry. Creation of an innovative product by developing a unique chromophore is a patentable and novel aspect in biotechnology. Considering all the inadequacies of synthetic, organic and plant-based pigments available in the market, researchers are now focusing on microbial-based biocolour, which could have a broad spectrum of activities. Prodigiosin ($C_{20}H_{25}N_3O$) is one of them. It is a membrane linked red pigment as well as a bioactive secondary metabolite, synthesized by both Gram positive and Gram negative bacteria like *Serratia marcescens*, *Streptomyces coelicolor*, *S. lividans*, *Pseudomonas magneslorubra*, *Vibrio psychroerythrous*, etc. (Khanafari et al. 2006; Shahitha and Poornima 2012; Kurbanoglu et al. 2015). *Serratia marcescens* is the largest producer of this vital pigment worldwide. Being an opportunistic human pathogen, frequency of spreading infection by pigmented *S. marcescens* is much lesser than non-pigmented one (Elkenawy et al. 2017). Prodigiosin is commercially exploited for several industrial applications such as antimicrobial, anti-oxidant, anti-malarial, anti-cancer, anti-protozoal, anti-neoplastic, anti-diabetic, non-steroidal, immunosuppressive, UV protective and antiproliferative activity (Khanafari et al. 2006; Borić et al. 2011; Lee et al. 2011; Montaner and Pérez-Tomás 2001; Park et al. 2012; Soto-Cerrato et al. 2004; Gulani et al. 2012). The same can also act as a bacteriostatic agent against *E. coli* (Danevčič et al. 2016). There are some isoforms of prodigiosin available in the literature, i.e. undecylprodigiosin, metacycloprodigiosin, nonylprodigiosin, norprodigiosin having the same clinical applications (Zang et al. 2014). It also plays a vital role in bacterial cells by acting as a protective cover during stress by reducing the fabrication of reactive oxygen species (ROS) (Gul et al. 2020).

With the growing solicitation of prodigiosin, reducing its manufacturing cost has become one of the most significant targets. It is very crucial to note that several aspects are still needed to be concerned about the large-scale production of prodigiosin. Moreover, the application of high-throughput technology and complex instrumentations with higher operating costs may, in turn, limit their commercialization and industrialization, especially in the third world countries. To develop efficient techniques for prodigiosin production, significant research has been directed towards strain improvement and the identification of dynamic systems and process conditions.

Improving the yield of prodigiosin cost-effectively is one of the critical challenges for its commercial viability. Recovery of prodigiosin from microorganism is a complex procedure. Therefore, enhancing the downstream recovery of the product with low processing cost is another challenge for the bioengineers (Acuña-Argüelles et al. 1995). Microbial prodigiosin production is inducible, and hence the yield could be considerably enhanced with the addition of inducers.

Pilot-scale production and solicitations of prodigiosin in developing countries do not seem to flourish dramatically due to excessive capital investment for industrial applications. Considering the current market demand, implementation of this potent biocolour is projected to increase logarithmically within the next few decades.

Considering the current mandate, usefulness and cost-effectiveness of the prodigiosin in the industry, it is imperative to focus on its enhanced yield using several existing tools of biotechnology. It will surely amplify the applications of prodigiosin in several industries, manufacturing processes and services for economic mobility through biotechnology and bioengineering.

The present book chapter attempts to address the knowledge gap surrounding efficient prodigiosin production by fermentation, nanobiotechnology and other essential methods for its effective commercialization. Bioengineering aspects that have been taken into consideration for its enhanced yield will also be covered in this chapter.

12.2 Natural Occurrence and Recombinant Sources of Prodigiosin

Prodigiosin is a naturally occurring pigment that can be obtained from both plants as well as microorganisms. Terrestrial plants are the primary sources of this pigment. There are numerous advantages for selecting microorganisms over plant sources for extracting prodigiosin. Some of the benefits like they are stable, their accessible cultivation technology, can be biosynthesized any time in the lab, a concise doubling time of bacteria thus fast specific growth rate, cost-effective culture media, optimal environmental parameters and their production is independent of weather condition (Chidambaram and Perumalsa 2009; Gulani et al. 2012). The pigment-producing microorganisms are found everywhere like air, water, soil, sewage and even in some household animals (Su et al. 2011). Nguyen et al. (2020) recently discovered that prodigiosin could also be synthesized by marine chitin (Nguyen et al. 2020). Among different chitin molecules, α chitin is the potent producer of prodigiosin by fermentation. However, *Serratia marcescens* is the dominant producer of this pigment to date; its large-scale production can be trimmed due to its durable infectious nature. On the contrary, the actinomycetes group of organisms, viz. *Streptomyces* spp. are on the safer side. Optimization and large-scale production can be easily carried out by these species (Luti and Mavituna 2011). Among the *Streptomyces* sp. *Streptomyces coelicolor* A3(2) have been reported to be capable of producing prodigiosin pigment (Hobbs et al. 1990). Besides *S. marcescens*, other extremophiles like *Hahellacea* and *Pseudoalteromonas* are also capable of synthesizing prodigiosin (Kim et al. 2007; Schloss et al. 2010).

Massive bioengineering studies were attempted to amplify the yield of prodigiosin by modifying the parent strain. Researchers have used classical mutational techniques along with molecular biology approaches to enhance the yield of this antibiotic. Prodigiosin biosynthesis genes from *S. marcescens* were randomly integrated into *Pseudomonas putida* KT2440 for making a potent strain (Domröse et al. 2015). Often, an isogenic mutant strain of *S. marcescens* SS-1 was also used for producing prodigiosin resembling pigments (Wei and Chen 2005).

12.3 Nanotechnological Aspects of Prodigiosin Synthesis

Nanobiotechnology is a rapidly growing field in medical science for its unique and promising features. 'Nano' (10^{-9}) is the smallest size that exists in the world having the capability of reflecting light. Nanoparticle or nanomaterials consist of a cluster of atoms in the range of 1–100 nm (Karthika et al. 2015; Al-Shabib et al. 2018; Prasad et al. 2016). They are mainly of two types, i.e. organic and inorganic. Inorganic nanoparticles have vast applications in the medical devices like biosensors; in environmental sample purification, in the pharmaceutical industry and also they can act as antimicrobials which can be integrated into optical fibres (Karthika et al. 2015; Prasad et al. 2017a,b, 2018a,b, 2020). Their potential biocidal activity is due to tiny size and relatively higher surface to volume ratio (Al-Shabib et al. 2018; Morones et al. 2005). Among inorganic nanoparticles, gold (AuNPs) and silver nanoparticles (AgNPs) have higher material property, functional diversity and stable in harsh environmental conditions (Karthika et al. 2015). Metal oxides can also be used as nanomaterial, i.e. Zinc oxides (ZnO). ZnO nanoparticles (ZnO-NPs) exhibit broad-spectrum antibacterial activity (Al-Shabib et al. 2018; Bhuyan et al. 2015). Prodigiosin driven biosynthesis of silver nanoparticles, conjugated gold/prodigiosin synthesis, its characterization and application in breast cancer will be highlighted in this chapter.

12.3.1 *Prodigiosin Mediated Biosynthesis of Silver Nanoparticles (AgNPs)*

Karthika et al. (2015) synthesized silver nanoparticles using culture supernatant containing prodigiosin from the bacterial strain of *S. marcescens* (Karthika et al. 2015). In brief, culture supernatant and 0.1 (M) $AgNO_3$ solution (equal volume) were mixed and incubated at 60^0C until the colour change was observed from pink to black. Reduction of Ag^+ was confirmed by the colour change and subsequently indicated the synthesis of silver nanoparticle. The resulting mixture was separated, followed by repeated washing and centrifugation at $10000\times g$ for 15 mins. The final suspension was dried and ready to use for further study. AgNPs were characterized by UV-visible spectroscopy, X-ray diffraction (XRD), scanning and transmission electron microscopy (SEM and TEM). Antibacterial activity against *Staphylococcus* sp., *E. coli* and *Pseudomonas* sp. was examined and observed by disc diffusion method. TEM images confirmed the spherical shape of AgNP and size less than 100 nm. Due to their tiny shape and size, AgNPs can easily reach to the genomic content of the bacteria, have the accessibility of the large surface area and subsequently destroy them.

12.3.2 Biosynthesis of Gold/prodigiosin Nanoparticles (AuNPs) and Its Application

Gold nanoparticles also have prospective applications in drug delivery, tissue analysis and in gene transfer (Dozie-Nwachukwu et al. 2017b). Dozie-Nwachukwu et al. (2017) biosynthesized gold/prodigiosin nanoparticles (AuNPs) from *Serratia marcescens* strain to combat triple-negative breast cancer (MDA-MB-231) cells (Dozie-Nwachukwu et al. 2017b; Dozie-Nwachukwu et al. 2017a). To target a specific cancer cell, AuNPs/PG drug was prepared under diverse pH environments.

Prodigiosin from *S. marcescens* was extracted and purified using size exclusion chromatography before nanomaterial synthesis. Unlike silver nanoparticle, AuNP/PG *conjugate* was prepared through physisorption, where particles were held together by weak Vander-walls force (Dozie-Nwachukwu et al. 2017a). Here, 1 mg of solid prodigiosin was dissolved in 2 ml of methanol and mixed to the gold nanoparticles *conjugate*. The resulting mixture was agitated at 500 rpm at 4°C for 30 min. AuNPs were characterized by UV-spectrophotometry, electron diffraction (ED), dynamic light scattering (DLS) and energy-dispersive X-ray spectroscopy (EDS). The adhesion between luteinizing hormone-releasing hormone (LHRH)-conjugated AuNP/PG drug and MDA-MB-231 breast cancer cells (nanocluster) was studied by atomic force microscopy (AFM) technique. It was observed that the adhesion force was higher in nanocluster than healthy breast cells. Greater adhesion leads to 5 fold increase overexpression/interaction between ligand and receptors on the membrane surface, which is very important to treat a specific target of triple-negative breast cancer.

In another study, cell-free extracts of prodigiosin solution was allowed to react with 2.5 mM $HAuCl_4$ (tetrachloroauric acid) to prepare gold/prodigiosin nanoparticle (Dozie-Nwachukwu et al. 2017b). The formation of nanoparticle was confirmed by colour change (bright red to pale pink). Such colour change occurred due to reduction of $HAuCl_4$ to Au^0 and the possible reaction is as follows, $HAuCl_4 + 3e^- \rightarrow Au + 4Cl^- + H^+$. However, the total colour change was observed after 6th day of incubation. Reduction of $HAuCl_4$ to Au^0 results from supersaturation solution and nucleation. Different analytical instruments were used to characterize gold nanoparticles. The reduction of Au^{3+} to Au^0 was observed using UV–vis spectrophotometer. The size and shape of nanoparticle was determined by TEM, higher resolution image of particle surface was taken with the help of Helium Ion Microscopy (HIM), and the lower concentration of AuNPs was detected with the help of DLS (Dozie-Nwachukwu et al. 2017b). Both biomass and cell-free extracts of prodigiosin were used for nanoparticle synthesis. Still, the cell-free extract is a better choice for lower reaction time. Final nanoparticle size was around 40–60 nm which is very efficient for cancer detection and treatment.

12.3.3 Biofabrication of Zinc Oxide Nanoparticle

Al-Shabib et al. (2018) synthesized zinc nanoparticles (ZnNPs) from leaf extract of *Ochradenus baccatus* (Al-Shabib et al. 2018). Zinc metal oxides (ZnO-NPs) were the nanoparticle of choice due to their broad-spectrum bactericidal activity, protein binding study, etc. 0.05 (M) Zinc nitrate solution and *Ochradenus baccatus* leaf extract were mixed and microwaved for 20 min. After cooling down, the precipitate was parted by centrifugation, followed by washing with de-ionized water and 100% C_2H_5OH. After that, they were allowed to dry at 80^0C for 24 hr. Lastly, the product was calcined at 800^0C for 2 hr. X-Ray Diffraction (XRD), Fourier Transform Infrared Spectroscopy (FTIR), Scanning Electron Microscopy (SEM) and EDX were used for ZnO-NPs characterization.

Effect of zinc oxide nanoparticle was analysed against prodigiosin synthesized by *S. marcescens* culture. Decrease of prodigiosin yield was documented at sub-MIC level ranging between 6.25 μg/ml and 50 μg/ml. It was thought that inhibition of prodigiosin would decrease the pathogenicity and virulence characteristics of *S. marcescens*.

Not only nanoparticles, nanogel is also an excellent choice to treat breast cancer cells (Obayemi et al. 2016b). Crosslinking between dextran and poly(lactide) was proved to be an effective conventional technique for preparing nanogel. It is injectable, multifunctional and biodegradable. The same was loaded with prodigiosin and was very effective against breast cancer cells.

12.4 Analytical Methods for Extraction, Estimation and Purification of Prodigiosin

Prodigiosin is principally an intracellular metabolite. It is bound to the cell membrane. During fermentation, only 17% of the total prodigiosin is released into the broth (Wang et al. 2004). Hence, an effective extraction method is necessary to recover the product. Well-defined separation and purification techniques are hugely recommended for its clinical application too. In this chapter, various analytical methods, i.e. extraction, estimation and purification of prodigiosin, will be discussed.

12.4.1 Extraction of Prodigiosin

More or less similar methods were adopted by the researchers to extract the prodigiosin from the culture broth. The two-step extraction procedure was followed for the efficient recovery of prodigiosin (Kurbanoglu et al. 2015; Williams 1973; Lin et al. 2019). For this purpose, culture broth (5 to 10 ml) from the stationary phase was taken in a test tube and centrifuged at 10000 rpm for 10 minutes to collect the cell pellet. After that, the supernatant was decanted, and acidified methanol was poured to the cell suspension. The mixture is vortexed and centrifuged under the same condition. Finally, the concentration of the prodigiosin was measured by spectrophotometer at 535 nm (Venil and Lakshmanaperumalsamy 2009; Kurbanoglu et al. 2015). Some researchers have used acetone in place of MeOH for its active recovery (Shahitha and Poornima 2012). Su et al. (2011) used de-ionized water for washing the cell pellet prior to adding acetone (Su et al. 2011). The crude pigment was obtained by concentrating it using a rotary evaporator (Lin et al. 2019). Wei and Chen (2005) extracted prodigiosin like pigment from the recombinant strain of *Serratia marcescens* SMΔR. They have used 3 (M) Chloroform to extract the pigment (Wei and Chen 2005). Park et al. (2012) opted nine organic solvents to find out the optimal and best solvent on the basis of extraction efficiency (Park et al. 2012). They found ethanol as the best solvent, whereas acetone was optimum for prodigiosin extraction from fermentation broth. However, by using acetone, the extraction process was 140 times faster than ethanol under the same conditions. Sometimes, a proper surfactant (Tween 80; 0.1% w/v) is useful for extracting prodigiosin from the bacterial cell envelope (Wang et al. 2004). Wang et al. (2004) have developed an efficient technique for in-situ recovery of prodigiosin from direct fermentation broth (Wang et al. 2004).

12.4.2 Estimation of Prodigiosin

There are different methods available for estimating prodigiosin after extraction from fermentation broth. Its concentration can be determined spectrophotometrically as well as by Liquid Chromatography (LC) technique. λ_{max} for prodigiosin is around 535 nm (Elkenawy et al. 2017). Its concentration can be calculated by taking its specific absorbancy of 51.5×10^3 litre per g per cm (Williams 1973). Haddix and Werner (2000) developed one formula to quantify the prodigiosin yield (Haddix and Werner 2000). The same has become one of the most useful equation to estimate prodigiosin concentration. It is shown below: Prodigiosin unit/cell = $\frac{[(OD_{499}-(1.381\times OD_{620}))]\times1000}{OD_{620}}$; whereas OD_{499}=is prodigiosin absorption, OD_{620}= bacterial culture absorption and 1.381 is $\frac{OD_{499}}{OD_{620}}$ quotient and taken as a constant. Williams et al. (1961) and Chen et al. (1993) added an alternative recipe to calculate the total prodigiosin (TP) yield, i.e. Total prodigiosin (mg/L) = $\frac{ADV_1}{7.07\times10^4 V2}$; whereas A= Absorbance at 535 nm, D = Dilution ratio, V_1= volume of methanol added, V_2= amount of fermentative culture and 7.07×10^4 is extinction coefficient of the antibiotic (Williams et al. 1961; Chen and Johns 1993). Lee et al. (2011) assayed the bioactivity of prodigiosin and cycloprodigiosin by *Agar Disc Diffusion* (ADD) method (Lee et al. 2011). They have checked the antimicrobial activity against different microorganisms like *Bacillus subtilis* KCTC 1914, *Escherichia coli* KCTC 1924, *Salmonella enterica* serovar Typhimurium KCTC 1926,

Staphylococcus aureus KCTC 1916 and *Candida albicans* KCTC 1940. Filter discs were poured with pure prodigiosin extract, and inhibition zones were recorded against test organisms. Finally, the concentration of the antibiotic was measured by measuring the zone diameter. ADD is a very primitive method for analysing the unknown concentration of many antibiotics, i.e. rapamycin, cyclosporine, etc. (Kojima et al. 1995; Lee et al. 1997; Dutta et al. 2014; Dutta et al. 2017).

12.4.3 Purification and Characterization

Many scientists are working on a laboratory scale for the production, purification and characterization of prodigiosin. However, the research terminates at the lab scale with a few progressing beyond it. This is due to problems associated with scale-up and downstream of the production process. LC-MS, thin-layer chromatography (TLC), HPLC, NMR, FTIR, etc., are the methods used for prodigiosin stability, purification and successive characterization study (Aruldass et al. 2014; Park et al. 2012; Wei and Chen 2005; Faraag et al. 2017; Rakh et al. 2017). Crude prodigiosin is purified by a silica gel column (Lin et al. 2019). In brief, crude prodigiosin extract was dissolved in methanol, and the subsequent mixture was allowed to pass through a hexane balanced silica gel packed column. The absorbed product was then eluted out from the column with the help of ethyl acetate. Finally, the orange colour eluate was dried in a rotary evaporator to obtain the red prodigiosin powder (Wei and Chen 2005). Wang et al. (2004) have developed an efficient technique by combining both static and column adsorption process for prodigiosin purification (Wang et al. 2004). They finally obtained 83% higher yield than conventional silica gel method.

Darshan and Manonmani (2013) determined the molecular mass of prodigiosin by LC-Mass spectroscopy (MS) technique as 324 D (Darshan and Manonmani 2013) and the structure was further confirmed by ^{1}H-NMR spectroscopy. Rakh et al. (2017) also characterized the chemical structure of purified prodigiosin by FTIR method (Rakh et al. 2017). Fourier transform mass spectrometry (FT-MS) technique was used for structural elucidation of prodigiosin like pigments produced by *Vibrio* sp. (Alihosseini et al. 2008). Characterization of prodigiosin by *Zooshikella rubidus* S1-1 was analysed by tandem mass spectroscopy (LC-MS/MS) (Lee et al. 2011). Researchers have identified two major peaks with '*m/z*' value of 322 and 324, respectively. Later they have concluded that two peaks were of cycloprodigiosin and prodigiosin.

12.5 Bioengineering Strategies Adopted for the Enhanced Yield of Prodigiosin

It was observed that there are two policies adopted to date for microbial prodigiosin production. It was either produced by wild type culture or recombinant or mutant microbial strain. The critical challenges behind using wild type microorganism include the availability of potent microbial strain and the solicitation of this biocatalyst for industrial-level production. Furthermore, isolating a dominant strain of anticipated product is tiresome and time-consuming. Hence, the robust approach will be to generate a mutant strain having the capability of producing a large amount of prodigiosin with the help of available biotechnological tools.

12.5.1 Contribution of Influential Parameters

Ryazantseva and Andreyeva (2014) experimented with *Serratia marcescens* 9986 for production of prodigiosin. It was reported that growth parameters controlled the synthesis of pigment. Prodigiosin yield was found to be 0.2–0.4 mg/L of culture medium in a batch process under aerobic environments. 0.1% of sodium dodecyl sulphate (SDS) was treated in biomass for extraction of prodigiosin (Ryazantseva and Andreyeva 2014). Locally isolated pigment-producing strain *Serratia marcescens* UMT-1 was used by Aruldass et al. (2014) for enhanced prodigiosin yield (Aruldass et al. 2014). They have used brown sugar as a nutrient for the development of bacteria. It was reported that not only culture conditions but also lactose and L-tryptophan supplementation were necessary for higher production of prodigiosin in both shake flask and 5-L bioreactor. The final output of prodigiosin was found to be 8000 mg/L. The study revealed that brown sugar could be used as cheap and potential media constituents for prodigiosin production. In another study, an isolated bacterial culture *Serratia marcescens* N10612 was used for the production of prodigiosin. In this experiment, a mixed culture media consisting of sucrose, peptone and yeast extract was used for maximum prodigiosin production. Media optimization was carried out through statistical experimental designs for maximum yield. The optimum concentrations of medium were yeast extract 5.6 g/L, peptone 6.9 g/L, sucrose 31.6 g/L and NaCl 1.0 g/L (Zang et al. 2014). Nakamura and Kitamura (1981) patented their research work on prodigiosin production from a novel *Serratia marcescens* R-2 strain. A fatty acid having 12 to 18 carbon atoms was employed as a sole carbon source in simulated culture media. It was observed that *Serratia marcescens* could assimilate the carbon and produce prodigiosin (Nakamura and Kitamura 1981). In another patent, production of prodigiosin was carried out in fermentor having media composition of 3% sorbitol, 1.5% soy flour, and 0.125% of $Mg(SO_4)_2$. The fermentation media was inoculated with a rough, nonmucoid strain of *Serratia marcescens*. The culture was transferred to a sizeable non-agitated fermenter having the capacity of 100 gallons and grown for 16 hours at

28^0C. The antifoaming agent used was 0.3% solution of octadecanol in mineral oil (Harned 1953).

In India, most of the works on prodigiosin production have been focused on either submerged fermentation where the pre-formulated synthetic medium was employed. Sundaramoorthy et al. (2009) investigated the production of prodigiosin from *Serratia marcescens*. Prodigiosin was characterized and reported that it has antifungal, immunosuppressive and antiproliferative activity. In this investigation, the factors, viz., temperature, pH, sugar and oil substrate were found to be the key parameters for prodigiosin production. Further, the optimization of influential factors was carried out to enhance the prodigiosin yield. It was also noticed that the maximum amount of prodigiosin was produced at a temperature of 30°C and pH 7.0. Besides, the maltose was found to be most suitable carbon source in the medium and yielded 425±40 mg/L of prodigiosin. It was also found that peanut oil considered to be the most appropriate oil substrate for production of prodigiosin (535±45 mg/L) (Sundaramoorthy et al. 2009).

12.5.2 Scale-up Study of Prodigiosin

Few scale-up studies were carried out for prodigiosin production by *Serratia marcescens* strain. There is a clear difference between scale-up and large-scale production of any metabolite. The scale-up study requires geometrical similarity, constant K_{La} (volumetric mass transfer coefficient), constant impeller tip speed (N_i) as well as constant H:D (H= height of the tank and D= tank diameter) ratio of the reactor. However, large-scale production of prodigiosin has been achieved by increasing the yield of metabolite through mutation, synthetic biology approach as well as molecular technique obviously from a small shake flask to reactor. Lack of scale-up study leads to higher in vivo experiment expenditure as well as an expensive downstream process. Naik et al. (2012) used peanut oil cake (POC) as a cheap substrate for improved prodigiosin production by *Serratia marcescens* CF-53 strain (Naik et al. 2012). They have also carried out scale-up study in a lab-scale 2 L bioreactor at temperature 30^0C, airflow rate of 2 *vvm*, agitation 200 rpm and pH 7. 40 mg/L prodigiosin yield was finally achieved. Chen et al. (2013) reported that they had increased the prodigiosin yield by 6.78 fold using statistical optimization and porous carrier addition strategy (Chen et al. 2013). *Serratia marcescens* C3 used in the study was a quorum sensing strain. A fractional factorial design was adopted for statistical analysis with four significant factors, i.e. $CaCl_2.2H_2O$, $FeSO_4.4H_2O$, $MgSO_4.7H_2O$ and $MnSO_4.4H_2O$. After optimization, around 7.05 g/L prodigiosin was obtained. Finally, they carried out immobilization study by adding calcium alginate beads as a porous carrier. Yield significantly increased from 7.05 g/L to 15.6 g/L. Incorporation of physical (UV ray) and chemical (EMS) mutagen for enhanced prodigiosin production was achieved by El-Bialy and El-Nour (2015) (El-Bialy and El-Nour 2015). The potent chemical mutant S_{26} strain of *Serratia marcescens* produced eightfold higher prodigiosin than that of parent culture. The said mutated culture was stable up to 80^0C and at alkaline pH.

12.5.3 Biokinetic Study

Prodigiosin is an idiolite. Its production mainly occurs during the transient of exponential phase and throughout the whole stationary period of bacterial growth. Hence, it is a non-growth associated product. A robust and deliberate kinetics study is very much necessary to observe the growth pattern of microorganism with its product formation strategy in the presence of different substrates available (Dutta et al. 2014). It is also very much crucial to develop and establish a proper mathematical model that can describe the kinetic behaviour of the organism easily.

Very few kinetics studies have been carried out by different researchers for prodigiosin production using diverse strains of *Serratia marcescens* in a lab-scale bioreactor (Mohammed and Luti 2020; Casullo de Araújo et al. 2010; Hobbs et al. 1990). Mohammed and Luti (2020) carried out the growth kinetics study of *S. marcescens* in a 7 L bioreactor using optimized media condition (Mohammed and Luti 2020). Preliminary they have optimized the culture media constituents, i.e. C/N ratio, inoculum size, etc., by Response Surface Methodology (RSM) based Central Composite Design (CCD) technique. It is globally used fractional factorial method for optimization studies. Cell growth kinetics calculation was based on dry cell weight (DCW), and product formation kinetics followed the Luedeking–Piret model. All the simulations were carried out by polymath 6 software. Biokinetic parameters were determined from logistic equation model. Stoichiometric and kinetic variables are shown in Table 12.1. Finally, they have achieved the maximum cell mass and prodigiosin yield of 14.4 mg/mL and 594.88 mg/L, respectively. Casullo de Araújo et al. (2010) carried out the kinetics study of prodigiosin production by *Serratia marcescens* UCP1459 using '*manipueira*' (cassava wastewater) supplemented with 2% mannitol as a renewable low-cost substrate (Casullo de Araújo et al. 2010). They were able to produce a very high level of prodigiosin, i.e. 43,000 mg/L in just 48 hr after fermentation. The specific growth rate (μ) and generation time (T_d) were found to be 0.36 h^{-1} and 1.88 hr. During kinetics study, they observed diauxic pattern of growth at 12 hr of fermentation. Williams (1973) have compared prodigiosin kinetics study from two different strains of *Serratia marcescens*, one is wild type and another is mutant (Williams 1973). He found that maximum prodigiosin production was achieved after 5 days of incubation apparently when the culture is in mid-stationary phase. From this study it has been again proved that prodigiosin is a secondary metabolite. Similar kinetics study was carried out for prodigiosin like pigments, i.e. actinorhodin and undecylprodigiosin by *Streptomyces coelicolor* A3(2) (Hobbs et al. 1990). Unlike prodigiosin, undecylprodigiosin is a growth-associated product, whereas actinorhodin was found to be a non-growth associated.

Table 12.1 Biokinetic parameters obtained from different study

Kinetic parameters	Microorganism	Reaction vessel	Experimental value	Reference
X_{max} (mg/ml)	*S. marcescens*	7 L bioreactor	14.42	(Mohammed and Luti 2020)
P_{max} (mg/L)	*S. marcescens*	7 L bioreactor	594.8	(Mohammed and Luti 2020)
$Y_{x/s}$ (g/g)	*S. marcescens*	7 L bioreactor	2.8	(Mohammed and Luti 2020)
$Y_{p/s}$ (mg/g)	*S. marcescens*	7 L bioreactor	129.1	(Mohammed and Luti 2020)
$Y_{p/x}$ (mg/g)	*S. marcescens*	7 L bioreactor	45.7	(Mohammed and Luti 2020)
q_p (mg_{prod}/ g_{cell}.h)	*S. marcescens*	7 L bioreactor	0.79	(Mohammed and Luti 2020)
μ_{max} (h^{-1})	*S. marcescens*	7 L bioreactor	0.0605	(Mohammed and Luti 2020)
	S. marcescens UCP1459	Shake flask	0.36	(Casullo de Araújo et al. 2010)
Td (hr.)	*S. marcescens*	7 L bioreactor	11.45	(Mohammed and Luti 2020)
	S. marcescens UCP1459	Shake flask	1.88	(Casullo de Araújo et al. 2010)

In another study, it was found that prodigiosin production is related to ATP synthesis (Haddix and Shanks 2018; Haddix et al. 2008). When the quantity of ATP within a cell is minimum, the cell population arrives at high-density exponential phase. Similarly, when the growth rate ceases, prodigiosin synthesis begins. The parallel result was also found during the kinetics analysis of prodigiosin production by *S. marcescens* Nima strain in batch culture (Haddix et al. 2008). Haddix et al. (2008) also suggested that prodigiosin production and ATP synthesis have an inverse relationship (Haddix et al. 2008). In the latest research published by Obayemi et al. (2016b) revealed the release kinetics of anti-cancer drug prodigiosin obtained from *Serratia marcescens* subsp. *Marcescens* (Obayemi et al. 2016a). They have analysed the drug release kinetics using the following model developed by Korsmeyer and Peppas, i.e. $F = K_p t^n$, where F is the amount of drug release at time t; K_p is the Peppas release rate constant and n is the release exponent. The release constant (k) and release coefficient (n) were found to be 0.93 min^{-1} and 0.85.

Dyeing kinetics of prodigiosin pigment and thermodynamical aspects of polyester fibre were carried out to get an overview to develop a suitable pigment dyeing technology (Shen and Yang 2013). Shen and Yang (2013) calculated the enthalpy and entropy change during dyeing kinetics study. Negative enthalpy and positive entropy value indicate that dyeing of prodigiosin is an exothermic process. The dyeing temperature was kept below 120^0C.

12.6 Conclusion

The present chapter has summarized more or less critical and holistic approaches that can be taken into consideration for enhanced prodigiosin yield. Some recent advancements on the nanotechnological synthesis, bioengineering aspects and relevant analytical techniques were concentrated. A lot of research has been conducted and is going on various parts of the globe on prodigiosin. However, in a country like India, researchers are still in their initial stages in the field of biocolour. The technological gaps reside on its efficient downstream processing and extraction hinders the solicitation of large scale prodigiosin in the food industry. Optimization of production media, incorporation of the robust mathematical model with the suitable molecular technique will help to trim the production cost of this potential multitasking component. After compelling purification study, the prodigiosin can assist as a promising drug for cancer treatment. This book chapter will serve as a bridge between the industry and the scientific institutes by helping them to develop a bench-scale technology that can be transferred to the industry and will further improve the country's economics.

References

Acuña-Argüelles M, Gutierrez-Rojas M, Viniegra-González G, Favela-Torres E (1995) Production and properties of three pectinolytic activities produced by *Aspergillus niger* in submerged and solid-state fermentation. Appl Microbiol Biotechnol 43(5):808–814

Al-Shabib NA, Husain FM, Hassan I, Khan MS, Ahmed F, Abul Qais F, Oves M, Rahman M, Khan RA, Khan A, Hussain A, Alhazza IM, Aman S, Noor S, Ebaid H, Al-Tamimi J, Khan JM, Al-Ghadeer ARM, Khan MKA, Ahmad I (2018) Biofabrication of zinc oxide nanoparticle from ochradenus baccatus leaves: broad-spectrum antibiofilm activity, protein binding studies, and in vivo toxicity and stress studies. J Nanomat 1:1. https://doi.org/10.1155/2018/8612158

Alihosseini F, Ju KS, Lango J, Hammock BD, Sun G (2008) Antibacterial colorants: characterization of prodiginines and their applications on textile materials. Biotechnol Prog 24(3):742–747

Aruldass CA, Venil CK, Zakaria ZA, Ahmad WA (2014) Brown sugar as a low-cost medium for the production of prodigiosin by locally isolated *Serratia marcescens* UTM1. Int Biodeteriorat Biodegrad 95:19–24

Bhuyan T, Mishra K, Khanuja M, Prasad R, Varma A (2015) Biosynthesis of zinc oxide nanoparticles from *Azadirachta indica* for antibacterial and photocatalytic applications. Mater Sci Semicond Process 32:55–61

Borić M, Danevčič T, Stopar D (2011) Prodigiosin from Vibrio sp. DSM 14379; a new UV-protective pigment. Microb Ecol 62(3):528

Casullo de Araújo HW, Fukushima K, Takaki GMC (2010) Prodigiosin production by *Serratia marcescens* UCP 1549 using renewable-resources as a low cost substrate. Molecules 15 (10):6931–6940

Chen MH, Johns MR (1993) Effect of pH and nitrogen source on pigment production by *Monascus purpureus*. Appl Microbiol Biotechnol 40(1):132–138

Chen W-C, Yu W-J, Chang C-C, Chang J-S, Huang S-H, Chang C-H, Chen S-Y, Chien C-C, Yao C-L, Chen W-M (2013) Enhancing production of prodigiosin from *Serratia marcescens* C3 by statistical experimental design and porous carrier addition strategy. Biochem Eng J 78:93–100

Chidambaram KV, Perumalsa L (2009) An insightful overview on microbial pigment, prodigiosin. eJBio 5(3):49–61

Danevčič T, Vezjak MB, Zorec M, Stopar D (2016) Prodigiosin-A multifaceted *Escherichia coli* antimicrobial agent. PLoS One 11(9):1

Darshan N, Manonmani H (2013) Production, purification and characterization of prodigiosin from *Serratia rubidaea* and evaluation of its antibacterial and antiproliferative potential. J Bacteriol Parasitol 1:1

Domröse A, Klein AS, H-H J, Thies S, Svensson V, Classen T, Pietruszka J, Jaeger KE, Drepper T, Loeschcke A (2015) Efficient recombinant production of prodigiosin in *Pseudomonas putida*. Front Microbiol 6:972

Dozie-Nwachukwu SO, Obayemi JD, Danyuo Y, Anuku N, Odusanya OS, Malatesta K, Soboyejo WO (2017a) A comparative study of the adhesion of biosynthesized gold and conjugated gold/prodigiosin nanoparticles to triple negative breast cancer cells. J Mater Sci Mater Med 28(9):143

Dozie-Nwachukwu SO, Obayemi JD, Danyuo YT, Etuk-Udo G, Chi Y, Hu J, Anuku N, Odusanya OS, Malatesta K, Soboyejo WO (2017b) Biosynthesis of gold nanoparticles and gold/prodigiosin nanoparticles with *Serratia marcescens* bacteria. Waste Biomass Valori 8 (6):2045–2059

Dutta S, Basak B, Bhunia B, Chakraborty S, Dey A (2014) Kinetics of rapamycin production by *Streptomyces hygroscopicus* MTCC 4003. 3 Biotech 4(5):523–531

Dutta S, Basak B, Bhunia B, Sinha A, Dey A (2017) Approaches towards the enhanced production of Rapamycin by *Streptomyces hygroscopicus* MTCC 4003 through mutagenesis and optimization of process parameters by Taguchi orthogonal array methodology. World J Microbiol Biotechnol 33(5):90

El-Bialy HA, El-Nour SAA (2015) Physical and chemical stress on *Serratia marcescens* and studies on prodigiosin pigment production. Ann Microbiol 65(1):59–68

Elkenawy NM, Yassin AS, Elhifnawy HN, Amin MA (2017) Optimization of prodigiosin production by *Serratia marcescens* using crude glycerol and enhancing production using gamma radiation. Biotechnol Rep 14:47–53

Faraag AH, El-Batal AI, El-Hendawy HH (2017) Characterization of prodigiosin produced by *Serratia marcescens* strain isolated from irrigation water in Egypt. Nat Sci 15:55–68

Gul S, Shad MA, Arshad R, Nawaz H, Ali A, Altaf A, Gul T, Iqbal W (2020) Response surface optimization of prodigiosin production by mutagen-treated *Serratia marcescens* in different growth media. Pharmacogn Mag 16(68):99

Gulani C, Bhattacharya S, Das A (2012) Assessment of process parameters influencing the enhanced production of prodigiosin from Serratia marcescens and evaluation of its antimicrobial, antioxidant and dyeing potentials. Malays J Microbiol 8(2):116–122

Haddix PL, Jones S, Patel P, Burnham S, Knights K, Powel JN, Laform A (2008) Kinetic analysis of growth rate, ATP, and pigmentation suggests an energy-spilling function for the pigment prodigiosin of *Serratia marcescens*. J Bacteriol 190(22):7453–7463

Haddix PL, Shanks RMQ (2018) Prodigiosin pigment of *Serratia marcescens* is associated with increased biomass production. Arch Microbiol 200:989–999. https://doi.org/10.1007/s00203-018-1508-0

Haddix PL, Werner TF (2000) Spectrophotometric assay of gene expression: *Serratia marcescens* pigmentation. Bioscene 26:3–13

Harned RL (1953) Method for production of prodigiosin. Google Patents

Hobbs G, Frazer CM, Gardner DCJ, Flett F, Oliver SG (1990) Pigmented antibiotic production by *Streptomyces coelicolor* A3(2) : kinetics and the influence of nutrients. J Gen Microbiol 136:2291–2296

Karthika D, Vadakkan K, Ashwini R, Shyamala A, Hemapriya J, Vijayanand S (2015) Prodigiosin mediated biosynthesis of silver nanoparticles and evaluation of its antibacterial activity. Int J Curr Microbiol Appl Sci 3(10):868–874

Khanafari A, Assadi MM, Fakhr FA (2006) Review of prodigiosin, pigmentation in Serratia marcescens. J Bio Sci 6(1):1–13

Kim D, Lee JS, Park YK, Kim JF, Jeong H, Oh TK, Kim BS, Lee CH (2007) Biosynthesis of antibiotic prodiginines in the marine bacterium *Hahella chejuensis* KCTC 2396. J Appl Microbiol 102(4):937–944

Kojima I, Cheng YR, Mohan V, Demain A (1995) Carbon source nutrition of rapamycin biosynthesis in *Streptomyces hygroscopicus*. J Ind Microbiol 14(6):436–439

Kurbanoglu EB, Ozdal M, Ozdal OG, Algur OF (2015) Enhanced production of prodigiosin by Serratia marcescens MO-1 using ram horn peptone. Braz J Microbiol 46(2):631–637

Lee JS, Kim YS, Park S, Kim J, Kang SJ, Lee MH, Ryu S, Choi JM, Oh TK, Yoon JH (2011) Exceptional production of both prodigiosin and cycloprodigiosin as major metabolic constituents by a novel marine bacterium, Zooshikella rubidus S1-1. Appl Environ Microbiol 77 (14):4967–4973

Lee M, Kojima I, Demain A (1997) Effect of nitrogen source on biosynthesis of rapamycin by *Streptomyces hygroscopicus*. J Ind Microbiol Biotechnol 19(2):83–86

Lin C, Jia X, Fang Y, Chen L, Zhang H, Lin R, Chen J (2019) Enhanced production of prodigiosin by *Serratia marcescens* FZSF02 in the form of pigment pellets. E J Bio 40:58–64

Luti KJK, Mavituna F (2011) Elicitation of *Streptomyces coelicolor* with dead cells of *Bacillus subtilis* and *Staphylococcus aureus* in a bioreactor increases production of undecylprodigiosin. Appl Microbiol Biotechnol 90(2):461–466

Mohammed SJ, Luti KJK (2020) A kinetic model for prodigiosin production by *Serratia marcescens* as a bio-colorant in bioreactor. In: AIP Conference Proceedings, vol 1. AIP Publishing LLC, Melville, p 020027

Montaner B, Pérez-Tomás R (2001) Prodigiosin-induced apoptosis in human colon cancer cells. Life Sci 68(17):2025–2036

Morones JR, Elechiguerra JL, Camacho A, Holt K, Kouri JB, Ramírez JT, Yacaman MJ (2005) The bactericidal effect of silver nanoparticles. Nanotechnol 16(10):2346

Naik C, Srisevita JM, Shushma KN, Farah N, Shilpa AC, Muttanna CD, Darshan N, Sannadurgappa D (2012) Peanut oil cake: a novel substrate for enhanced cell growth and prodigiosin production from *Serratia marcescens* CF-53. J Res Biol 2(6):549–557

Nakamura K, Kitamura K (1981) Process for preparation of prodigiosin. Google Patents,

Nguyen VB, Chen S-P, Nguyen TH, Nguyen MT, Tran TTT, Doan CT, Tran TN, Nguyen AD, Kuo YH, Wang S-L (2020) Novel Efficient Bioprocessing of Marine Chitins into Active Anticancer Prodigiosin. Mar Drugs 18(1):15

Obayemi JD, Danyuo Y, Dozie-Nwachukwu S, Odusanya OS, Anuku N, Malatesta K, Yu W, Uhrich KE, Soboyejo WO (2016a) PLGA-based microparticles loaded with bacterial-synthesized prodigiosin for anticancer drug release: effects of particle size on drug release kinetics and cell viability. Mater Sci Eng C 66:51–65

Obayemi JD, Malatesta KA, Odusanya OS, Yiporo D, Yu W, Uhrich KE, Soboyejo WO (2016b) Abstract C60: injectable, biodegradable micro-and nano-particles loaded with prodigiosin-based drug for localized anticancer drug delivery. AACR,

Park H, Lee SG, Kim TK, Han SJ, Yim JH (2012) Selection of extraction solvent and temperature effect on stability of the algicidal agent prodigiosin. Biotechnol Bioproc E 17(6):1232–1237

Prasad R, Pandey R, Barman I (2016) Engineering tailored nanoparticles with microbes: quo vadis. WIREs Nanomed Nanobiotechnol 8:316–330

Prasad R, Kumar M, Kumar V (2017a) Nanotechnology: an agriculture paradigm. Springer Nature Singapore Pte Ltd., Singapore. ISBN 978-981-10-4573-8

Prasad R, Kumar V, Kumar M (2017b) Nanotechnology: food and environmental paradigm. Springer Nature Singapore Pte Ltd., Singapore. ISBN 978-981-10-4678-0

Prasad R, Jha A, Prasad K (2018a) Exploring the realms of nature for nanosynthesis. Springer International Publishing, Cham. ISBN 978-3-319-99570-0. https://www.springer.com/978-3-319-99570-0

Prasad R, Kumar V, Kumar M, Wang S (2018b) Fungal nanobionics: principles and applications. Springer Nature Singapore Pte Ltd., Singapore. ISBN 978-981-10-8666-3. https://www.springer.com/gb/book/9789811086656

Prasad R, Siddhardha B, Dyavaiah M (2020) Nanostructures for antimicrobial and antibiofilm applications. Springer International Publishing, Cham. ISBN 978-3-030-40336-2. https://www.springer.com/gp/book/9783030403362

Rakh RR, Dalvi SM, Musle BB, Raut LS (2017) Production, Extraction and Characterization of Red Pigment Produced by *Serratia rubidaea* JCM 1240T isolated from Soil. Int J Curr Microbiol App Sci 6(1):143–154

Ryazantseva I, Andreyeva I (2014) Application of prodigiosin as a colorant for polyolefines. Adv Bio Chem 2014:1

Schloss PD, Allen HK, Klimowicz AK, Mlot C, Gross JA, Savengsuksa S, McEllin J, Clardy J, Ruess RW, Handelsman J (2010) Psychrotrophic strain of *Janthinobacterium lividum* from a cold Alaskan soil produces prodigiosin. DNA Cell Biol 29(9):533–541

Shahitha S, Poornima K (2012) Enhanced production of prodigiosin production in Serratia marcescens. J Appl Pharmaceut Sci 2(8):138

Shen JJ, Yang Y (2013) Kinetics and thermodynamics studies of prodigiosin dyeing on polyester. In: Adv Mat Res. Trans Tech Publ, Stafa-Zurich, pp 156–160

Soto-Cerrato V, Llagostera E, Bz M, Scheffer GL, Perez-Tomas R (2004) Mitochondria-mediated apoptosis operating irrespective of multidrug resistance in breast cancer cells by the anticancer agent prodigiosin. Biochem Pharmacol 68(7):1345–1352

Su W-T, Tsou T-Y, Liu H-L (2011) Response surface optimization of microbial prodigiosin production from *Serratia marcescens*. J Taiwan Inst Chem E 42(2):217–222

Sundaramoorthy N, Yogesh P, Dhandapani R (2009) Production of prodigiosin from *Serratia marcescens* isolated from soil Ind. J Sci Technol 2(10):32–34

Venil CK, Lakshmanaperumalsamy P (2009) Application of statistical design to the optimization of culture medium for prodigiosin production by *Serratia marcescens* SWML08. Malays J Microbiol 5(1):55–61

Wang X, Tao J, Wei D, Shen Y, Tong W (2004) Development of an adsorption procedure for the direct separation and purification of prodigiosin from culture broth. Biotechnol Appl Biochem 40:277–280

Wei Y-H, Chen W-C (2005) Enhanced production of prodigiosin-like pigment from *Serratia marcescens* SMΔR by medium improvement and oil-supplementation strategies. J Biosci Bioeng 99(6):616–622

Williams RP (1973) Biosynthesis of Prodigiosin, a Secondary Metabolite of *Serratia marcescens*. Appl Microbiol 25(3):396–402

Williams RP, Gott CL, Green JA (1961) Studies on Pigmentation of *Serratia Marcescens* V.: Accumulation of Pigment Fractions with Respect to Length of Incubation Time1. J Bacteriol 81 (3):376

Zang C-Z, Yeh C-W, Chang W-F, Lin C-C, Kan S-C, Shieh C-J, Liu Y-C (2014) Identification and enhanced production of prodigiosin isoform pigment from Serratia marcescens N10612. J Taiwan Inst Chem E 45(4):1133–1139

Chapter 13
Applications of Nanomaterials for Water Disinfection

Guhankumar Ponnusamy and Jayaprakash Saththasivam

Abstract The presence of harmful pathogens in water is a major threat to human health when it comes to drinking water and hygiene. In addition to that, the emergence of new pathogens and antimicrobial resistance microbes will further challenge the operation of water treatment plants. Hence, there is a pressing need to develop cost-effective treatment technologies that will ensure safe and clean water supply for drinking and sanitation requirements. In recent years, nanomaterials have been extensively studied to inactivate waterborne pathogens. This chapter provides a short review of the applications of nanomaterials for disinfection against bacteria, protozoa, and viruses. The first section of the chapter briefly discusses the type of nanomaterials, synthesis protocols, and disinfection mechanisms. The subsequent section discusses some of the experimental studies related to the inactivation of bacteria, protozoa, and viruses using various nanomaterials. Challenges and limitations of nanomaterials in water treatment applications are also discussed towards the end of the chapter.

Keywords Pathogen · Disinfection · Nanomaterial · Water treatment · Drinking water

13.1 Introduction

Access to safe and clean drinking water is of paramount importance to humankind. One of the continuous threats to the supply of safe drinking water is the presence of microbial pathogens that could seriously affect public health and safety. The emerging and re-emerging nature of the pathogens can be attributed to the changes in demographics and human behavior, poor and improper maintenance of public health

G. Ponnusamy · J. Saththasivam (✉)
Qatar Environment and Energy Research Institute, Hamad Bin Khalifa University, Qatar Foundation, Doha, Qatar
e-mail: jsaththasivam@hbku.edu.qa

N. R. Maddela et al. (eds.), *Nanotechnology for Advances in Medical Microbiology*, Environmental and Microbial Biotechnology,
https://doi.org/10.1007/978-981-15-9916-3_13

utilities, microbial adaptation, and also by the impact of climate change and extreme weather events (Jones et al. 2008). A study analyzing the failure of the conventional water systems in developed countries revealed that malfunction of infrastructure and institutional practices led to frequent recurrence of microbial contamination in water supplies (Onyango et al. 2015). The effect of climate change on the rise of waterborne diseases can be directly correlated as extreme weather conditions such as floods and storms could pollute water bodies. For instance, there were reported cases of drinking water reservoir contamination by *Cryptosporidium* after heavy rainfall events (Saxena 2019). A *Campylobacter* outbreak in New Zealand affected nearly 5500 people as the groundwater supply was contaminated by seepage of sheep feces into a nearby aquifer following heavy rainfall (Gilpin et al. 2020). The contamination of water bodies by wastewater is another cause of concern for waterborne disease outbreaks. Numerous studies have shown that wastewater is the major source for fecal microorganisms, including emerging pathogens which when improperly handled contaminates the surface waters. As wastewater production and efficiency of treatment facilities vary across different regions, the presence of nutrients, chemicals, and pathogenic microorganisms needs to be adequately addressed to prevent environmental and health issues. Wastewater from household hosts a myriad of pathogenic microorganisms such as bacteria, viruses, and protozoa (Herschy 2012). Also, excess nutrition in the water bodies resulting from the industrial discharge of untreated wastewater can lead to algal bloom and eutrophication. This in turn causes the proliferation of pathogenic microorganisms and deteriorates the overall water quality (Roy et al. 2011). The presence of natural organic matter (NOM) in water supplies increases the microbial contamination risk as NOM shields microorganisms from conventional disinfection technologies and serves as a breeding medium for a large variety of other microorganisms (Gledhill 1987). Apart from these factors, the ability of the pathogen itself to evolve and develop antibiotic-resistant genes is posing innumerable challenges to the conventional disinfection process (Dungeni et al. 2010).

The United States Environmental Protection Agency Contaminant Candidate List 3 (CCL 3) identifies many waterborne pathogens that could compromise the safety and quality of the drinking water ("Final Contaminant Candidate List 3 Microbes: Screening to the PCCL" 2009). Common symptoms related to waterborne diseases are diarrhea, vomiting as well as eye, ear, skin, and/or respiratory infections. Some of the common pathogens that are found in water supplies are *Cryptosporidium, Legionella, Escherichia coli O157, Rotavirus, Hepatitis E virus, and Norovirus*. A survey by Moreira and Bondelind (2017) indicated that *Cryptosporidium, Norovirus, Giardia, Campylobacter*, and *Rotavirus* are among the most reported waterborne pathogens. *Shigella flexneri* is another common food and waterborne pathogen that has been reported in many different regions and is known to be resistant to most of the antibiotics (Nisa et al. 2020). The emergence of more new pathogens such as *Helicobacter pylori* (*H. pylori*) that could be potentially transmitted through water is also a growing concern in the water treatment sectors. Other pathogens that belong to the *Enterobacteriaceae* family such as *Klebsiella spp.* and

Salmonella spp. are also among the common cause of water supply contamination (Mehrad et al. 2015).

According to the commitments to the Sustainable Development Goals, governments should take measures to considerably increase recycling and safe reuse of the wastewater by 2030 (Malik et al. 2015). In addition to that, there is a need for the transition to decouple the economic growth and development essentially with the consumption of finite resources like water, in achieving a circular economy (Shan et al. 2016). The United States Environmental Protection Agency (USEPA) has permitted the use of copper as an antibacterial agent due to its ability to generate reactive oxygen species to disinfect pathogens through chromosomal aberrations and oxidative damage to DNA (Environmental Protection Agency Federal Facilities Restoration and Office 2017). The efficiency of the water treatment process in removing pathogens is measured using Log Reduction Value (LRV). LRVs are generally monitored in water treatment plants from the entry point and after successive treatments,

$$\mathrm{LRV} = \log_{10}\left(\mathrm{C_{in}}/\mathrm{C_{out}}\right)$$

where C_{in} is the influent pathogen concentration and C_{out} is the effluent pathogen concentration. (Bennett 2008).

In general, WHO recommends the LRVs for pathogens as follows: (1) enteric viruses (9.5-$\log_{10}$) (2) bacteria (8.5-$\log_{10}$), and (3) protozoa (8.5-$\log_{10}$). The determination of LRV for viruses is more complicated and only several countries have regulations to control the indicator viruses and on the LRVs (Ahmed et al. 2020).

Many different treatment processes are deployed in a water treatment plant as a multiple-barrier to remove and inactivate pathogens. This includes (1) pretreatment (such as roughing filters, micro strainers, riverbank filtration) (2) coagulation, flocculation, and sedimentation/flotation, and (3) filtration (granular media filtration, sand filtration, microfiltration, and ultrafiltration). These treatment processes are followed by disinfection steps that can be achieved using oxidizing compounds such as chlorine, chlorine dioxide, and ozone or by Ultraviolet light (UV). Conventional disinfection processes have their own merits and disadvantages. For instance, chlorination is effective in deactivating different types of pathogens and able to provide positive residual protection for a longer period. However, chlorination effectiveness is pH-dependent and could potentially create toxic disinfection by-products such as trihalomethanes. UV treatment, despite being excellent in eliminating pathogens by breaking the bonds of microbial nucleic acids, requires frequent maintenance and does not work well in turbid water. Therefore, there is a need for developing efficient processes for disinfection and microbial control. Nanomaterials with antimicrobial properties have huge potentials as viable technologies to inactivate pathogens in water supplies.

13.2 Application of Nanomaterials in Disinfection

In recent years, nanomaterials have been explored and utilized in different water treatment processes such as oxidation, adsorption, filtration, and disinfection. The exceptional properties of nanomaterials owing to its different shape and size along with others features such as (1) high efficiency, (2) high surface area, (3) durability, (4) ease of functionalization, and (5) antimicrobial structures have led to the development of various innovative water treatment technologies (Ibrahim et al. 2016; Bora and Dutta 2014; Gautam et al. 2019; Ul-Islam et al. 2017; Guerra et al. 2018; Punniyakotti et al. 2020).

There are a plethora of scientific studies related to the application of nanomaterials for disinfection purposes. One of the studies demonstrated excellent removal of *E. coli* K12, *P. mendocina,* and virus using polysulfonate ultrafiltration membranes impregnated with silver nanoparticles (Zodrow et al. 2009). In another study, carbon nanotubes were found effective in removing *E. coli* and *Staphylococcus aureus* and *Poliovirus sabin 1* (Srivastava et al. 2004). A photocatalytic material made of Al-doped $BiVO_4$ has a superior photocatalytic antibacterial property and was effective in controlling the growth of Methicillin-resistant *S. aureus* MRSA (Vicas et al. 2019). More studies related to microbial inactivation will be in the subsequent sections.

13.3 Types of Nanomaterials

Carbon nanotubes (CNT), graphene oxide (GO), and fullerenes are among the popular carbon-based nanomaterials that are used in water disinfection processes due to their excellent antimicrobial properties. Studies have confirmed that single-wall carbon nanotubes (SWCNT) inactivate *E. coli* by damaging its cell wall (Kang et al. 2007). Figure 13.1 shows the morphological degradation of *E. coli* after being exposed to SWCNT for 2 hours in saline solution (Ibrahim et al. 2016).

Another study claimed that the antimicrobial properties of SWCNT could be attributed to its -OH and -COOH surface groups (Arias and Yang 2009). Graphene Oxide (GO), made up of a monolayer of tightly packed carbon treated with hydroxyl, epoxyl, and carboxyl groups, has been widely studied in the field of disinfection. GO inactivates microbes by forming cell-GO aggregates and disrupting the membrane of microbes using its sharp nanowalls edges (Dizaj et al. 2015). Fullerenes, made up of carbon atoms, have excellent antimicrobial activity against *E. coli, Salmonella*, and *Streptococcus spp*. The antimicrobial mechanism of fullerenes involves the inhibition of the bacterial energy metabolism by disrupting the oxygen uptake as well as by damaging the cell membrane (Tegos et al. 2005). C60 and C70 fullerenes have shown good results against viruses and bacteria by cleaving their genetic material (Chae et al. 2014). Fullerol was studied extensively for water treatment applications, especially against MS2 bacteriophage, which has similar morphology of *hepatitis A*

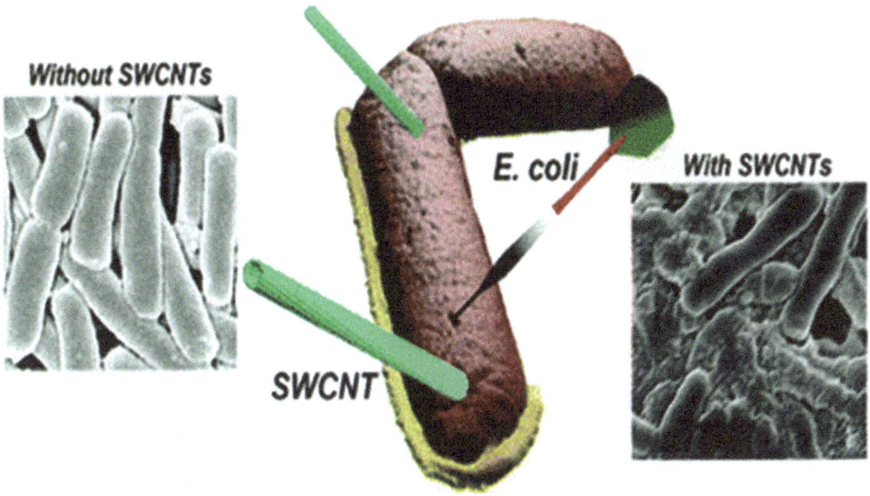

Fig. 13.1 SEM images of punctured *E. Coli* after exposure to SWCNT

virus and *poliovirus*. In the presence of fullerol and UV light, excellent inactivation of MS2 bacteriophage was achieved (Badireddy et al. 2007).

Metal oxide-based nanomaterials have been attracting major interest in recent decades due to their numerous applications, especially in the environmental remediation field. Titanium dioxide, TiO_2 is among the popular metal oxides that are well known for its antimicrobial properties. Under UV/visible light exposure, TiO_2 generates reactive oxygen species that destroy cellular membranes, DNA, and proteins. Other excellent candidates are Zinc oxide (ZnO) and Copper (II) oxide (CuO) and Magnesium oxide (MgO). ZnO has shown 100% inhibition, antibacterial and antibiofilm activities against *Staphylococcus aureus* and *Pseudomonas aeruginosa* (Akhil et al. 2016). CuO nanosheets showed antibacterial activity against *Enterococcus faecalis* and *Micrococcus luteus* along with its good reusability and photo corrosion and inhibition properties (Fakhri et al. 2018). Ag@TiO_2 nanoparticles under UV light irradiation showed complete disinfection of *E. coli* at a 0.4 g/L catalyst loading rate (Sreeja and Vidya Shetty 2016).

13.4 Nanomaterial Synthesis Methods

Chemical methods such as coprecipitation, hydrothermal, facile hydrothermal, and microwave-assisted approaches are widely adopted in synthesizing various types of nanomaterials. This method is generally simple, tunable, and can be scaled up for mass production (Hyeon 2003). The hydrothermal technique is one of the commonly used chemical methods for the synthesis of nanomaterials. It promotes chemical reaction under elevated temperature and pressure where the properties and

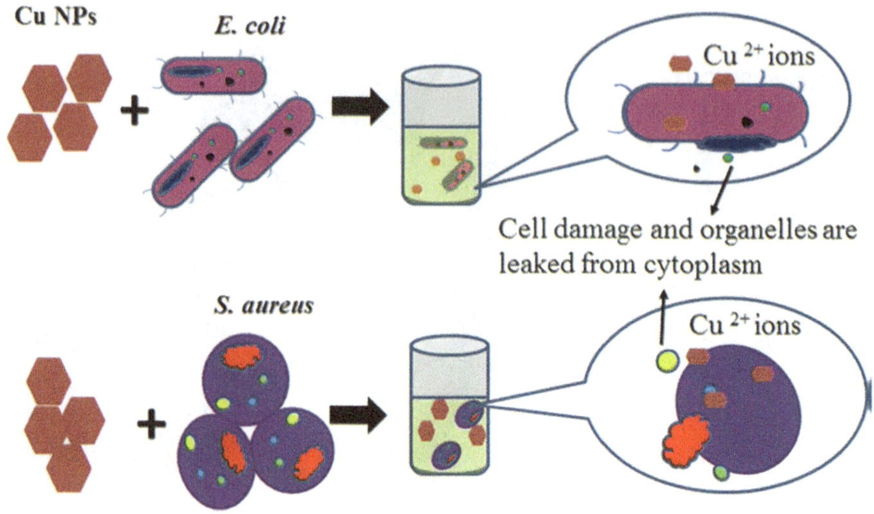

Fig. 13.2 Schematic representation of Cu NPs on Gram-positive and Gram-negative bacteria

characteristics of nanomaterials can be tuned by varying the reaction pressure and temperature of an autoclave reactor. It is a highly preferred method due to its capability to operate across different solvent, operating pressure, and temperature to manipulate the critical point of the substances used in the reaction (Rane et al. 2018).

As the chemical synthesis methods have its drawbacks such as (1) requiring expensive chemicals, (2) hazardous solvent, (3) and waste disposal issues, there is a need to develop more environment-friendly synthesis methods (Chen et al. 2007). Green chemistry principles are now being explored to synthesize nanomaterials using extracts of plants, fruits, and fungi (Prasad 2014, 2016, 2017; Prasad et al. 2016, 2018a,b; Srivastava et al. 2021). This method is non-toxic, low cost, and reduces the usage of harmful substances. For instance, copper nanoparticles can be synthesized in a greener way by mixing copper chloride solution and leaf extracts of *Cardiospermum halicacabum,* (commonly known as balloon vine). The green-synthesized Cu NPs liberate their cations (Cu+) into the growth medium stick on the cell membrane of cells because of electrostatic attraction and leads to cell death as shown in Fig. 13.2 (Punniyakotti et al. 2020).

Silver nanoparticles (AgNPs) have also been synthesized using a facile method with the extracts of *Excoecaria agallocha* (*E. agallocha*), a mangrove tree leaf (Bhuvaneswari et al. 2017). The synthesized AgNPs were pure, stable, and displayed excellent cytotoxic properties. In another study, Ag NPs were synthesized using Ives cultivar (*Vitis labrusca*) pomace as reducing and stabilizing compounds and were capable of reducing 47% of *E. coli* in the raw wastewater (Raota et al. 2019).

13.5 Mechanism of Inactivation by Nanomaterial

Inactivation of pathogens by nanomaterials involves the combination of different mechanisms. MS-2 phage and *E. coli* were inactivated differently when photocatalytic TiO_2 was used (Cho et al. 2005). MS-2 phage was mainly inactivated by the formation of free hydroxyl radicals in the bulk phase, while *E. coli* was targeted by the surface and free hydroxyl radicals as well as by other reactive oxygen species. Ag NPs inactivate pathogen using several mechanisms. Apart from releasing silver ions that are capable of altering enzyme activities and interfering with DNA replications, Ag NPs can also accomplish its activation by forming free radicals on the cell wall of the pathogens that would result in the damage of cellular membranes (Saleh 2017; Prasad and Swamy 2013; Aziz et al. 2014, 2015, 2016). Similarly, ZnO nanoparticle can achieve its antimicrobial activities using its superior photoconductivity properties resulting from UV illumination as well as by damaging the cell walls using its morphology (Sirelkhatim et al. 2015; Bhuyan et al. 2015). Like other nanomaterials, carbon nanotubes also inactivate microbes using several mechanisms such as by damaging DNA, disrupting the cell wall, and transmembrane electron transfer (Liu et al. 2018).

13.6 Bacterial Inactivation Using Nanomaterials

The current disinfection technology such as chlorine can effectively control pathogens but forms disinfection by-products, which causes harmful effects on human health. Therefore, there is a need for an ideal disinfectant with broad antimicrobial property. Therefore, NPs and nanomaterials could be a better alternative for pathogen removal.

Copper NPs (CuNPs) are among the excellent antibacterial agents against various pathogens such as *E. coli, P. aeruginosa,* and *S. aureus* species (Yadav et al. 2017). In one of the studies, the antimicrobial performance of a cellulosic paper coated with CuNP against *E. coli* was evaluated. Their study showed that CuNP embedded paper achieved up to 8.8 log reduction value. It was also shown that the leaching of copper from the paper substrate was only 0.14%, thus indicating the robustness of their facile deposition method. The direct contact between CuNP and *E. coli* during the filtration process was postulated as the main mechanism of inactivation (Dankovich and Smith 2014). In a recent study, copper oxide nanoparticles were synthesized using *Citrus aurantifolia* leaves to evaluate the antimicrobial performance against *S. aureus* and *E. coli.* The results indicated that *E. coli* was more susceptible to copper oxide nanoparticles and the inactivation mechanism was attributed to the presence of amine groups and large surface area of the nanomaterial (Rafique et al. 2020).

Inactivation of mixed *E. coli* and *Bacillus spp.* culture using TiO_2 Degussa P-25 in a coaxial photocatalytic reactor illuminated with neon lamps showed that *E. coli* was more susceptible than the *Bacillus sp.* It also showed that higher microbe inactivation was achieved at elevated dissolved oxygen levels (Rincón and Pulgarin

2005). Another study investigated the inactivation of *M. smegmatis* using nine different TiO_2 nanotubes electrodes impregnated with Ag nanoparticles. It demonstrated that inactivation efficiency as high as 99.6% was achieved after 30 minutes of visible light exposure while full activation was attained within three minutes of using UV light (Brugnera et al. 2014). In another study, Ag/TiO_2 nanofiber membranes developed showed excellent filtration and antimicrobial properties (Liu et al. 2012). The membrane was able to inactivate 99.9% of *E. coli* within 30 min of solar irradiation. Bacteria regrowth test using the Kirby–Bauer approach confirmed the antimicrobial nature of the membrane where no regrowth of *E. coli* was observed. Similar findings were observed, where the addition of silver improved the photocatalytic performance of TiO_2. However, excessive addition of silver could be detrimental as this could reduce the active site available for photocatalytic reactions (Taylor et al. 2011).

AgNP decorated GO nanocomposite synthesized using a facile and green method showed good antibacterial property when tested against Gram-negative *E. coli* ATCC 25922 and Gram-positive *S. aureus* ATCC 6538. The synthesis of AgNP–GO was made by suspending GO in deionized water along with $AgNO_3$ with further addition of 1 mM glucose and 4% starch as a reducing agent and stabilizer, respectively. The undesired agglomeration of silver nanoparticles was overcome in this study by using GO as a support media. SEM microscopy analyses confirmed the substantial deformation in the shape and size of *E. coli* and *S. aureus after* being exposed to AgNP–GO, hence suggesting that the inactivation of the pathogen was likely due to the damage of cell walls (Shao et al. 2015).

In another investigation, the effect of capping agents such as ethylene glycol (EG), gelatin, polyvinyl alcohol (PVA), and polyvinylpyrrolidone (PVP) on the microbial activities of ZnO NPs was evaluated. The experiments were conducted using *Staphylococcus aureus* and *Pseudomonas aeruginosa* where the bacterial cells were exposed to ZnO NPs synthesized using different capping agents for four hours under dark and light conditions. The results indicated that most of the capping agents except for PVP yielded in lower inactivation when compared with pristine ZnO NPs (synthesized without any capping agent). Their results were also in agreement with many other findings that confirmed the higher inactivation activity of ZnO with the presence of light (Akhil et al. 2016).

Cerium oxide (CeO_2) nanoparticle is another promising nanomaterial that is being actively studied due to its photocatalytic and antimicrobial properties. CeO_2 nanoparticles, synthesized using *Calotropis Procera* flower extract, showed efficient antimicrobial activity against Gram-positive and Gram-negative bacteria such as *Bacillus subtilis, Staphylococcus saprophyticus, E. coli,* and *Pseudomonas aeruginosa*. This study showed that the prepared CeO_2 NP was effective against these bacteria and it was noted that higher inhibition activity was achieved against Gram-negative bacteria. The antimicrobial characteristics of CeO_2 NP were attributed to the huge surface area and surface reactivity of the nanomaterial (Muthuvel et al. 2020).

The antimicrobial property of monoclinic bismuth vanadate (m-$BiVO_4$) octahedral nanostructures was investigated using *E. coli* as an indicator of bacteria (Sharma et al.

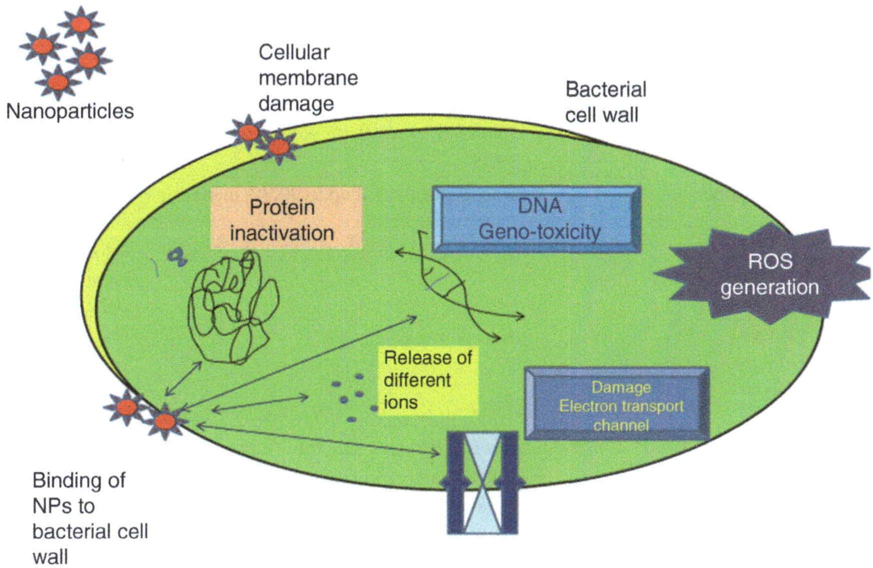

Fig. 13.3 Antimicrobial mechanism of nanoparticle

2016). It was found that exposure to m-$BiVO_4$ led to cellular disintegration where complete inactivation of *E. coli* was achieved after 8 hours of exposure to 80 ppm of m-$BiVO_4$. In another study, a one-pot hydrothermal approach was adopted to synthesize Al:$BiVO_4$ by adding 1% aluminum oxide powder (w/v) to the precursor solutions that comprised of bismuth nitrate and ammonium vanadate. The resultant nanoparticles exhibited higher bacteria inactivation than the pristine $BiVO_4$ even at the lowest Al concentration of 0.02 g/L. It was postulated that the addition of Al lowered the bandgap energy and improved the charge transfer characteristics of the material, thus leading to the enhancement of photocatalytic efficiency (Vicas et al. 2019). Figure 13.3 shows the various inactivation mechanisms of bacteria by nanoparticles. Nanoparticles bind to the surface of the bacterial cell wall and deploy various mechanisms to inactivate cells. This includes (1) releasing of ions to disrupt the cell function (2) puncturing of the cell wall (3) binds with DNA and causes genotoxicity, and (4) generation of reactive oxygen species (ROS) that led to oxidative stress (Chaudhary et al. 2020; Prasad 2019a,b; Prasad et al. 2020).

In a very recent study, photocatalytic activity of a WO_3/ZrO_2 was improved using ruthenium (Ru), a rare transition metal. By coating the $Ru/WO_3/ZrO_2$ on aluminum plates, they were able to reuse the synthesized materials for extended disinfection cycles where the catalyst inactivated more than 90% of Gram-negative (e.g., *Shigella, Salmonella, Vibrio parahaemolyticus,* and *Vibrio cholerae*) and Gram-positive bacteria (e.g., *Enterococcus*) within 120 min of exposure (Fouad et al. 2021). They reported that the holes and hydroxyl radicals played a prominent role in activating the microbes with negligible contribution from superoxide radicals. This study also reported that a high concentration of NOM (e.g., 28.7 mg/L) in water

Table 13.1 Nanoparticle or nanomaterial used for the removal of bacteria

Nanoparticle type / nanomaterial used	Concentration used	Susceptible pathogens and mechanism	Refs.
Zinc oxide	100 mg/L	*S. aureus*, *P. aeruginosa*, and *E. coli* inhibited to 100% by the production of ROS	(Akhil et al. 2016)
Ag/TiO_2 nanofiber membrane	–	*E. coli* inactivated up to 99.9% in 30 min using solar light	(L. Liu et al. 2012)
TiO_2 photocatalysis	–	*E. coli, S. typhimurium*, and *S. sonnei*	(Moncayo-Lasso et al. 2012)
TiO_2, Pt–TiO_2, and Ag–TiO_2—photocatalysis	–	*E. coli*	(Taylor et al. 2011)
Ti/TiO_2–Ag—photocatalysis	–	100% inactivation of *Mycobacterium smegmatis* using UV for 3 min	(Brugnera et al. 2014)
TiO_2 nanowires on fabrics	–	100% inactivation of *E. coli* and *S. epidermidis* under visible light for 15 min	(Xu et al. 2018)
Films of TiO_2 and TiO_2/Ag	4% w/w	100% inactivation of fecal coliform using UV irradiation for 6 min	(Domínguez-Espíndola et al. 2017)
SnO_2-doped nanomaterials	–	*E. coli, P. aeruginosa, S. aureus, L. monocytogenes, B. subtilis, S. typhi, and T. viride*	(Pandiyan et al. 2019)
Modified core shell Fe_3O_4–SiO_2–NH_2	–	93.4% *S. aureus*, 97.4% *B. subtilis*, 95.1% *E. coli*, and 90.1% of *P. aeruginosa* and *Salmonella*	(Zhan et al. 2014)
Fe_3O_4@CTAB	–	99% inactivation of *E. coli* and *B. subtilis* within 60 min	(Jin et al. 2015)
AgNPs-loaded Clay	0.1 mg/L	Removal up to 90% *Salmonella spp*, 80% of *E. coli, Klebsiella pneumonia*, and *Shigella flexneri*, and 70% *Klebsiella aerogenes* at 2 h exposure Time	(Kassem et al. 2019)
Hybrid polyaniline/graphene/CNT		Up to 99.5% removal of *S. aureus* and 99.2% for *E. coli*	(Hussein et al. 2018)
$BiVO_4$ QDs/g-C_3N_4 and $AgVO_3$ QDs/g-C_3N_4	–	Removal up to 87.5% and 96.4% against *Salmonella* under visible light for 10 min	(Wang et al. 2018)

led to poor inactivation of microbes. This was due to the absorbance of light energy by NOM, adsorption of NOM on the catalyst surface that affects the production of reactive oxygen species (ROS) and finally scavenging of ROS by the NOM itself. Some of the other types of nanomaterials and nanoparticles used in the investigation of pathogenic bacterial removal along with their mechanism are provided in Table 13.1.

13.7 Inactivation of Protozoa by Nanomaterials

Some of the common waterborne protozoa that affect the water industries are *Cryptosporidium, Giardia, Cyclospora, Acanthamoeba,* and *Isospora*. Inactivation of *E. histolytica* and *C. parvum* cysts using silver and copper oxide nanoparticles was investigated and the study indicated that high inactivation was achieved for both the cysts after three hours of exposure (Saad et al. 2015). The inactivation of these parasites also increased with the concentration of NPs and exposure time. Post-treated SEM images of the parasites confirmed the structural damage in the cell wall of *C. parvum.* A study that investigated the efficacy of Ag NPs in inactivating *C. parvum* concluded that lower nanoparticles and ions concentration affect the cell viability while higher concentration led to oocyst rupture (Cameron et al. 2016). Another study investigated the effect of different dosages of silver nanoparticles (Ag NPs) (0.05, 0.1, and 1 mg/L) on the mortality of *Cryptosporidium parvum* (CP). The result indicated that a 97.2% drop in oocyst count was obtained when *C. parvum* was exposed to 1 ppm Ag NPs for 30 minutes. The inactivation of these protozoa was also possible at a much lower concentration of 0.05 ppm Ag NPs. However, at least four hours of exposure time was required to achieve over 90% inactivation efficiency (Hassan et al. 2019). In one of the studies, the performance of TiO_2 slurry under simulated solar radiation in inactivating *Cryptosporidium parvum* oocysts was compared between distilled water and simulated sewage effluent (Abeledo-Lameiro et al. 2016). This study reported that more than 95% of *C. parvum* can be inactivated in distilled water after five hours of exposure to 100 mg/L TiO_2 under simulated solar irradiation. In contrast, the inhabitation was significantly lower in the simulated wastewater effluent under similar operating conditions. The presence of scavengers such as carbonates, bicarbonates, and organic carbon in the wastewater effluent could potentially reduce the photocatalytic activity of TiO_2 and led to the poor inactivation efficiency. In another study, oocysts viability was studied by exposing them to solar radiation in the presence of 100 mg/L of TiO_2 and/or 50 mg/L of H_2O_2. The results showed a strong reduction in oocytes in the water samples with TiO2/ H_2O_2 up to 99.5% after exposure to sunlight, thus proving TiO_2 as one of the best photocatalytic disinfection agents against *C. parvum* (Abeledo-Lameiro et al. 2017). A study on the inactivation of *C. cayetanensis* oocyst at different doses of Magnesium oxide nanoparticles (MgO NPs) (1.25–25 mg/mL) revealed that exposure to MgO Nps induced morphological changes to the protozoa where cell wall damage/ rupture and leakage of oocysts content were observed. This study also claimed that antimicrobial properties of MgO Nps were controlled by the size of the nanomaterials where smaller dimension led to high inactivation efficiency. Longer exposure time also led to higher mortality of *C. cayetanensis* oocyst (Hussein et al. 2018).

13.8 Inactivation of Virus by Nanomaterials

Apart from bacteria and protozoa, the presence of viruses in water bodies poses a great danger to public health. The World Health Organization (WHO) recommends monitoring the microbial quality of wastewater and food using indicator organisms (e.g., fecal indicator bacteria (FIB) and bacteriophages) and reference pathogens (e.g., HNoV) in the position of measuring all of the human pathogens that can be possibly present. Some of the common nanomaterials used for disinfection include carbon nanotube, graphene, nano ZnO, nano Ni, nano Fe_3O_4, and nano TiO_2–anatase where MS2 phage is commonly used as a virus indicator.

Fullerol nanoparticles under UVA irradiation were reported to have achieved 4-log removal of MS2 bacteriophage. It was confirmed that the generation of superoxide and singlet oxygen during the photosensitization reaction played a key role in MS2 inactivation (Badireddy et al. 2007). A part of the study conducted by Brady-Estévez et al. (2008) investigated the removal of MS2 bacteriophage using single-walled carbon nanotube (SWNT) filter. This study demonstrated that the viral particles were completely removed using a 6-μm thick SWNT layer at a loading of 0.8 mg/cm^2. The researchers suggested that the high filtration efficiency was due to the effective filtration depth of the SWNT layer. In another investigation, the effectiveness of polymeric graphitic carbon nitride (g-C_3N_4) in inactivating MS2 bacteriophage was evaluated (Li et al. 2016). The photocatalytic experiments conducted under the visible light irradiation showed that at a concentration of 150 mg/L, the material was able to achieve an approximately 8-log reduction within six hours of exposure. However, g-C_3N_4 loading needs to be optimized as an excessive amount could negatively affect the photocatalytic reactions. Viral regrowth studies also confirmed that MS2 bacteriophage was completely inactivated as no regrowth was observed after 72 hours of incubation. The photocatalytic performance of Cu–TiO_2 nanofibers in inactivating bacteriophage f2 showed more than a 5-log reduction of bacteriophage f2 within four hours of exposure under visible light (Zheng et al. 2018). The inactivation efficiency was relatively unaffected when the pH of bulk water was varied between 6 and 9. Bacteriophage f2 inactivation increased with the addition of catalyst but tapered off as the catalyst loading reached 75 mg/L. It was mentioned that excessive addition of catalyst would result in turbid water and eventually led to a poor photocatalytic reaction. Some of the other types of nanomaterials and nanoparticles used in the investigation of viral removal along with their mechanism are provided in Table 13.2 below.

13.9 Limitation of Nanomaterial

Despite showing huge potentials in solving many water treatment problems, the risk associated with nanotoxicity needs to be effectively assessed. The toxicity of nanomaterials is governed by its shape, size, concentration, and reactivity (Sarkar

Table 13.2 Nanoparticle or nanomaterial used for the removal of virus

Nanoparticle type/ nanomaterial used	Susceptible pathogens and mechanism	Refs.
Cu–TiO_2 nanofibers/ photocatalysis	Bacteriophage f2 and its host *E. coli* 285 under visible light irradiation	(Zheng et al. 2018)
Magnetic Fe_3O_4–SiO_2–NH_2 nanoparticles	76.7% of bacteriophage f2 and 81.5% Poliovirus-1	(Zhan et al. 2014)
g–C_3N_4/EP composite nanomaterial	Complete inactivation of 8-log *E. coli* and MS2 within 3 and 4 h of visible light irradiation	(Zhang et al. 2020)

et al. 2019). Exposure to nanomaterials can lead to severe health and environmental problems. Owing to their nanoscale nature, nanomaterials can penetrate and disrupt cell walls, interact and bind with proteins, cause organ damage as well as triggering undesirable immune responses (Naqvi et al. 2018). CNT and metal oxide-based nanomaterials are known to create oxidative stress to cells by generating reactive oxygen species (Metzel 1990). Fullerenes have been reported to cause ecotoxicity to bacteria, daphnia, earthworms, fish, and human cell lines (Hlongwane et al. 2019). It was also reported that the growth of marine algae was reduced by 50 to 75% due to the exposure to ZnO NPs (Miller et al. 2010). A toxicology study on the effect on CuO and ZnO NPs on crustaceans and protozoans indicated that the presence of solubilized ion was likely the root cause for toxicity (Blinova et al. 2010). Another study demonstrated that flat shape Ag NPs is more toxic to embryos of zebrafish than Ag NP of spherical shape (Abramenko et al. 2018). A geno and ecotoxicities study on *D. magna* revealed that Ag Nps caused damage to DNA and could have genotoxic potential towards *Daphnia* (Park and Choi 2010). The utilization of nanomaterials in water treatment processes requires more extensive testing and investigations to gain a deeper understanding of the nanomaterial leaching and its and toxicology effects.

Apart from toxicity, recovery of nanomaterial from the treated water matrices is another practical limitation that needs to be overcome for successful implementation in the field. Other key challenges include aggregation and leaching of nanoparticles. The stability of nanomaterials could be also compromised by the water quality parameters such as pH, turbidity, temperature, dissolved organic carbon, and alkalinity.

13.10 Conclusion

Pathogens are usually found in wastewater and have a detrimental effect on human health. Unlike chemical emerging pollutants, whose effects on humans at low concentrations are yet to be established, most pathogenic contaminants are known to cause a wide range of waterborne diseases that have adverse effects on humans. Despite the advancements in monitoring and microbiological detection technologies, waterborne pathogens continue to pose risks. This is evident from the increasing number of infectious disease cases caused by unidentified or known microorganisms in the last few decades worldwide (Kot et al. 2015). It is of great importance that water treatment plants are equipped with the best available technologies and regularly monitored to ensure that they are free from harmful microorganisms. Nanomaterials with antimicrobial properties have huge potential in activating waterborne pathogens as demonstrated by various studies. However, there are some concerns and limitations that need to be addressed. It is essential to develop nanomaterials with a wide range of properties that are capable of treating various kinds of pathogens and emerging chemicals at different water matrices. Extensive studies related to the stability and toxicity of nanomaterials at a large pilot scale are required to completely de-risk the technology.

References

Abeledo-Lameiro MJ, Ares-Mazás E, Gómez-Couso H (2016) Evaluation of solar Photocatalysis using TiO_2 slurry in the inactivation of Cryptosporidium parvum Oocysts in water. J Photochem Photobiol B Biol 163:92–99. https://doi.org/10.1016/j.jphotobiol.2016.08.016

Abeledo-Lameiro MJ, Reboredo-Fernández A, Polo-López MI, Fernández-Ibáñez P, Ares-Mazás E, Gómez-Couso H (2017) Photocatalytic inactivation of the waterborne protozoan parasite Cryptosporidium parvum using TiO_2/H_2O_2 under simulated and natural solar conditions. Catal Today 280:132–138. https://doi.org/10.1016/j.cattod.2016.05.046

Abramenko NB, Demidova TB, Abkhalimov EV, Ershov BG, Krysanov EY, Kustov LM (2018) Ecotoxicity of different-shaped silver nanoparticles: case of Zebrafish embryos. J Hazard Mater 347:89–94. https://doi.org/10.1016/j.jhazmat.2017.12.060

Ahmed W, Kitajima M, Tandukar S, Haramoto E (2020) Recycled water safety: current status of traditional and emerging viral indicators. Current Opinion in Environmental Science and Health 16:62–72. https://doi.org/10.1016/j.coesh.2020.02.009

Akhil K, Jayakumar J, Gayathri G, Khan SS (2016) Effect of various capping agents on photocatalytic, antibacterial and antibiofilm activities of ZnO nanoparticles. J Photochem Photobiol B Biol 160:32–42. https://doi.org/10.1016/j.jphotobiol.2016.03.015

Arias LR, Yang L (2009) Inactivation of bacterial pathogens by carbon nanotubes in suspensions. Langmuir 25(5):3003–3012. https://doi.org/10.1021/la802769m

Aziz N, Fatma T, Varma A, Prasad R (2014) Biogenic synthesis of silver nanoparticles using *Scenedesmus abundans* and evaluation of their antibacterial activity. J Nanoparticles, Article ID 689419. https://doi.org/10.1155/2014/689419

Aziz N, Faraz M, Pandey R, Sakir M, Fatma T, Varma A, Barman I, Prasad R (2015) Facile algae-derived route to biogenic silver nanoparticles: synthesis, antibacterial and photocatalytic properties. Langmuir 31:11605–11612. https://doi.org/10.1021/acs.langmuir.5b03081

Aziz N, Pandey R, Barman I, Prasad R (2016) Leveraging the attributes of *Mucor hiemalis*-derived silver nanoparticles for a synergistic broad-spectrum antimicrobial platform. Front Microbiol 7:1984. https://doi.org/10.3389/fmicb.2016.01984

Badireddy AR, Hotze EM, Chellam S, Alvarez P, Wiesner MR (2007) Inactivation of bacteriophages via photosensitization of fullerol nanoparticles. Environ Sci Technol 41(18):6627–6632. https://doi.org/10.1021/es0708215

Bennett A (2008) Drinking water: pathogen removal from water—technologies and techniques. Filtration and Separation 45(10):14–16. https://doi.org/10.1016/S0015-1882(08)70495-6
Bhuvaneswari R, Xavier RJ, Arumugam M (2017) Facile synthesis of multifunctional silver nanoparticles using mangrove plant Excoecaria Agallocha L. for its antibacterial, antioxidant and cytotoxic effects. J Parasit Dis 41(1):180–187. https://doi.org/10.1007/s12639-016-0773-6
Bhuyan T, Mishra K, Khanuja M, Prasad R, Varma A (2015) Biosynthesis of zinc oxide nanoparticles from *Azadirachta indica* for antibacterial and photocatalytic applications. Mater Sci Semicond Process 32:55–61
Blinova I, Ivask A, Heinlaan M, Mortimer M, Kahru A (2010) Ecotoxicity of nanoparticles of CuO and ZnO in natural water. Environ Pollut 158(1):41–47. https://doi.org/10.1016/j.envpol.2009.08.017
Bora T, Dutta J (2014) Applications of nanotechnology in wastewater treatment—a review. J Nanosci Nanotechnol 14(1):613–626. https://doi.org/10.1166/jnn.2014.8898
Brady-Estévez AS, Kang S, Elimelech M (2008) A single-walled-carbon-nanotube filter for removal of viral and bacterial pathogens. Small 4(4):481–484. https://doi.org/10.1002/smll.200700863
Brugnera MF, Miyata M, Fujimura Leite CQ, Zanoni MVB (2014) Silver ion release from electrodes of nanotubes of TiO_2 impregnated with Ag nanoparticles applied in photoelectrocatalytic disinfection. J Photochem Photobiol A Chem 278(1–8). https://doi.org/10.1016/j.jphotochem.2013.12.020
Cameron P, Gaiser BK, Bhandari B, Bartley PM, Katzer F, Bridle H (2016) Silver nanoparticles decrease the viability of Cryptosporidium parvum Oocysts. Appl Environ Microbiol 82(2):431–437. https://doi.org/10.1128/AEM.02806-15
Chae SR, Hotze EM, Wiesner MR (2014) Possible applications of fullerene nanomaterials in water treatment and reuse. Nanotechnology applications for clean water: solutions for improving water quality: Second Edition. Elsevier Inc. https://doi.org/10.1016/B978-1-4557-3116-9.00021-4
Chaudhary P, Fatima F, Kumar A (2020) Relevance of nanomaterials in food packaging and its advanced future prospects. J Inorg Organomet Polym Mater. https://doi.org/10.1007/s10904-020-01674-8
Chen M, Feng YG, Wang X, Li TC, Zhang JY, Qian DJ (2007) Silver nanoparticles capped by oleylamine: formation, growth, and self-organization. Langmuir 23(10):5296–5304. https://doi.org/10.1021/la700553d
Cho M, Chung H, Choi W, Yoon J (2005) Different inactivation behaviors of MS2 phage and. Appl Environ Microbiol 71(1):270–275. https://doi.org/10.1128/AEM.71.1.270
Dankovich TA, Smith JA (2014) Incorporation of copper nanoparticles into paper for point-of-use water purification. Water Res 63:245–251. https://doi.org/10.1016/j.watres.2014.06.022
Dizaj SM, Mennati A, Jafari S, Khezri K, Adibkia K (2015) Antimicrobial activity of carbon-based nanoparticles. Advanced Pharmaceutical Bulletin 5(1):19–23. https://doi.org/10.5681/apb.2015.003
Domínguez-Espíndola RB, Varia JC, Álvarez-Gallegos A, Ortiz-Hernández ML, Peña-Camacho JL, Silva-Martínez S (2017) Photoelectrocatalytic inactivation of fecal coliform bacteria in urban wastewater using nanoparticulated films of TiO_2 and TiO_2/Ag. Environmental Technology (United Kingdom) 38(5):606–614. https://doi.org/10.1080/09593330.2016.1205148
Dungeni M, van Der Merwe RR, Momba MNB (2010) Abundance of pathogenic bacteria and viral indicators in chlorinated effluents produced by four wastewater treatment plants in the Gauteng Province, South Africa. Water SA 36(5):607–614. https://doi.org/10.4314/wsa.v36i5.61994
Environmental Protection Agency Federal Facilities Restoration, Us, and Reuse Office (2017) EPA technical fact sheet—nanomaterials (November 9) https://www.epa.gov/sites/production/files/2014–03/documents/ffrrofactsheet_emergingcontaminant_nanomaterials_jan2014_final.pdf
Fakhri A, Azad M, Fatolahi L, Tahami S (2018) Microwave-assisted photocatalysis of neurotoxin compounds using metal oxides quantum dots/nanosheets composites: photocorrosion inhibition,

reusability and antibacterial activity studies. J Photochem Photobiol B Biol 178(October 2017):108–114. https://doi.org/10.1016/j.jphotobiol.2017.10.038

"Final Contaminant Candidate List 3 Microbes: Screening to the PCCL" (2009) In *Office of Water (4607M)*, EPA 815-R-. United States Environmental Protection Agency

Fouad M, Alalm MG, El-Etriby HK, Boffito DC, Ookawara S, Ohno T, Fujii M (2021) Visible-light-driven photocatalytic disinfection of raw surface waters (300–5000 CFU/ML) using reusable coated Ru/WO3/ZrO2. J Hazard Mater 402(April):123514. https://doi.org/10.1016/j.jhazmat.2020.123514

Gautam PK, Singh A, Misra K, Sahoo AK, Samanta SK (2019) Synthesis and applications of biogenic nanomaterials in drinking and wastewater treatment. J Environ Manag 231(June 2018):734–748. https://doi.org/10.1016/j.jenvman.2018.10.104

Gilpin BJ, Walker T, Paine S, Sherwood J, Mackereth G, Wood T, Hambling T et al (2020) A large scale waterborne campylobacteriosis outbreak, Havelock North, New Zealand. J Infect 81 (3):390–395. https://doi.org/10.1016/j.jinf.2020.06.065

Gledhill WE (1987) Microbial toxicity and degradation test methodology: an industrial perspective. Toxicity Assessment 2(1):89–96. https://doi.org/10.1002/tox.2540020108

Guerra FD, Attia MF, Whitehead DC, Alexis F (2018) Nanotechnology for environmental remediation: materials and applications. Molecules 23(7):1–23. https://doi.org/10.3390/molecules23071760

Hassan D, Farghali M, Eldeek H, Gaber M, Elossily N, Ismail T (2019) Antiprotozoal activity of silver nanoparticles against Cryptosporidium parvum oocysts: new insights on their feasibility as a water disinfectant. J Microbiol Methods 165(May):105698. https://doi.org/10.1016/j.mimet.2019.105698

Herschy, RW (2012) Water quality for drinking: WHO guidelines Encyclopedia of Earth Sciences Series, 876–883. https://doi.org/10.1007/978-1-4020-4410-6_184

Hlongwane GN, Sekoai PT, Meyyappan M, Moothi K (2019) Simultaneous removal of pollutants from water using nanoparticles: a shift from single pollutant control to multiple pollutant control. Sci Total Environ 656:808–833. https://doi.org/10.1016/j.scitotenv.2018.11.257

Hussein EM, Ahmed SA, Mokhtar AB, Elzagawy SM, Yahi SH, Hussein AM, El-Tantawey F (2018) Antiprotozoal activity of magnesium oxide (MgO) nanoparticles against cyclospora cayetanensis oocysts. Parasitol Int 67(6):666–674. https://doi.org/10.1016/j.parint.2018.06.009

Hyeon T (2003) Chemical synthesis of magnetic nanoparticles. Chem Commun 3(8):927–934. https://doi.org/10.1039/b207789b

Ibrahim RK, Hayyan M, AlSaadi MA, Hayyan A, Ibrahim S (2016) Environmental application of nanotechnology: air, soil, and water. Environ Sci Pollut Res 23(14):13754–13788. https://doi.org/10.1007/s11356-016-6457-z

Jin Y, Deng J, Liang J, Shan C, Tong M (2015) Efficient bacteria capture and inactivation by cetyltrimethylammonium bromide modified magnetic nanoparticles. Colloids Surf B: Biointerfaces 136:659–665. https://doi.org/10.1016/j.colsurfb.2015.10.009

Jones KE, Patel NG, Levy MA, Storeygard A, Balk D, Gittleman JL, Daszak P (2008) Global trends in emerging infectious diseases. Nature 451(7181):990–993. https://doi.org/10.1038/nature06536

Kang S, Pinault M, Pfefferle LD, Elimelech M (2007) Single-walled carbon nanotubes exhibit strong antimicrobial activity. Langmuir 23(17):8670–8673. https://doi.org/10.1021/la701067r

Kassem A, Ayoub GM, Malaeb L (2019) Antibacterial activity of chitosan nano-composites and carbon nanotubes: a review. Sci Total Environ 668:566–576. https://doi.org/10.1016/j.scitotenv.2019.02.446

Kot M, Castleden H, Gagnon GA (2015) The human dimension of water safety plans: a critical review of literature and information gaps. Environ Rev 23(1):24–29. https://doi.org/10.1139/er-2014-0030

Li Y, Zhang C, Shuai D, Naraginti S, Wang D, Zhang W (2016) Visible-light-driven photocatalytic inactivation of MS2 by metal-free g-C_3N_4: virucidal performance and mechanism. Water Res 106:249–258. https://doi.org/10.1016/j.watres.2016.10.009

Liu D, Mao Y, Ding L (2018) Carbon nanotubes as antimicrobial agents for water disinfection and pathogen control. J Water Health 16(2):171–180. https://doi.org/10.2166/wh.2018.228

Liu L, Liu Z, Bai H, Sun DD (2012) Concurrent filtration and solar photocatalytic disinfection/degradation using high-performance Ag/TiO_2 Nanofiber membrane. Water Res 46 (4):1101–1112. https://doi.org/10.1016/j.watres.2011.12.009

Malik OA, Hsu A, Johnson LA, De Sherbinin A (2015) Science direct a global indicator of wastewater treatment to inform the Sustainable Development Goals (SDGs). *Environmental Science and Policy* 48:172–185. https://doi.org/10.1016/j.envsci.2015.01.005

Mehrad B, Clark NM, Zhanel GG, Lynch JP (2015) Antimicrobial resistance in hospital-acquired gram-negative bacterial infections. Chest 147(5):1413–1421. https://doi.org/10.1378/chest.14-2171

Metzel E (1990) Neurotraumatologie Im Alten Agypten. Neurochirurgia 33(3):78–80

Miller RJ, Lenihan HS, Muller EB, Tseng N, Hanna SK, Keller AA (2010) Impacts of metal oxide nanoparticles on marine phytoplankton. Environ Sci Technol 44(19):7329–7334. https://doi.org/10.1021/es100247x

Moncayo-Lasso A, Mora-Arismendi LE, Rengifo-Herrera JA, Sanabria J, Benítez N, Pulgarin C (2012) The detrimental influence of bacteria (*E. Coli, Shigella* and *Salmonella*) on the degradation of organic compounds (and vice versa) in TiO_2 photocatalysis and near-neutral photo-Fenton processes under simulated solar light. Photochem Photobiol Sci 11(5):821–827. https://doi.org/10.1039/c2pp05290c

Moreira NA, Bondelind M (2017) Safe drinking water and waterborne outbreaks. J Water Health 15 (1):83–96. https://doi.org/10.2166/wh.2016.103

Muthuvel A, Jothibas M, Mohana V, Manoharan C (2020) Green synthesis of cerium oxide nanoparticles using Calotropis procera flower extract and their Photocatalytic degradation and antibacterial activity. Inorg Chem Commun 119:108086. https://doi.org/10.1016/j.inoche.2020.108086

Naqvi S, Vinay K, Gopinath P (2018) Nanomaterial Toxicity: A Challenge to End Users. In Appl Nanomat 315–343. https://doi.org/10.1016/b978-0-08-101971-9.00012-0

Nisa I, Qasim M, Yasin N, Ullah R, Ali A (2020) Shigella flexneri: an emerging pathogen. Folia Microbiol 65(2):275–291. https://doi.org/10.1007/s12223-020-00773-w

Onyango LA, Quinn C, Tng KH, Wood JG, Leslie G (2015) A study of failure events in drinking water systems as a basis for comparison and evaluation of the efficacy of potable reuse schemes. Environmental Health Insights 9(s3):11–18. https://doi.org/10.4137/EHI.S31749

Pandiyan R, Mahalingam S, Ahn YH (2019) Antibacterial and photocatalytic activity of hydrothermally synthesized SnO_2 doped GO and CNT under visible light irradiation. J Photochem Photobiol B Biol 191(November 2018):18–25. https://doi.org/10.1016/j.jphotobiol.2018.12.007

Park S-Y, Choi J-H (2010) Geno- and ecotoxicity evaluation of silver nanoparticles in freshwater crustacean Daphnia magna. Environmental Engineering Research 15(1):23–27. https://doi.org/10.4491/eer.2010.15.1.428

Prasad R (2014) Synthesis of silver nanoparticles in photosynthetic plants. J Nanoparticles, Article ID 963961. https://doi.org/10.1155/2014/963961

Prasad R (2016) Advances and applications through fungal nanobiotechnology. Springer International Publishing, Cham. ISBN 978-3-319-42989-2

Prasad R (2017) Fungal nanotechnology: applications in agriculture, industry, and medicine. Springer Nature Singapore Pte Ltd., Singapore. ISBN 978-3-319-68423-9

Prasad R (2019a) Microbial nanobionics: basic research and applications. Springer International Publishing, Cham. ISBN 978-3-030-16534-5. https://www.springer.com/gp/book/9783030165338

Prasad R (2019b) Microbial nanobionics: state of art. Springer International Publishing, Cham. ISBN 978-3-030-16383-9. https://www.springer.com/gp/book/9783030163822

Prasad R, Swamy VS (2013) Antibacterial activity of silver nanoparticles synthesized by bark extract of *Syzygium cumini*. J Nanoparticles. https://doi.org/10.1155/2013/431218

Prasad R, Pandey R, Barman I (2016) Engineering tailored nanoparticles with microbes: quo vadis. WIREs Nanomed Nanobiotechnol 8:316–330. https://doi.org/10.1002/wnan.1363

Prasad R, Jha A, Prasad K (2018a) Exploring the realms of nature for nanosynthesis. Springer International Publishing, Cham. ISBN 978-3-319-99570-0. https://www.springer.com/978-3-319-99570-0
Prasad R, Kumar V, Kumar M, Wang S (2018b) Fungal nanobionics: principles and applications. Springer Nature Singapore Pte Ltd., Cham. ISBN 978-981-10-8666-3. https://www.springer.com/gb/book/9789811086656
Prasad R, Siddhardha B, Dyavaiah M (2020) Nanostructures for antimicrobial and antibiofilm applications. Springer International Publishing, Cham. ISBN 978-3-030-40336-2. https://www.springer.com/gp/book/9783030403362
Punniyakotti P, Panneerselvam P, Perumal D, Aruliah R, Angaiah S (2020) Anti-bacterial and anti-biofilm properties of green synthesized copper nanoparticles from Cardiospermum halicacabum leaf extract. Bioprocess Biosyst Eng., no. 0123456789. https://doi.org/10.1007/s00449-020-02357-x
Rafique M, Tahir MB, Irshad M, Nabi G, Gillani SSA, Iqbal T, Mubeen M (2020) Novel Citrus aurantifolia leaves based biosynthesis of copper oxide nanoparticles for environmental and wastewater purification as an efficient photocatalyst and antibacterial agent. Optik 219:165138. https://doi.org/10.1016/j.ijleo.2020.165138
Rane, AV, Krishnan K, Abitha VK, Thomas S (2018) Methods for synthesis of nanoparticles and fabrication of nanocomposites. Synthesis of inorganic nanomaterials Elsevier Ltd https://doi.org/10.1016/b978-0-08-101975-7.00005-1
Raota CS, Cerbaro AF, Salvador M, Delamare APL, Echeverrigaray S, Da Silva Crespo J, Silva TBD, Giovanela M (2019) Green synthesis of silver nanoparticles using an extract of Ives cultivar (Vitis Labrusca) pomace: characterization and application in wastewater disinfection. J Environ Chem Eng 7(5):103383. https://doi.org/10.1016/j.jece.2019.103383
Rincón, AG, Pulgarin C (2005) Use of coaxial photocatalytic reactor (CAPHORE) in the TiO_2 photo-assisted treatment of mixed *E. Coli* and *Bacillus Sp.* and bacterial community present in wastewater Catalysis Today 101(3–4 spec. iss.):331–44. https://doi.org/10.1016/j.cattod.2005.03.022
Roy A, Das BK, Bhattacharya J (2011) Development and validation of a spectrophotometric method to measure sulfate concentrations in mine water without interference. Mine Water Environ 30. https://doi.org/10.1007/s10230-011-0140-x
Saad HA, Soliman MI, Azzam AM, Mostafa B (2015) Antiparasitic activity of silver and copper oxide nanoparticles against Entamoeba histolytica and Cryptosporidium parvum cysts. J Egypt Soc Parasitol 45(3):593–602. https://doi.org/10.12816/0017920
Saleh TA (2017) An overview of nanomaterials for water technology. Advanced Nanomaterials for Water Engineering, Treatment, and Hydraulics 2016:1–12. https://doi.org/10.4018/978-1-5225-2136-5.ch001
Sarkar A, Sarkar D, Poddar K (2019) Nanotoxicity: sources and effects on environment. Springer, Cham, pp 169–179. https://doi.org/10.1007/978-3-030-16534-5_9
Saxena SK (2019) Water-associated infectious diseases. https://doi.org/10.1007/978-981-13-9197-2
Shan Y, Guan D, Liu J, Liu Z, Liu J, Schroeder H, Chen Y, Shao S, Mi Z, Zhang Q (2016) CO_2 emissions inventory of Chinese cities. Atmos. Chem. Phys. Discuss.:1–26. https://doi.org/10.5194/acp-2016-176
Shao W, Liu X, Min H, Dong G, Feng Q, Zuo S (2015) Preparation, characterization, and antibacterial activity of silver nanoparticle-decorated graphene oxide nanocomposite. ACS Appl Mater Interfaces 7(12):6966–6973. https://doi.org/10.1021/acsami.5b00937
Sharma R, Uma SS, Verma A, Khanuja M (2016) Visible light induced bactericidal and photocatalytic activity of hydrothermally synthesized $BiVO_4$ nano-octahedrals. J Photochem Photobiol B Biol 162(June):266–272. https://doi.org/10.1016/j.jphotobiol.2016.06.035
Sirelkhatim A, Mahmud S, Seeni A, Kaus NHM, Ann LC, Bakhori SKM, Hasan H, Mohamad D (2015) Review on zinc oxide nanoparticles: antibacterial activity and toxicity mechanism. Nano-Micro Letters 7(3):219–242. https://doi.org/10.1007/s40820-015-0040-x

Sreeja S, Vidya Shetty K (2016) Microbial disinfection of water with endotoxin degradation by photocatalysis using Ag@TiO_2 core Shell nanoparticles. Environ Sci Pollut Res 23 (18):18154–18164. https://doi.org/10.1007/s11356-016-6841-8

Srivastava A, Srivastava ON, Talapatra S, Vajtai R, Ajayan PM (2004) Carbon nanotube filters. Nat Mater 3(9):610–614. https://doi.org/10.1038/nmat1192

Srivastava S, Usmani Z, Atanasov AG, Singh VK, Singh NP, Abdel-Azeem AM, Prasad R, Gupta G, Sharma M, Bhargava A (2021) Biological nanofactories: using living forms for metal nanoparticle synthesis. Mini-Rev Med Chem 21(2):245–265

Taylor, P, Suri RPS, Thornton HM, Muruganandham M (2011) Disinfection of water using Pt- and Ag-doped TiO_2 photocatalysts September 2012;37–41

Tegos GP, Demidova TN, Arcila-Lopez D, Lee H, Wharton T, Gali H, Hamblin MR (2005) Cationic fullerenes are effective and selective antimicrobial photosensitizers. Chem Biol 12 (10):1127–1135. https://doi.org/10.1016/j.chembiol.2005.08.014

Ul-Islam M, Ullah MW, Khan S, Manan S, Khattak WA, Ahmad W, Shah N, Park JK (2017) Current advancements of magnetic nanoparticles in adsorption and degradation of organic pollutants. Environ Sci Pollut Res 24(14):12713–12722. https://doi.org/10.1007/s11356-017-8765-3

Vicas CS, Keerthiraj N, Byrappa N, Byrappa K (2019) Hydrothermally synthesized Al-doped $Bivo_4$ as a potential antibacterial agent against methicillin-resistant Staphylococcus aureus. Environmental Engineering Research 24(4):566–571. https://doi.org/10.4491/eer.2018.228

Wang R, Kong X, Zhang W, Zhu W, Huang L, Wang J, Zhang X et al (2018) Mechanism insight into rapid photocatalytic disinfection of salmonella based on vanadate QDs-interspersed g-C_3N_4 heterostructures. Appl Catal B Environ 225:228–237. https://doi.org/10.1016/j.apcatb.2017.11.060

Xu Y, Wen W, Wu JM (2018) Titania nanowires functionalized polyester fabrics with enhanced photocatalytic and antibacterial performances. J Hazard Mater 343:285–297. https://doi.org/10.1016/j.jhazmat.2017.09.044

Yadav L, Tripathi RM, Prasad R, Pudake RN, Mittal J (2017) Antibacterial activity of Cu nanoparticles against *E. coli*, *Staphylococcus aureus* and *Pseudomonas aeruginosa*. Nano Biomed Eng 9(1):9–14. https://doi.org/10.5101/nbe.v9i1.p9-14

Zhan S, Yang Y, Shen Z, Shan J, Li Y (2014) Efficient removal of pathogenic bacteria and viruses by multifunctional amine-modified magnetic nanoparticles. 274:115–123. https://doi.org/10.1016/j.jhazmat.2014.03.067

Zhang T, Yu H, Li J, Song H, Wang S, Zhang Z, Chen S (2020) Green light–triggered antimicrobial cotton fabric for wastewater disinfection. Materials Today Physics 15:100254. https://doi.org/10.1016/j.mtphys.2020.100254

Zheng X, peng Shen Z, Cheng C, Shi L, Cheng R, hai Yuan D (2018) Photocatalytic disinfection performance in virus and virus/bacteria system by cu-TiO_2 nanofibers under visible light. Environ Pollut 237:452–459. https://doi.org/10.1016/j.envpol.2018.02.074

Zodrow K, Brunet L, Mahendra S, Li D, Zhang A, Li Q, Alvarez PJJ (2009) Polysulfone ultrafiltration membranes impregnated with silver nanoparticles show improved biofouling resistance and virus removal. Water Res 43(3):715–723. https://doi.org/10.1016/j.watres.2008.11.014

Chapter 14
The Evolving Role of Nanoparticles in Bacteria Mediated Cancer Therapy

Swapnil C. Kamble, Farhan F. Shaikh, and Joyita Sarkar

Abstract Effectiveness of bacteria mediated cancer therapy is gaining momentum as a re-discovered method of cancer treatment. It is principled on homing nature of certain anaerobic bacteria towards tumor microenvironment induced by its inherent hypoxic conditions. However, simple homing phenomenon is of low utility for tumor management, considering the toxicity and infection caused by these very same bacteria to the body. Recently, researchers have developed the ways of conjugation of nanoparticles with selected bacteria towards improving the treatment efficacy. Cargo-laden, expression-based, and antibody-guided approaches have emerged as major routes of application. Cargo-based technique potentiates delivery of chemotherapeutic drug as per tumor type. Genetically engineered microbes can deliver similar results, though long-term efficacy is unseen. Antibody-guided methods exercise specificity of drug-conjugated antibody against antigen-expressed over bacterial wall (located in tumor niche). Nonetheless, multiple challenges and executional limitations remain. In this review, we have focused on enhancement of bacterial functionality by nanoparticles for cancer management. We attempt to identify the challenges ahead and future perspectives of this emerging science. Considering the limited literature, we hope that this review will give its reader the enthusiasm in this field and possibly explore new avenues for precision bacteria-based cancer therapy.

Keywords Bacteria mediated cancer therapy · Cancer theranostics · Nanoparticle-bacteria conjugates · Tumor regression

---

S. C. Kamble
Department of Technology, Savitribai Phule Pune University, Pune, Maharashtra, India

F. F. Shaikh · J. Sarkar (✉)
Institute of Chemical Technology Marathwada Campus, Jalna, Maharashtra, India
e-mail: j.sarkar@marj.ictmumbai.edu.in

N. R. Maddela et al. (eds.), *Nanotechnology for Advances in Medical Microbiology*, Environmental and Microbial Biotechnology,
https://doi.org/10.1007/978-981-15-9916-3_14

14.1 Introduction

Cancer, regardless of primary or metastatic, tumorigenic or non-tumorigenic, is detrimental to human health and therefore development of its treatment strategies has gained much importance. Transformation of normal cells into tumorigenic may be stimulated by various risk factors—external agents (like smoking, alcohol, pollutants, specific food or high energy waves) or internal agents (like genetic/ hereditary causes, age, body weight). Though these external factors may induce cancer, an antagonist role by similar action has been attempted for therapeutic value. Over recent years, association of microbial infection leading to cancer and therapeutic effects of bacteria on cancer have been elucidated (Rai et al. 2020; Ashu et al. 2019; Song et al. 2018).

Microbial role in cancer treatment has been in existence since the nineteenth century. The serendipity, in 1813, of the clostridial infection induced tumor regression made space in treatment approaches of certain cancers (Minton 2003; Vautier 1813). This therapy route gained traction in 1890 when William Coley perfected the therapy, leading to dawn of Coley's Toxins (Coley 1991). It is principled on tendency of anaerobic bacteria to lodge themselves inside the tumor due to the favorable conditions of hypoxia and availability of nutrients (Dang et al. 2001; Zhao et al. 2005). Unlike Gram-positive bacteria, Gram-negative bacteria elicit the secretion of cytokines like tumor necrosis factor α (TNFα) owing to their endotoxins-lipopolysaccharide in the cell wall (Shear et al. 1943; Wiemann and Starnes 1994). TNFα plays an important role, but itself has limited efficacy for tumor regression (Wiemann and Starnes 1994). This finding brought T-cells into the forefront of active host immune cell-dependent tumor regression. However, as a pre-requisite, the T-cell must be made to respond to tumor before the therapy for effective regression of tumor (Berendt et al. 1978).

Bacteria after attaching on the certain tumor sites, can stimulate the production of Interleukin-12 (IL-12) (D'Andrea et al. 1992; Macatonia et al. 1995). IL-12 exhibits tumor regression better than TNFα. However, this treatment works well in some cases only. This is due to the requirement of Th1 response to be present in the body for IL-12 to exert its effects, *i.e.* the patient should have "pre-existing immunity" (Berendt et al. 1978; Iwasaki et al. 2000; Le et al. 2001). The removal of $CD4^{+}CD25^{+}$T-cells can enable this immunity (Ghiringhelli et al. 2004; Onizuka et al. 1999; Shimizu et al. 1999). Thereon, bacteria mediated cancer therapy has made tremendous progress with development of multitude of approaches (Fig. 14.1). "Targeting of tumor cells" approaches have traditionally involved injection of pro-drugs and bacteria in tumor hosts. However, limitations of routine chemotherapy, *i.e.* obstruction caused by hypoxic and acidic environment, render the purpose less fruitful.

Unfortunately, these bacteria, although cause regression of tumors, have toxic side effects due to their virulity on the body. Recent studies have shown reduced toxicity by removing certain virulent genes from the bacteria (Berendt et al. 1978). Also, antibiotics can be used to control the effects of bacteria (D'Andrea et al. 1992).

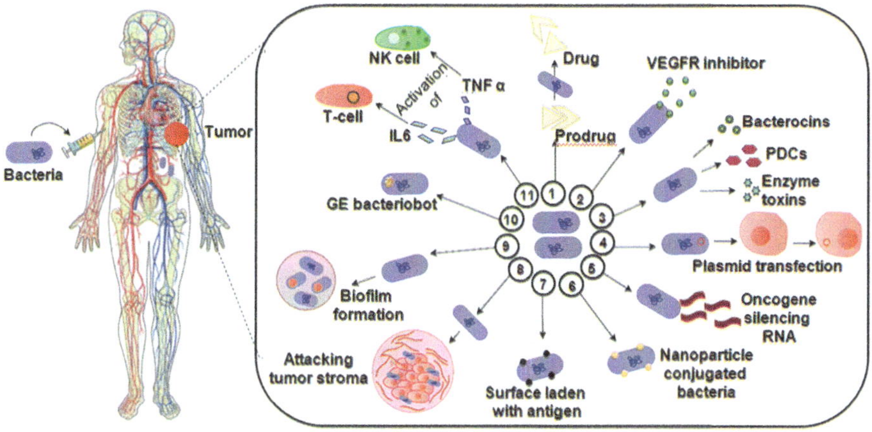

Fig. 14.1 Some of the major approaches involving bacteria for cancer therapy. (1) Bacteria expressing enzymes to activate anticancer "prodrugs" at tumor sites; (2) Bacteria expressing anticancer agents directly (e.g. anti-angiogenic like VEGFR inhibitor); (3) through bacteria released substances (toxins/enzymes—Bacteriocins/Phenazine 1,6-di-carboxylic acid (PDC)/bacterial enzyme toxin/bacterial spores and vectors as tumoricidal agents); (4) Bacteria transferring eukaryotic expression vectors into tumor cells (plasmid transfection); (5) Bacteria expressing oncogene silencing RNA; (6) delivery of tumor killing nanoparticle conjugated bacteria based therapies; (7) Bacteria expressing tumor specific antigens and antibodies; (8) Targeting of tumor stroma; (9) Bacteria as anticancer agents through biofilms; (10) Genetically Engineered (GE) Bacteria-based microrobot (Bacteriobot); (11) Bacteria as anticancer agents through enhancing immunity (Activating inflammasome pathway/CD4, CD25, CD8 antitumor effector cell response/TNF-α innate immune system in bacteria-based tumor necrosis/Activation of Immune cells)

The earlier studies thus summarized, exhibit that though the bacteria are efficient in targeting tumors but their efficacy in tumor regression is disputed and bacteria elicits different response in different individuals. This signifies that if external aid is provided, these bacteria can be tuned and manipulated for the treatment of cancer.

Simultaneous and independent development in utility of nanoparticles in cancer diagnosis and management has seen fruitful scientific progress. Application in therapeutics of nanoparticle delivery systems are considered important due to

1. improved targeting of cancer cells,
2. minimal systemic toxicity due to lower dosage,
3. improvement in regulation of circulation time of the intended therapeutic, and
4. conservation of therapeutics' bioavailability.

Standardizations of nanoparticles for therapy have been inclusive of vesicle dimensions, interaction with cells, binding efficacy, and its sustained release (Yoo et al. 2011).

Despite the above listed and many more advantages, various aspects hinder immediate clinical application. For example, uncertainties in the physical structure of the tumor microenvironment (due to organ of residence), and interstitial fluid

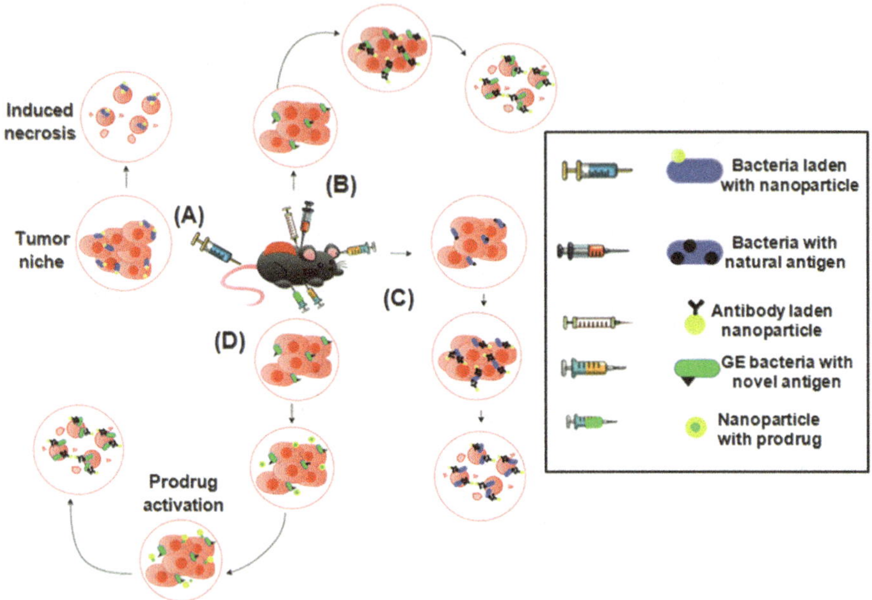

Fig. 14.2 Selected strategies for combinatorial approaches. (**a**) Cargo-laden approach. (**b**) Antibody-guided approach. (**c**) Genetically engineered (GE) approach. (**d**) Co-delivery approach

pressure loss due to any lymphatic drainage, etc. (Dewhirst and Secomb 2017; Minchinton and Tannock 2006). The complexity is increased with the presence of stromal cells, nature of extracellular matrix of the tumor microenvironment, physical stress generated by growing number of cells within confined space, and expanded intercapillary spaces that impede delivery and/or functionality of chemotherapeutic agents (Jain and Stylianopoulos 2010; Jang et al. 2003; Nia et al. 2017; Saggar et al. 2013). Thus, extermination of cancer cell requires an effective disbursement of optimally medicated therapeutic macromolecules within the tumor.

An approach to combine hypoxic environment targeting bacterial activity with functionality of nanoparticles has been a work that was waiting to happen. Both domains have been critical in targeting and managing tumors either directly (killing tumor cells) or in-directly (targeting stroma, activating immune cells, etc.). However, limited progress made in clinical setting overreaches the limitations these methods have. Hence a potential angle generated by an overlay of these two approaches may have a more impact than either of them alone (Fig. 14.2). Since nanoparticles have been widely utilized for targeted cancer theranostics, evidently, it has also been employed in conjunction with bacteria for cancer therapy. In this chapter, we aim to consolidate the research works carried out on cancer therapy mediated by nanoparticles in concurrence with bacteria mediated tumor regression and targeting.

14.2 Bacteria Associated with Cancer Management

As already summarized in Fig. 14.1, bacteria mediate tumor regression by different ways. Bacteria cause tumor regression by inducing tumor apoptosis and autophagy either by production of toxic substances or depleting nutrients supply to the tumor cells. The bacteria can also activate certain pro-drugs to kill a tumor (Kim et al. 2013). Further, certain bacterial spores have antitumor cytotoxic enzymes which are expressed after spore germination (Bettegowda et al. 2006; Kubiak et al. 2015). Besides, host immunity also plays a crucial role in inducing antitumor activity (Agrawal et al. 2004). The specific mechanisms of tumor regression by some bacteria are listed as follows.

14.2.1 Salmonella

Salmonella shows tropism towards tumor sites due to the presence of receptors of aspartate, serine, ribose/galactose on its surface. These three receptors have characteristic function in the chemotaxis of *Salmonella* towards the tumor. The aspartate receptor starts the chemotaxis, the serine receptor helps in penetrating the tumor cell wall, and the ribose/galactose receptor guides the bacteria towards necrotic sites. Once settled at the tumor site, the bacteria continue to proliferate leading to overgrowth of bacteria around the tumor. It has been determined that *S. typhimurium* cells number more than 1×10^{10} CFU/g of tumor tissues after 3 days of administration (Ganai et al. 2011; Uchugonova et al. 2015). *Salmonella* spp. also induces autophagy of tumor cells by downregulating the AKT/mTOR pathway, which in turn negatively regulates autophagy (Lee et al. 2014; Liu et al. 2016). It induces apoptosis of the tumor cells by increasing the levels of caspase-1, which cleaves pro-IL-β and pro-IL-18, yielding their active forms at tumor sites (Phan et al. 2015). *Salmonella* upregulates the protein Connexin 43, promoting formation of gap junction between tumor and dendritic cells, and giving access to tumor antigen. Dendritic cells containing antigen peptides are then passed through these junctions into the tumor cell (Shilling et al. 2007). Once inside, these antigen peptides activate $CD8^+$ T-cells for tumor regression. This also lead to reduced production of immunity suppressing enzyme indoleamine2,3-dioxygenase in T-cells causing increased activation of T-cells (Chang et al. 2013; Saccheri et al. 2010). *Salmonella* also leads to tumor regression by inducing an autoimmune response. The signalling of Flagellin-derived Toll-like receptor 5 not only reduces tumor proliferation but also decreases $CD4^+$ $CD25^+$ T-cells, which stimulates the pre-existing immunity of the body (Cai et al. 2011; Leigh et al. 2014). The Flagellin helps in increasing the levels of interferons, CXCL9 and CXCL10 in *Salmonella* inhabited tumor sites. It is these cytokines that further recruit natural killer cells and cytotoxic T-cells surrounding the tumor site (Kupz et al. 2014).

14.2.2 Clostridium

The spores of *Clostridium* find their way to the tumor sites by the undeveloped vasculature of tumor owing to enhanced permeability and retention (EPR) effect that allows entry of macromolecules and nutrients of fixed size (usually in nm to μm) in tumor parenchyma (Fang et al. 2003; Matsumara and Maeda 1986). Also, the hypoxic levels of tumor site provide favorable environment for this anaerobic bacteria to grow and proliferate (Van Mellaert et al. 2006). Once the bacterial spores reach the tumor sites and germinate, they can self-propel around the tumor vasculature (Forbes 2010). The injection of the virgin clostridium bacteria in mice caused tumor lysis but also the death of the mice due to its toxic nature; few mice which survived owing to high antibiotic doses, later died due to regrowth of the tumor (Malmgren and Flanigan 1955). The recent engineered strain of clostridium, *C. Novyi*-NT, has shown more promising result and is undergoing human clinical trial (Dang et al. 2001; Roberts et al. 2014).

14.2.3 Bifidobacteria

Apart from competing for nutrients with tumor cells, another interesting tumor regression mechanism of this bacterium is biotransformation, *i.e.* transforming certain compounds into tumor regression compounds. Anticancer drugs like Lapachol and 5-fluorocytosine are converted to active antitumor compounds (Bae et al. 2000; Hidaka et al. 2007; Nakamura et al. 2002; Oliveira Silva et al. 2014). The bacteria also alter the expression of cancerous genes and cytokines, either by suppressing or enhancing them by increasing levels of Interferon-gamma secreting cells (Gu et al. 2016; Reddy 1999; Wu et al. 2016).

14.3 Selected Strategies of Combinatorial Approaches

Tumor is very difficult to penetrate, be it for nutrients or the T-cells. However, in the previous section it is seen that the anaerobic bacteria infiltrate into tumor region with ease. This has led to the rise of the idea to use this phenomenon as a Trojan horse. The conjugation of bacteria with nano-sized particles is of much interest (Lee et al. 2013). The conjugation with the bacteria is done by either modifying receptors on the bacterial cell wall to recognize the nanoparticles or by ligand exchange (Wang et al. 2010; Zhai et al. 2017). The teaming up of bacteria with nanoparticles provides us with advanced imaging and therapeutic strategies, and thereby can be utilized for theranostic purposes. This conjugation opens up various therapeutic avenues for tumor regression. Based on the mode of conjugation and action, nano-bacteria therapy can be broadly classified into cargo-laden, genetically engineered, antibody-guided and co-delivery approaches (Fig. 14.2).

14.3.1 Cargo-Laden Approach

The cargo-laden approach is the most common method in which the bacterium is the carrier and the nanoparticle, quantum dots or anticancer drugs are the cargo. The natural affinity of some anaerobic bacteria, for example, *Salmonella*, *Clostridium*, *Bifidobacteria*, etc. are exploited to carry the cargo either on the surface or inside the tumor. The bacteria may also itself produce the cargo. Liu and colleagues demonstrated the quantum dot-internalized *Bifidobacterium bifidum* having folic acid on its surface can target the folate receptor expressing tumor cells (Liu et al. 2012). Here, the cadmium-selenium-sulfur quantum dots were encapsulated in lipid layer and internalized in the bacteria by electroporation. To further enhance the targeting efficiency, the bacterial surface was modified with folic acid. The formulation was found to be efficient in targeting the core of solid tumors in mice model and may prove to be an excellent tool for tumor imaging (Fig. 14.3a). However, the side effects and fate of the quantum dots in the body have not been elaborated and are subject to further research. Luo et al. used upconversion nanorods (UCNRs) and bioimaging and photothermal ablation of tumors, respectively (Luo et al. 2016). Considering the deeper penetration using near-infrared (NIR) light in biological tissues, both UCNRs are susceptible of NIR light excitation. The ligand-free UCNRs (LF-UCNRs) were deposited electrostatically on negatively charged surface of *Bifidobacterium breve*. It was observed that the bacteria-conjugated UCNRs showed significantly better targeting efficiency as compared to the LF-UCNRs (Fig. 14.3b). In a similar study, liposomes loaded with doxorubicin were inserted in *Salmonella* by electroporation to generate nanoswimmers termed BADOX (Zoaby et al. 2017). The BADOX exhibited an enhanced speed in targeted delivery of doxorubicin to cancer cells. The doxorubicin upon reaching to the cancer cells were released from the liposomes in response to ammonia secreted by the cancer cells and hence creating an osmotic misbalance. Doxorubicin, being both cancer chemotherapeutic and antibiotic drug, kills the tumor cells as well as the bacteria, thereby lowering the probability of any side effects that could be generated by the bacteria. However, the ability of the fabricated nanoswimmers to invade solid tumors has to be clinically proved.

In another interesting study, *E. coli* attached with a complex drug system was designed (Park et al. 2017). The cargo was of doxorubicin-loaded polyelectrolyte multilayer (PEM) microparticles including embedded magnetic nanoparticles. The 1 μm wide PEM loaded *E. coli* were directed using an external magnetic guidance system to 4T1 breast cancer cells. This is unlike the obligate anaerobes like *Salmonella* and *Clostridium* which have tendency to migrate towards hypoxic tumor. The microswimmers were easily guided to the desired site and the presence of bacteria resulted in lowering of local pH that led to enhanced uptake of anticancer drugs by the cancer cells. Overall, the platform is highly tunable but needs to be perfected in vivo. In another similar approach, *Magnetococcus marinus* strain MC-1 was used that naturally produced magnetosome (chain of iron oxide nanoparticles) (Felfoul et al. 2016). Anticancer drug loaded nanoliposomes were covalently attached to the

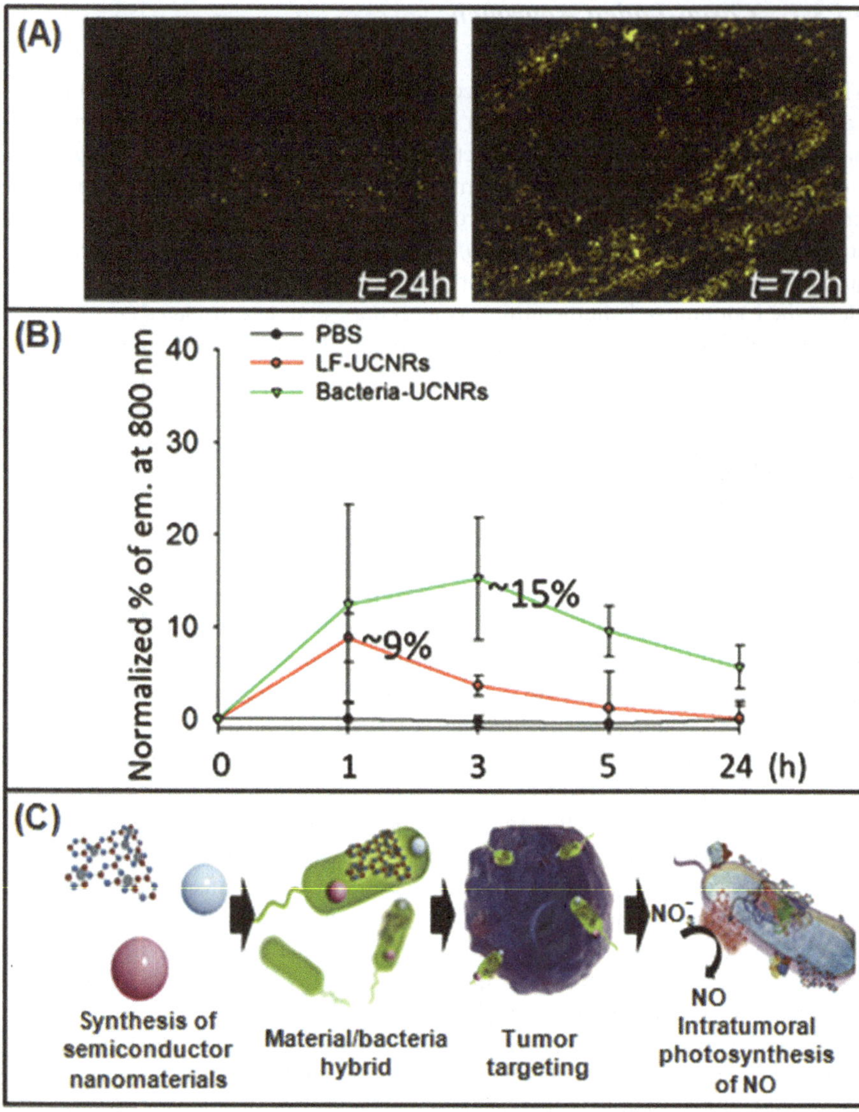

Fig. 14.3 *Cargo-laden approach*. (**a**) Fluorescent image showing internalization of quantum dot-*B. bifidum*-folic acid conjugate by solid tumor at 24 and 72 h post intravenous injection (Liu et al. 2012). (Reproduced with permission). (**b**) Near infrared (NIR) intensities of tumor site after injection of cargo-laden bacteria-nanoparticle conjugate (Luo et al. 2016). (Reproduced with permission). *PBS* Phosphate buffer saline, *LF-UCNRs* Ligand-free upconversion nanorods. *Genetically engineered approach*. (**c**) Schematic showing synthesis of nanomaterial-bacteria hybrid, tumor targeting and intratumoral photosynthesis of NO to engender photo-controlled bacterial metabolite therapy (Zheng et al. 2018). (Reproduced with permission)

surface of MC-1 bacterial cells and the formulation was guided under magnetic field to HCT116 colorectal xenografts in mice model. Approximately 70 nanoliposomes per MC-1 cells were attached and around 55% of the bacteria-nanoliposome conjugate were able to penetrate the hypoxic region of the xenograft. Apart from imaging and chemotherapeutic studies, the bacteria-nanoparticle conjugate has also been used as a vaccine for cancer immunotherapy. Cationic polymer, β-cyclodextrin-Polyethyleneimine600 (CP), nanoparticles containing plasmid DNA encoding vascular endothelial growth factor receptor 2 (VEGFR2) were coated on liver attenuated *Salmonella* bacteria (Hu et al. 2015). This coating enabled the bacteria to be administered orally by escaping phagosomes and imparting resistance to the acidic pH of stomach. Upon oral administration of the bacteria-nanoparticle construct, successful T-cell and cytokine activation along with suppression of angiogenesis in tumor was observed leading to tumor necrosis.

14.3.2 Genetically Engineering Bacteria for Expressing Specific Biomolecule

In this approach, a secondary molecule having high affinity to a bacteria-expressed biomolecule is either delivered simultaneously with the bacteria or later. The bacteria itself does not carry the active drug. In one such approach, Park and co-workers modified the facultative anaerobic *Salmonella typhimurium* to express biotin that enables interaction with streptavidin conjugated microbeads (Park et al. 2014). The "bacteriobot" was seen with the increase of the tumor targeting by the observation of the fluorescent microbeads. Fan and co-workers developed an *E. coli* MG1655 based vehicle for oral administration (Fan et al. 2018). This non-invasive thermally sensitive programmable bacterial (TPB) therapeutic system expressed TNF-α, and bio-mineralized gold nanoparticles (AuNPs). TPB is transported into internal microcirculation by microfold cells (of Peyer's patches) followed by homing to tumor microenvironments. Once the TBPs accumulated at the tumor sites, its irradiation by NIR induced expression of TNF-α, eventually inducing apoptotic cell death in tumor treatment. In a very intriguing approach, a metabolic pathway by "charging" facultative anaerobe *E. coli* with a nano-photocatalyst that strengthens their routine metabolic activities has been proposed (Zheng et al. 2018). Carbon nitride (C_3N_4), when metabolized by nitric oxide (NO) generation enzymes, enables development of photo-controlled bacterial metabolite therapy (PMT). In the presence of light treatment, C_3N_4 produced photoelectrons and transferred to *E. coli* induce enzymatic reduction of cellular NO_3^- to cytotoxic NO (Fig. 14.3c). Preclinical study involving a tumor-bearing mouse model, C_3N_4 loaded bacteria collected within the tumor microenvironment and the PMT led to inhibition of tumor progression by 80%. A similar work in non-toxin strains of genetically engineered *Salmonella* or *Clostridium* may be used to increase therapeutic efficiency.

14.3.3 Antibody-Guided Approach

As the name suggests, in this approach antibody acts as a guide to deliver the cargo rather than directly loading on the bacteria. Customarily, the bacteria are injected first, allowed to colonize the tumor site and then the antibody conjugated nanoparticles or drugs are injected, leading to regression of tumor. The antibody is against the bacterial antigen; and the microorganisms are not genetically engineered. Apart from guiding the cargo, the approach has also been used to guide bacteria to the tumor specific antigen for augmented colonization and invasion of solid tumor. Probably, the first instance of antibody-guided bacteria mediated delivery of nanoparticles was demonstrated by Akin and co-workers (Akin et al. 2007). Streptavidin-coated polystyrene nanoparticle was first attached to biotinylated antibody against a surface antigen of *Listeria monocytogenes*; the nanoparticle was then loaded with green fluorescent protein. The cargo was then docked on to bacterial surface giving rise to "microbots." The microbots were then targeted to desired cells where they were able to deliver the nanoparticle containing plasmid that was further successfully expressed in the target cell. The bacteria were then killed by treatment with antibiotics. Later Kojima and co-workers used a similar biotin-streptavidin approach to fabricate bacteria driven liposomes by raft domain binding method (Kojima et al. 2012). *Vibrio alginolyticus* mutant strain VIO5 with polar flagella associated with biotinylated antibody was used. The bacteria interacted with biotin-modified liposome *via* streptavidin.

In a different study discussed in cargo-laden approach by Luo et al., core shell-UCNRs modified with PEG polymers and bound to *Clostridium* polyclonal antibodies that target the vegetative *C. difficile* were introduced into mice model post-colonization of tumor sites by *C. difficile* (Luo et al. 2016). This approach was more efficient than the cargo-laden approach in which nanoparticles were loaded on *B. beveri* (Fig. 14.4A). Suh and co-workers utilized bioconjugation method based on streptavidin–biotin interaction (Suh et al. 2019). *Salmonella enterica* serovar Typhimurium VNP20009 surface was modified using a biotinylated antibody targeting tumor. The bacterium with streptavidin-coated poly(lactic-co-glycolic acid) (PLGA) nanoparticles to the outer membrane constituted the NanoBEADS agent (Fig. 14.4B). As the authors did not genetically engineer the bacteria for the construction of NanoBEADS of any specific cargo, the same system can be utilized as a platform for delivery of required drug. Inherent to its nature, *S. enterica* Typhimurium VNP20009 homes to tumor region through intercellular translocation mode, skipping any external requirement for guidance towards tumor site. Effective tumor targeting and accumulation of therapeutic agent was observed both in vitro and in vivo. The NanoBEADS platform thereby provides a versatile and simple approach to enhance the bacterial colonization and invasion in the tumor and simultaneous delivery of anticancer cargo.

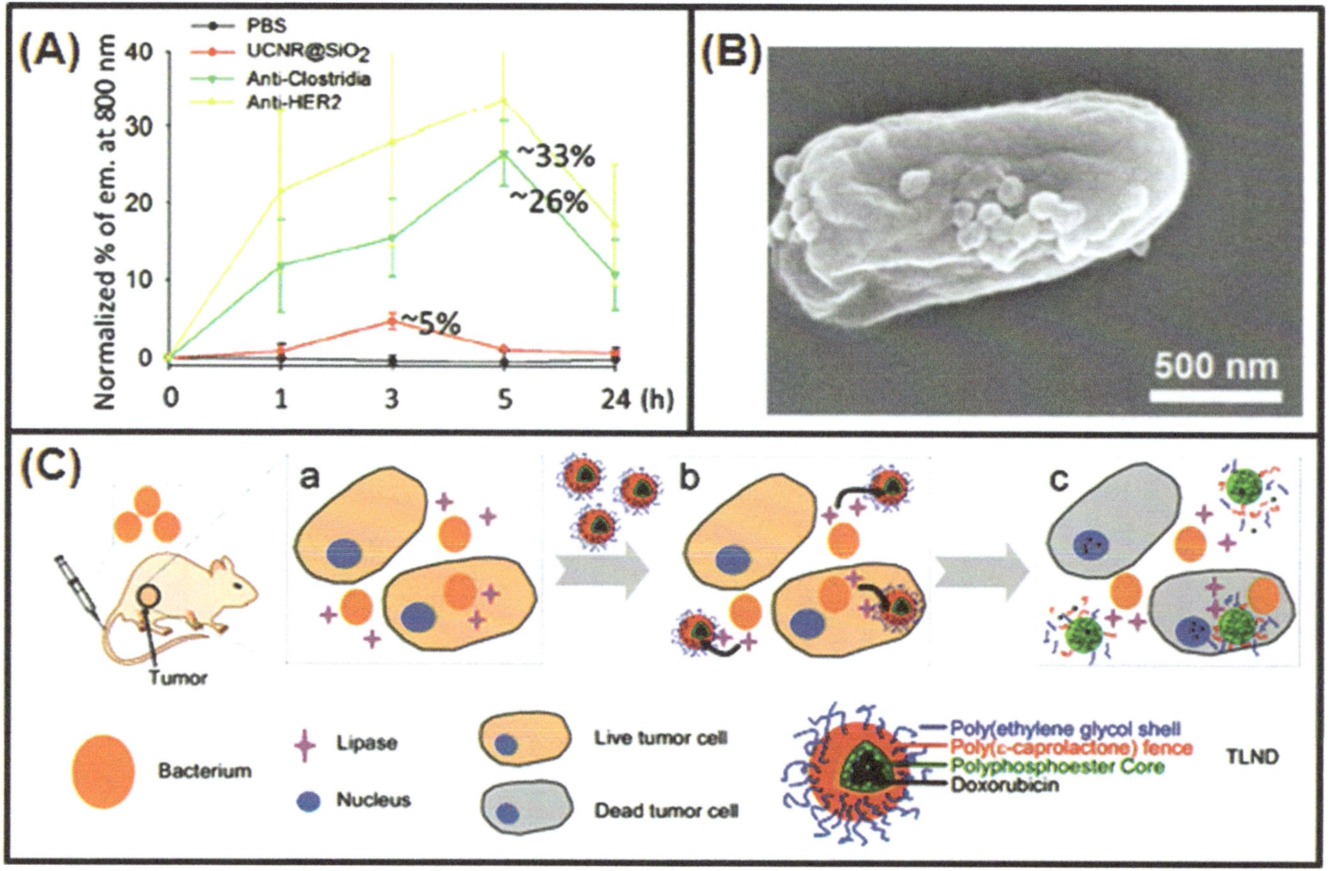

Fig. 14.4 *Antibody-guided approach.* (**A**) Near infrared (NIR) intensities of tumor site after injection of antibody-guided bacteria and nanoparticle formulation (Luo et al. 2016). (Reproduced with permission). *PBS* Phosphate buffer saline, *UCNRs* upconversion nanorods. (**B**) Scanning electron microscope image of

14.3.4 Co-delivery

In this case, a pro-drug and microbe are injected in body independently. Pro-drug is activated either by the microbe or by an external stimulant like light (for example, in photo-dynamic light therapy). Using this approach, Park and colleagues attempted to overcome limitation of remnant tumor cells of primary tumor targeted by any therapy (Park et al. 2020). They formulated an emulsion having two components—multifunctional nanoscintillators (NSs) and *C. novyi* spores coated with branched gold nanoparticle (for synergistic image-guided combinational treatment). NSs were composed of $NaGdF_4$:Tb,Ce@$NaGdF_4$ core/shell structure that allowed MRI visibility. The emulsion injected into the tumor allowed MRI/CT image guidance and photoactivation of NSs for photo-dynamic therapy (PDT). In an approach similar to pro-drug activation method, attenuated bacteria SBY1 were co-delivered with bacteria-sensitive triple-layered nanogel (TLN) (Xiong et al. 2013). Initially administered SBY1 exclusively accumulated in tumor site after certain duration. The ensuing administration of doxorubicin-loaded TLN was actively degraded by SBY1 to release doxorubicin within the tumor niche. Thus, targeted delivery of doxorubicin was achieved (Fig. 14.4C).

14.4 Conclusion

One can conclude that under careful manifestation, bacterial agents can be employed to manage tumor or cancer. Many of the innovative therapies have excellent efficacy at in vitro and animal model levels but have not been translated to clinical practice due to either incomplete tumor killing or recurrence of treated primary cancer in a short term. An effective combinatorial therapy using advances in nanoparticle-based and bacteria mediated systems for cancer has minimized remnant tumor cells. Many of these concepts can be extended, and cross improvised to generate an entirely new entity that can overcome limitations of many of above systems. Here we have attempted to present the existing systems that have potential for translational studies. We hope that the reader will be encouraged to foresee and undertake research in this domain.

Fig. 14.4 (continued) NanoBEADS agent (Suh et al. 2019). (Reproduced with permission). *Co-delivery approach.* (**C**) Schematic showing doxorubicin-loaded triple layered nanogel; (a) Accumulation of bacteria at tumor site; (b) Accumulation of nanoparticles at tumor site; (c) Bacteria degrading the poly(ε-caprolactone) fence triggering release of doxorubicin (Xiong et al. 2013). (Reproduced with permission)

References

Agrawal N, Bettegowda C, Cheong I, Geschwind J-F, Drake CG, Hipkiss EL, Tatsumi M, Dang LH, Diaz LA, Pomper M, Abusedera M, Wahl RL, Kinzler KW, Zhou S, Huso DL, Vogelstein B (2004) Bacteriolytic therapy can generate a potent immune response against experimental tumors. Proc Natl Acad Sci U S A 101:15172–15177. https://doi.org/10.1073/pnas.0406242101

Akin D, Sturgis J, Ragheb K, Sherman D, Burkholder K, Robinson JP, Bhunia AK, Mohammed S, Bashir R (2007) Bacteria-mediated delivery of nanoparticles and cargo into cells. Nat Nanotechnol 2:441–449. https://doi.org/10.1038/nnano.2007.149

Ashu EE, Xu J, Yuan ZC (2019) Bacteria in cancer therapeutics: a framework for effective therapeutic bacterial screening and identification. J Cancer 10:1781–1793. https://doi.org/10.7150/jca.31699

Bae EA, Park SY, Kim DH (2000) Constitutive beta-glucosidases hydrolyzing ginsenoside Rb1 and Rb2 from human intestinal bacteria. Biol Pharm Bull 23:1481–1485. https://doi.org/10.1248/bpb.23.1481

Berendt MJ, North RJ, Kirstein DP (1978) The immunological basis of endotoxin-induced tumor regression: requirement for a pre-existing state of concomitant anti-tumor immunity*. J Exp Med 148:1560–1569. https://doi.org/10.1084/jem.148.6.1560

Bettegowda C, Huang X, Lin J, Cheong I, Kohli M, Szabo SA, Zhang X, Diaz LA, Velculescu VE, Parmigiani G, Kinzler KW, Vogelstein B, Zhou S (2006) The genome and transcriptomes of the anti-tumor agent Clostridium novyi-NT. Nat Biotechnol 24:1573–1580. https://doi.org/10.1038/nbt1256

Cai Z, Sanchez A, Shi Z, Zhang T, Liu M, Zhang D (2011) Activation of toll-like receptor 5 on breast cancer cells by flagellin suppresses cell proliferation and tumor growth. Cancer Res 71:2466–2475. https://doi.org/10.1158/0008-5472.CAN-10-1993

Chang W-W, Lai C-H, Chen M-C, Liu C-F, Kuan Y-D, Lin S-T, Lee C-H (2013) Salmonella enhance chemosensitivity in tumor through connexin 43 upregulation. Int J Cancer 133:1926–1935. https://doi.org/10.1002/ijc.28155

Coley WB (1991) The treatment of malignant tumors by repeated inoculations of erysipelas. Clin Orthop Relat Res 262:3–11

D'Andrea A, Rengaraju M, Valiante NM, Chehimij J, Kubin M, Aste M, Chan SH, Kobayashi M, Young D, Nickbarg E, Chizzonite R, Wolf SF, Trinchieri G (1992) Production of natural killer cell stimulatory factor (interleukin 12) by peripheral blood mononuclear cells. J Exp Med 176:1387–1398. https://doi.org/10.1084/jem.176.5.1387

Dang LH, Bettegowda C, Huso DL, Kinzler KW, Vogelstein B (2001) Combination bacteriolytic therapy for the treatment of experimental tumors. Proc Natl Acad Sci U S A 98:15155–15160. https://doi.org/10.1073/pnas.251543698

Dewhirst MW, Secomb TW (2017) Transport of drugs from blood vessels to tumour tissue. Nat Rev Cancer. https://doi.org/10.1038/nrc.2017.93

Fan JX, Li ZH, Liu XH, Zheng DW, Chen Y, Zhang XZ (2018) Bacteria-mediated tumor therapy utilizing photothermally-controlled TNF-α expression via Oral administration. Nano Lett 18:2373–2380. https://doi.org/10.1021/acs.nanolett.7b05323

Fang J, Sawa T, Maeda H (2003) Factors and mechanism of "EPR"; effect and the enhanced antitumor effects of macromolecular drugs including SMANCS. Adv Exp Med Biol 519:29–49. https://doi.org/10.1007/0-306-47932-X_2

Felfoul O, Mohammadi M, Taherkhani S, de Lanauze D, Zhong Xu Y, Loghin D, Essa S, Jancik S, Houle D, Lafleur M, Gaboury L, Tabrizian M, Kaou N, Atkin M, Vuong T, Batist G, Beauchemin N, Radzioch D, Martel S (2016) Magneto-aerotactic bacteria deliver drug-containing nanoliposomes to tumour hypoxic regions. Nat Nanotechnol 11:941–947. https://doi.org/10.1038/nnano.2016.137

Forbes NS (2010) Engineering the perfect (bacterial) cancer therapy. Nat Rev Cancer 10:785–794. https://doi.org/10.1038/nrc2934

Ganai S, Arenas RB, Sauer JP, Bentley B, Forbes NS (2011) In tumors Salmonella migrate away from vasculature toward the transition zone and induce apoptosis. Cancer Gene Ther 18:457–466. https://doi.org/10.1038/cgt.2011.10

Ghiringhelli F, Larmonier N, Schmitt E, Parcellier A, Cathelin D, Garrido C, Chauffert B, Solary E, Bonnotte B, Martin F (2004) CD4+CD25+ regulatory T cells suppress tumor immunity but are sensitive to cyclophosphamide which allows immunotherapy of established tumors to be curative. Eur J Immunol 34:336–344. https://doi.org/10.1002/eji.200324181

Gu Q-Y, Zhang J, Feng Y-C (2016) Role of NLRP3 inflammasome in Bifidobacterium longum-regulated visceral hypersensitivity of postinfectious irritable bowel syndrome. Artif Cells Nanomed Biotechnol 44:1933–1937. https://doi.org/10.3109/21691401.2015.1111238

Hidaka A, Hamaji Y, Sasaki T, Taniguchi S, Fujimori M (2007) Exogenous cytosine deaminase gene expression in Bifidobacterium breve I-53-8w for tumor-targeting enzyme/prodrug therapy. Biosci Biotechnol Biochem 71:2921–2926. https://doi.org/10.1271/bbb.70284

Hu Q, Wu M, Fang C, Cheng C, Zhao M, Fang W, Chu PK, Ping Y, Tang G (2015) Engineering nanoparticle-coated bacteria as Oral DNA vaccines for cancer immunotherapy. Nano Lett 15:2732–2739. https://doi.org/10.1021/acs.nanolett.5b00570

Iwasaki M, Yu WG, Uekusa Y, Nakajima C, Yang YF, Gao P, Wijesuriya R, Fujiwara H, Hamaoka T (2000) Differential IL-12 responsiveness of T cells but not of NK cells from tumor-bearing mice in IL-12-responsive versus - unresponsive tumor models. Int Immunol 12:701–709. https://doi.org/10.1093/intimm/12.5.701

Jain RK, Stylianopoulos T (2010) Delivering nanomedicine to solid tumors. Nat Rev Clin Oncol. https://doi.org/10.1038/nrclinonc.2010.139

Jang SH, Wientjes MG, Lu D, Au JLS (2003) Drug delivery and transport to solid tumors. Pharm Res. https://doi.org/10.1023/A:1025785505977

Kim K-A, Jung I-H, Park S-H, Ahn Y-T, Huh C-S, Kim D-H (2013) Comparative analysis of the gut microbiota in people with different levels of ginsenoside Rb1 degradation to compound K. PLoS One 8:e62409. https://doi.org/10.1371/journal.pone.0062409

Kojima M, Zhang Z, Nakajima M, Fukuda T (2012) High efficiency motility of bacteria-driven liposome with raft domain binding method. Biomed Microdevices 14:1027–1032. https://doi.org/10.1007/s10544-012-9711-2

Kubiak AM, Poehlein A, Budd P, Kuehne SA, Winzer K, Theys J, Lambin P, Daniel R, Minton NP (2015) Complete genome sequence of the nonpathogenic soil-dwelling bacterium clostridium sporogenes strain NCIMB 10696. Genome Announc 3. https://doi.org/10.1128/genomeA.00942-15

Kupz A, Curtiss R, Bedoui S, Strugnell RA (2014) In vivo IFN-γ secretion by NK cells in response to Salmonella typhimurium requires NLRC4 inflammasomes. PLoS One 9:e97418. https://doi.org/10.1371/journal.pone.0097418

Le HN, Lee NC, Tsung K, Norton JA (2001) Pre-existing tumor-sensitized T cells are essential for eradication of established tumors by IL-12 and cyclophosphamide plus IL-12. J Immunol 167:6765–6772. https://doi.org/10.4049/jimmunol.167.12.6765

Lee J-M, Yoon T-J, Cho Y-S (2013) Recent developments in nanoparticle-based siRNA delivery for cancer therapy. Biomed Res Int 2013:782041. https://doi.org/10.1155/2013/782041

Lee C-H, Lin S-T, Liu J-J, Chang W-W, Hsieh J-L, Wang W-K (2014) Salmonella induce autophagy in melanoma by the downregulation of AKT/mTOR pathway. Gene Ther 21:309–316. https://doi.org/10.1038/gt.2013.86

Leigh ND, Bian G, Ding X, Liu H, Aygun-Sunar S, Burdelya LG, Gudkov AV, Cao X (2014) A flagellin-derived toll-like receptor 5 agonist stimulates cytotoxic lymphocyte-mediated tumor immunity. PLoS One 9:e85587. https://doi.org/10.1371/journal.pone.0085587

Liu Y, Zhou M, Luo D, Wang L, Hong Y, Yang Y, Sha Y (2012) Bacteria-mediated in vivo delivery of quantum dots into solid tumor. Biochem Biophys Res Commun 425:769–774. https://doi.org/10.1016/j.bbrc.2012.07.150

Liu B, Jiang Y, Dong T, Zhao M, Wu J, Li L, Chu Y, She S, Zhao H, Hoffman RM, Jia L (2016) Blockage of autophagy pathway enhances Salmonella tumor-targeting. Oncotarget 7:22873–22882. https://doi.org/10.18632/oncotarget.8251

Luo CH, Huang CT, Su CH, Yeh CS (2016) Bacteria-mediated hypoxia-specific delivery of nanoparticles for tumors imaging and therapy. Nano Lett 16:3493–3499. https://doi.org/10.1021/acs.nanolett.6b00262

Macatonia SE, Hosken NA, Litton M, Vieira P, Hsieh CS, Culpepper JA, Wysocka M, Trinchieri G, Murphy KM, O'Garra A (1995) Dendritic cells produce IL-12 and direct the development of Th1 cells from naive CD4+ T cells. J Immunol 154(10):5071–5079

Malmgren RA, Flanigan CC (1955) Localization of the vegetative form of Clostridium tetani in mouse tumors following intravenous spore administration. Cancer Res 15:473–478

Matsumara Y, Maeda H (1986) A new concept for macromolecular therapeutics in cancer chemotherapy: mechanism of tumoritropic accumulation of proteins and the antitumor agent smancs. Cancer Res 46:6387–6392

Minchinton AI, Tannock IF (2006) Drug penetration in solid tumours. Nat Rev Cancer. https://doi.org/10.1038/nrc1893

Minton NP (2003) Clostridia in cancer therapy. Nat Rev Microbiol 1:237–242. https://doi.org/10.1038/nrmicro777

Nakamura T, Sasaki T, Fujimori M, Yazawa K, Kano Y, Amano J, Taniguchi S (2002) Cloned cytosine deaminase gene expression of Bifidobacterium longum and application to enzyme/prodrug therapy of hypoxic solid tumors. Biosci Biotechnol Biochem 66:2362–2366. https://doi.org/10.1271/bbb.66.2362

Nia HT, Liu H, Seano G, Datta M, Jones D, Rahbari N, Incio J, Chauhan VP, Jung K, Martin JD, Askoxylakis V, Padera TP, Fukumura D, Boucher Y, Hornicek FJ, Grodzinsky AJ, Baish JW, Munn LL, Jain RK (2017) Solid stress and elastic energy as measures of tumour mechanopathology. Nat Biomed Eng 1. https://doi.org/10.1038/s41551-016-0004

Oliveira Silva E, Cruz de Carvalho T, Parshikov IA, Alves dos Santos R, Silva Emery F, Jacometti Cardoso Furtado NA (2014) Cytotoxicity of lapachol metabolites produced by probiotics. Lett Appl Microbiol 59:108–114. https://doi.org/10.1111/lam.12251

Onizuka S, Tawara I, Shimizu J, Sakaguchi S, Fujita T, Nakayama E (1999) Tumor rejection by in vivo administration of anti-CD25 (interleukin-2 receptor α) monoclonal antibody. Cancer Res 59:3128–3133

Park D, Park SJ, Cho S, Lee Y, Lee YK, Min J-J, Park BJ, Ko SY, Park J-O, Park S (2014) Motility analysis of bacteria-based microrobot (bacteriobot) using chemical gradient microchamber. Biotechnol Bioeng 111:134–143. https://doi.org/10.1002/bit.25007

Park BW, Zhuang J, Yasa O, Sitti M (2017) Multifunctional bacteria-driven microswimmers for targeted active drug delivery. ACS Nano 11:8910–8923. https://doi.org/10.1021/acsnano.7b03207

Park W, Cho S, Kang D, Han JH, Park JH, Lee B, Lee J, Kim DH (2020) Tumor microenvironment targeting nano–bio emulsion for synergistic combinational X-ray PDT with oncolytic bacteria therapy. Adv Healthc Mater 9. https://doi.org/10.1002/adhm.201901812

Phan TX, Nguyen VH, Duong MT-Q, Hong Y, Choy HE, Min J-J (2015) Activation of inflammasome by attenuated Salmonella typhimurium in bacteria-mediated cancer therapy. Microbiol Immunol 59:664–675. https://doi.org/10.1111/1348-0421.12333

Rai N, Singh AK, Keshri PK, Barik S, Kamble SC, Singh SK, Kumar R, Mishra P, Kotiya D, Gautam V (2020) Probiotics for Management of Gastrointestinal Cancers. In Probiotic Research in Therapeutics, edited by Indu Pal Kaur and Parmeet Kaur Deol, 1st ed. 373. Singapore: Springer Singapore. https://doi.org/10.1007/978-981-15-8214-1

Reddy BS (1999) Possible mechanisms by which pro- and prebiotics influence colon carcinogenesis and tumor growth. J Nutr 129:1478S–1482S. https://doi.org/10.1093/jn/129.7.1478S

Roberts NJ, Zhang L, Janku F, Collins A, Bai R-Y, Staedtke V, Rusk AW, Tung D, Miller M, Roix J, Khanna KV, Murthy R, Benjamin RS, Helgason T, Szvalb AD, Bird JE, Roy-Chowdhuri S, Zhang HH, Qiao Y, Karim B, McDaniel J, Elpiner A, Sahora A,

Lachowicz J, Phillips B, Turner A, Klein MK, Post G, Diaz LA, Riggins GJ, Papadopoulos N, Kinzler KW, Vogelstein B, Bettegowda C, Huso DL, Varterasian M, Saha S, Zhou S (2014) Intratumoral injection of Clostridium novyi-NT spores induces antitumor responses. Sci Transl Med 6:249ra111. https://doi.org/10.1126/scitranslmed.3008982

Saccheri F, Pozzi C, Avogadri F, Barozzi S, Faretta M, Fusi P, Rescigno M (2010) Bacteria-induced gap junctions in tumors favor antigen cross-presentation and antitumor immunity. Sci Transl Med 2:44ra57. https://doi.org/10.1126/scitranslmed.3000739

Saggar JK, Yu M, Tan Q, Tannock IF (2013) The tumor microenvironment and strategies to improve drug distribution. Front Oncol 3:154. https://doi.org/10.3389/fonc.2013.00154

Shear BMJ, Turner FC, Perrault A, Shovelton T (1943) Chemical treatment of tumors. V. Isolation of the hemorrhage-producing fraction from Serratia marcescens (Bacillus prodigiosus) Culture Filtratel. J Natl Cancer Inst 4:81–97

Shilling DA, Smith MJ, Tyther R, Sheehan D, England K, Kavanagh EG, Redmond HP, Shanahan F, O'Mahony L (2007) Salmonella typhimurium stimulation combined with tumour-derived heat shock proteins induces potent dendritic cell anti-tumour responses in a murine model. Clin Exp Immunol 149:109–116. https://doi.org/10.1111/j.1365-2249.2007.03393.x

Shimizu J, Yamazaki S, Sakaguchi S (1999) Induction of tumor immunity by removing CD25 +CD4+T cells: a common basis between tumor immunity and autoimmunity. J Immunol 163:5211–5218

Song S, Vuai MS, Zhong M (2018) Role of bacteria in cancer therapy. Infect Agent Cancer 13:1–7. https://doi.org/10.1186/s13027-018-0180-y

Suh S, Jo A, Traore MA, Zhan Y, Coutermarsh-Ott SL, Ringel-Scaia VM, Allen IC, Davis RM, Behkam B (2019) Nanoscale bacteria-enabled autonomous drug delivery system (NanoBEADS) enhances intratumoral transport of nanomedicine. Adv Sci 6:1801309. https://doi.org/10.1002/advs.201801309

Uchugonova A, Zhang Y, Salz R, Liu F, Suetsugu A, Zhang L, Koenig K, Hoffman RM, Zhao M (2015) Imaging the different mechanisms of prostate Cancer cell-killing by tumor-targeting Salmonella typhimurium A1-R. Anticancer Res 35:5225–5229

Van Mellaert L, Barbé S, Anné J (2006) Clostridium spores as anti-tumour agents. Trends Microbiol 14:190–196. https://doi.org/10.1016/j.tim.2006.02.002

Vautier AH (1813) General perspectives on the cancer illness. The Áse du Paris 43

Wang S, Chen K-J, Wu T-H, Wang H, Lin W-Y, Ohashi M, Chiou P-Y, Tseng H-R (2010) Photothermal effects of supramolecularly assembled gold nanoparticles for the targeted treatment of cancer cells. Angew Chem Int Ed Engl 49:3777–3781. https://doi.org/10.1002/anie.201000062

Wiemann B, Starnes CO (1994) Coley's toxins, tumor necrosis factor and cancer research: a historical perspective. Pharmacol Ther 64:529–564. https://doi.org/10.1016/0163-7258(94)90023-X

Wu B-B, Yang Y, Xu X, Wang W-P (2016) Effects of Bifidobacterium supplementation on intestinal microbiota composition and the immune response in healthy infants. World J Pediatr 12:177–182. https://doi.org/10.1007/s12519-015-0025-3

Xiong MH, Bao Y, Du XJ, Tan ZB, Jiang Q, Wang HX, Zhu YH, Wang J (2013) Differential anticancer drug delivery with a nanogel sensitive to bacteria-accumulated tumor artificial environment. ACS Nano 7:10636–10645. https://doi.org/10.1021/nn403146t

Yoo JW, Irvine DJ, Discher DE, Mitragotri S (2011) Bio-inspired, bioengineered and biomimetic drug delivery carriers. Nat Rev Drug Discov. https://doi.org/10.1038/nrd3499

Zhai Y, Su J, Ran W, Zhang P, Yin Q, Zhang Z, Yu H, Li Y (2017) Preparation and application of cell membrane-camouflaged nanoparticles for cancer therapy. Theranostics 7:2575–2592. https://doi.org/10.7150/thno.20118

Zhao M, Yang M, Li XM, Jiang P, Baranov E, Li S, Xu M, Penman S, Hoffman RM (2005) Tumor-targeting bacterial therapy with amino acid auxotrophs of GFP-expressing Salmonella typhimurium. Proc Natl Acad Sci U S A 102:755–760. https://doi.org/10.1073/pnas.0408422102

Zheng DW, Chen Y, Li ZH, Xu L, Li CX, Li B, Fan JX, Cheng SX, Zhang XZ (2018) Optically-controlled bacterial metabolite for cancer therapy. Nat Commun 9. https://doi.org/10.1038/s41467-018-03233-9

Zoaby N, Shainsky-Roitman J, Badarneh S, Abumanhal H, Leshansky A, Yaron S, Schroeder A (2017) Autonomous bacterial nanoswimmers target cancer. J Control Release 257:68–75. https://doi.org/10.1016/j.jconrel.2016.10.006

Chapter 15
Recent Progress on Nanostructured Materials for Biomedical Applications

Sudip Mondal, Sumin Park, Jaeyeop Choi, and Junghwan Oh

Abstract The focus of this chapter is to explore the progress of nanostructured materials and their potential biomedical applications. Nanotechnology offers incredible opportunities in manipulating chemical and biological entities at the nano scale level. The nanoscience plays a key role in technological development for biomedical applications, especially in the areas of preclinical diagnosis, non- or minimal invasive biomedical imaging, drug discovery, and drug delivery. This research work discusses a brief history of nanotechnology, different synthetic routes, characterizations, fundamental concepts regarding morphologies, characteristics, biological interactions, and clinical applications. A few nanoparticles such as metal nanoparticles (Au, Ag, Pt, etc.), magnetic and metal oxide nanoparticles (Fe_3O_4, Fe_2O_3, ZnO, TiO_2, etc.), quantum dots (CdTe, Cds, etc.), mesoporous silica nanoparticles, carbon nano tubes (CNT, SWNT, etc.), ceramics nano materials (apatite, hydroxyapatite, bio-glass, etc.), polymeric nanoparticles (polypyrrole,

S. Mondal
Department of Biomedical Engineering, Pukyong National University, Busan, Republic of Korea

S. Park
Industry 4.0 Convergence Bionics Engineering, Pukyong National University, Busan, Republic of Korea

J. Choi
Industry 4.0 Convergence Bionics Engineering, Pukyong National University, Busan, Republic of Korea

Ohlabs Corp, Busan, Republic of Korea

J. Oh (✉)
Department of Biomedical Engineering, Pukyong National University, Busan, Republic of Korea

Industry 4.0 Convergence Bionics Engineering, Pukyong National University, Busan, Republic of Korea

Ohlabs Corp, Busan, Republic of Korea
e-mail: jungoh@pknu.ac.kr

N. R. Maddela et al. (eds.), *Nanotechnology for Advances in Medical Microbiology*, Environmental and Microbial Biotechnology,
https://doi.org/10.1007/978-981-15-9916-3_15

β-cyclodextrin, chitosan, fucoidan, etc.) are widely used in biomedical field. In this article, we present the recent trend and challenges in the advances of nanomaterials for clinical applications. This review might be considered as a general guide and will help the readers to find key information regarding the recent advances in nanomedicine.

Keywords Nanomaterials · Metal nanoparticles · Ceramics nanoparticles · Polymeric nanoparticles · Cancer treatment · Tissue engineering · Biomedical application

15.1 A Brief History of Nanotechnology

The advances in nanomaterials research and applications with pharmaceutical and biomedical applications have been increased remarkably. To realize the importance of nanotechnology, the chronological achievements of this science need to be considered always. The journey of nanoscience and technology begins since the fourth century (The Lycurgus Cup stained with colloidal gold silver), ninth–seventeenth century which evidenced on in the of Europe's medieval stained windows glass in cathedrals. With progress in time, the development of nanotechnology tremendously accelerated to a new dimension where a new era of nanotechnology started. Nanotechnology represents an escalating research area, including materials, objects, and systems with improved characteristics and functions due to the special tailored preparation in 1–100 nm scale usually. A chronological development of nanoscience and technology has been represented in a tabular form (Table 15.1). The nanoscience and technology grow its arms to multifunctional application starting from energy, healthcare, environment, space, oceans, information and communication, optical, materials, and applied field of research. Recently, a steadily growth observed in nanotechnology for different multidisciplinary healthcare applications such as targeted drug delivery, hyperthermia, photothermal and photodynamic therapy, bioimaging, biosensors, and so on (Mondal and Oh 2019).

15.2 Introduction to Nanoparticles for Biomedical Applications

Nanotechnology is subject to contribute in every field of science, including physics, chemistry, materials science, biology, medical science, computer science, and engineering. With the swift advancement of nanotechnology, drug delivery study extends its opportunities for enhanced therapeutic delivery. Due to their exclusive structure and superior properties, nanomaterials can successfully deliver therapeutics such as drugs, proteins, biomolecules, peptides, genes, nucleic acids, etc. In the field

Table 15.1 A chronological development of nanoscience and nanotechnology (https://www.nano.gov/timeline)

Year	Progress on nanoscience and nanotechnology
Fourth century	The Roman Lycurgus cup found stained in this ancient era with colloidal gold silver nanoparticles.
Ninth¯seventeenth century	The vibrant colors of stained glass in the windows of Europe's medieval cathedrals are due to metallic nanoparticles.
1857	Michael Faraday discovered nano ruby gold colloidal solution.
1936	Invention of the field emission microscope by Erwin Müller, which explore the future of nano.
1947	The semiconductor transistor by J. Bardeen, W. Shockley, and W. Brattain.
1950	The process for monodisperse colloidal materials reported by Victor La Mer and R. Dinegar.
1951	Discovery of the field ion microscope by Erwin Müller.
1958	Design and fabrication of the first integrated circuit by Kilby (Nobel Prize winner 2000).
1959	Richard Feynman gave the first lecture on atomic scale, "There's Plenty of Room at the Bottom" at an American Physical Society meeting at Caltech.
1974	Professor Norio Taniguchi coined the term nanotechnology.
1981	Gerd Binnig and Heinrich Rohrer (Nobel prize winner 1986) at IBM's Zurich lab invented the scanning tunneling microscope.
1981	Alexei Ekimov discovered nanocrystalline, semiconductor quantum dots.
1985	Rice University researchers Harold Kroto, Sean O'Brien, Robert Curl, and Richard Smalley (Nobel prize winner 1996) discovered the Buckminsterfullerene (C60).
1985	Louis Brus discovered colloidal semiconductor quantum dots.
1986	The atomic force microscope invented by G. Binnig, C. Quate, and C. Gerber.
1990	Nanotechnology companies starts. To name a few Nanophase Technologies (1989), Helix Energy Solutions Group (1990), Zyvex (1997), Nano-Tex (1998) etc. and so on.
1991	Discovery of the carbon nanotube (CNT) by Sumio Iijima.
1992	Invention of nanostructured catalytic materials MCM-41 and MCM-48.
1993	Controlled synthesis of quantum dots was reported by Moungi Bawendi.
1998	National Science and Technology Council established the Interagency Working Group on Nanotechnology (IWGN).
1999	Dip-pen nanolithography® (DPN®) was invented by Chad Mirkin. The application of this tool was meant for "writing" electronic circuits for biomedical research.
2000	National Nanotechnology Initiative (NNI) was initiated by President Clinton.
2004	Britain published "Nanoscience and Nanotechnologies: Opportunities and Uncertainties" by Royal Society and the Royal Academy of Engineering.
2005	DNA-based computation was proposed by Erik Winfree and Paul Rothemund.

(continued)

Table 15.1 (continued)

Year	Progress on nanoscience and nanotechnology
2006	First nanoscale car was made by James Tour and colleagues built with oligo (phenylene ethynylene), alkynyl axles, and four spherical C60 fullerene wheels.
2007	First report of non-harmful virus mediated lithium-ion battery by Angela Belcher and colleagues at MIT.
2008	Nanotechnology-Related first Environmental, Health, and Safety (EHS) Research was published by NNI.
2010	DNA-like robotic nanoscale assembly devices was fabricated by Nadrian Seeman and colleagues.
2012	Nanotechnology Signature Initiatives (NSIs) Nanosensors, and the Nanotechnology Knowledge Infrastructure (NKI) was initiated by NNI.
2014	The updated 2014 strategic plan was released by NNI on environmental, health, and safety.
2015–2020	An era of tremendous development and improvement of techniques for medical diagnosis and imaging, targeted drug delivery, and hyperthermia (magnetic/photothermal) for biomedical application. Application of nanotechnology for military aid such as battle suits with advanced material as bulletproof or resistant to bacterial chemical attacks. Environmental technology by empowering catalysis mediated reactions which might reduce the toxic materials as effluents. In space technology by building heat proof tiles, optoelectronic instruments for communication, etc.
Future of nano	Converging technologies with multidisciplinary fields to initiate nanoscale or nano-system based foundation for future requirements. To explain such plan, we can consider nanorobot mediated molecular level therapeutic agent which can pre-diagnose and cure the problems. Such ideas are non-exhaustive, for example, carbon nanotube cables for space elevator, advanced nano-opto electronics device for faster communication, nano level replicators help to build exact atomic level mimicry, etc.

of nanotechnology research many nanoparticles already approved by FDA for clinical applications. To name a few superparamagnetic iron oxide nanoparticles (SPION), gadolinium nanoparticles, etc. are already approved for MRI contrast agents. Gold, silver, palladium nanoparticles with tailored shapes and sizes have been extensively studied for bioimaging, biosensing, drug targeting, photothermal therapy applications for cancer treatment (Fig. 15.1) (Bharathiraja et al. 2018; Manivasagan et al. 2019a, b; Phan et al. 2020; Prasad et al. 2016, 2020).

The definition of nanoparticles is depending upon its application and properties. But the most common characteristics for all nanomaterial are its size range. According to the National Nanotechnology Initiative (NNI) nanoparticles sizes are considered in between the range of 1 and 100 nm in general. Whereas many researchers claimed up to 1000 nm might be considered in the range of nano. The foremost benefits of nanostructured materials are its enhanced surface-to-volume ratio. For the last two decades, a steady development in biomedical application of nanoparticles specifically in photoablation therapy, targeted drug delivery, and

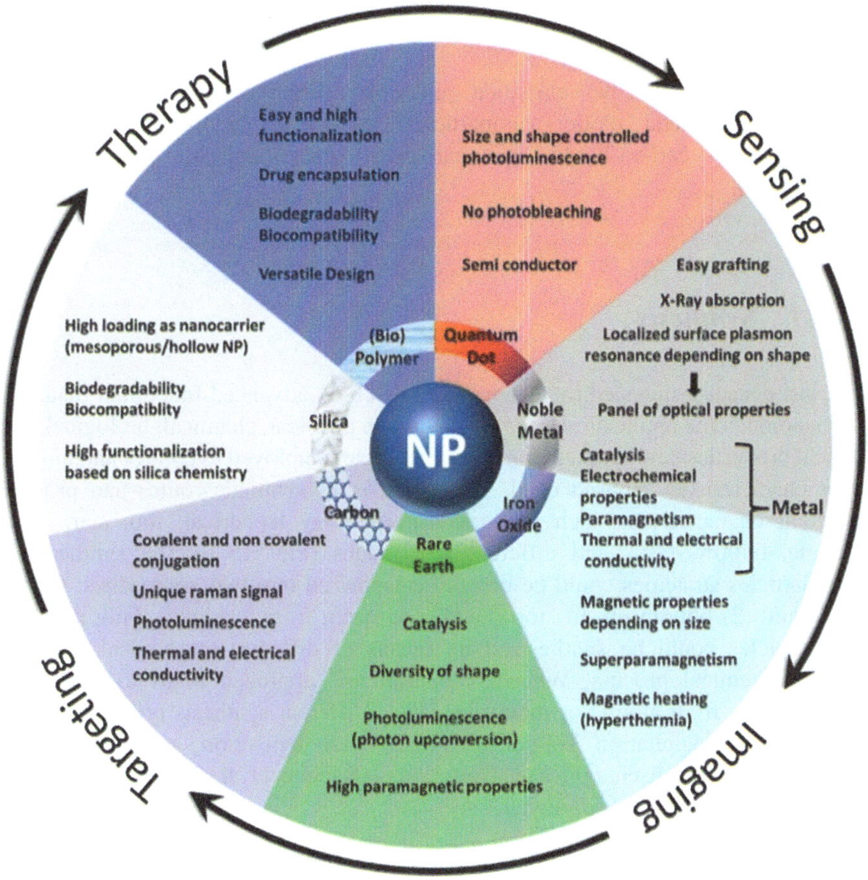

Fig. 15.1 Nanostructured materials for biomedical applications (De Crozals et al. 2016)

bioimaging purposes is observed (Thangadurai et al. 2020a, b). Different ceramics, metallic, and polymeric nanoparticles have shown successful promising results with excellent biocompatibility, chemical stability, non-toxicity, and other associated properties (Prasad et al. 2017). The reliability of nanostructured inorganic materials is much better due to its efficiency, low toxicity, targeted delivery, and tailored morphology. Till date any gene delivery agent has not approved by Food and Drug Administration (FDA) due to its uncertain long-term toxicity. Different synthesis techniques such as chemical, physical, biological approaches were adopted to synthesize tailored nanoparticles (Prasad et al. 2016, 2018; Srivastava et al. 2021). Although researchers have already reported a large number of synthetic routes but most of the techniques lack of reproducibility, and low yield. Moreover, difficult synthesis routes are often making limitations to get better nanoparticles. In this article we have classified nanoparticles in six groups and discussed their different synthesis routes.

15.3 Classification of Nano-systems

Nanoparticles could be classified into six different nano-systems metal nanoparticles, metal oxide nanoparticles, polymeric nanoparticles, ceramics nanoparticles, carbon-based nanoparticles, and composite nanoparticles (Table 15.2).

15.3.1 Synthesis

The last decade evidenced the extensive research on advanced functional materials for nanomedicine application. Due to the unique physical, chemical, biological, and optical properties new synthetic methods have been employed to control and tailored their characteristics (Prasad et al. 2016). Different synthetic routes can produce different characteristics such as varied morphology (spherical, rod, star, cube, triangle, hollow, etc.), and different size ranges (Fig. 15.2). The synthesis of nanoparticles strategies could be broadly categorized into two approaches: (1) top down and (2) bottom up. In "top down" approach, from a primary bulk material, nanoparticles could be synthesized by means of different mechanical, thermal, optical, chemical process. Whereas "bottom up" approach helps to synthesize nanoparticles from atomic or molecular level. Different synthesis procedures such as chemical precipitation, sol-gel, chemical vapor deposition, atomic molecular condensation, arc discharge, and evaporation (Bachilo et al. 2002; Hafner et al. 1998) were successfully employed.

Table 15.2 Classifications of nanoparticles

Nanoparticle type	Examples	Reference
Metal nanoparticles	Au, Ag, Pt, Fe, Zn, Cu, Mn, Co, etc.	Kim et al. (2018a), McNamara and Tofail (2017)
Metal oxide nanoparticles	TiO_2, Al_2O_3, SiO_2, ZnO, Fe_3O_4, Fe_2O_3, etc.	McNamara and Tofail (2017), Qin et al. (2011)
Polymeric nanoparticles	Polypyrrole, chitosan, dendrimer, liposome, etc.	Bharathiraja et al. (2018), Manivasagan et al. (2018), Mondal et al. (2020a) Phan et al. (2018)
Ceramics nanoparticles	Calcium phosphate, hydroxyapatite, bio-glass, etc.	Mondal et al. (2018a, 2020c)
Carbon-based nanoparticles	Carbon nanotube, SWNT, MWNT, etc.	Cherukuri et al. (2004), Mondal et al. (2019a)
Composite nanoparticles	Composite of above materials together	McNamara and Tofail (2017), Mondal et al. (2016b)

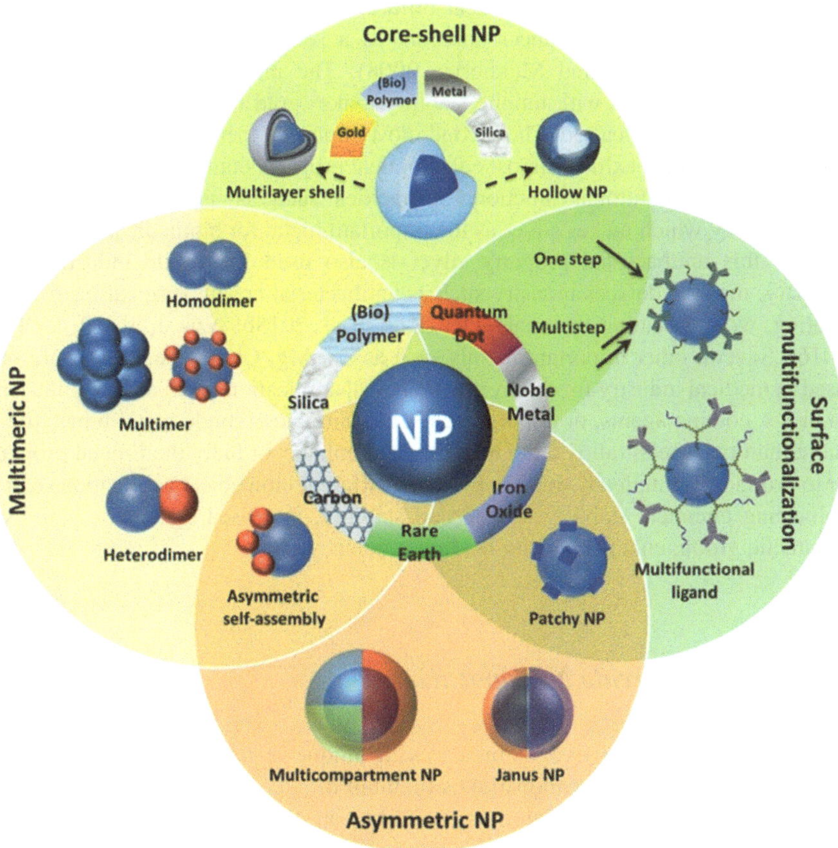

Fig. 15.2 Different morphologies for nanostructured materials and their multifunctional applications (De Crozals et al. 2016)

15.3.2 Metallic Nanoparticles

As long as for 5000 years ago gold served as novel metal for different medicinal purposes. Gold (Au) categorized under metallic nanoparticles exhibits exclusive electronic and optical properties along with strong chemical stability. Gold also possess most active surface functionality which allows different positively charged materials to bind on its surface by strong ionic interactions (Manivasagan et al. 2019c, d). For fabricating contrast agents strong optical absorption of noble metals with surface plasmon resonance (SPR) property is important (Liao et al. 2006). The fundamental phenomenon of metal based nanoparticle for biomedical application is its surface plasmon resonance (Mondal et al. 2017b). With different morphologies (rod, cube, caged, star, triangle, spherical, etc.) the absorption spectra also changed, and due to this property, the application of gold nanoparticle is always demanding.

The absorption in the visible and near-infrared regions already utilized for several photothermal therapy for cancer ablation, as a sensor, and biomedical imaging purpose (Bhattacharya and Mukherjee 2008). The most important factor is its versatile morphologies with unique stability makes gold is a first choice for metal based medicinal application. The second most important metal nanoparticle is silver nanoparticles which exhibit almost equal optical property compared to similar group of metals and additional functional properties such as antibacterial and anti-inflammatory which makes silver as the important metal for biomedical application. Due to this antibacterial property silver is also used for textile industry, food industry, medical devices industry such as antibacterial cream, surgical instruments coating, wound healing materials, etc. (Kim et al. 2018b; Aziz et al. 2014, 2015, 2016). Several other important metals such as Zn, Mg, Cu, Pt, Fe, Co, Cr are also used in medical industry for different biomedical applications such as micronutrients in drugs, coating agents, or doping agents, etc. Similar to single metal bimetallic or more than two metal (alloys) are also used for enhancing individual metal property or to produce a completely new property for different biomedical application such as enhancing chemical stability, or thermal stability, increased enzymatic activity a prosthetic group, enhancing drug loading efficiency etc.

15.3.3 Metal Oxide Nanoparticles

Metal oxides are derivative of metals with unique chemical properties. Among widely used metal oxide nanoparticles titanium dioxide (TiO_2), zinc oxide (ZnO), mesoporous silica (SiO_2), cerium oxide (CeO_2) are the most important. The application of those material is not only restricted in a particular field, whereas a wide application is observed including food color agent, sunscreen, tooth paste, pharmaceutical and cosmetics industry. Recently, CeO_2 nanoparticles are as potential biological antioxidant and anticancer agent along with its extended application in biosensor (Alpaslan et al. 2015). Another important nanoparticle is silica-based hybrid nanomaterials.

The use of SiO_2 nanoparticles is well known for greater surface area which could accommodate large number of drug molecules. This type of nanoparticles also could be useful for triggered and controlled drug delivery application due to its tailored morphology including variable pore structure, particle size, and tunable biocompatibility (Moorthy et al. 2018, 2019). Research study suggested that silica particles can successfully carry and deliver the therapeutic or imaging cargoes (Moorthy et al. 2017). As a biomolecule carrier agent ZnO is also in the priority list. Its unique electronic properties make this molecule a bioimaging contrast enhancer. Though its only drawback is its solubility in biological fluids. To overcome such limitations different coating materials are employed to coat the ZnO surface to protect them in biological system (Ngo et al. 2009).

15.3.4 Quantum Dots

Quantum dots (QDs) are tiny fluorescent semiconductor nanocrystals (specifically 1–10 nm in range) with exclusive optical properties (Choi et al. 2007). QDs possess more stability compared with organic dyes and fluorescent proteins, while their brightness could be controlled by different synthetic process. The other exclusive characteristics are narrow linewidth emission spectra, comparatively long (5 to >100 ns) fluorescence lifetime (1–5 ns for organic dyes), and negligible photobleaching over minutes to hours (Resch-Genger et al. 2008). The luminescence property of QDs is widely used as a tagging molecule for medical imaging applications. Synthetic techniques directly control the absorption and emission characteristics and optical properties of QDs. Different synthetic routes were reported by researchers for the synthesis of cadmium selenide (CdSe), cadmium sulfide (CdS), or cadmium telluride (CdTe) QDs (Murray et al. 1993). QDs photoluminescence efficiency could be enhanced by different core shell techniques such as CdSe core and a ZnS shell that shields the core (Talapin et al. 2004). Use of ZnS capping also enhances the stability of QDs by decreasing the oxidative photobleaching.

15.3.5 Iron Oxide Nanoparticles

One of the most commonly used nanoparticles in biomedical research is iron oxide nanoparticles. Iron oxide is a common name, whereas different oxidation states of the nanoparticles may change its property and corresponding applications. To discuss the oxidation levels, we can categorize including iron (II) oxide (FeO), iron (III) oxide Fe_2O_3 and Fe_3O_4. Different crystalline polymorphs are observed for iron (III) oxide (Fe_2O_3) [α-Fe_2O_3, β-Fe_2O_3, γ-Fe_2O_3 and ε-Fe_2O_3]. For biomedical application maghemite (γ-Fe_2O_3) and magnetite (Fe_3O_4) are the most suitable material, whereas in terms of super paramagnetic iron oxide (SPION) nanomaterial magnetite (Fe_3O_4) is the best choice as its facile synthesis process and high magnetic saturation with low remanence and coercivity (Mondal et al. 2017a; Abd-Elsalam et al. 2019). Different doping activities reported to enhance the magnetic as well as biological properties. Such doping is accompanied with magnetically susceptible elements such as cobalt (Co), manganese (Mn), and nickel (Ni) (McNamara and Tofail 2017).

15.3.6 Carbon-Based Nanoparticles

The main constituent of organic molecule is carbon. So, for biomedical application there is no substitute of carbon in respect of favorable biochemical properties. The advanced nanotechnology research helps us to tailor the morphology of carbon nanomaterials with superior characteristics. Different morphologies with carbon nanoparticles such as carbon nanotubes (CNTs) have expand their application in multidisciplinary fields. The advanced research reports different surface modification of carbon-based nanoparticles with biological molecules has increased their use in drug delivery, and bioimaging research. CNTs are the most discussed and promising nanostructure which is basically a monolayered graphite sheet rolled into tubes. Single and multi-walled CNTs structure could be useful for different applications. The CNTs are the most important carbon-based nanostructure due to its exclusive high electrical and thermal conductivity and enhanced tensile strength properties.

15.3.7 Liposomes

Bangham and Horne (Bangham and Horne 1964) first reported about small artificial lipid-bilayer spherical vesicles called "Liposomes." Liposomes are a promising drug delivery carrier especially used in ophthalmic therapy. The synthesis of liposome is also highly tunable compared to size, surface charge, shape, etc. By changing different lipid molecules liposome fluidity and rigidity could easily be controllable. There are several classifications already reported for different liposomes. The classification is primarily based on their size (diameter). The size of liposomes ranges between 10 and 100 nm diameter considered as small uni-lamellar vesicles (SUVs) and ranges above 100–1000 nm considered as large uni-lamellar vesicles (LUVs). There are other classifications also considered for liposomes based on the number of lipid bilayers. Different multiple phospholipid bilayers could be used for multilamellar liposome vesicles in aqueous media with a comparatively bigger size range of 400–3500 nm in diameter (Akbarzadeh et al. 2013). Liposomes have been engineered to deliver a wide range of therapeutic agent or target molecules which could not be possible for conventional nanoparticle system. For example, liposomes can transport a wide range of compounds, such as aqueous soluble or insoluble drugs, antibiotics, enzymes, antioxidants, proteins etc. for different vaccination, therapeutic purposes. Due to protective lipid bilayers the functionality of these carrier molecules is well preserved and the efficiency or circulating time enhanced dramatically even in harsh conditions (Akbarzadeh et al. 2013).

15.3.7.1 Dendrimers

In modern drug delivery application polymeric biomaterials are most promising (Fig. 15.3). Among different polymeric materials dendrimers are the new carriers with highly branched molecules and variable morphologies. The molecular structure could be tailored according to the demand with low polydispersity. The wide range molecular cargo trafficking with tunable size and shape makes dendrimers as a promising biomolecule. The other advantage of this molecule is to carry more than

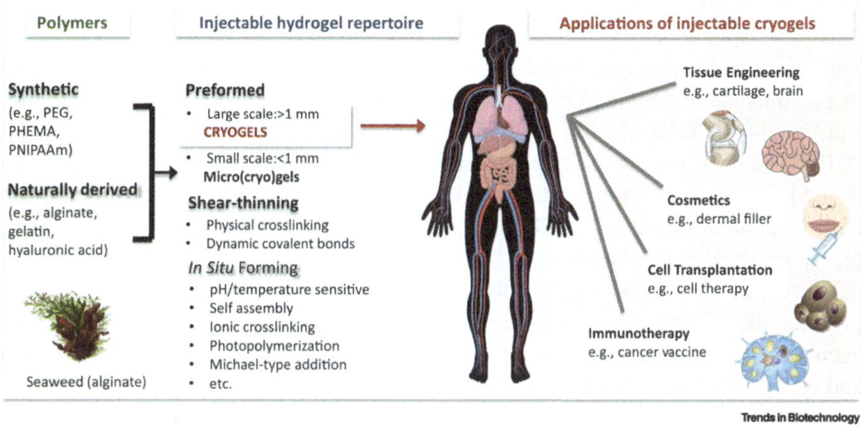

Fig. 15.3 Different biomedical applications of polymeric nanoparticles (Eggermont et al. 2020)

one type of target molecules in specific site. Gadolinium-polyamidoamine (PAMAM) starburst dendrimers were first reported by Wiener et al. which could be useful for MRI contrast agents (Wiener et al. 1994). The long-term effects of dendrimers in body system were studied by Kobayashi et al., who reported that the different sizes of dendrimer show variable blood retention, circulation, and clearance rates (Kobayashi et al. 2003).

15.3.7.2 Lipid-Based Nanoparticles

Most of the oral drugs are based on lipid-based nanoparticles such as liposomes and micelles (Muchow et al. 2008). Lipids contain both hydrophobic and hydrophilic parts with a unique property of self-assembling in aqueous environments. Lipid-based nanoparticle is comparatively a new field of research for bioimaging application. There are very few and small number of successful research work reported till date. To address this relatively new technique Muller et al. first reported the lipid-based nanoparticles for contrast enhancing agent for MRI application (Mulder et al. 2006). Another study was performed by Koole et al. on silica nanoparticles coated with paramagnetic lipid molecule for multimodal imaging purpose. The fabricated nanostructure contains a core quantum dot which acts as a contrast agent for MRI and fluorescence imaging (Koole et al. 2008). Among the few successful study Cressman et al. reported lipid incorporated liposomal nanoparticles labeled with RGD on its surface for imaging biomolecular movements in endothelial cells (Cressman et al. 2009).

15.3.7.3 Apatite Nanoparticles

The most reliable and widely used nanoparticle is calcium phosphate material which received its attention due to extremely low or no toxicity, excellent biocompatibility and bio-absorbability (Liu et al. 2005; Mondal et al. 2016a, 2018a; Mondal and Pal 2019). The most widely used apatite material as bone tissue engineering application is hydroxyapatite. Other different types of apatites are calcium phosphate (CP) dicalcium phosphate (DCP), tri calcium phosphate (TCP), tetra calcium phosphate, etc. Calcium phosphate is mostly used for biomedical purposes such as delivery vehicle for drugs, biotherapeutics, gene, proteins, peptide, DNA, etc. (Maitra 2005). The study with calcium phosphate is not exhaustive as delivery agent only, whereas recent studies suggested its potential use for biomedical imaging and contrast agent also (Mondal et al. 2020d, e).

Mondal et al. reported a rapid microwave assisted facile synthetic technique to synthesize gold loaded HAp nanoparticles (Au-HAp). Further the Au-HAp nanoparticles were coated with collagen and used for doxorubicin drug delivery applications. Though the authors reported different concentration Au loaded HAp but the DOX loading and releasing study was performed with optimized for 0.1 wt% Au–HAp–Col nanoparticles. A high drug loading efficiency of ~58.22% and a pH responsive releasing of ~53% (at pH 4.5) were observed. To evaluate the cytotoxicity osteoblast-like MG-63 cells were studied for AO/PI and MTT assay. The promising nontoxic results extend with scaffold fabrication and cellular attachment study (Fig. 15.4). The overall result qualifies the Au-HAp nanoparticles for drug delivery and tissue engineering application (Mondal et al. 2019b).

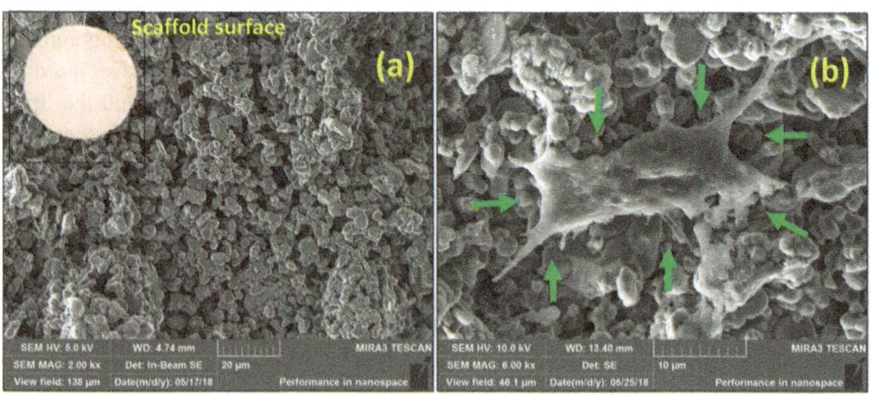

Fig. 15.4 FE-SEM study of (a) 0.1 wt.% Au-HAp-Col scaffold prior to MG-63 cell treatment (b) attached MG-63 cells (green arrow) on 0.1 wt.% Au-HAp-Col fabricated scaffold (Mondal et al. 2019b)

15.4 Biomedical Applications

15.4.1 Sensing

Due to its variable color property with different shape and morphology gold nanoparticles are widely used as a colorimetric sensing agent (Zhao et al. 2008). Different biochemical reactions associated with colloidal gold solution produce different color intensity due to gold nanoparticles unique plasmon resonance effects and that help to clue the sensing property of Au nanoparticles. In simple words the assay is directly depend upon the formation of color due to surface plasmon resonance of Au nanoparticles. The plasmon resonance frequency is controlled by different factors such as shape, morphology, sizes, average distance between gold particles and this might change the color from red to purple or even blue. Mirkin and co-workers first reported the DNA-gold sensors which rely the same color changing phenomenon (Mirkin 2000). This biosensing property is used for detecting various biomolecules, including different proteins, peptides, metals, enzymes, nucleic acids, etc. (Liu and Lu 2004; Singh et al. 2020).

15.4.2 Imaging

Kee and Danila et al. demonstrated that gold nanoparticle is an excellent blood pool contrast agent and targeted imaging of myocardial scar in a rat model of myocardial infarction with CNA35-gold nanoparticles (Fig. 15.5). A molecular imaging approach was taken by gold nanoparticles contrast agent which successfully detects the CT-based specific imaging for myocardial scar.

Magnetic resonance imaging is a noninvasive extremely efficient tool for imaging deep tissues inside body system. Nowadays clinician highly depends upon imaging modules for diagnosis and treatment. With the advances of therapeutic approach deep tissue penetration imaging could be possible by using nanoparticles. Noninvasive is a non-painful technique which could visualize the structure and morphologies of tissues. To better visualize nowadays different nanoparticles have been used as contrast in noninvasive imaging. Among all different nanoparticles iron oxide is the most widely used and FDA approved due to its superparamagnetic property with adequate biocompatibility. To enhance its biocompatibility surface modification could be performed by different polymeric coating agents with antibody, peptide, or small molecule conjugation for active targeting in affected tissues (Wunderbaldinger et al. 2002). Hainfeld et al. reported 1.9 nm size gold nanoparticles mediated contrast agent for detecting tumors in mice by X-ray CT (Hainfeld et al. 2006). After 24 h post injection the gold nanoparticles were not traced in the blood, whereas significant accumulation observed in kidney. Due to extreme small size of gold nanoparticles there was no accumulation evidence found inside liver or spleen.

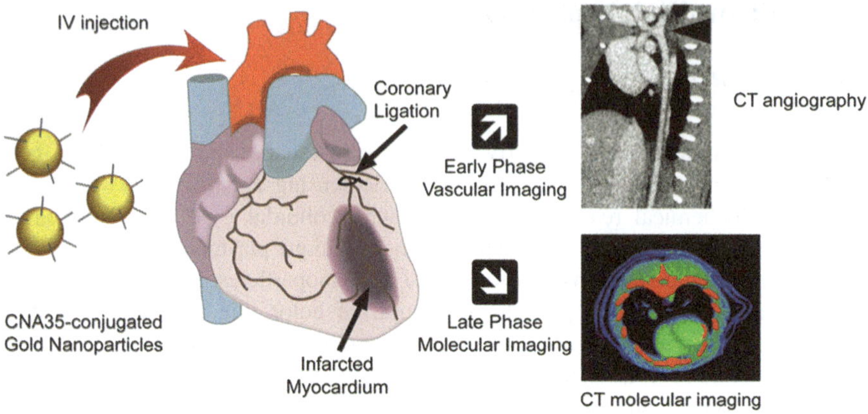

Fig. 15.5 Management of patients suffering from myocardial infarction is based on the extent of coronary artery stenosis and myocardial scar burden. (Upper panel) Gold nanoparticles (AuNPs) functionalized with collagen-binding adhesion protein 35 (CNA35) perform vascular imaging at early phase and molecular imaging at late phase. (Lower panel) Myocardial infarction imaging at 6 hours after injection. Reproduced with permission. Copyright 2018. Elsevier. (Kee and Danila 2018)

For cancer treatment and diagnosis purposes carbon nanotubes successfully synthesized and applied. Several researchers have reported CNTs based bioimaging system applied for detection and destruction for cancer cells. Different fluoroprobes have already been successfully linked with single wall nanotubes (SWNTs) by means of covalent bonding which make possible visible wavelength imaging (Kam et al. 2004). Kam et al. conjugated SWNTs with streptavidin to detect human T cells and promyelocytic leukemia cells by confocal microscopy. Another study reported by Pantarotto et al. synthesized amino-modified SWNTs with FITC in dimethylformamide (Pantarotto et al. 2004) *to detect cancer cells*. The capacity of the FITC-labeled SWNTs confirms after successful images captured by confocal microscopy. Cherukuri et al. also reported the cytotoxicity study of pristine SWNTs. In the reported article NIR imaging was performed by fluorescence microscope and spectrofluorometer (Cherukuri et al. 2004). The fluorescence property of SWNTs has been opened the paves for nanoparticle based promising imaging techniques which could be useful as a powerful tool for tracing diseased or damaged tissues conjugated with CNTs.

15.4.3 Drug Delivery

Drug delivery is an important characteristic for nanomaterials in biomedical application. The affinity between nanoparticles and drug molecule is very important for drug loading and releasing purposes. The strong attraction between drug molecules and the nanocarriers is mainly due to covalent and ionic interactions. With strong interactions drug loading and stability will enhance, whereas for drug releasing purpose these interactions may hamper the releasing efficiency. On the other hand, carrier molecules could give stability and long circulating time inside body system. Drug loading and releasing could be externally controlled by different characteristics such as for releasing purposes external stimuli by ultrasound, heat, magnetic effects, etc. and for loading purposes pH, heat, morphologies, etc. In targeted drug delivery a special type of carrier molecule is necessary such as magnetic nanoparticles. To be specific superparamagnetic nanoparticles are most important. When the external stimuli are applied by magnetic fields the SPIONs will be activated and targeted to specific direction, whereas in absence of the external magnetic field the nanoparticle will behave like nonmagnetic particles. This strategy could help nanoparticles to target in specific location. Also, iron oxide nanoparticles sometime cause toxic effect when circulating inside body system, and to overcome such unwanted toxic effects, nanoparticles are coated with biomolecules such as polymers, ceramics, etc. In general iron oxide nanoparticles are coated with biopolymers such as collagen, poly ethylene glycol, polyvinylpyrrolidone, chitosan, fucoidan, dextran, etc. Moreover, the nanoparticles are also loaded with specific drug to treat and target specific disease to control. Several research studies demonstrate pH dependent drug release capability to target specific tumor cells. The tumor cells have acidic pH compared to normal healthy cells. In this respect mesoporous silica, nano HAp, mediated drug delivery agents are investigated and reported. The release efficiency is also studied by different enzymatic activity, changes in temperature, or in different osmolality environment (Chatterjee et al. 2014). It is always recommended that specific target-oriented drug delivery is most beneficial to manage disease control and treatment. For favorable tumor targeting nanoparticles are always conjugated with target ligands, or receptor molecules complementary to target site. A large number of different targeting receptor and ligand molecules are already studied and reported by several researchers. Nowadays, selection and modification of nanoparticle surface with specific target-oriented biological molecules are well practiced and pharmaceutical companies are running phase II and III clinical trials on this drugs (De Crozals et al. 2016).

15.4.4 Magnetic Hyperthermia

Hyperthermia is a heat mediated therapeutic approach where heat is applied to destroy affected cells and tissues without damaging the healthy cells. The fundamental phenomenon of hyperthermia relies on 41–46 °C mediated cell apoptosis of cancerous cells when heated specifically. Thermal ablation is also another type of hyperthermia where temperature rises more than 50 °C for treatment of affected tissues. In general, three different types (local, regional, and whole-body hyperthermia) were performed to treat cancer. In local hyperthermia, radio frequency, microwave, or ultrasound mediated heat is generated to specific small target site. Large tissue areas are being treated with regional hyperthermia, whereas whole-body hyperthermia is rarely used for treat cancers to encounter metastatic phase. For all this hyperthermia mediated treatment the use of nanoparticle enhances the success rate of treatment. Different magnetic nanoparticles generate different heat energy, so application specific nanomaterials need to be selected for treatment purpose. Nanoparticles are sometime coated with different polymeric materials to enhance its biocompatibility. Magnetic nanoparticles might be injected directly to the tumor or cancer region or might be targeted to specific region by binding with specific target molecules such as specific antibody or peptides.

15.4.5 *Photoablation Therapy: Photodynamic and Photothermal Therapy*

Different approaches are invented to inhibit cancer cells activity and photoablation therapy is one of the most promising techniques to counter act. Photodynamic therapy (PDT) and photothermal therapy (PTT) are the two classification of photoablation therapy. PDT uses photosensitizers a nontoxic light sensitive material or complex which become toxic to targeted cells or tissues upon light exposure. To activate the sensitizer a specific excitation wavelength is required. This therapeutic approach is mainly used for cancer cells management. In the PDT procedure photosensitizers (TiO_2, Ce6, etc., nanoparticles) are exposed to excitation wavelength, which results in formation of photo-induced electrons and holes. The subsequent electron and hole react with system surrounding water molecules which generates hydroxyl ions, oxidative radicals [reactive oxygen species (ROS)], and singlet oxygen. The final step of this PDT is free radical mediated cell organelles damage to targeted or affected cells which results in cell death.

Photothermal therapy uses a suitable near-infrared (NIR-I: 700–900 nm, NIR-II: 1000–1700 nm) light source to irradiate cancer cells. The exposure of this light energy subsequently converted to heat energy due to surface plasmon resonance (SPR) effect which can cause hyperthermia mediated cell death. Manivasagan et al. reported a dual modal multifunctional theragnostic agent for imaging-guided photothermal treatment (Fig. 15.6). In this study gold nano shell was wrapped with chitosan conjugated with paclitaxel drug and anti-EGFR antibody for specific targeting tumor cells. The promising in vivo results show successful thermal ablation of tumors from nude mice model.

With gold the second most widely used nano metal is silver (Ag). There are many promising characteristics of Ag nanoparticles and the most important is its near equal optical characteristics like gold. The other advantage is its low cost and antibacterial activity and easy synthesis process. In spite of having excellent opportunities the main limitation of Ag is its low stability. Though with different approaches such as coating the stability of Ag nanoparticles enhanced. Therefore, Ag could act as a promising material for different therapeutic applications. Different research activities

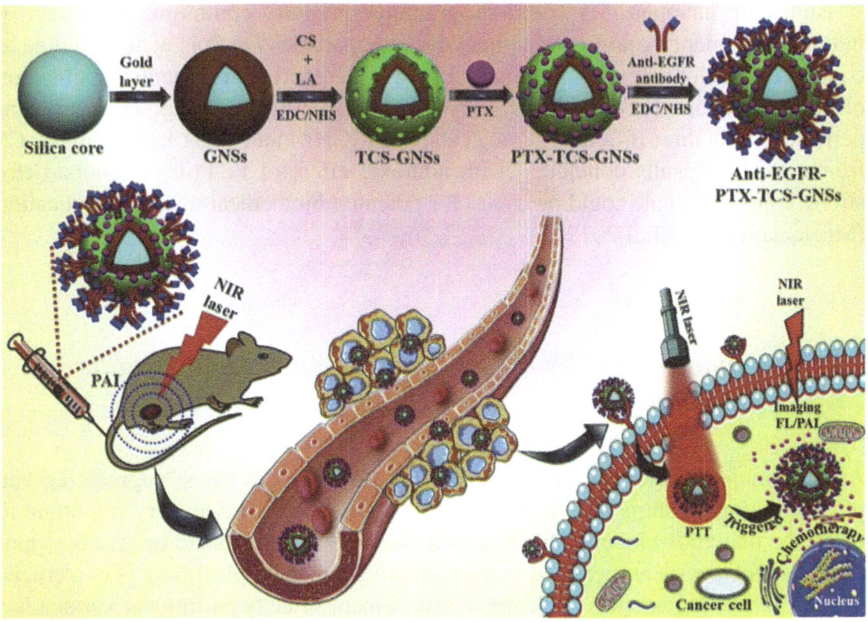

Fig. 15.6 Representation of anti-EGFR-PTX-TCS-GNSs mediated near-infrared fluorescence/photoacoustic imaging-guided chemo-photothermal therapy for cancer treatment (Manivasagan et al. 2019d)

have reported the promising performance of Ag nanoparticles in different clinical activities such as drug delivery, diagnosis, imaging, antibacterial activity, etc. (Prasad 2014). Franco-Molina et al. reported the anticancer activity of colloidal silver nanoparticles with MCF-7 human breast cancer cells (Franco-Molina et al. 2010). MCF-7 cells were incubated with different concentrations colloidal Ag nanoparticles to determine its cytotoxic efficacy. The trypan blue cell exclusion assay was performed to determine cell viability. The results suggested that colloidal Ag had a concentration dependent cytotoxic effect on MCF-7 cell line. The apoptotic cell death confirms the anticancer activity of the colloidal Ag nanoparticles (Franco-Molina et al. 2010; Aziz et al. 2019).

Qin et al. (Qin et al. 2011) reported two different methods to load TiO_2 nanoparticles with doxorubicin (DOX). The two methods non-covalent complexation (TiO_2/DOX) and the covalent conjugation (TiO_2–DOX) were employed to load drugs. Further studies were performed to evaluate the best outcomes in respect with cellular uptake, cytotoxicity, and glioma (C6) cell line mediated intracellular distribution. TiO_2 nanoparticles are widely used for sunscreen agent. The UV light is harmful for human skin due to its mutagenic effect. Fortunately, TiO_2 nanoparticles have unique absorption capability which could be helpful for protecting from direct sunlight. On the other hand, researchers are using this unique property of nanoparticles as photosensitizers for PDT treatment to treat cancer.

Bimetallic nanoparticles could show unique property combining the effect of individual element. Among these bimetallic nanoparticles Fe–Pt is widely studied as contrast agents for diagnosis of tumors. Liang et al. reported the potential application of Fe–Pt nanoparticles coated with L-cysteine for MRI/CT imaging. The study was performed with three different glioma cell lines (SGH44 and U251 from humans, C6 from rat). The results concluded with nontoxic effect of Fe-Pt-Cys nanoparticles strong contrast signals could be useful for potential biomedical imaging application (McNamara and Tofail 2017).

15.4.6 Nanotechnology to Engineer the Surface of Metallic Implants

Another major application of nanotechnology is found in tissue engineering and prosthetic implantation. The surface modification of implants is very important to make the materials biocompatible and favorable to accommodate on its host environment. Nanoparticle mediated surface modification of implant may play a crucial role for tissue engineering application. Attachment of cells on implant surface is a big challenge in tissue engineering. For example, poly lactic acid (PLA) is an example of excellent biomaterials but its major drawback is its smooth surface which cannot allow cells to attach or grow. To overcome such situation, we need to modify the surface of the scaffold or implant by means of chemical, physical, or structural modifications (Mondal et al. 2020b). The enhanced surface activity may facilitate the cells to attach and grow firmly. Till date titanium (Ti) and its alloys are considered one of the best materials for bone replacement applications due to its adequate mechanical properties, high resistance to corrosion, and bioinert property. The limitation of Ti implant is a thin fibrous layer which separates the scaffold from the host bone causing loose bonding and subsequent implant failure. In this situation, for more successful bone implant surgery, a better bone–scaffold interaction is necessary by improving scaffold surface (Le Guéhennec et al. 2007). Mondal et al. reported a composite biomaterial formulation with HAp, Al_2O_3, bio glass, and starch for enhanced mechanical support and better biocompatibility (Fig. 15.7). The scaffold was fabricated by gel-casting and achieved compressive strength of ~157 ± 2 MPa and tensile strength of ~83 ± 2 MPa with 20–25% porosity when sintered after 1200 °C for 2 h. The study extends with successful MG-63 mediated cytotoxicity evaluation and cell attachment and proliferation study on its surface. The result confirms the nontoxic behavior of the scaffold formulation which facilitates cell attachment and proliferation on its surface (Mondal et al. 2018b).

Due to excellent bioactivity CaP nanoparticles were used as a coating material over different implant surface. Kokubo et al. reported that simulated body fluid (SBF) is most commonly used biomimetic solution (Kokubo et al. 1990) which consists of a precursor Ca and P ionic formulation that mimic human body fluids. This biomimetic procedure helps to promote immobilization of apatite

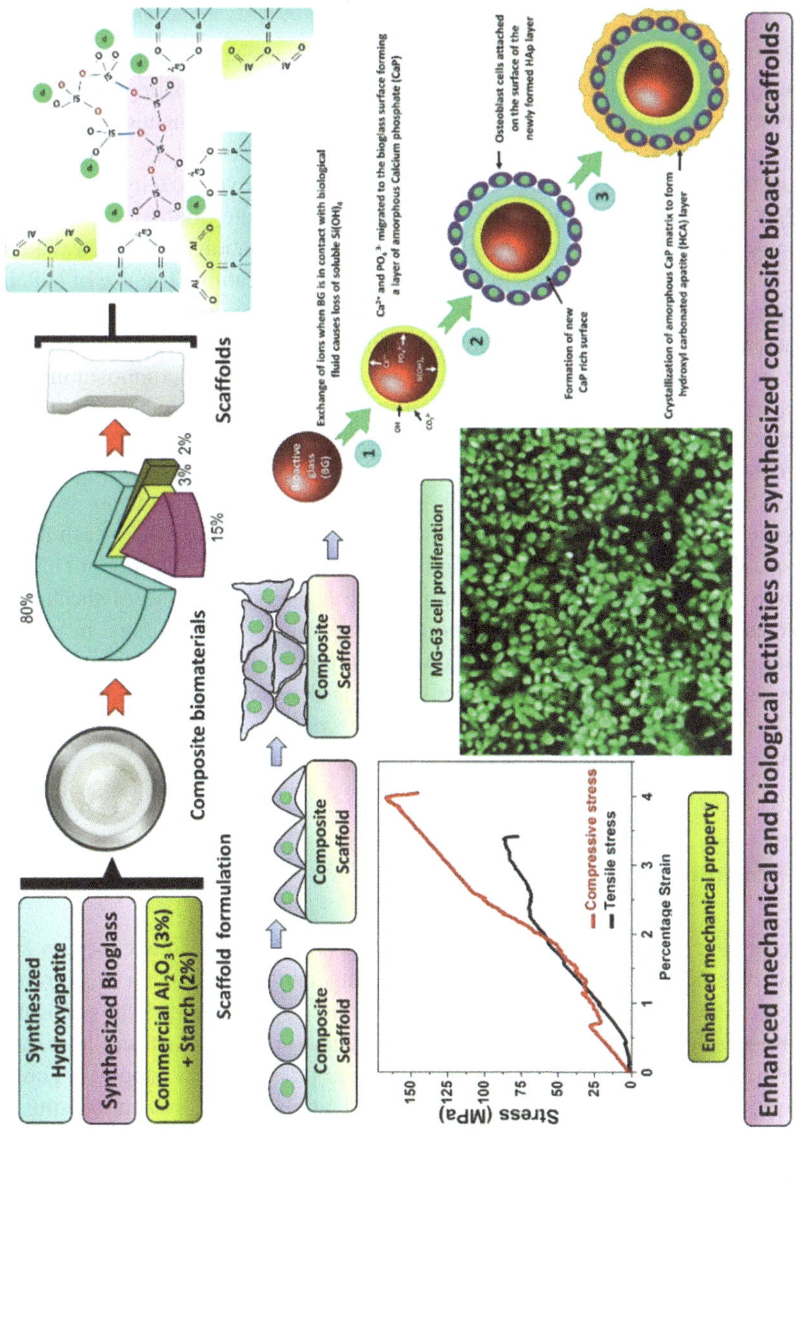

Fig. 15.7 Fabrication of composite scaffold with enhanced mechanical and biological performance for tissue engineering application (Mondal et al. 2018b)

structure, growth factors, therapeutic agents, drugs, proteins, etc. on its surface along with paving its way to promote new cell generation.

15.5 Limitations and Challenges

Several materials, including ceramics, polymers, metal, metal oxide nanoparticles, bio glasses, and polymers, are in top priorities for biomedical applications. However, each of these materials also has exclusive limitations. Poor shape holding, low mechanical strength, inadequate cell adhesion, and in vivo toxicity are the key features for experimental failure. To date, no such absolute nontoxic biomaterials can be used without further modifications. The prime limitations associated with the failure of present approaches are toxicity for in vivo application. The use of suitable dose of nanoparticles as drug needs to be prudently measured. The toxicity of nanomaterials is associated with multiple factors such as morphology, surface charge, sizes, chemical structure, and obviously on dose and composition. Low contrast efficiency for bioimaging application, synthesis reproducibility is another concern regarding synthesis of nanoparticles. Therefore, it is always be prime concern to choose the biomaterials according to the specific target region with tailored shape and biological properties. The limitations of nanoparticles as a drug delivery agent depend upon its payload capacity, drug intake, and release efficiency. Also, specific targeting (to avoid killing healthy cells) is the most important characteristics of nanoparticles. Therefore, targeted drug delivery is one of the biggest challenges. Nanoparticle colloid stability, aggregations, and storage in clinical locations remain a big challenge to consider. Reproducible large-scale production for clinical application with a cost-effective way is also a big challenge. Establishing the real consequences of nanomaterials in body system is an extremely important, challenging, and in the similar way interesting task.

15.6 Summary

Nanotechnology, a multidisciplinary field of research brings all the concepts from physics, chemistry, biology, medicine, engineering, and others under a single roof of science and technology. The future of nanotechnology for biomedical application must integrate a multiple factors-based technology to promote a strong host materials interaction with all chemical, biological, and physiological supports. Though the current trends of nano-research in biomedical application appear promising, till there are many miles to go to achieve the direct benefits. This study recapitulates emerging research and their promising results for healthcare applications. Different nanoparticles with diverse characteristics are aimed to employ for noninvasive imaging, drug delivery, photothermal/photodynamic therapy for different cancer treatment. The developed nanoparticle will cross the limits for conventional materials which restricts their application for therapeutic approaches. Till date many nanomaterials such as quantum dots, gold, silver nano particles are most promising for fluorescence, photoacoustic, X-ray, CT imaging applications. The rich surface chemistry and precise control over size and composition of iron oxide nanoparticles are useful and FDA approved contrast agents for MRI. As a most reliable bioactive biomaterial hydroxyapatite nanoparticles win the race since 1970s and till going on. Nonetheless, in order to take advantage of the potential application of nanomaterials, extensive safety and toxicology studies will need to consider along with its clinical trials.

Conflicts of Interests The author declares no conflict of interest.

References

Abd-Elsalam K, Mohamed AA, Prasad R (2019) Magnetic nanostructures: environmental and agricultural applications. Springer International Publishing, Cham. ISBN 978-3-030-16438-6. https://www.springer.com/gp/book/9783030164386

Akbarzadeh A, Rezaei-Sadabady R, Davaran S et al (2013) Liposome: classification, preparation, and applications. Nanoscale Res Lett 8(1):102

Alpaslan E, Yazici H, Golshan NH, Ziemer KS, Webster TJ (2015) pH-dependent activity of dextran-coated cerium oxide nanoparticles on prohibiting osteosarcoma cell proliferation. ACS Biomater Sci Eng 1(11):1096–1103

Aziz N, Fatma T, Varma A, Prasad R (2014) Biogenic synthesis of silver nanoparticles using *Scenedesmus abundans* and evaluation of their antibacterial activity. J Nanoparticles, Article ID 689419. https://doi.org/10.1155/2014/689419

Aziz N, Faraz M, Pandey R, Sakir M, Fatma T, Varma A, Barman I, Prasad R (2015) Facile algae-derived route to biogenic silver nanoparticles: synthesis, antibacterial and photocatalytic properties. Langmuir 31:11605–11612. https://doi.org/10.1021/acs.langmuir.5b03081

Aziz N, Pandey R, Barman I, Prasad R (2016) Leveraging the attributes of *Mucor hiemalis*-derived silver nanoparticles for a synergistic broad-spectrum antimicrobial platform. Front Microbiol 7:1984. https://doi.org/10.3389/fmicb.2016.01984

Aziz N, Faraz M, Sherwani MA, Fatma T, Prasad R (2019) Illuminating the anticancerous efficacy of a new fungal chassis for silver nanoparticle synthesis. Front Chem 7:65. https://doi.org/10.3389/fchem.2019.00065

Bachilo SM, Strano MS, Kittrell C, Hauge RH, Smalley RE, Weisman RB (2002) Structure-assigned optical spectra of single-walled carbon nanotubes. Science 298(5602):2361–2366

Bangham AD, Horne R (1964) Negative staining of phospholipids and their structural modification by surface-active agents as observed in the electron microscope. J Mol Biol 8(5):660–IN10

Bharathiraja S, Bui NQ, Manivasagan P et al (2018) Multimodal tumor-homing chitosan oligosaccharide-coated biocompatible palladium nanoparticles for photo-based imaging and therapy. Sci Rep 8(1):1–16

Bhattacharya R, Mukherjee P (2008) Biological properties of "naked" metal nanoparticles. Adv Drug Deliv Rev 60(11):1289–1306

Chatterjee K, Sarkar S, Rao KJ, Paria S (2014) Core/shell nanoparticles in biomedical applications. Adv Colloid Interf Sci 209:8–39

Cherukuri P, Bachilo SM, Litovsky SH, Weisman RB (2004) Near-infrared fluorescence microscopy of single-walled carbon nanotubes in phagocytic cells. J Am Chem Soc 126 (48):15638–15639

Choi HS, Liu W, Misra P et al (2007) Renal clearance of quantum dots. Nat Biotechnol 25 (10):1165–1170

Cressman S, Dobson I, Lee JB, Tam YYC, Cullis PR (2009) Synthesis of a labeled RGD– lipid, its incorporation into liposomal nanoparticles, and their trafficking in cultured endothelial cells. Bioconjug Chem 20(7):1404–1411

De Crozals G, Bonnet R, Farre C, Chaix C (2016) Nanoparticles with multiple properties for biomedical applications: a strategic guide. Nano Today 11(4):435–463. https://doi.org/10.1016/j.nantod.2016.07.002

Eggermont LJ, Rogers ZJ, Colombani T, Memic A, Bencherif SA (2020) Injectable cryogels for biomedical applications. Trends Biotechnol 38(4):418–431

Franco-Molina MA, Mendoza-Gamboa E, Sierra-Rivera CA et al (2010) Antitumor activity of colloidal silver on MCF-7 human breast cancer cells. J Exp Clin Cancer Res 29(1):148

Hafner JH, Bronikowski MJ, Azamian BR et al (1998) Catalytic growth of single-wall carbon nanotubes from metal particles. Chem Phys Lett 296(1–2):195–202

Hainfeld J, Slatkin D, Focella T, Smilowitz H (2006) Gold nanoparticles: a new X-ray contrast agent. Br J Radiol 79(939):248–253

Kam N, Jessop T, Wender P, Dai H (2004) Nanotube molecular transporters: internalization of carbon nanotube-protein conjugates into mammalian cells. J Am Chem Soc 126(22):6850–6851

Kee PH, Danila D (2018) CT imaging of myocardial scar burden with CNA35-conjugated gold nanoparticles. Nanomedicine 14(6):1941–1947. https://doi.org/10.1016/j.nano.2018.06.003

Kim H, Mondal S, Bharathiraja S, Manivasagan P, Moorthy MS, Oh JJCI (2018a) Optimized Zn-doped hydroxyapatite/doxorubicin bioceramics system for efficient drug delivery and tissue engineering application. Ceram Int 44(6):6062–6071

Kim H, Mondal S, Jang B, Manivasagan P, Moorthy MS, Oh J (2018b) Biomimetic synthesis of metal–hydroxyapatite (Au-HAp, Ag-HAp, Au-Ag-HAp): structural analysis, spectroscopic characterization and biomedical application. Ceram Int 44(16):20490–20500. https://doi.org/10.1016/j.ceramint.2018.08.045

Kobayashi H, Kawamoto S, Jo S-K, Bryant HL, Brechbiel MW, Star RA (2003) Macromolecular MRI contrast agents with small dendrimers: pharmacokinetic differences between sizes and cores. Bioconjug Chem 14(2):388–394

Kokubo T, Kushitani H, Sakka S, Kitsugi T, Yamamuro T (1990) Solutions able to reproduce in vivo surface-structure changes in bioactive glass-ceramic A-W3. J Biomed Mater Res 24 (6):721–734

Koole R, van Schooneveld MM, Hilhorst J et al (2008) Paramagnetic lipid-coated silica nanoparticles with a fluorescent quantum dot core: a new contrast agent platform for multimodality imaging. Bioconjug Chem 19(12):2471–2479

Le Guéhennec L, Soueidan A, Layrolle P, Amouriq Y (2007) Surface treatments of titanium dental implants for rapid osseointegration. Dent Mater 23(7):844–854

Liao H, Nehl CL, Hafner JH (2006) Biomedical applications of plasmon resonant metal nanoparticles. Nanomedicine 1(2):201–208

Liu J, Lu Y (2004) Accelerated color change of gold nanoparticles assembled by DNAzymes for simple and fast colorimetric Pb2+ detection. J Am Chem Soc 126(39):12298–12305

Liu T, Tang A, Zhang G et al (2005) Calcium phosphate nanoparticles as a novel nonviral vector for efficient transfection of DNA in cancer gene therapy. Cancer Biother Radiopharm 20 (2):141–149

Maitra A (2005) Calcium phosphate nanoparticles: second-generation nonviral vectors in gene therapy. Expert Rev Mol Diagn 5(6):893–905

Manivasagan P, Bharathiraja S, Santha Moorthy M et al (2018) Marine natural pigments as potential sources for therapeutic applications. Crit Rev Biotechnol 38(5):745–761

Manivasagan P, Hoang G, Santha Moorthy M et al (2019a) Chitosan/fucoidan multilayer coating of gold nanorods as highly efficient near-infrared photothermal agents for cancer therapy. Carbohydr Polym 211:360–369. https://doi.org/10.1016/j.carbpol.2019.01.010

Manivasagan P, Jun SW, Truong NTP et al (2019b) A multifunctional near-infrared laser-triggered drug delivery system using folic acid conjugated chitosan oligosaccharide encapsulated gold nanorods for targeted chemo-photothermal therapy. J Mater Chem B 7(24):3811–3825

Manivasagan P, Khan F, Hoang G et al (2019c) Thiol chitosan-wrapped gold nanoshells for near-infrared laser-induced photothermal destruction of antibiotic-resistant bacteria. Carbohydr Polym 225:115228. https://doi.org/10.1016/j.carbpol.2019.115228

Manivasagan P, Nguyen VT, Jun SW et al (2019d) Anti-EGFR antibody conjugated thiol chitosan-layered gold nanoshells for dual-modal imaging-guided cancer combination therapy. J Control Release 311-312:26–42. https://doi.org/10.1016/j.jconrel.2019.08.007

McNamara K, Tofail SA (2017) Nanoparticles in biomedical applications. Advances in Physics: X 2(1):54–88

Mirkin CA (2000) Programming the assembly of two-and three-dimensional architectures with DNA and nanoscale inorganic building blocks. Inorg Chem 39(11):2258–2272

Mondal S, Oh J (2019) Nanostructured materials and their biomedical application. In: Biomedical engineering and its applications in healthcare. Springer, pp 205–227

Mondal S, Pal U (2019) 3D hydroxyapatite scaffold for bone regeneration and local drug delivery applications. Journal of Drug Delivery Science and Technology 53:101131. https://doi.org/10.1016/j.jddst.2019.101131

Mondal S, Dey A, Pal U (2016a) Low temperature wet-chemical synthesis of spherical hydroxyapatite nanoparticles and their in situ cytotoxicity study. Advances in Nano Research 4(4):295

Mondal S, Pal U, Dey A (2016b) Natural origin hydroxyapatite scaffold as potential bone tissue engineering substitute. Ceram Int 42(16):18338–18346. https://doi.org/10.1016/j.ceramint.2016.08.165

Mondal S, Manivasagan P, Bharathiraja S et al (2017a) Hydroxyapatite coated iron oxide nanoparticles: a promising nanomaterial for magnetic hyperthermia cancer treatment. Nano 7 (12):426

Mondal S, Reyes MEDA, Pal U (2017b) Plasmon induced enhanced photocatalytic activity of gold loaded hydroxyapatite nanoparticles for methylene blue degradation under visible light. RSC Adv 7(14):8633–8645

Mondal S, Dorozhkin SV, Pal U (2018a) Recent progress on fabrication and drug delivery applications of nanostructured hydroxyapatite. WIREs Nanomedicine and Nanobiotechnology 10(4):e1504. https://doi.org/10.1002/wnan.1504

Mondal S, Hoang G, Manivasagan P et al (2018b) Nano-hydroxyapatite bioactive glass composite scaffold with enhanced mechanical and biological performance for tissue engineering application. Ceram Int 44(13):15735–15746. https://doi.org/10.1016/j.ceramint.2018.05.248

Mondal S, Hoang G, Manivasagan P, Kim H, Oh J (2019a) Nanostructured hollow hydroxyapatite fabrication by carbon templating for enhanced drug delivery and biomedical applications. Ceram Int 45(14):17081–17093. https://doi.org/10.1016/j.ceramint.2019.05.260

Mondal S, Hoang G, Manivasagan P et al (2019b) Rapid microwave-assisted synthesis of gold loaded hydroxyapatite collagen nano-bio materials for drug delivery and tissue engineering application. Ceram Int 45(3):2977–2988. https://doi.org/10.1016/j.ceramint.2018.10.016

Mondal S, Manivasagan P, Oh J (2020a) Marine polysaccharide-based nanomaterials. Encyclopedia of Marine Biotechnology 2:1231–1248

Mondal S, Nguyen TP, Hoang G et al (2020b) Hydroxyapatite nano bioceramics optimized 3D printed poly lactic acid scaffold for bone tissue engineering application. Ceram Int 46 (3):3443–3455

Mondal S, Nguyen TP, Pham VH et al (2020c) Hydroxyapatite nano bioceramics optimized 3D printed poly lactic acid scaffold for bone tissue engineering application. Ceram Int 46 (3):3443–3455. https://doi.org/10.1016/j.ceramint.2019.10.057

Mondal S, Nguyen VT, Park S et al (2020d) Rare earth element doped hydroxyapatite luminescent bioceramics contrast agent for enhanced biomedical imaging and therapeutic applications. Ceram Int. https://doi.org/10.1016/j.ceramint.2020.08.099

Mondal S, Nguyen VT, Park S et al (2020e) Bioactive, luminescent erbium-doped hydroxyapatite nanocrystals for biomedical applications. Ceramics international 46(10, part B):16020-16031. https://doi.org/10.1016/j.ceramint.2020.03.152

Moorthy MS, Subramanian B, Panchanathan M et al (2017) Fucoidan-coated core–shell magnetic mesoporous silica nanoparticles for chemotherapy and magnetic hyperthermia-based thermal therapy applications. New J Chem 41(24):15334–15346

Moorthy MS, Bharathiraja S, Manivasagan P et al (2018) Synthesis of Fe 3 O 4 modified mesoporous silica hybrid for pH-responsive drug delivery and magnetic hyperthermia applications. J Porous Mater 25(4):1251–1264

Moorthy MS, Hoang G, Manivasagan P et al (2019) Chitosan oligosaccharide coated mesoporous silica nanoparticles for pH-stimuli responsive drug delivery applications. J Porous Mater 26 (1):217–226

Muchow M, Maincent P, Müller RH (2008) Lipid nanoparticles with a solid matrix (SLN®, NLC®, LDC®) for oral drug delivery. Drug Dev Ind Pharm 34(12):1394–1405

Mulder WJ, Strijkers GJ, van Tilborg GA, Griffioen AW, Nicolay K (2006) Lipid-based nanoparticles for contrast-enhanced MRI and molecular imaging. NMR in Biomedicine: An International Journal Devoted to the Development and Application of Magnetic Resonance In vivo 19(1):142–164

Murray C, Norris DJ, Bawendi MG (1993) Synthesis and characterization of nearly monodisperse CdE (E= sulfur, selenium, tellurium) semiconductor nanocrystallites. J Am Chem Soc 115 (19):8706–8715

Ngo C, Yoon S, Chua S (2009) Ambient temperature dependence on emission spectrum of InAs quantum dots. Phys Status Solidi B 246(4):799–802

Pantarotto D, Briand J-P, Prato M, Bianco A (2004) Translocation of bioactive peptides across cell membranes by carbon nanotubes. Chem Commun 10(1):16–17

Phan TTV, Bui NQ, Cho S-W et al (2018) Photoacoustic imaging-guided photothermal therapy with tumor-targeting HA-FeOOH@ PPy nanorods. Sci Rep 8(1):1–13

Phan TTV, Huynh T-C, Manivasagan P, Mondal S, Oh J (2020) An up-to-date review on biomedical applications of palladium nanoparticles. Nano 10(1):66

Prasad R (2014) Synthesis of silver nanoparticles in photosynthetic plants. J Nanoparticles, Article ID 963961. https://doi.org/10.1155/2014/963961

Prasad R, Pandey R, Barman I (2016) Engineering tailored nanoparticles with microbes: quo vadis. WIREs Nanomed Nanobiotechnol 8:316–330. https://doi.org/10.1002/wnan.1363

Prasad R, Pandey R, Varma A, Barman I (2017) Polymer based nanoparticles for drug delivery systems and cancer therapeutics. In: Kharkwal H, Janaswamy S (eds) Natural polymers for drug delivery. CAB International, London, pp 53–70

Prasad R, Jha A, Prasad K (2018) Exploring the realms of nature for nanosynthesis. Springer International Publishing, Cham. ISBN 978-3-319-99570-0. https://www.springer.com/978-3-319-99570-0

Prasad R, Siddhardha B, Dyavaiah M (2020) Nanostructures for antimicrobial and antibiofilm applications. Springer International Publishing, Cham. ISBN 978-3-030-40336-2. https://www.springer.com/gp/book/9783030403362

Qin Y, Sun L, Li X et al (2011) Highly water-dispersible TiO_2 nanoparticles for doxorubicin delivery: effect of loading mode on therapeutic efficacy. J Mater Chem 21(44):18003–18010

Resch-Genger U, Grabolle M, Cavaliere-Jaricot S, Nitschke R, Nann T (2008) Quantum dots versus organic dyes as fluorescent labels. Nat Methods 5(9):763

Singh S, Kumar V, Dhanjal DS, Datta S, Prasad R, Singh J (2020) Biological biosensors for monitoring and diagnosis. In: Singh J, Vyas A, Wang S, Prasad R (eds) Microbial biotechnology: basic research and applications. Springer Nature, Singapore, pp 317–336

Srivastava S, Usmani Z, Atanasov AG, Singh VK, Singh NP, Abdel-Azeem AM, Prasad R, Gupta G, Sharma M, Bhargava A (2021) Biological nanofactories: using living forms for metal nanoparticle synthesis. Mini-Rev Med Chem 21(2):245–265

Talapin DV, Mekis I, Götzinger S, Kornowski A, Benson O, Weller H (2004) CdSe/CdS/ZnS and CdSe/ZnSe/ZnS core – shell – shell nanocrystals. J Phys Chem B 108(49):18826–18831

Thangadurai D, Sangeetha J, Prasad R (2020a) Nanotechnology for food, agriculture, and environment. Springer International Publishing, Cham. ISBN 978-3-030-31937-3. https://www.springer.com/gp/book/9783030319373

Thangadurai D, Sangeetha J, Prasad R (2020b) Functional bionanomaterials. Springer International Publishing, Cham. ISBN 978-3-030-41464-1. https://www.springer.com/gp/book/9783030414634

Wiener E, Brechbiel M, Brothers H et al (1994) Dendrimer-based metal chelates: a new class of magnetic resonance imaging contrast agents. Magn Reson Med 31(1):1–8

Wunderbaldinger P, Josephson L, Weissleder R (2002) Crosslinked iron oxides (CLIO): a new platform for the development of targeted MR contrast agents. Acad Radiol 9(2):S304–S306

Zhao W, Brook MA, Li Y (2008) Design of gold nanoparticle-based colorimetric biosensing assays. Chembiochem 9(15):2363–2371

Chapter 16
Nanotechnology-A New Frontier in Medical Microbiology

Silpa Somavarapu, Bellamkonda Ramesh, Ch. Venkatrayulu, and M. Subhosh Chandra

Abstract Nanotechnology relates to microbiology at a number of levels as the microbial entities are nano-machines. In the second half of this decade, nanotechnology expanding its applications in the field of medical microbiology. Nanotechnology is clinically appropriate and retains the potential to be valuable in the diagnosis of general and microbial infections. The rapid detection of pathogenic microbes at the point of care is extremely critical. The application of nanoparticles permits for the detection of infectious pathogens in small sample volumes directly in a sensitive, specific, and rapid format at lower costs than current in-use technologies. A bio-conjugated nanoparticle-based bioassay for in situ pathogen quantification can detect a single microbe. The waveguide technology is an emergent area in the medical microbiology for the fast and successful diagnosis of infectious diseases. Nanotechnology is demonstrated for the detection of Avian influenza virus H5N1, Respiratory Syncytial Virus (RSV), HIV, and Severe acute respiratory syndrome (SARS) Coronavirus in clinical samples with a great degree of sensitivity. Nanoparticle-based bio-barcode amplification (BCA) assay is being applied for early detection of HIV-1 capsid antigen. The gold nanoparticle interferometer sensor has been validated for detection of Herpes simplex virus (HSV) and silver nanorod array substrates can detect spectral differences between the viral strains. A nanoparticle label technology with highly fluorescent chelated nanoparticle label has been developed for Adenovirus and Human papillomavirus (HPV). The nano-gold labelled amplification is a novel technique for the detection of Hepatitis B virus, Hepatitis C virus, and Hepatitis E virus in patient's samples. Norovirus is a leading cause of gastroenteritis and nanospray mass spectrometry is evaluated for norovirus detection. With the manifestation and intensification of microbes resistant to antibiotics, silver nanoparticle antiseptics have been evaluated for the antimicrobial

S. Somavarapu (✉) · B. Ramesh · C. Venkatrayulu
Department of Food Technology, Vikrama Simhapuri University, Nellore, Andhra Pradesh, India

M. Subhosh Chandra
Department of Microbiology, Yogi Vemana University, Kadapa, Andhra Pradesh, India

N. R. Maddela et al. (eds.), *Nanotechnology for Advances in Medical Microbiology*, Environmental and Microbial Biotechnology,
https://doi.org/10.1007/978-981-15-9916-3_16

activity against Gram-positive and Gram-negative bacteria. All these technologies would have to be assessed in clinical settings prior to their complete admission is highly recommended.

Keywords Nanotechnology · Microorganisms · Virus · Bacteria · Infectious diseases

16.1 Introduction

Microbiology is the science that deals with microscopic analysis of microorganisms, which are not visible to the naked human eye. Microscopic invention by Antony van Leeuwenhoek in the year of 1676 marked the beginning of microbiology.

Advent of light microscopy by Ernst Abbe and Carl Zeiss (1880s) and electron microscopy by Ernst Ruska (1931) produced high resolution images for visual characterization and released a new avenue in the study of biological structures. Microorganisms were the primitive life recorded on earth. Microbes occupy a place as significant component of biota with peculiar cell structure and ability to survive under extreme environmental conditions. They exist as unicellular, multicellular, and acellular organisms. Microorganisms are beneficial as well as harmful to human beings (Minocheherhomji 2016). They can be divided into six major types: bacteria, archaea, fungi, protozoa, algae, and viruses.

Nanotechnology is a highly promising field of research in the modern era of scientific research. Eric Drexler popularized the word 'nanotechnology' in 1986 in his book 'Engines of creation: The Coming Era of Nanotechnology (Shinde 2012). It is an interdisciplinary science, whose potential has been widely touted in various domains of lifecycle with its development from submissive nanomaterial to active nanotechnology (e.g., drug delivery system) and nanosystems (e.g., robotics). Nanotechnology, in other forms, means building things from the bottom up, with atomic precision. This theoretical capability was envisioned as early as 1959 by the renowned physicist Richard Feynman in his talk 'There's **Plenty** of Room at the Bottom' at Caltech, USA (Feynman 1960) and promoted by Eric Drexlerby his book *Engines of Creation: The Coming Era of Nanotechnology* in the 1980s.

Nanotechnology is the study and application of nanoparticles. Nanoparticles are particles that exist on a nanometre scale (i.e., below 100 nm in at least one dimension). They possess physical properties such as uniformity, conductance or special optical properties and are of pronounced technical interest (Prasad et al. 2016). They form a bridge between bulk materials and atomic or molecular structures. Nanoparticles were used by artisans as far back as the ninth century in Mesopotamia for generating a glittering effect on the surface of pots. In the fourth and fifth centuries BC it was used by traditional medical practitioners in the world with the name 'Swarna Bhasma' for therapeutic purposes in treating cognitive disorders and syphilis (Paul and Chugh 2011; Balzani 2005; Drexler and Peterson 1989; Dykman and Khlebtsov 2011). The worldwide market of nanomaterials used in the biomedical, pharmaceutical, and cosmetic industries showed an elevation from \$170.17 million in 2006 to \$684.4 million in 2012.

16.2 Role of Nanotechnology in Medical Microbiology

The boom of Nanotechnology revolutionized microbiology. Nanotechnology proved significant in medical microbiology in the detection of pathogens. Microbial nanotechnology has gained prominence with the advancement of chip-based DNA control, quantum dots, and carbon nanotubes (CNTs). Secure and cost viable novel materials are designed utilizing the viral like protein frames for biomedical imaging, Bioengineering, Drug Design, Enzyme Technology, etc. Additionally, synthesis of nanomaterials utilizing the energy of microorganisms also opens a new-fangled opportunity for 'green' synthesis, which is eco-friendly cost-effective process (Ahmad et al. 2016; Prasad et al. 2016, 2018; Prasad 2019a, b; Srivastava et al. 2021). Multidrug resistance is the present day concern (Andersson and Hughes 2010; Magiorakos et al. 2012). Biofilms offer resistance to antimicrobial agents while present in the human body and oblige as a reservoir of bacteria that can cause continuous and chronic infections (Watnick and Kolter 2000; Inamuddin et al. 2021). A 80% of infections in the hospitals are reported due to biofilms (Costerton 1999; Beyth et al. 2015).There is an immense need for substitute to antibacterial and anti-biofilm agents (Prasad et al. 2020). The nanotechnology can provide novel diagnostics and treatments for the bacterial infections and encouraged extensive investigation. Nanotechnology marks an innovative tool to address this challenge. Antibacterial nanodrugs are active against MDR bacteria and biofilms. A nanoporous polymer matrix composed of sodium dodecyl sulfate proved to have excellent anti-biofilm activity against *E. coli*. Vitamin E-conjugated cationic polymer cross-linked hydrogels were found to have good bactericidal and antifungal effects (Lee et al. 2013).

16.2.1 Broad Spectrum of Nanoparticles

Liposomes are the nanoparticles comprised of lipid bilayer membranes, which is surrounded by an aqueous interior. Liposomes can be widely applied for delivery of antibiotics to the target sites (Drulis-Kawa et al. 2009). Liposomes exhibited biofilm inhibition and minimum sensing distraction on clinical variants of *E. coli*, *Acinetobacter lwoffii*, *A. baumannii*, *Bordetella bronchiseptica*, *Klebsiella pneumoniae*, and *P. aeruginosa* (Drulis-Kawa et al. 2009). Dendrimers with lower molecular weight peptides are proven to be efficient antimicrobials against *E.coli* and *S. aureus* without any further antibiotics (Johansson et al. 2008). Fucose-specific lectins (LecB) from fucose peptide dendrimers prevented the formation of biofilm in *P. aeruginosa*. A lipid dendrimer hybrid nanoparticle (LDHN) delivered vancomycin against methicillin resistant *Staphylococcus aureus* (MRSA) infections

(Sonawane et al. 2016). Nano-emulsions with antimicrobial activity against bacteria such as *E. coli*, *Salmonella sps*, *S. aureus*, enveloped viruses (Human immunodeficiency virus and Herpes simplex), fungi (Candida), and spore forms of *B. anthracis* have been reported.

Bio-responsive smart nanoparticles, which are comprised with peripheral energy and energy absorbing NPs having the therapeutic properties towards respective antimicrobial infections (Ramasamy et al. 2016). Quantum dots (Qdots) are nanoscale semiconducting nanoparticles that can transport electrons. Qdots emit various colors (wavelength) when the light is excited in different wavelengths according to their size. Qdots have been extensively used in the Fluorescent Resonance Energy Transfer (FRET) based immunoassays for fast and accurate detection of *Aspergillus sps*. Qdot barcodes have been widely applied in the detection of HIV (Kaittanis et al. 2010; Kattke et al. 2011).

16.2.2 Nanoparticles with Intrinsic Antibacterial Properties

Silver is a renowned potent antimicrobial agent since prehistoric times (Rai et al. 2009; Reidy et al. 2013; Duran et al. 2015; Franci et al. 2015; Aziz et al. 2014, 2015, 2016). Silver based nanoparticles (AgNPs) exhibit antimicrobial activity against *Pseudomonas aeruginosa*, one of the important opportunistic pathogens triggering nosocomial infections, *Mycobacterium tuberculosis* (Mapara et al. 2015; Singh et al. 2015; El-Zahry et al. 2015; Pal et al. 2015) and *Staphylococcus aureus* (Actis et al. 2015). Silver nanoparticles are also inhibited the formation of biofilm by means of *P. aeruginosa* and *S. epidermidis* by more than 95% (Sinha et al. 2011).

Gold nanoparticles (AuNPs) adapted with various surfaces exhibit various enzyme-like accomplishments including peroxidase, glucose oxidase, superoxide dismutase, and catalase mimetics (He et al. 2013). MSN-AuNPs can inhibits the biofilm formation in *Bacillus subtilis* and even be able to collision existing biofilm (Tao et al. 2015). They proved effective against enteropathogenic *Escherichia coli* (EPEC), *Enterococcus faecium*, *Enterococcus faecalis* (including vancomycin-resistant strains) (Kaittanis et al. 2010). Gold and silver nanoparticles have been applied in resonance scattering confocal microscopy or two-photon luminescence confocal microscopy and as a transporters for drugs (Dykman and Khlebtsov 2011). ZnO NPs have better antibacterial activities and low toxicities in mammalian cells and is effective against *E. faecalis*, *S. aureus*, *S. epidermidis*, *Bacillus subtilis*, and *E. coli* (Lee et al. 2014).

Gold nanoparticle interferometric approach is a sensitive and quantitative detection method for sensing single particle, which is calibration-free allowing molecular counting and capable of directly processing with complex biological samples. Gold nanoparticles are used for sensing due to localized surface plasmon resonance (LSPR). The detection is typically based on the shifting of wavelength of the Localized surface plasmon resonance by scattering of light through a single gold nanoparticle. The experiment utilizes a nonlinear confocal microscope and interferometer arm similar to a Michelson interferometer. The modifications to the local RI (refractive index) signal or absorption of molecules on the nanoparticle surface are captured. Gollmer et al. (2014) described the Gold nanotriangle (GNT) arrays synthesized by e-beam lithography. A biosensor designed from electrochemically produced with Gold nanoparticles-modified screen-printed carbon electrode coupled with the thiolated aptamer AG3 might achieve a limit of detection of around 180 virus particles of MNV (Giamberardino et al. 2013). Gold nanoparticles that have enzyme-like catalytic activity (i.e., NanoZyme activity) are immobilized with the AG3 aptamer (Kd of 18.5 nM) for Mouse norovirus (MNV) recognition.

16.2.3 A Bio-Conjugated Nanoparticle-based Bioassay

Cancer is a deadly disease caused by the abnormal cell growth, which affects any part of the human body. Early diagnosis of cancer (prior to metastasis) is a critical aspect in its treatment. Nanomaterials demonstrated as excellent materials for the early stage diagnose of many cancers. Gold nanoparticles (AuNps) comprised with biomolecules corresponding to target molecule are used for early detection of cancer by means of colorimetric detection. Kang et al. (2010) established a colorimetric procedure using anionic citrate coated AuNps for the diagnosis of cancer. Zhang et al. (2016) proposed Gold nanoparticles with telomerase primer on its surface and treated with human cell lines of leukaemia (HL-60, K562), hepatocellular carcinoma (HepG2), embryonic kidney (293 T), and normal skin fibroblasts (HSF). Guo et al. (2014) conjugated folic acid to Au nanorods (FGNRs) and produced optical detecting sensor with fluorescence and Localized surface plasmon resonance absorption feature, which facilitated in the cancer diagnosis.

The biomolecule-conjugated AgNPs shows exciting applications by challenging the clinical complexities such as multidrug resistance, designing biocompatible nanopharmaceutics, cancer therapy superior drug delivery transporters, fluorescence biosensors, and the next-generation antibiotics. The bio-conjugated Gold nanoparticles clenched unique optical and plasmonic features, which are exploited for developing diagnostic sensors and for the detection of biomarkers using immunoassays. Liu et al. (2013) reported a highly sensitive technique for the quantitative detection of HIV nucleic acids. Abbaspour et al. (2015) reported an assay for the detection of Gram-positive bacteria *Staphylococcus aureus* by dual-aptamers bio-conjugated AgNPs sandwich immunosensor. Kurdekar et al. (2017) explored the application of fluorescent AgNPs bio-conjugated with streptavidin for early diagnosis of HIV infection.

16.2.4 Nanotechnology for Viral Detection

Development of a variety of novel isolates of pathogenic viruses presently marked universal fatality challenging human health and demanding urge to develop suitable detection methods, nanovaccines, and therapeutic nano-based options. Most respiratory tract infections are caused by frequently Influenzae virus, Respiratory syncytial virus (RSV), Rhinovirus (RV), and Severe/acute respiratory syndrome (SARS) Coronavirus. Respiratory viruses affect infants, children, elderly people, and immune compromised patients. Nanomaterials are considered to be suitable aspirants against many viral infections, especially Coronaviruses due to their capability to move into cells effortlessly and interact with viruses and interfere with viral genome replication. Gold NPs and quantum dots (QDs) form new nanotechnology-based detection procedures for several respiratory viruses. Au NPs comprised with silver staining and employed for the recognition of HPV (Human papillomavirus) in a cancer type called cervical carcinoma cell lines (Zehbe et al. 1997). Nanostructures like metal NPs, graphene oxide (GO), Quantum Dots, carbon nanotubes are used for virus testing and detection (Wang et al. 2009). Gold or silica based nanoparticles synthesized on thin silicon membranes form nanochips, which are used to screen clinical samples. This method enabled the diagnosis of HIV 1 virus in the plasma samples (Lee et al. 2004).

Mass spectrometry is employed in the detection of nuclear NoV(CCN3) protein and this was first evidenced by Colquhoun et al. (2006) using matrix-assisted laser desorption ionization coupled with a time-of-flight(MALDI-TOF) and nano-electrospray ionization mass spectrophotometry (ESI-MS) in the detection of NoV virus like particles (VLPs) in clinically significant mediums. Khoris et al. (2019) demonstrated a silver-enhanced nanozyme-based immunoassay for the analysis of NoV with unaided-eye. Moreover, a LSPR-based fluorescence nanobiosensor reported by Takemura et al. (2017), which is capable to accomplish a limit of detection of 0.4 pg/mL of NoV VLPs, through incorporating antibody-mediated AuNPs and quantum dots.

Bio-barcode amplification assay (BCA) with nanoparticles have been reported for the detection of nucleic acid (DNA or RNA) and proteins at ultra-low level with short oligonucleotides as surrogate targets, which can be measured with light scattering, calorimetry, fluorescence or gel based assay (Singh et al. 2018; Draz and Shafiee 2018). NP-I-PCR is a modification of BCA, where real-time PCR is used for detection. It helps in identification of infectious microorganisms and can be used for biomarker discovery for different infectious diseases. In general, gold nanoparticles are being used in the BCA. Kim et al (Kim et al. 2008) developed 235 NP-based BCA for the detection of HIV-1 Gag p24 protein in blood samples, and after the conjugation of anti-HIV 1 p24 Gag pAbs (coated on MMPs) to p24 Gag present in samples, the biotinylated anti-HIV p24 Gag mAbs were added followed by the addition of streptavidin coated GNPs and the biotinylated oligonucleotides. Perez et al. (2011) developed Nanoparticle-amplified I-PCR technique for the rapid detection of RSV surface protein with the help of synthesized GNPs.

16.2.5 Nano-based Antimicrobial and Anti-Biofilm Coatings

Catheters are coated with copper (Cu) and silver (Ag) nanoparticles to accelerate killing of *E. coli* (Rtimi et al. 2016). Silicon coated urinary devices killed different bacterial strains that commonly cause urinary tract infections (UTI) such as *E. coli, Staphylococcus aureus, Staphylococcus epidermidis*, and *Pseudomonas aeruginosa*. ZnO and MgF2 nanoparticles proved to prevent biofilm formation of *Staphylococcus aureus* and *Streptococcus pneumonia*. Catheters coated iron NPs inhibited biofilm in *S. aureus* and *P. aeruginosa* (Anghel et al. 2012) iron NPs eliminated *S. epidermis* infection on orthopaedic implants.

Nanocantilevers consist of tiny pieces of silicon-based materials that can recognize proteins and detect pathogenic bacteria and viruses (Kumar 2006). A technique was developed using 16S rRNA gold nanoprobe-nucleic acid sequence-based amplification (NASBA), for the detection of *Salmonella* (Mollasalehi and Yazdanparast 2013). Surface Enhanced Raman Scattering (SERS) nanoparticles help in sensitive detection of pathogens in complex samples (Weidemaier et al. 2015)

16.2.6 Nanosystems for Food Borne Pathogen Detection and Biofilm Inhibition

Surface enhanced Raman spectroscopy (SERS) is applied as a nano-biosensing technique for the detection of highly sensitive infectious microorganisms directly in a rapid and accurate way (Chandra et al. 2011). Different types of nano-biosensors have been employed to identify foodborne pathogens (bacteria and virus) (Thakur and Ragavan 2013; Li et al. 2004). Silver nanocolloids are commonly used in SERS (Baranwal et al. 2016) to detect microbes as silver nanocolloids increase Raman signals. In addition to silver nanocolloids, graphene oxide (rGO), magnetic beads, carbon nanotubes, and plasmonic gold are most commonly used nanomaterials for the detection of foodborne pathogens. Furthermore, synthetic DNA molecular beacon probes labelled with colour-codes are used as nanobarcodes to detect food pathogens (Li et al. 2004). Therefore, surfaces of refrigerators and storage containers are coated with silver nanoparticles to prevent growth of foodborne pathogens and food spoilage bacteria.

Nanofibers are efficiently employed in the prevention of biofilm-associated infections (Zhang et al. 2011). Furthermore, nickel oxide nanoparticles (NiO-NPs) are proposed as potential antibacterial and antitumor agents. Over and done with a green approach and using *Eucalyptus globulus* leaf extract, researchers (Sallem et al. 2017) synthesized NiO-NPs, which are in the size of 10–20 nm and evaluated their anti-biofilm activity.

16.2.7 Nanotechnology for Designing Vaccines

Nanoparticles/nanocarriers act as vaccine adjuvants and nanovaccine delivery platforms as they retain physicochemical characteristics that enhance their immunogenicity. This offers novel opportunities of novel nano-therapeutics and diagnostics which is the urge of present drug resistance age. These vaccines are more effective, safe, and convenient over conventional vaccines. Silver Nanoparticles inhibited the viral entry into host cells, for HIV-1 virus by interacting with the cell receptors (Kerry et al. 2019; Zhu et al. 2019). Gold Nanoparticles stabilized by biocompatible polymers showed antiviral activity against HIV-1 and influenza virus (e.g., H1N1, H3N2, H5N1) (Seo et al. 2020; Rauch et al. 2018). Multidrug nanoparticles (with great biocompatibility and drug loading) for the reduction of uncontrolled inflammation is very promising, particularly in the case of COVID-19 (Callaway 2020; Le et al. 2020; Martinez 2020).

16.3 SARS-COVID

Severe acute respiratory syndrome (SARS) is a respiratory disease caused by novel coronavirus called SARS-associated corona virus (SARS-CoV) that caused the major pandemic of this decade (Drosten et al. 2003). For the first time, SARS has been reported in Asia in 2003 and then the infection was transmitted to more than 12 countries including America, Europe, and other Asian countries before the SARS global outbreak of 2003 is contained.

Recently, the global outbreak has outstretched by the novel pathogenic viral transmission caused by the severe acute respiratory syndrome coronavirus 2 (SARS-CoV-2) which is identified on January 2020 by Chinese scientists and originated in Wuhan, China and spread around the world (Pandemic). The World Health Organization (WHO) officially declared the COVID-19 epidemic as a public health emergency of global concern. Infection with SARS-CoV-2 virus can cause sickness, ranging from common cold to more severely, respiratory diseases, such as SARS and MERS. The virus is highly transmitted human-to-human with rapid rate. The beginning of SARS-CoV-2, since the SARS-CoV in 2002 and Middle East respiratory syndrome coronavirus (MERS-CoV) in 2012, manifests the third introduction of a highly pathogenic and large-scale epidemic coronavirus into the human population in the twenty-first century. Coronavirus is a betacoronavirus, enveloped, non-segmented, single-stranded, positive-sense RNA virus. Coronavirus resembles a crown and the electron microscopic image of Coronavirus is illustrated in Fig. 16.1.

Progressive resources are the principal basis of various technologies that can reduce the effect of Coronavirus. Nanovaccines are the influential proxies that could inhibit viral contaminations and nanosystems with antiviral activity might diminish the virus for its inactivation. Nanotechnology is a versatile technology that can serve

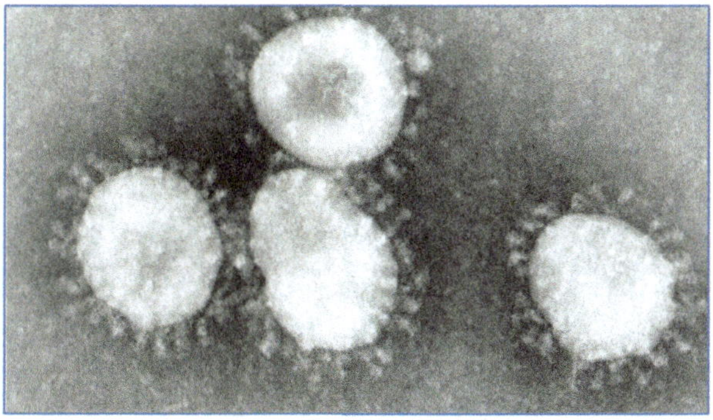

Fig. 16.1 Electronic microscopic image of Coronavirus (Courtesy: The COVID-19 Pandemic: A Summary. A special supplement from Thepathologist.com)

the solution to the pandemic situations such as Coronavirus outbreaks. Nanodiagnostics resemble nanoparticles, which encounter molecules of interest to generate a signal by permitting the recognition of infective viruses (Jackson et al. 2017). Nanotechnology provides significant benefits including rapid analysis, sensitive, highly accurate, and reproducible results. Nanotechnology-based biosensors can greatly reduce the use of chemical reagents needed for the analysis. A biosensor is a device that assembles with a profound biological detection module and a physical transducer to recognize viruses in the body fluids.

The characteristics of nanomaterial based biosensors include stability in fluids, suitable surface chemical properties, more surface energy and high amplification effect to generate measurable signals. Biosensors designed with the Metallic nanoparticles (MNPs), carbon-graphene based nanotubes, and photonic crystals (PCs) are widely used in the field of viral diagnostics. Figure 16.2 illustrates the various types of nanoparticle-based biosensors for SARS-CoV-1 and MERS-CoV detection. The principle involved in the technology is specific adsorption. The developed biosensors are the prominent modules for the development of SARS-CoV-2 biosensors. Table 16.1 presents the Biosensors developed for the detection of closely related COVID19 viruses.

16.4 SARS-CoV-1, MERS-CoV and SARS-CoV-2

16.4.1 Nucleic Acid-based Biosensors

Researchers designed a biosensor based on the surface plasmon resonance (SPR) on a chip with nucleotides for targeting the general respiratory viruses such as SARS-CoV. The test RNA samples were collected and subjected to the RT-PCR and hybridization has been carried out on the nanochip, which intensifies the signal and enhances the biosensor accuracy. The multi-target biosensor for the identification of target genetic materials from different infectious pathogens such as MERS-CoV was evaluated. The arch-shaped biosensors are applied to detect the other

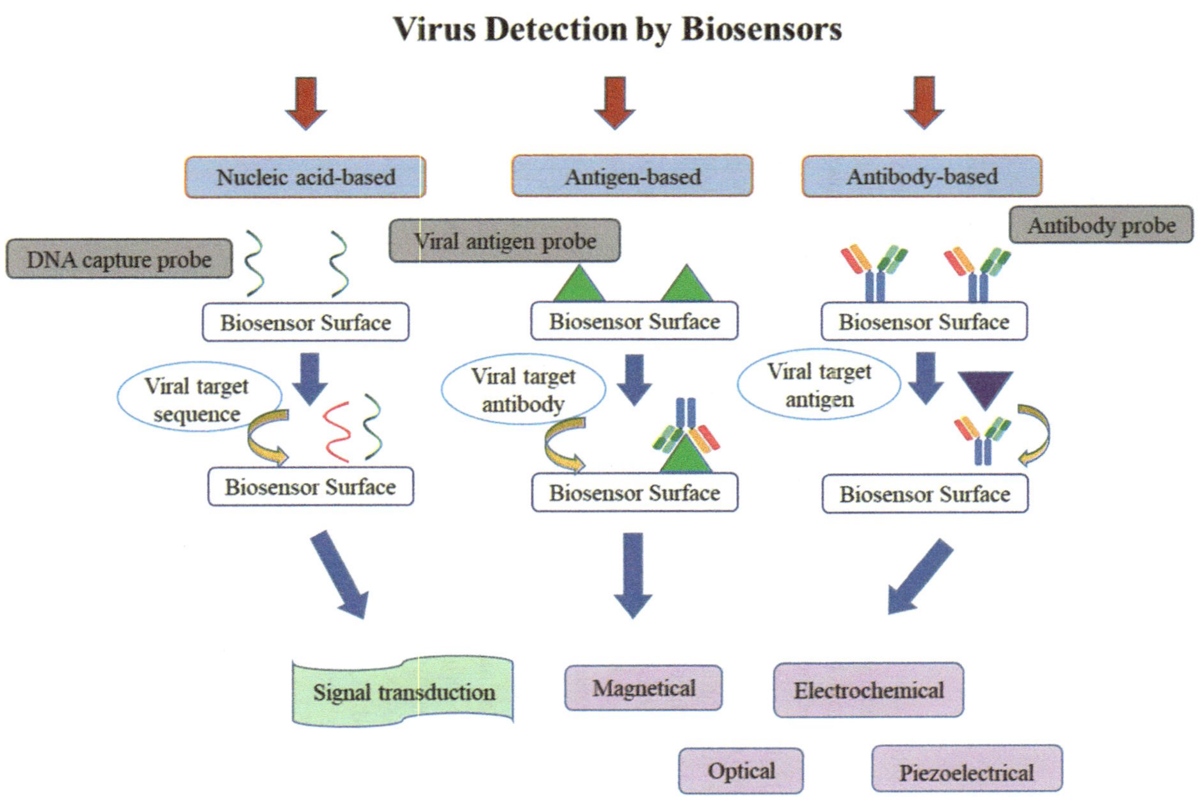

Fig. 16.2 Different types of Biosensors and their principle involved in the detection of virus infections

Table 16.1 Detection of COVID19 viruses (SARS-CoV-1 and MERS-CoV) by Biosensors

Biosensor	Virus	Detection Probe	Detection mechanism	Major findings	Reference
SPR chip with immobilized specific oligonucleotides	SARS-CoV-1	Oligonucleotide (DNA)	RI signal	Simultaneous identification of SARS-CoV-1, influenza A and B, H1N1, PIV-1, PIV-2, PIV-3, RSV, Adenovirus	ShiL et al.
Photo crystals based biosensor chip	SARS-CoV-1	Surface envelope protein	RI signal	Detection of protein-protein, DNA-DNA and protein-metal interactions	Park et al.
Carbon nanotube field-effect transistor (FET)-based biosensor	SARS-CoV-1	Engineered antibody mimic protein (AMP)	Conductance	Efficient detection of SARS nucleocapsid protein	Ishikawa et al.
Arch-shaped multi-target sensor	MERS-CoV	Oligonucleotide (DNA > 50 bp)	Resonance wavelength	Rapid detection of MERS-CoV	Koo et al.
RCA microfluidic device	MERS-CoV	Oligonucleotide (ssDNA)	Naked eye	Multiple detection of pathogens	Na et al.
ESPS microfibers (electrospun polystyrene)	MERS-CoV	His-MERS NP antigen protein	Fluorescence	MERS NP detection	Hoy et al.
Electrochemical immunosensor based on an array of carbon electrodes (DEP) modified with gold nanoparticles	MERS-CoV	S1 protein	Square wave voltammetry	Detection in spiked nasal samples	Layqah et al.
S. aureus nanobioparticles	MERS-CoV	MER-CoV nucleoprotein	Agglutination test	MERS NP detection	Qiao et al.
Dual-functional plasmonic biosensor	SARS-CoV-2	Oligonucleotide (DNA)	Plasmonic resonance wavelength	High sensitivity towards selected SARS-CoV-2 sequences	Qiu et al.
FET-based biosensor (COVID-19 FET)	SARS-CoV-2	Specific antibody against SARS-CoV-2Â S protein	Electrical response	SARS-CoV-2 in medium culture and in nasopharyngeal swab samples from COVID-19 patients	Seo et al.

infectious pathogens including Zika virus, Ebola virus, and SARS-CoV-1.Recently, the plasmonic photo-thermal (PPT) and localized surface plasmon resonance (LSPR) sensing transduction based biosensors are developed to detect the RNA from SARS-CoV-2. The sensor is assimilated on a chip by two-dimensional gold nanoislands (AuNIs). The developed device was also validated for the detection of various genome sequences from both the SARS-CoV-2 and CoV-1 and the significant outcomes are exhibited.

16.4.2 Antigen-based Biosensors

A photosensitive based biosensor is designed for the detection of SARS-CoV-1 S antigen. The biosensor comprised of photonic crystals (PCs) synthesized with chemical carboxyl groups (aldehyde or ketones) for the attachment of S protein, which endorsed identifying SARS-CoV-1 S antibodies. Researchers proposed biosynthesized nanoparticles inside the *Staphylococcus aureus* cells, where the MERS and Ebola virus nucleoproteins ((MERS NP and EBOV NP) are coupled with the help of a cell wall binding domain (CBD) from a bacteriophage lysin PlyV12. The nanoparticles are designed using responding the monotetrazolium redox dye at room temperature for 15 min to produce insoluble formazan crystals inside the cells followed by inactivation at high temperature (65°C). *S. aureus* nanoparticles are used to execute an agglutination test to detect the IgG antibodies of Ebola virus and MERS nanoparticles.

16.4.3 Antibody-based Biosensors

Ishikawa et al., invented biosensors with carbon nanotubes which work based on the field-effect transistor. The designed nanotubes are capable to detect the nucleocapsid (N) protein of SARS-CoV. A novel biosensor based on antibody for the detection of S protein from SARS-CoV-2 is developed. The sensor containing grapheme bundles of the field-effect transistor (FET), which are smeared with CoV-2 antibodies. Furthermore, the performance of biosensor was evaluated for the detection of nasopharyngeal swab samples from COVID-19 patients.

16.4.4 Virus Inhibition Using Nanosystems

Virus inhibition can be done with Nanosystems against SARS-CoV-2 and worked based on the detection of inactivate enveloped viruses. The hydrophobic regions present on the nanosystems interact with the fatty acids on the surface portion of the virus endorsing its denaturation, which causes the inhibition of virus. Researchers developed various types of nanoparticle-based nanosystems to detect the viruses by following the inhibition mechanism. The graphene oxide and their derivatives act as antiviral agents against Porcine epidemic diarrhea virus (PEDV) and Pseudorabies virus. Graphene oxide-silver nanosystems (GO-Ag) are verified against enveloped and non-enveloped virus including feline Coronavirus and infectious bursal disease virus. The magnetic hybrid colloids (MHC) comprised with Ag nanoparticles have a vital role in inactivation of viral pathogens. Functionalized carbon quantum dots (CQD) are the most effective therapy against human coronavirus (HCoV-229E). The carbon quantum dots are produced using an aqueous carbonization process. Ag_2S nanoclusters (NC) having quantum dots with glutathione are the good inhibition properties against coronavirus proliferation. The curcumin based carbon dots are developed with the antiviral activity against intestinal Cononavirus.

Nanosystems can be effectively applied to destroy the virus as well. The electromagnetic irradiation is engaged by producing reactive species (RS) or photo-thermal heating. Hence, the viruses might be destroyed by the redox or denaturation processes. The carbon nanohorns (CNH) synthesized with a polyethylene glycol (PEG) derivative on one side and a T7 promoter tag antibody on the other side PEG is effective approach to kill the virus.

16.4.5 Nanovaccine Models Against COVID-19

Nanotechnology proved effective in combating COVID-19 with the production of nanovaccines. Nanovaccines consist of nanoparticles functioning as delivery vectors for antigens that activate defensive immunity. Various nanoparticles have the inherent immune-stimulatory characteristics that favour vaccine activity. Hence, nanosystems can be effectively used for the designing of vaccines. Virus like Particles (VLPs) mark significant in nanovaccine production. Most of the recombinant vaccines available in the healthcare industry are, indeed, based on VLPs (HBsAg and HPV). Additionally, other nanoparticles based vaccines are being explored including liposomes and particles comprised of gold and chitosan. Additionally, gold nanoparticles (AuNPs) based nanosystems have been demonstrated in the development of nanovaccines against Coronaviruses.

16.5 Conclusion

Microbial nanotechnology is the present day promising field. Nanotechnology-based advancements are commonly denoted to as systematically extracted information and open new avenues of research and applications that will progressively affect all portions of the world. Hence, there is an immense requirement from the scientific community for a complete and comprehensive understanding of the toxicity of the particle if any, their interactions within the ecosystem and ultimately the fate of the biosynthesized nanoparticles so that this powerful technology could be adapted for the welfare of the human and animal welfare as well in the environmental conservation. The lack of exhaustive and complete toxicological data is also due to the actual difficulty to characterize, detect, and measure nanoparticles alone and in complex matrices like food/feed and biological samples. Concerns are elevated as no administration has established a monitoring structure that reveals the nanoparticles. Therefore, extensive research is prerequisite to reach strong and dependable risk assessment procedures for nanomaterials to be applied in medical the field of microbiology and disease diagnosis.

References

Abbaspour A, Norouz-Sarvestani F, Noori A, Soltani N (2015) Aptamer-conjugated silver nanoparticles for electrochemical dual-aptamer-based sandwich detection of *Staphylococcus aureus*. Biosens Bioelectron 68:149–155. https://doi.org/10.1016/j.bios.2014.12.040

Actis L, Srinivasan A, Lopez-Ribot JL et al (2015) Effect of silver nanoparticle geometry on methicillin susceptible and resistant *Staphylococcus aureus*, and osteoblast viability. J Mater Sci Mater Med 26:215. https://doi.org/10.1007/s10856-015-5538-8

Ahmad A, Qureshi AS, Li L et al (2016) Antibacterial activity of grapheme supported FeAg bimetallic nanocomposites. Colloid Surface B 143:490–498. https://doi.org/10.1016/j.colsurfb.2016.03.065

Andersson DI, Hughes D (2010) Antibiotic resistance and its cost: is it possible to reverse resistance? Nat Rev Microbiol 8:260–271. https://doi.org/10.1038/nrmicro2319

Anghel LC, Grumezescu AM, Anghel AG, Bleotu G, Chifiriuc MC (2012) *In vitro* evaluation of antipathogenic surface coating nanofluid, obtained by combining Fe3O4/C12 nanostructures and 2-((4-ethylphenoxy) methyl)- N-(substituted phenylcarbamothioyl)-benzamides. Nanoscale Res Lett 7:1–10. https://doi.org/10.1186/1556-276X-7-513.

Aziz N, Fatma T, Varma A, Prasad R (2014) Biogenic synthesis of silver nanoparticles using *Scenedesmus abundans* and evaluation of their antibacterial activity. J Nanoparticles, Article ID 689419. https://doi.org/10.1155/2014/689419

Aziz N, Faraz M, Pandey R, Sakir M, Fatma T, Varma A, Barman I, Prasad R (2015) Facile algae-derived route to biogenic silver nanoparticles: synthesis, antibacterial and photocatalytic properties. Langmuir 31:11605–11612. https://doi.org/10.1021/acs.langmuir.5b03081

Aziz N, Pandey R, Barman I, Prasad R (2016) Leveraging the attributes of *Mucor hiemalis*-derived silver nanoparticles for a synergistic broad-spectrum antimicrobial platform. Front Microbiol 7:1984. https://doi.org/10.3389/fmicb.2016.01984

Balzani V (2005) Nanoscience and nanotechnology: A personal view of chemist. Small 1 (3):278–283. https://doi.org/10.1002/smll.200400010

Baranwal A, Mahato K, Srivastava A, Maurya PK, Chandra P (2016) Phytofabricated metallic nanoparticles and their clinical applications. RSC Adv 6:105996–106010. https://doi.org/10.1039/C6RA23411A

Beyth N, Houri-Haddad Y, Domb A et al (2015) Alternative antimicrobial approach: nano-antimicrobial materials. Evid-Based Compl Alt 2015:1–16. https://doi.org/10.1155/2015/246012

Callaway E (2020) The race for coronavirus vaccines: a graphical guide. Nature 580:576–577. https://doi.org/10.1038/d41586-020-01221-y

Chandra P, Noh HB, Won MS, Shim YB (2011) Detection of daunomycin using phosphatidylserine and aptamercoimmobilized on Au nanoparticles deposited conducting polymer. Biosens Bioelectron 26:4442–4449. https://doi.org/10.1016/j.bios.2011.04.060

Colquhoun DR, Schwab KJ, Cole RN, Halden RU (2006) Detection of norovirus capsid protein in authentic standards and in stool extracts by matrix-assisted laser desorption ionization and nanospray mass spectrometry. Appl Environ Microbiol 72:2749–2755. https://doi.org/10.1128/AEM.72.4.2749-2755.2006

Costerton JW (1999) Bacterial biofilms: a common cause of persistent infections. Science 284:1318–1322. https://doi.org/10.1126/science.284.5418.1318

Draz MS, Shafiee H (2018) Applications of gold nanoparticles in virus detection. Theranostics 8:1985. https://doi.org/10.7150/thno.23856

Drexler KE, Peterson C (1989) Foresight briefing. Nanotechnology and enabling technologies. http://www.foresight.org/updates/Briefing2.html

Drosten C, Gunther S, Preiser W, van der Werf S, Brodt HR, Becker S, Rabenau H, Panning M, Kolesnikova L, Fouchier RA, Berger A, Burguiere AM, Cinatl J, Eickmann M, Escriou N, Grywna K, Kramme S, Manuguerra JC, Muller S, Rickerts V, Sturmer M, Vieth S, Klenk HD, Osterhaus AD, Schmitz H, Doerr HW (2003) Identification of a novel coronavirus in patients with severe acute respiratory syndrome. N Engl J Med 348:1967–1976. https://doi.org/10.1056/NEJMoa030747

Drulis-Kawa Z, Dorotkiewicz-Jach A, Gubernator J, Gula G, Bocer T, Doroszkiewicz W (2009) The interaction between *Pseudomonas aeruginosa* cells and cationic PC: Chol: DOTAP liposomal vesicles versus outer-membrane structure and envelope properties of bacterial cell. Int J Pharm 367(1):211–219. https://doi.org/10.1016/j.ijpharm.2008.09.043

Duran N, Duran M, de Jesus MB et al (2015) Silver nanoparticles: a new view on mechanistic aspects on antimicrobial activity. Nanomedicine 12:789–799. https://doi.org/10.1016/j.nano.2015.11.016

Dykman LA, Khlebtsov NG (2011) Gold nanoparticles in biology and medicine. Recent Advances and Prospects Acta Naturae 3(2):34–55. https://doi.org/10.1039/C1CS15166E

El-Zahry MR, Mahmoud A, Refaat IH et al (2015) Antibacterial effect of various shapes of silver nanoparticles monitored by SERS. Talanta 138:183–189. https://doi.org/10.1016/j.talanta.2015.02.022

Feynman RP (1960) There's plenty of room at the bottom. Eng Sci 23(5):22–36. ISSN 0013-7812

Franci G, Falanga A, Galdiero S et al (2015) Silver nanoparticles as potential antibacterial agents. Molecules 20:8856–8874. https://doi.org/10.3390/molecules20058856

Giamberardino A, Labib M, Hassan EM, Tetro JA, Springthorpe S, Sattar SA, Berezovski MV, DeRosa MC (2013) Ultrasensitive norovirus detection using DNA aptasensor technology. PLoS One 8:e79087. https://doi.org/10.1371/journal.pone.0079087

Gollmer D, Walter F, Lorch C, Novák J, Banerjee R, Dieterle J et al (2014) Fabrication and characterization of combined metallic nanogratings and ITO electrodes for organic photovoltaic cells. Microelectron Eng 119:122–126. https://doi.org/10.1016/j.mee.2014.03.042

Guo YJ, Sun GM, Zhang L, Tang YJ, Luo JJ, Yang PH (2014) Multifunctional optical probe based on gold nanorods for detection and identification of cancer cells. Sens Actuators B 191:741–749. https://doi.org/10.1016/j.snb.2013.10.027

He W, Zhou YT, Wamer WG et al (2013) Intrinsic catalytic activity of Au nanoparticles with respect to hydrogen peroxide decomposition and superoxide scavenging. Biomaterials 34:765–773. https://doi.org/10.1016/j.biomaterials.2012.10.010

Inamuddin, Ahamed MI, Prasad R (2021) Advanced antimicrobial materials and applications. Springer, Singapore. ISBN 978-981-15-7098-8. https://www.springer.com/gp/book/9789811570971

Jackson TC, Patani BO, Ekpa DE (2017) Nanotechnology in diagnosis: a review. Adv Nanoparticles 6(3):93–102. https://doi.org/10.4236/anp.2017.63008

Johansson EMV, Crusz SA, Kolomiets E et al (2008) Inhibition and dispersion of Pseudomonas aeruginosa biofilms by glycopeptides dendrimers targeting the fucose-specific lectin LecB. Chem Biol 15(12):1249–1257. https://doi.org/10.1016/j.chembiol.2008.10.009

Kaittanis C, Santra S, Perez JM (2010) Emerging nanotechnology-based strategies for the identification of microbial pathogenesis. Adv Drug Deliv Rev 62(4–5):408–423. https://doi.org/10.1016/j.addr.2009.11.013

Kang JH, Asami Y, Murata M, Kitazaki H, Sadanaga N, Tokunaga E, Shiotani S, Okada S, Maehara Y, Niidome T (2010) Gold nanoparticle-based colorimetric assay for cancer diagnosis. Biosens Bioelectron 25(8):1869–1874. https://doi.org/10.1021/ac702037y

Kattke MD, Gao EJ, Sapsford KE, Stephenson LD, Kumar A (2011) FRET-based quantum dot immunoassay for rapid and sensitive detection of *Aspergillus amstelodami*. Sensors 11:6396–6410. https://doi.org/10.3390/s110606396

Kerry RG, Malik S, Redda YT, Sahoo S, Patra JK, Majhi S (2019) Nano-based approach to combat emerging viral (NIPAH virus) infection. Nanomed Nanotechnol Biol Med 18:196–220. https://doi.org/10.1016/j.nano.2019.03.004

Khoris IM, Takemura K, Lee J, Hara T, Abe F, Suzuki T, Park EY (2019) Enhanced colorimetric detection of norovirus using in-situ growth of Ag shell on Au NPs. Biosens Bioelectron 126:425–432. https://doi.org/10.1016/j.bios.2018.10.067

Kim EY, Stanton J, Korber BT et al (2008) Detection of HIV-1 p24 Gag in plasma by a nanoparticle-based bio-barcode-amplification method. Nanomedicine 3:293–303. https://doi.org/10.2217/17435889.3.3.293

Kumar CSSR (2006) Nanomaterials for biosensors. Wiley-VCH Weinheim. https://doi.org/10.1016/C2015-0-04697-4

Kurdekar AD, Chunduri LAA, Chelli SM et al (2017) Fluorescent silver nanoparticle based highly sensitive immunoassay for early detection of HIV infection. RSC Adv 7:19863–19877. https://doi.org/10.1039/C6RA28737A

Le T, Andreadakis Z, Kumar A, Gómez Román R, Tollefsen S, Saville M, Mayhew S (2020) The COVID-19 vaccine development landscape. Nat Rev Drug Discov 19:305–306. https://doi.org/10.1038/d41573-020-00073-5

Lee K-B, Kim E-Y, Mirkin CA, Wolinsky SM (2004) The use of nanoarrays for highly sensitive and selective detection of human immunodeficiency virus type 1. Nano Lett 10:1869–1872. https://doi.org/10.1021/nl049002y

Lee ALZ, Ng VWL, Wang W, Hedrick JL, Yang YY (2013) Block copolymer mixtures as antimicrobial hydrogels for biofilm eradication. Biomaterials 34(38):10278–10286. https://doi.org/10.1016/j.biomaterials.2013.09.029

Lee JH, Kim YG, Cho MH, Lee J (2014) ZnO nanoparticles inhibit Pseudomonas aeruginosa biofilm formation and virulence factor production. Microbiol Res 169(12):888–896. https://doi.org/10.1016/j.micres.2014.05.005

Li Y, Tseng YD, Kwon SY, Despaux L, Bunch JS, Mceuen PL (2004) Controlled assembly of dendrimer-like DNA. Nat Mater 3:38–42. https://doi.org/10.1038/nmat1045

Liu K, Qu S, Zhang X, Tan F, Wang Z (2013) Improved photovoltaic performance of silicon nanowire/organic hybrid solar cells by incorporating silver nanoparticles. Nanoscale Res Lett 8:88. https://doi.org/10.1186/1556-276X-8-88

Magiorakos AP, Srinivasan A, Carey RB et al (2012) Multidrug-resistant, extensively drug-resistant and pandrug-resistant bacteria: an international expert proposal for interim standard definitions for acquired resistance. Clin Microbiol Infect 18:268–281. https://doi.org/10.1111/j.1469-0691.2011.03570.x

Mapara N, Sharma M, Shriram V et al (2015) Antimicrobial potentials of Helicteresisora silver nanoparticles against extensively drugresistant (XDR) clinical isolates of *Pseudomonas aeruginosa*. Appl Microbiol Biotechnol 99:10655–10667. https://doi.org/10.1007/s00253-015-6938-x

Martinez MA (2020) Compounds with therapeutic potential against novel respiratory 2019 Coronavirus. Antimicrob Agents Chemother 64:e00399–e00320. https://doi.org/10.1128/AAC.00399-20

Minocheherhomji FP (2016) Microorganisms in environment: boon and bane. Int J Adv Res 4(10):826–830. https://doi.org/10.21474/IJAR01/1864

Mollasalehi H, Yazdanparast R (2013) An improved non-crosslinking gold nanoprobe-NASBA based on 16S rRNA for rapid discriminative biosensing of major salmonellosis pathogens. Biosens Bioelectron 47:231–236. https://doi.org/10.1016/j.bios.2013.03.012

Pal S, Tak YK, Song JM (2015) Does the antibacterial activity of silve nanoparticles depend on the shape of the nanoparticle? study of the gram-negative bacterium *Escherichia coli*. J Biol Chem 290:1712–1720. https://doi.org/10.1128/AEM.02218-06

Paul S, Chugh A (2011) Assessing the role ayurvedic bhasmas as ethano-nanomedicine in the metal based nanomedicine patent regime. J Intellect Prop Rights 16:509–515

Perez JW, Vargis EA, Russ PK, Haselton FR, Wright DW (2011) Detection of respiratory syncytial virus using nanoparticle amplified immuno-polymerase chain reaction. Anal Biochem 410:141–148. https://doi.org/10.1016/j.ab.2010.11.033

Prasad R (2019a) Microbial nanobionics: basic research and applications. Springer International Publishing, Cham. ISBN 978-3-030-16534-5. https://www.springer.com/gp/book/9783030165338

Prasad R (2019b) Microbial nanobionics: state of art. Springer International Publishing, Cham. ISBN 978-3-030-16383-9 https://www.springer.com/gp/book/9783030163822

Prasad R, Pandey R, Barman I (2016) Engineering tailored nanoparticles with microbes: quo vadis. WIREs Nanomed Nanobiotechnol 8:316–330. https://doi.org/10.1002/wnan.1363

Prasad R, Jha A, Prasad K (2018) Exploring the realms of nature for nanosynthesis. Springer International Publishing, Cham. ISBN 978-3-319-99570-0. https://www.springer.com/978-3-319-99570-0

Prasad R, Siddhardha B, Dyavaiah M (2020) Nanostructures for antimicrobial and antibiofilm applications. Springer International Publishing, Cham. ISBN 978-3-030-40336-2. https://www.springer.com/gp/book/9783030403362

Rai M, Yadav A, Gade A (2009) Silver nanoparticles as a new generation of antimicrobials. Biotechnol Adv 27:76–83. https://doi.org/10.1016/j.biotechadv.2008.09.002

Ramasamy M, Kim S, Lee SS, Yi DK (2016) Recyclable photothermalnano-aggregates of magnetic nanoparticle conjugate gold nanorods for effective pathogenic bacterialysis. J Nanosci Nanotechnol 16(1):555–561. https://doi.org/10.1166/jnn.2016.10603

Rauch S, Jasny E, Schmidt KE, Petsch B (2018) New vaccine technologies to combat outbreak situations. Front Immunol 9:1963. https://doi.org/10.3389/fimmu.2018.01963

Reidy B, Haase A, Luch A et al (2013) Mechanisms of silver nanoparticle release, transformation and toxicity: a critical review of current knowledge and recommendations for future studies and applications. Materials (Basel) 6:2295–2350. https://doi.org/10.3390/ma6062295

Rtimi S, Sanjines R, Pulgarin C et al (2016) Microstructure of Cu-Ag uniform nanoparticulate films on polyurethane 3D catheters: surface properties. ACS Appl Mater Interfaces 8:56–63. https://doi.org/10.1021/acsami.5b09738

Sallem S, Ahmed B, Khan MS, Al-Shaeri S, Musarrat J (2017) Inhibition of growth and biofilm formation of clinical bacterial isolates by NiO nanoparticles synthesized from *Eucalyptus globulus* plants. Microb Pathog 111:375–387. https://doi.org/10.1016/j.micpath.2017.09.019

Seo G, Lee G, Kim MJ, Baek SH, Choi M, Ku KB, Lee CS, Jun S, Park D, Kim HG et al (2020) Rapid detection of COVID-19 causative virus (SARS-CoV-2) in human nasopharyngeal swab specimens using field-effect transistor-based biosensor. ACS Nano 14:5135–5142. https://doi.org/10.1021/acsnano.0c02823

Shinde NC (2012) Nanoparticles: advances in drug delivery systems. Res J Pharm Biol Chem Sci 3 (1):922–929. https://doi.org/10.33887/rjpbcs

Singh R, Nawale LU, Arkile M (2015) Chemical and biological metal nanoparticles as antimycobacterial agents: a comparative study. Int J Antimicrob Agents 46:183–188. https://doi.org/10.1016/j.ijantimicag.2015.03.014

Singh N, Dahiya B, Radhakrishnan VS, Prasad T, Mehta PK (2018) Detection of *Mycobacterium tuberculosis* purified ESAT-6 (Rv3875) by magnetic bead-coupled gold nanoparticle-based immuno-PCR assay. Int J Nanomedicine 13:8523–8535. https://doi.org/10.2147/IJN.S181052

Sinha R, Karan R, Sinha A, Khare SK (2011) Interaction and nanotoxic effect of ZnO and Ag nanoparticles on mesophilic and halophilic bacterial cells. Bioresour Technol 102 (2):1516–1520. https://doi.org/10.1016/j.biortech.2010.07.117

Sonawane SJ, Kalhapure RS, Rambharose S, Mocktar C, Vepuri SB, Soliman M, Govender T (2016) Ultrasmall lipid-dendrimer hybrid nanoparticles as a promising strategy for antibiotic delivery: in vitro and in silico studies. Int J Pharm 504(1–2):1–10. https://doi.org/10.1016/j.ijpharm.2016.03.021

Srivastava S, Usmani Z, Atanasov AG, Singh VK, Singh NP, Abdel-Azeem AM, Prasad R, Gupta G, Sharma M, Bhargava A (2021) Biological nanofactories: using living forms for metal nanoparticle synthesis. Mini-Rev Med Chem 21(2):245–265

Takemura K, Adegoke O, Takahashi N, Kato T, Li TC, Kitamoto N, Tanaka T, Suzuki T, Park EY (2017) Versatility of a localized surface plasmon resonance-based gold nanoparticle-alloyed quantum dot nanobiosensor for immunofluorescence detection of viruses. Biosens Bioelectron 89:998–1005. https://doi.org/10.1016/j.bios.2016.10.045

Tao Y, Ju E, Ren J et al (2015) Bifunctionalized mesoporous silica supported gold nanoparticles: intrinsic oxidase and peroxidase catalytic activities for antibacterial applications. Adv Mater 27:1097–1104. https://doi.org/10.1002/adma.201405105

Thakur MS, Ragavan KV (2013) Biosensors in food processing. J Food Sci Technol 50:625–641. https://doi.org/10.1007/s13197-012-0783-z

Wang X, Liu LH, RamstroemO YM (2009) Engineering nanomaterial surfaces for biomedical applications. Exp Biol Med 234:1128–1139. https://doi.org/10.3181/0904-MR-134

Watnick P, Kolter R (2000) Biofilm, city of microbes. J Bacteriol 182:2675–2679. https://doi.org/10.1128/JB.182.10.2675-2679.2000

Weidemaier K, Carruthers E, Curry A, Kuroda M, Fallows E, Thomas J, Sherman D, Muldoon M (2015) Real-time pathogen monitoring during enrichment: a novel nanotechnology based approach to food safety testing. Int J Food Microbiol 198:19–27. https://doi.org/10.1016/j.ijfoodmicro.2014.12.018

Zehbe I, Hacker GW, Su H, Hauser-Kronberger C, Hainfeld JF, Tubbs R (1997) Sensitive in situ hybridization with catalyzed reporter deposition, streptavidin-nanogold, and silver acetate autometallography: detection of single-copy human papillomavirus. Am J Pathol 150:1553

Zhang L, Luo J, Menkhaus TJ, Varadaraju H, Sun Y, Fong H (2011) Antimicrobial nano-fibrous membranes developed from electrospun polyacrylonitrile nanofibers. J Membr Sci 369:499–505. https://doi.org/10.1016/j.memsci.2010.12.032

Zhang S, Pan W, Liang Q, Song X (2016) Exonuclease I manipulating primer modified gold nanoparticles for colorimetric telomerase activity assay. Biosens Bioelectron 77:144–148. https://doi.org/10.1016/j.bios.2015.08.045

Zhu S, Li J, Huang AG, Huang JQ, Huang YQ, Wang GX (2019) Anti-betanodavirus activity of isoprinosine and improved e_cacy using carbon nanotubes based drug delivery system. Aquaculture 512:734377. https://doi.org/10.1016/j.aquaculture.2019.734377

Chapter 17
Chitosan Nanoparticles: An Overview on Preparation, Characterization and Biomedical Applications

Arundathi Mesa, Grace Sugandha Sowjanya Mythatha, Rathna Silviya Lodi, Sudheer Ravuri, and Ramesh Balli

Abstract Chitosan (CS) and Chitosan nanoparticles (CNPs) have multifaceted applications in medicine, agriculture, pharmaceutics, tissue engineering, waste water treatment and food industries. CS is recognized as a less or non-toxic, biocompatible polymer by US Food and Drug Administration (FDA) for wound dressing as well as in dietary application. The properties of CS have upgraded by making their nanoparticles. Due to their exceptional properties including nanosize with large surface area to volume ratio, presence of reactive groups ($-NH_2$ and $-OH$), cationic nature (NH_3^+), bioadhesivity, biocompatibility, bioavailability and biodegradable nature; CNPs are explored in many ways in biomedical filed as an antimicrobial agent, wound healing agent, scaffolds for tissue engineering, anti-tumour agent in cancer therapy, carriers for gene and drug delivery, etc. In this chapter we highlight on CNPs preparation, characterization and certain important biomedical applications.

Keywords Chitosan (CS) · Chitosan nanoparticles (CNPs) · Scanning electron microscopy (SEM) · Transmission electron microscopy (TEM) · Tripolyphosphate (TPP)

A. Mesa · G. S. S. Mythatha · R. Balli (✉)
Department of Genetics and Genomics, Yogi Vemana University, Kadapa, Andhra Pradesh, India

R. S. Lodi
International Genome Center, Jiangsu University, Zhenjiang, Jiangsu, China

S. Ravuri
Center for Regenerative Sports Medicine (CRSM), Steadman Philippon Research Institute (SPRI), Vail, CO, USA

N. R. Maddela et al. (eds.), *Nanotechnology for Advances in Medical Microbiology*, Environmental and Microbial Biotechnology,
https://doi.org/10.1007/978-981-15-9916-3_17

Abbreviations

AFM	Atomic force microscopy
CNPs	Chitosan nanoparticles
CS	Chitosan
DA	Degree of acetylation
DD	Degree of deacetylation
DLS	Dynamic light scattering
ELS	Electrophoretic light scattering
FTIR	Fourier-transform infrared spectroscopy
HMW	High molecular weight
IBV	Infectious bronchitis virus
LMW	Low molecular weight
MMW	Medium molecular weight
nAg	Silver nanoparticles
nCu	Copper nanoparticles
NDV	New castle disease virus
NPs	Nanoparticles
$nTiO_2$	Titanium dioxide nanoparticles
nZnO	Zinc oxide nanoparticles
PDI	Polydispersity index
PEC	Polyelectrolyte complex
SEM	Scanning electron microscopy
TEM	Transmission electron microscopy
TPP	Tripolyphosphate
XRD	X-ray diffraction

17.1 Introduction

Marine products have been in the forefront of natural materials used in the therapeutic applications against several human diseases (Venugopal 2008). Majority of marine products are derived from exoskeletons of crustaceans such as crabs, shrimps, krills and lobsters. Chitin (polysaccharide) is one of the biopolymers majorly extracts from shells of crustaceans, cell walls of fungi and certain insects. Since chitin is insoluble in many common solvents, it is not widely used for fabrication of products or a food commodity (Tsigos et al. 2000; Crini 2006). The deacetylated chitin, i.e., chitosan has remarkable properties including biocompatibility, biodegradability, mucoadhesive, non-antigenic, non-toxic, solubility in weak acids, cationic nature, and hence, used in many biomedical applications (Gupta et al. 2019; Onsosyen and Skaugrud 1990; Felt et al. 1998; Han et al. 1999; Zhang and Zhang 2002; Kmiec et al. 2017; Guo and DiPietro 2010). The United States Food and Drug Administration (USFDA) has recognized chitosan as a GRAS (Generally

Recognized as Safe) material, and approved to use in food, agriculture and biomedicine (Kumar et al. 2019). For the last two decades, extensive studies have been conducted on fabrication and application of chitosan based nanocomposites in various fields including medicine, pharmaceutics and agriculture. Most of the research studies revealed that chitosan nanoparticles have superior physiochemical and biocompatible properties over CS, and have significant attraction in tissue engineering, biomedicine, drug delivery and cancer therapy. Due to small size, CNPs possess larger surface area to volume ratio, and hence, used as delivery vehicles for anti-cancer drugs, anti-inflammatory drugs, vaccines, antibiotics, peptides (arginine-glycine-aspartate; RGD), aptamers, folate, glycoproteins, polysaccharides, genes, growth factors, etc. (Jayasuriya 2017). However, due to the larger surface area and charge, CNPs can be readily absorb impurities from the medium, and hence, to overcome this limitation, selection of appropriate method for their fabrication and further characterization is an important key parameter. This chapter describes various methods and their principles for fabrication and characterization of CNPs for biomedical applications.

17.1.1 Chitin

Chitin is the most abundant linear biopolymer and structural polysaccharide widely occurring in the nature after cellulose. It is found in the exoskeletons of crustaceans, cell walls of fungi and in certain invertebrates. Various crystalline allomorphs of chitin are α, β and γ, which differ in orientation of microfibrils. The α-forms are more abundant and stable than β and γ forms. The α-forms are mainly present in cell walls of fungi, shells of crustaceans and arthropods, whereas β and γ forms are found in in squid pens, *Ptinus* beetles and Loligo squids (Jang et al. 2004; Carlstrom 1957). Insolubility of chitin in solvents such as water, organic solvents and basic solutions is due to the generation of intra- and intermolecular hydrogen bonds with acetyl, amino and hydroxyl groups of its polysaccharide chain. Insolubility of chitin affects the production of chitin based products. Crustacean's wastes from fish processing industries are the main source of chitin.

17.1.2 Chitosan

Chitosan is a natural linear polysaccharide synthesized by deacetylation of chitin. It is composed of D-glucosamine and N- acetyl D-glucosamine sub-units. The polycationic nature of chitosan is remarkable as most of the polysaccharides in the acidic solutions are either neutral (or) negatively charged. Because of this specific property, it forms electrostatic complexes with negatively charged polymers, lipids, proteins and DNA (Venkatesan and Kim 2010; Pavinatto et al. 2010; Madihally and Matthew 1999; Takahashi et al. 1990; Kim et al. 2007). CS has exceptional properties such as non-toxicity, low allergenicity, biocompatibility, hydrophilicity, antimicrobial activity, bioactivity (Kumar et al. 2004; Gällstedt and Hedenqvist 2006; Pillai et al. 2009), and hence, it is widely used in various applications include, among other, wound healing (Chandy and Sharma 1990), waste water treatment (Onsosyen and Skaugrud 1990; Kumar et al. 2019), drug carrier (Felt et al. 1998), treatment for obesity (Han et al. 1999), and as a scaffolds for tissue engineering (Zhang and Zhang 2002).

CS is insoluble in aqueous solutions, however, it is soluble in dilute aqueous acidic solutions such as acetic acid ($pH < 6.3$). CS solubility decreases as pH increases. At lower pHs, CS becomes protonated and shows stronger antimicrobial activity. Solubility of CS in aqueous solutions can be improved by chemical modifications, such as quaternization of nitrogen atoms of the amino groups (Goy et al. 2009).

17.2 Chitin to Chitosan Nanoparticles

Fabrication of CNPs from exoskeletons of crustaceans or composite chitin involves many methodologies including, extraction of pure chitin, chitin deacetylation for chitosan, preparation and characterization of CNPs. Further, these CNPs can be used for various biomedical applications and drug delivery. Overall strategies for CNPs fabrication are represented in the pictorial diagram (Fig. 17.1)

Marine bio-wastes such as shells of crustaceans used as a principal raw material for industrial production of chitin. The exoskeleton of crustaceans is composed of chitin (15–40%), proteins (20–40%), calcium carbonate (20–50%), pigments and lipids (Yan and Chen 2015). This composite chitin is much harder than the pure chitin. Two main extraction methods including conventional chemical extraction and biological extraction were followed to extract the pure chitin from composite chitin in industrial processing. Further, the chitin is converted to chitosan by deacetylation (chemical or enzymatic deacetylation). Chemical extraction of chitin is suitable for large scale production, however, it has many disadvantages like high energy consumption, high environmental pollution and difficulty in recovering waste products (pigments and proteins) (Gortari and Hours 2013; Cheung et al. 2015; Manni et al. 2010). Proper washing, drying, grinding and sieving up to 1 mm size of shells of crustaceans are the common initial steps in both extraction methods.

17.2.1 Chemical Extraction

In this extraction method, the composite chitin under goes demineralization, deproteinization and decoloration/bleaching to produce pure chitin. Demineralization majorly involves separation of minerals such as calcium carbonate and calcium phosphate by concentrated/diluted acid (HCl, HNO_3, H_2SO_4 and CH_3COOH)

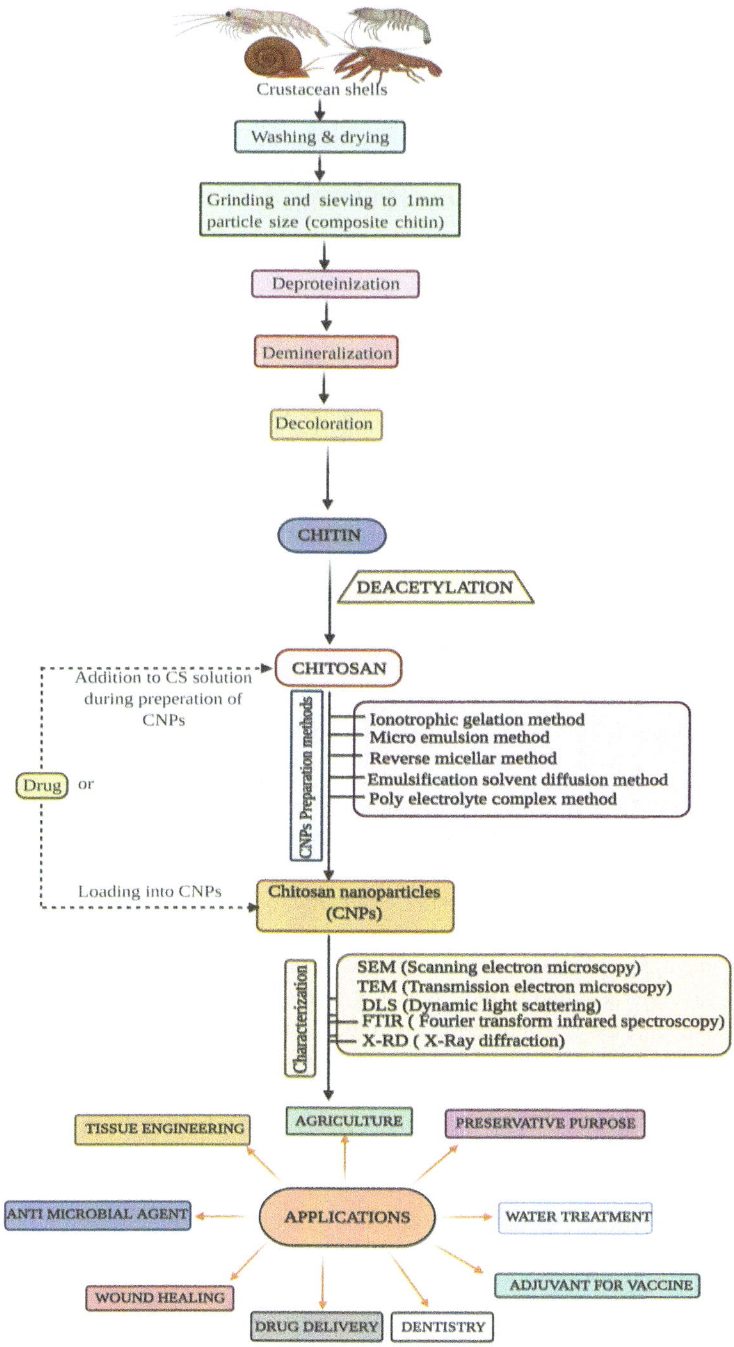

Fig. 17.1 Diagram representing overall fabrication methodologies of CNPs from chitin and their applications

treatment at room temperature. Then the alkaline (NaOH/KOH) treatment at high temperature (100°C to 120°C) separates proteins (deproteinization). Order of these two methods is interchangeable based on the source and proposed use of chitin. The third phase involves separation of pigments such as carotenoids (mainly astaxanthin and its esters) by NaOCl, H_2O_2 or $KMnO_4$ treatment. The fine white powder of pure chitin is subsequently deacetylates and converts into chitosan by alkaline treatment (25–50% NaOH) at high temperature (80°–140°).

17.2.2 Biological Extraction

In this method, demineralization and subsequent deproteinization are carried out by means of microbial (bacteria: *Lactobacillus* sp. *Pseudomonas* sp. *Bacillus* sp./Fungi, *Aspergillus* sp.) fermentation and proteolytic enzymes (chymotrypsin, trypsin alcalase, pepsin, papain, devolvase and pancreatin) treatment, respectively. Demineralization must be the prior step to deproteinization, since the minerals can inhibit the activity of proteases. The resultant pure chitin is then deacetylated by chemical (NaOH)/enzyme (chitin deacetylase) treatment to convert into chitosan.

17.3 Preparation of Chitosan Nanoparticles

Nanoparticles (NPs) are small solid colloidal particles ranging from 10 to 1000 nm, provide large surface area to volume ratio and unique physiochemical properties that allows them in enormous applications (Du et al. 2009; Prasad et al. 2016, 2017). NPs shows more specialized characteristics compared to their bulk materials, because, as the size decreases, the percentage of surface atoms increases (Gupta et al. 2007). Studies have revealed that CNPs can be acquired unique physiochemical and biological properties than to their bulk CS form. Therefore, CNPs are widely used in drug delivery, tissue engineering and other biomedical fields.

CNPs were first synthesized in 1994 by Ohya and co-workers through emulsification and cross-linking method for intravenous delivery of anti-cancer drug 5-fluorouracil (Grenha 2012). Since then, many methods have been employed for synthesis of CNPs, these include, Ionotropic gelation, microemulsion, emulsification solvent diffusion, polyelectrolyte complex and reverse micellar method (Tiyaboonchai 2003). All these methods comprise bottom-up fabrication processes, which involves the assembly of molecules in solution to form defined structures (Chan and kwok 2011).

17.3.1 Ionotropic Gelation Method

This method is relatively simple and mild, and conducted at aqueous conditions without the use of any organic solvent (Fig. 17.2). It was first reported by Calvo et al. (1997). The main strategy of this method is to establish electrostatic interactions between the cationic chitosan polymer and polyanion like tripolyphosphate (TPP) or sodium sulphate with or without stabilizing agent such as poloxamer. In this method, firstly, chitosan is dissolved in acetic acid aqueous solution to become cationic polymer and further the solution is allowed to react with polyanions. It leads to the formation of CNPs under constant stirring at room temperature. Physiochemical properties of nanoparticles (size and surface charge) could be modulated by changing the ratio of chitosan and TPP, and the pH value of the solution (Calvo et al. 1997).

17.3.2 Microemulsion Method

In this method, Chitosan nanoparticles are prepared using surfactant, for example, AOT (sodium bis (2-ethylhexyl) sulfosuccinate) and a cross linker, glutaraldehyde. Initially, surfactant/hexane mixture is prepared by dissolving surfactant into n-hexane, thereafter, the chitosan solution and glutaraldehyde are added into the above mixture by continuous stirring at room temperature. Overnight stirring allows establishing cross link between free amine group of chitosan and glutaraldehyde.

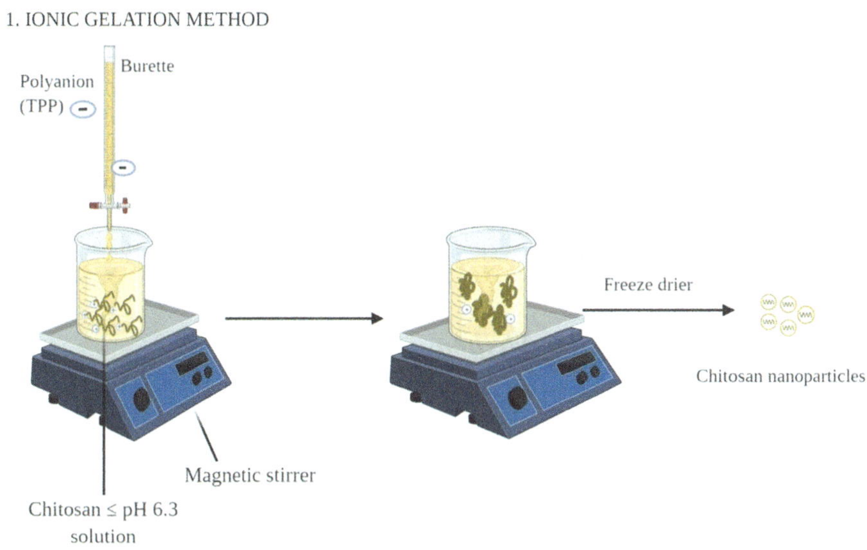

Fig. 17.2 Schematic diagram representation of Ionotropic gelation method

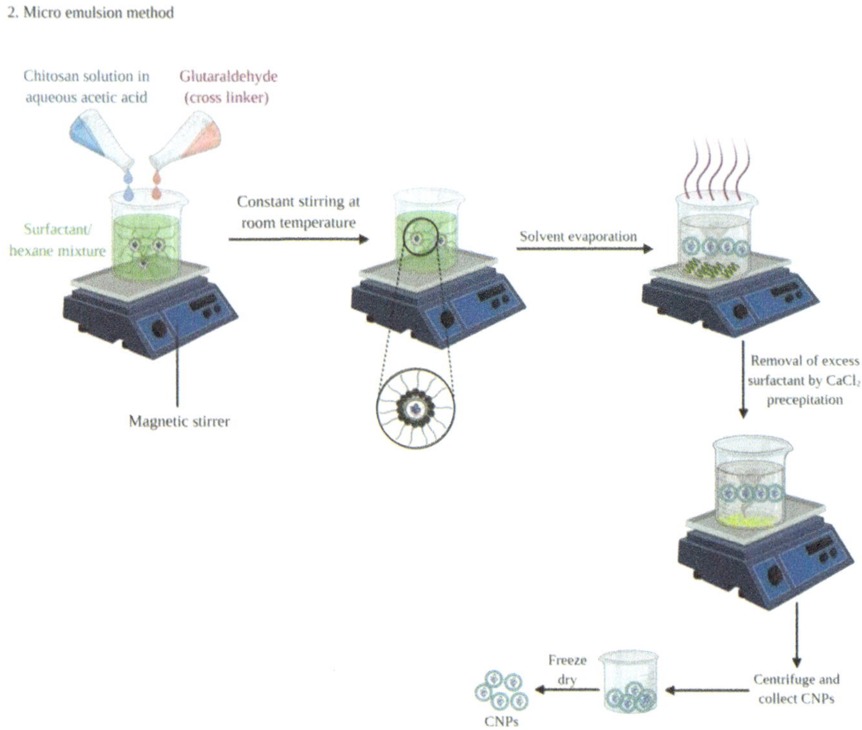

Fig. 17.3 Schematic diagram representation of microemulsion method

Further, the organic solvent and excess surfactant are removed by evaporation under low pressure and precipitation with $CaCl_2$ followed by centrifugation, respectively. The resultant nanoparticle suspension is then dialyzed and lyophilized (Maitra et al. 1997). In this method less than 100 nm sized nanoparticles can be produced, and further the size could be altered by varying the concentration of glutaraldehyde (Sailaja et al. 2011).Usage of glutaraldehyde (toxic agent) and cumbersome process could be the disadvantages of this method (Fig. 17.3).

17.3.3 Reverse Micellar Method

This method was reported by Brunel et al. (2008), it is an adoption of microemulsion method. The method is free from addition of cross linker and toxic organic solvents. In brief, the surfactant is dissolved in organic solvent, to which added chitosan aqueous solution under constant stirring to obtain reverse micelles. By this method extremely thin nanoparticles can be formed (Fig. 17.4).

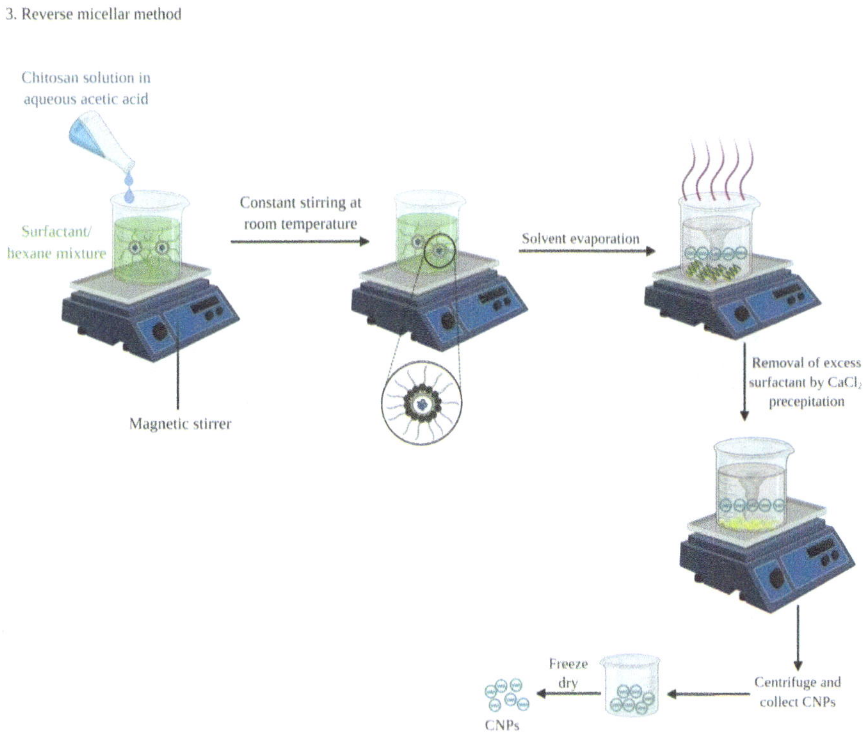

Fig. 17.4 Schematic diagram representation of reverse micellar method

17.3.4 Emulsification Solvent Diffusion Method

This method was first discovered for the fabrication of poly D, L-lactide/glycolide (PLGA) nanoparticles (Niwa et al. 1993), and later it was adapted to prepare chitosan nanoparticles by El-Shabouri (El-Shabouri 2002). In this method, an organic phase (e.g., methylene chloride and acetone) was injected into chitosan solution containing stabilizing agent (e.g., poloxamer and lecithin) under high shearing force, followed by high-pressure homogenization (Fig. 17.5). The resultant emulsion is then diluted with more water to overcome the organic solvent miscibility in water. Polymer precipitation occurs upon the diffusion of organic solvent into water, which subsequently leads to the formation of NPs. High percentage of hydrophobic drug entrapment could be achieved by this approach, however, use of organic solvents and high shearing forces are again the major drawbacks of this method (Mohammed et al. 2017).

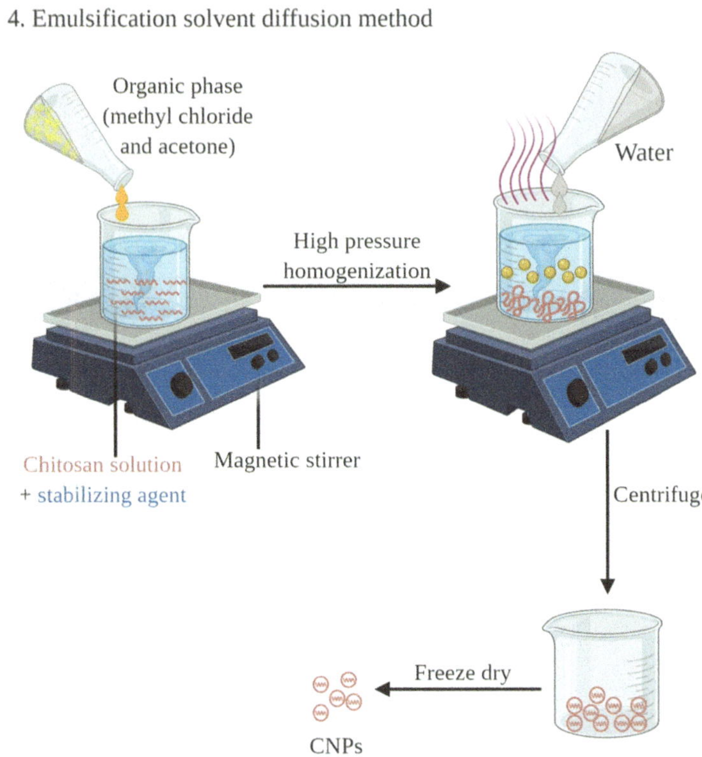

Fig. 17.5 Schematic diagram representation of emulsification solvent diffusion method

17.3.5 Polyelectrolyte Complex Method (PEC)

It is quite simple and not requires any catalysts/ initiators or toxic organic solvents for preparation of nanoparticles. PECs are self-assembled NPs, resulted from electrostatic interactions between cationic polymer and polyanions. The cationic chitosan polymer ($pH < 6.0$) is spontaneously associated with polyanions of chondroitin sulphate and hyaluronate (Denuzier et al. 1998), dextran sulphate (Chen et al. 2003; Chen et al. 2007), carboxymethyl cellulose (Ichikawa et al. 2005), heparin (Liu et al. 2001; Tan Tang et al. 2011), and DNA (Erbacher et al. 1998) in solutions to form PECs (Fig. 17.6).

17.4 Characterization of Chitosan Nanoparticles

Characterization of NPs tells about their physiochemical properties include, among others, particle size and size distribution, surface morphology and surface charge, etc. Particle morphology and surface characteristics can be studied by Scanning

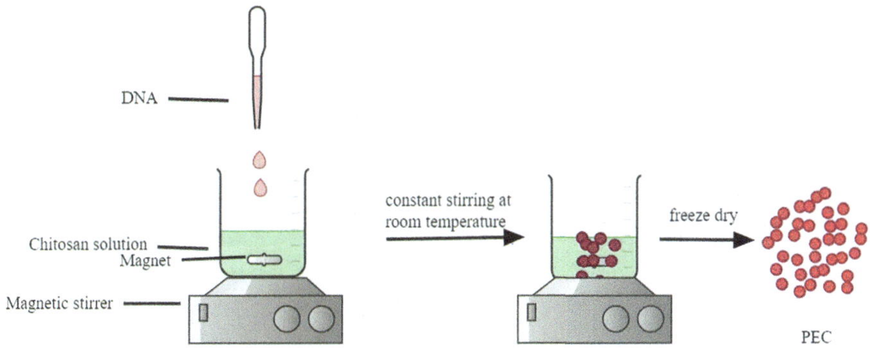

Fig. 17.6 Schematic diagram representation of formation of PEC

electron microscopy (SEM), Transmission electron microscopy (TEM) and Atomic force microscopy (AFM). Under these microscopic studies, most often spherical shaped CNPs are observed, however, some authors also reported a mixture of globular, toroids, rod-like particles (Danielsen et al. 2004; MacLaughlin et al. 1998; Köping-Höggård et al. 2001; Huang et al. 2005). The instrument, Zetasizer provides other particle characteristics using Dynamic light scattering (DLS), poly-dispersity index (PDI) and zeta potential (ζ).

17.4.1 Scanning Electron Microscopy (SEM)

SEM is one of the microscopic techniques to study the surface morphology and size of the nanoparticles. In principle, a high energy beam of electrons scan over the surface of nanoparticles and emit signals such as low-energy secondary electrons, backscattered electrons and X-rays, etc., that are detected by a detector and generates their three dimensional image. Samples to be studied under SEM are mounted on a metal stub and coated with a thin film of gold or other conducting material under vacuum. In our laboratory we have prepared chitosan nanoparticles by ionic gelation method found to have spherical shape with 141 nm in size Fig. 17.7 (Unpublished data). Spherical chitosan nanoparticles with an average size of 200 nm have observed under SEM (Jingou et al. 2011).

17.4.2 Transmission Electron Microscopy (TEM)

The size and surface morphology of nanoparticles can be studied by using TEM. It has much higher spatial resolution over SEM and it provides 2-dimensional image of the sample. The working principle of TEM is, when a high-energy electron beam pass through the sample, generates transmitted and diffracted electron beams. The

Fig. 17.7 SEM image of CNPs prepared by Ionotropic gelation method

interference between the two beams forms an image on the fluorescent screen which is coupled with a charge-coupled device (CCD) detector. Samples are to be examined under TEM, dispersed on carbon coated copper grid and allowed to dry at room temperature. TEM images of chitosan nanoparticles showed an average size distribution of 25–30 nm (Phan et al. 2019). Deng et al. (2006) observed, lysosome loaded CNPs as spherical structures with 50–280 nm in diameter. In a study, polymeric CNPs are resulted from reverse micellar method, which are spherical in shape with smooth surface and narrow size distribution of about 90 nm (Manchanda and Nimesh 2010).

17.4.3 Zetasizer

The instrument Zetasizer can be used to study the particle characteristics such as particle size, size distribution, surface charge, etc., using Dynamic light scattering (DLS), polydispersed index (PDI) and zeta potential (ζ).

17.4.3.1 Dynamic Light Scattering

Dynamic light scattering (DLS), also known as Photon correlation spectroscopy or Quasi-Elastic light scattering is one of the popular light scattering analytical

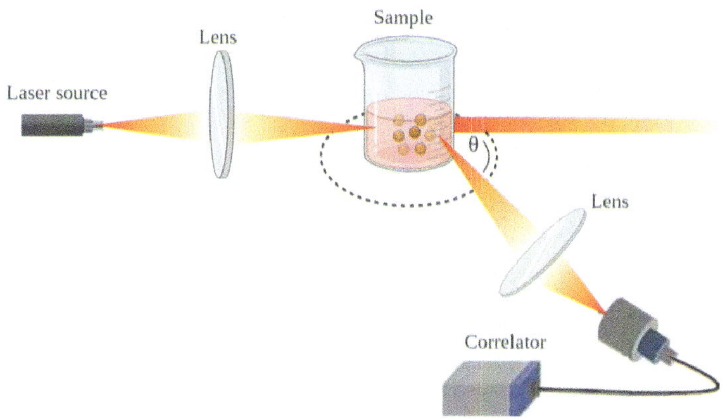

Fig. 17.8 Diagram representing dynamic light scattering principle

techniques used to study size distribution of nanoparticles or submicron particles include, among others, micelles, polymers, emulsions and proteins. The basic principle is simple: particles (in solution) to be studied are illuminated with monochromatic light beam, which in turn generates fluctuations in scattered light intensity are subsequently detected at a known scattered angle (θ) by a photo detector (Fig. 17.8). When a particle sample is dispersed in a solution, particles move randomly (Brownian motion) due to the collision of the solvent molecules around them (Choudhary et al. 2017a). Smaller particles move with greater velocity than the larger particles, hence, the distance between the particles is constantly varying. When moving particles in solution are exposed to a single frequency laser beam over a period of time, generates time dependent fluctuations in scattered light intensity depending on their sizes. These time dependent fluctuations can be related to particle speed by autocorrelation function. The autocorrelation function is used to determine the diffusion coefficient. The Stokes–Einstein equation can be used to convert the diffusion coefficient to the hydrodynamic diameter (Barth 1984). The size of CNPs in water measured using DLS ranges from 40 to 374 nm with an average size of ~ 250 nm (Saharan et al. 2013, 2015; Choudhary et al. 2017a, b). The CNPs obtained from ionotropic gelation method have shown size ranges from 68–75 nm by DLS (Fig. 17.9) (unpublished data).

17.4.3.2 Polydispersity Index Value (PDI)

The polydispersity index value (PDI) is a measure of the heterogeneity of a sample based on size. It means, the particles exhibit either monodispersed or polydispersed distribution in solutions, which can be studied by the instrument that use dynamic light scattering (DLS). According to the International standards organizations (ISOs), the PDI value < 0.05 represents that the particles are monodispersed in nature, whereas the value > 0.7 indicates larger polydispersed particle distribution.

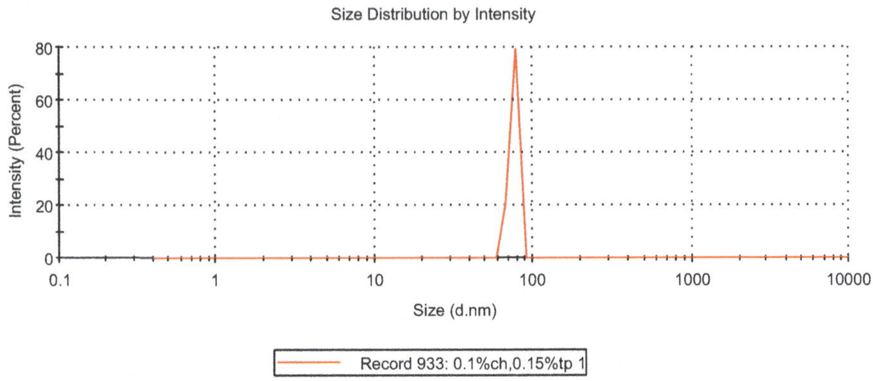

Fig. 17.9 Size distribution of CNPs

The PDI value of chitosan nanoparticles varies from 0.1 to 0.4 which indicates uniformity and stability of particles in suspension. Othman et al. (2018) reported that Chitosan nanoparticles loaded with L-Ascorbic Acid and Thymoquinone (CNP-LAA-TQs) had PDI values of 0.207 ± 0.013. PDI values of Methotrexate-chitosan–polyanion NPs, i.e. MTX-DCH-PAM-18Na NPs and MTX-DCH-PAM-18K NPs ranging from 0.238 to 0.485 and 0.247 to 0.339, respectively (Ciro et al. 2020).

17.4.3.3 Zeta Potential (ζ)

Zeta potential is an important parameter which tells about degree of electrostatic repulsion between charged groups present on particle surface (Saharan et al. 2013, 2015). It is an indicator of the stability of colloidal dispersion, which can be detected by electrophoretic light scattering (ELS). Formation of aggregation is not observed when the particles in the suspension have either high negative or positive zeta potentials as they tend to repel each other, whereas aggregation is observed at low zeta potentials due to less repulsion forces of particles. In general, nanoparticles with zeta potentials of $> +30$ to < -30 have high stability (Kumar et al. 2017). The zeta potential value of nanoparticles is affected by surface chemistry, particle concentration, size of particle, pH of the medium, temperature, solvent, and ionic strength (Mudalige et al. 2019). The optimised zeta potential values for stability of colloidal dispersion were given in the Table 17.1 (Kumar et al. 2017). In different studies, the zeta potential values of CNPs ranged between +21 mV and +50 mV, indicating CNPs are highly stable (Rampino et al. 2013; Li et al. 2018; Kheiri et al. 2017; Ali et al. 2011).

Table 17.1 The optimized zeta potential values for stability of colloidal dispersion (Kumar et al. 2017)

Zeta potential (mV)	Stability behaviour of colloidal Dispersion
0 to ±5	Rapid coagulation or flocculation
±10 to ±30	Incipient instability
±30 to ±40	Moderate stability
±40 to ±60	Good stability
>61	Excellent stability

17.4.4 Fourier-Transform Infrared (FTIR) Spectroscopy

FTIR is one of the absorption spectroscopic techniques widely used in nanoparticles characterization. It measures all of the infrared frequencies simultaneously, rather than individually, and provides sufficient information about the functional groups of a compound. When a sample is illuminated with infrared radiation, it absorbs and transmits certain amount of radiation, from which a detector generates interpretable spectrum that provide structural insights of the sample. The spectrum consists stretching (symmetric and asymmetric stretching) and bending (scissoring, rocking, wagging and twisting) vibrations. Stretching vibrations changes the bond length, whereas bending vibrations change the angle between two bonds of the molecules. In FTIR spectra of chitosan nanoparticles, the peak at 3447 cm^{-1} is attributed to –NH_2 and –OH groups stretching vibration. These peaks shift hypsochromically to 1639 and 1557 cm^{-1} in the FTIR spectra of CN which is caused by the interaction between NH_3^+ groups of chitosan and phosphate groups of TPP. The peaks at 1657 cm^{-1} and 1598 cm^{-1} are attributed to the $CONH_2$ and NH_2 groups, respectively (Lustriane et al. 2018; Qi and Xu 2004; Bhumkar and Pokharkar 2006; Sarkar et al. 2013). Similar FTIR spectra have been obtained for CNPs in our laboratory (Fig. 17.10).

17.5 Biomedical Applications

17.5.1 Antimicrobial Activity

Chitosan is a versatile biopolymer has many biomedical applications. Its effective role on microbes majorly depends on its molecular weight, pH and degree of deacetylation (DD). Many studies have revealed antimicrobial activity of chitosan against bacteria, fungi and yeasts; however, exact mode of action is still not fully understood. As of now, the following possible theories have somehow made an attempt to explain the mode of antimicrobial action of chitosan.

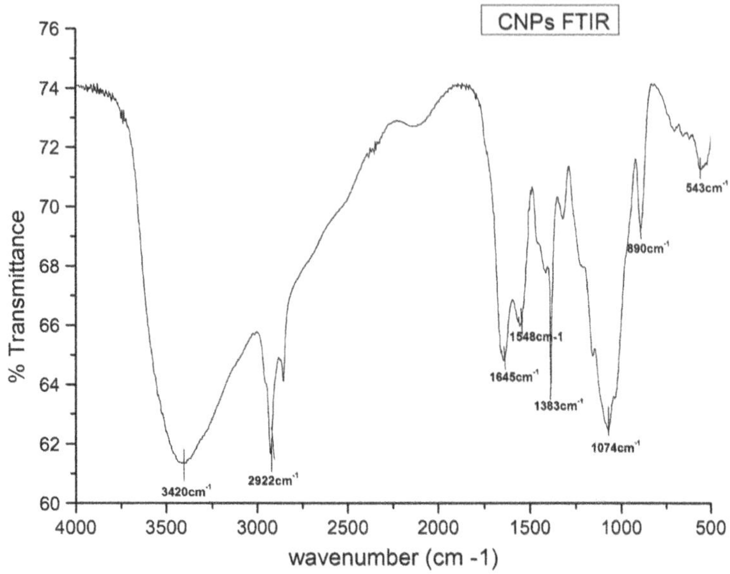

Fig. 17.10 FTIR spectrum of CNPs, shown specific peak at 3420 cm^{-1} due to the overlap of O-H and N-H stretching, similarly, peaks at 1645 cm^{-1}, 1548 cm^{-1} and 1383 cm^{-1} are attributed to the $CO\text{-}NH_2$, NH_2, C-H groups, respectively (unpublished data)

17.5.1.1 Polycationic Nature of Chitosan

As per this hypothesis antimicrobial activity presumably depends on the alteration of membrane permeability of microbes such as bacteria, fungi and viruses. The cell walls of Gram-positive bacteria is largely composed of peptidoglycan layer with certain composition of teichoic acids (wall teichoic acid and lipoteichoic acid), which gives negative charge to the bacterial surface, while the Gram-negative bacteria cell wall outer membrane possess lipopolysacharides (LPS), which provides surface negative charge. In acidic aqueous solutions, the NH_2 groups at C2 position of chitosan protonates to yield NH_3^+, which in turn forms electrostatic interactions with negatively charged groups (mostly phosphate groups of teichoic acids and LPS) located on the bacterial cell surfaces, leads to enhance the cell wall/ cell membrane permeability followed by leakage of intracellular constituents and death of the cell (Fig. 17.11) (Tsai and Su 1999; Aziz et al. 2014, 2015, 2016; Inamuddin et al. 2021).

In a study, in vitro assays, killing kinetics, cellular leakage measurements, membrane potential estimation, electron microscopy and transcriptional response analysis have given a speculation that antimicrobial activity of chitosan (LMW) is due to electrostatic binding of protonated amine groups with negatively charged teichoic acids (predominantly with lipoteichoic acids) of bacterial cell wall of Gram-positive bacteria *Staphylococcus aureus* (Raafat et al. 2008).

In several studies, CS shown more bactericidal effect in Gram +ve bacteria (*Listeria monocytogenes*, *Bacillus megaterium*, *B. cereus*, *S. aureus*, *Lactobacillus*

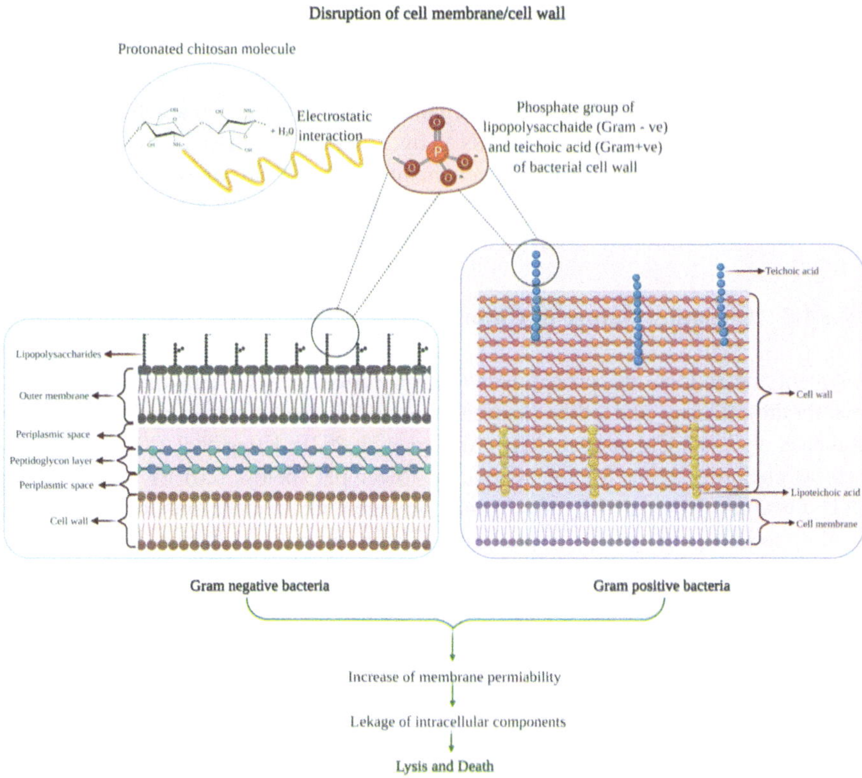

Fig. 17.11 Antimicrobial effect of CS by Polycationic nature

plantarum, *L. brevis*, *L. bulgaricus*) than in Gram –ve bacteria (*E. coli*, *Psedomonas fluorescens*, *Salmonella typhimurium*, *Vibrio parahaemolyticus*) (No et al. 2002).

Similar mechanisms might also be applicable to fungal pathogens due to presence of phospholipids in their membrane. At low pH, the protonated amine groups of chitosan interact with the negatively charged phosphate groups of carbohydrate side chains of fungal cell wall proteins and decrease the negative charge that leads to alteration of important metabolic pathways (Ing et al. 2012; Pena et al. 2013). In a study, LMW CS showed antifungal activity against the pathogenic yeast *Candida albicans* by decreasing the cell surface negative charge. Sialic acid, a constituent of cell wall glycoprotein of *C. albicans*, which provides negative charge to the cell wall (Soares et al. 2000; Tronchin et al. 2008). Positively charged groups of chitosan increase *E.coli* membrane permeability and lysis of membranes (Li et al. 2015).

17.5.1.2 Chitosan Interaction with Nucleic Acid

Low molecular weight (LMW), micro and nano size chitosan particles can able to penetrate into the cytoplasm of microorganisms and bind to the negatively charged biomolecules such as DNA and RNA by electrostatic interactions and subsequently effect on the downstream mechanisms such as transcription and translation (Fig. 17.12) (Jarmila and Vavrikova 2011; Sudarshan et al. 1992).

17.5.1.3 Chelating Nature of Chitosan

Divalent metal ions are prerequisite for microbial growth, enzymatic functions, membrane integrity and other (Varma et al. 2004; Hosseinnejad and Jafari 2016; Rabea et al. 2003; Chien et al. 2016; Matica et al. 2019; Kong et al. 2008). Chitosan acts as chelating agent for metal ions, at lower pH (below 6.0) its amine groups (NH_2) become protonated ($-NH_3^+$) and compete with divalent ions for electrostatic binding to phosphate groups of teichoic acid and lipopolysaccharide (LPS) of Gram +ve and Gram −ve bacteria, respectively (Fig. 17.13). On other hand, chitosan at higher pH value (above pKa 6.3) also chelates many metal ions. Less availability of essential metal ions leads to enhancement of cell wall permeability and sensitive to several chemicals or antibiotics (Clifton et al. 2015).

17.5.1.4 Cell Surface Blocking Nature

High molecular weight (HMW) chitosan molecules deposit as dense polymer layer on cell surface of microbes and became a barrier to uptake of essential nutrients, minerals and oxygen (aerobic microbes) as well as excretion of their metabolic products, leads to death of cells (Fig. 17.14) (Yuan et al. 2016; Devlieghere et al. 2004).

17.5.2 Factors Affecting the Antimicrobial Activity of Chitosan

The main factors that affect antimicrobial activity of chitosan and its derivatives are pH, temperature, molecular weight and degree of acetylation.

Higher antimicrobial activity is observed at low pH, i.e. <6.3 while the inhibitory efficiency is decreased with increased pH (Muzzarelli 1996; Helander et al. 2001).

Temperature may affect the chitosan viscosity, molecular weight and antimicrobial activity during storage (No et al. 2006). In a study, it was observed that CS was remain stable and showed antimicrobial activity against *Listeria monocytogenes*,

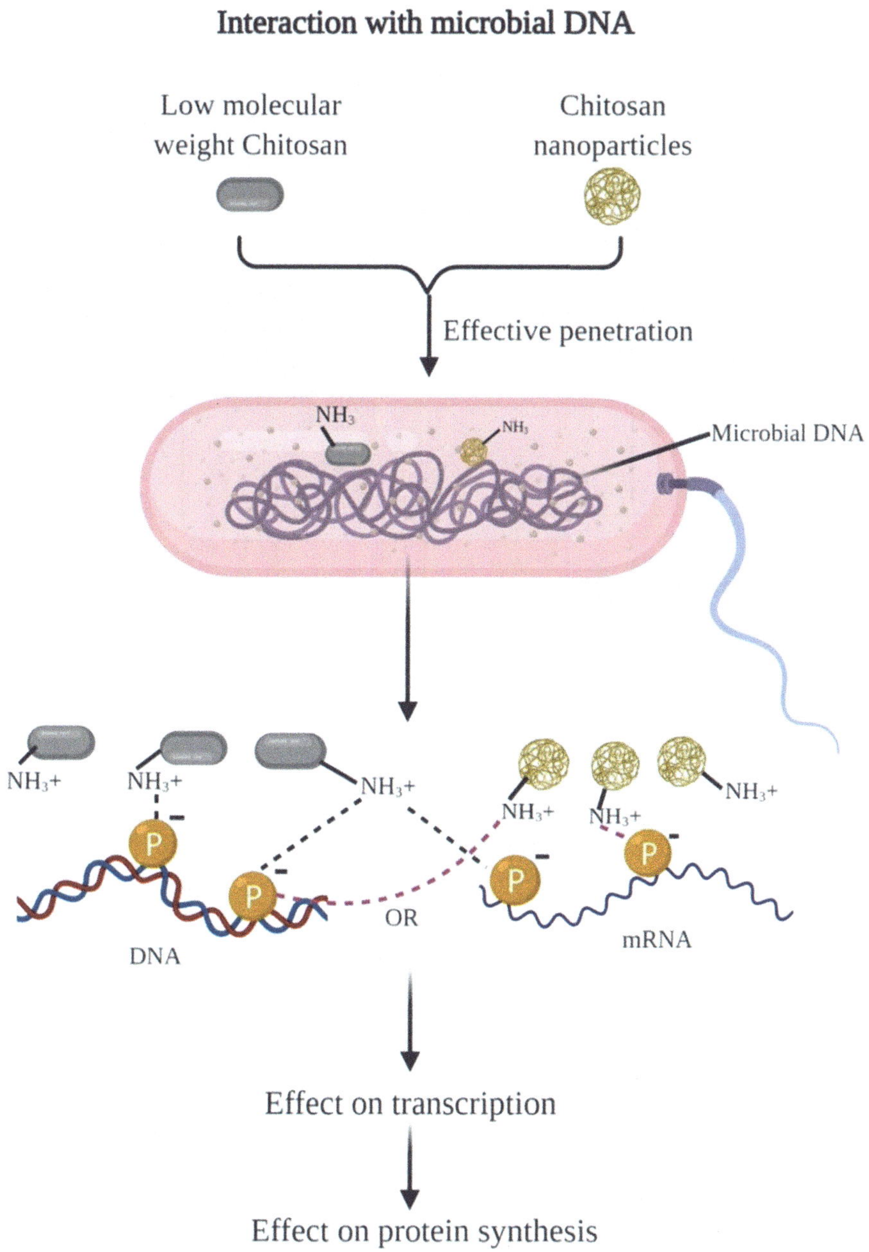

Fig. 17.12 Antimicrobial effect of CS and CNPs by polycationic nature

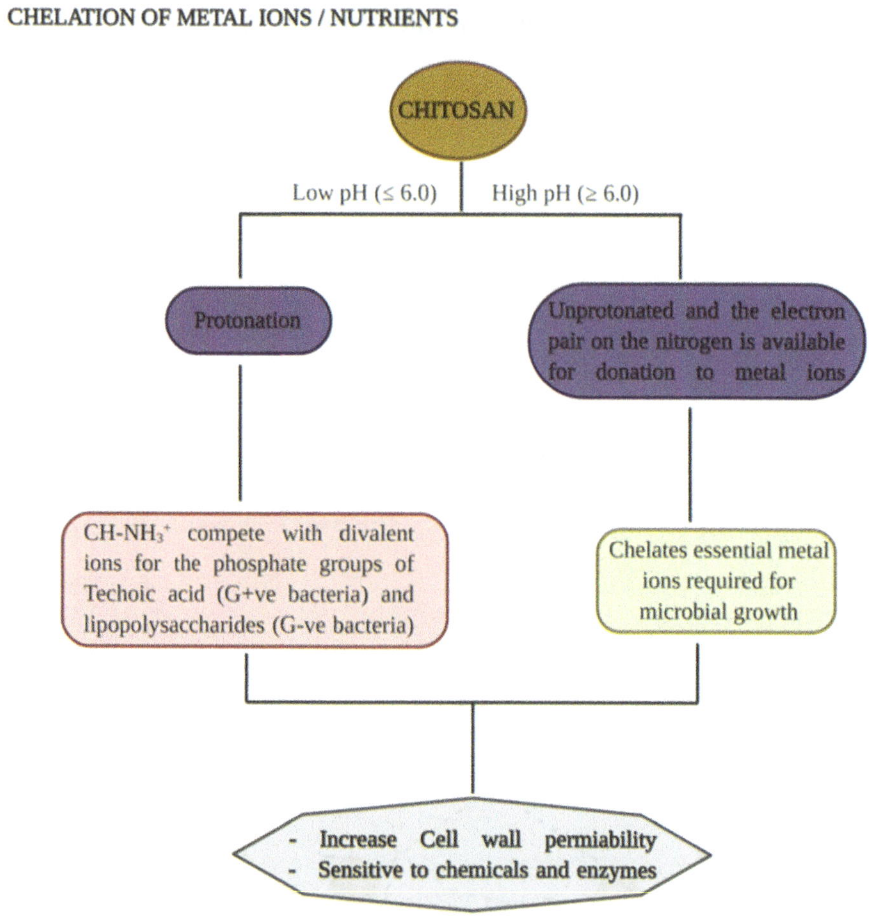

Fig. 17.13 Antimicrobial effect of CS by chelating nature

Salmonella enteritidis, *Staphylococcus aureus* and *E.coli* even after 15 weeks storage at 4°C than 25°C (No et al. 2006).

Molecular weight and degree of acetylation will affect the chitosan antimicrobial activity. Based on the molecular weight, CS is often classified as high molecular weight (HMW, 64.8 kDa to 375 kDa), medium molecular weight (MMW, 250 kDa to 310 kDa), low molecular weight (LMW, 10 kDa to 150 kDa) (Matica et al. 2019). Relation between antimicrobial activity and molecular weight depends on the type of microorganisms. High molecular weight (HMW) chitosan accumulate on the surface of bacterial membrane and inhibit the nutrient transport, resulting in cell death (Li et al. 2010), whereas low molecular weight (LMW) chitosan could pierce into the bacterial surface membranes and bind with DNA, thus blocking mRNA and protein synthesis. (Kulikov et al. 2015). When chitosan molecular weight is

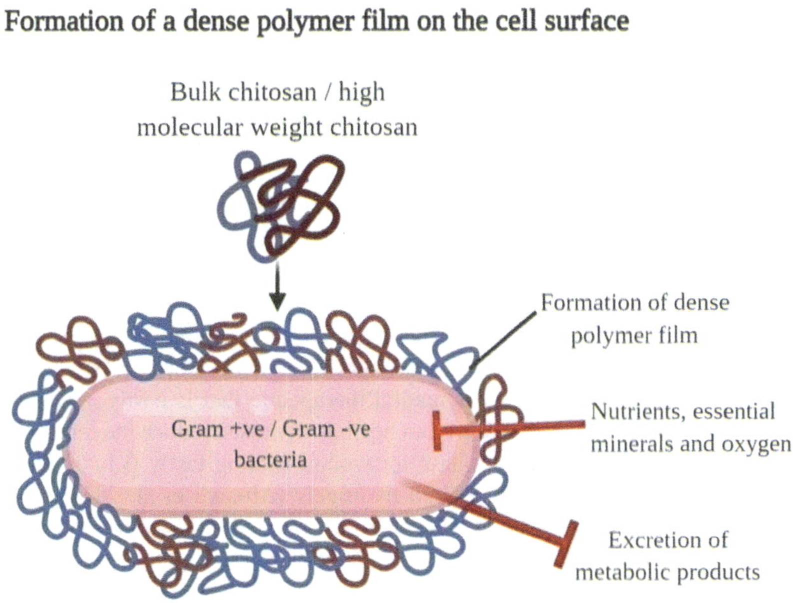

Fig. 17.14 Antimicrobial effect of CS by cell surface blocking nature

decreased bactericidal effect of Gram-negative bacteria was enhanced (Younes et al. 2014).

In the same way the degree of acetylation also influence the chitosan antimicrobial activity. Chitosan antimicrobial activity was increased with the decrease in the degree of acetylation (Goy et al. 2009). More positively charged cations of chitosan are associated with the degree of acetylation (Kong et al. 2008). Highest antibacterial activity against *S. aureus* and *E. coli* was observed at 30–40% of degree of acetylation (Takahashi et al. 2008). Chitosan with a higher degree of deacetylation have more cationic positive amino free groups which influence the antimicrobial activity.

17.5.3 Wound Healing Activity of Chitosan Nanoparticles

Chitosan plays vital role in wound healing because of its antimicrobial, haemostatic, film forming and analgesic and anti-inflammatory properties (Gupta et al. 2019). Chitosans have the similar structure of glycosaminoglycans (GAGs) which are constituents of extra cellular matrix (ECM), hence, used in skin tissue engineering (Chen et al. 2008). Chitosan is biocompatible and non-toxic to living cells and tissues and has been proved in vitro with different types of cells like fibroblasts, keratinocytes, hepatocytes, myocardial and endothelial cells (Dash et al. 2011). Wound healing is a natural response to injury, patients with non-healing disorders due to factors like age, sex hormones, stress diabetes, obesity, alcoholism, smoking and nutrition makes them stressed and inconvenient or discomfort (Guo and DiPietro 2010).

General programmed phases for wound healing are haemostasis, inflammation, proliferation and remodelling. Many factors include age, sex hormones, infection, oxygenation, medication, nutrition, alcoholism, smoking, diabetes, stress and obesity involved in wound healing by disturbing the programmed phases leads to delay in wound healing (Guo and DiPietro 2010). Impaired healing of wounds enters in to pathological inflammation which leads to chronic wounds like ulcers, diabetes mellitus and venous stasis disease. Non-healing wounds result in immense healthcare expenditures. Thus many studies lead to therapeutics that promote tissue repair, improve impaired wound healing and at the same time inexpensive easily available to sufferers. Chitosan is a natural therapeutic that is easily available and promote tissue repair and improve wound healing. Chitosan reduce the inflammatory phase and accelerates proliferate phase for wound healing fastly (Liu et al. 2018). Chitosan can easily mould into desired hydrogels (Ahmadi et al. 2015), sponges (Huang et al. 2015), membranes (Mi et al. 2001) and films without hazardous chemicals. Hydrogels moistens the infected area by storing high capacity of water (Hoffman 2012). Sponges give prefect matrixes to most wound healing areas due to its open porosity and swelling properties (Mori et al. 2016). Membranes fabricate the three-dimensional matrices with high surface-volume ratio for nutrient supply and cell proliferation. Films should be resistant to pathogenic bacteria in biomedical applications (Zhang et al. 2015).

Chitosan based antimicrobial wound dressing can be incorporated with antibiotics (ciprofloxacin, gentamicin, sulfadiazine or tetracycline), metallic antimicrobial nanoparticles (e.g. nAg, nCu, nZnO and $nTiO_2$) and natural compounds and extracts (honey, *Aloe vera, Juglena regia,* etc.) or fabricated alone with native molecules (Simões et al. 2018; Yang et al. 2016; Ahmadi et al. 2015; Huang et al. 2015; Mi et al. 2001; Coma et al. 2002).

Chitosan is associated with antibiotics to evoke the antimicrobial effect by interfering with bacterial metabolic pathways (Bermingham and Derrick 2002) bacterial structure, cell wall biosynthesis (Patrick 2003), protein synthesis (Hong et al. 2014). Genotoxic, oxidative and cytotoxic effects with metallic nanoparticles can be reduced by using chitosan based biomaterials as carriers (Travan et al. 2009).

In recent studies silver nanoparticles (nAg) owed much interest as a potent antimicrobial agent and in clinical studies in wound dressing to nAg coated medical equipment (Madhumathi et al. 2010). nAg is the metallic nanoparticle showed broad inhibitory activity against many antibiotic resistant bacteria (Zewde et al. 2016).

CNPs are used as stabilizing materials as they are having more permeability towards aqueous solution and its mechanical strength, biofilm formation, liable to chemical modifications and cost-effectiveness (Javid et al. 2013). CNPs and its derivatives make it possible for versatile applications in the medical fields in blood clotting, wounds healing and skin tissue engineering, skin burns, blood lipid cholesterol control, membrane and scaffolds, surgical sutures, etc. (Baghdan et al. 2018; Li et al. 2018; Mohebbi et al. 2019; Gupta et al. 2019).

In the recent studies CS-polyvinyl alcohol (PVA)-silver nanoparticles were used to provoke the wound healing process as it is involved in the wound healing dressing (Hajji et al. 2019).

In another in vivo study on albino rats CS-nanosilver dressings showed enormous and best wound healing activities when compared with intra dermal injection of mesenchymal stem cells injections (Ghannam et al. 2018).

Chitosan-entrapped metallic nanoparticles are safe to use to living cells and have the properties of anti-bacterial effects; interacts with cell wall composition and inhibits the membrane of mitochondrial organelle; enhance the mechanical support and provoke the regrowth of granulation tissue. Chitosan/sodium alginate-Cu (hydrogel), Chitosan-nAu(film), Quaternized chitosan-nAg (film), Chitosan/algetic acid-nZnO (sponge), Chitosan/ECM/n-TiO_2, Chitosan/gelatine-nFe_3O_4 (composite)are the chitosan based nanoparticles used in the recent studies against the microbial activity of *Staphylococcus aureus*, *Escherichia coli*, *Pseudomonas aeruginosa*, *Candida albicans* (Wichai et al. 2019; Rahimi et al. 2019; Regiel-Futyra et al. 2015; Bal-Ozturk et al. 2019; Cai et al. 2016; Woo et al. 2015).

17.5.4 Chitosan Based Nanoparticles in Vaccine Delivery

As the nanoparticles size is small they are easily incorporated into antigen presenting cells and are used as adjuvant in vaccines (Kreuter 1995). Properties of chitosan like polycationic, non-reactivity and high affinity for metals make it to be used as carrier molecule. As the size of pathogens met during proliferation of immune system is almost same that of nanoparticle carrier it is easily taken up by the antigen presenting cells (Xiang et al. 2006). Nanoparticle carriers provoke the mucosal uptake of vaccines and stimulate the mucosal immune response (IgA). Chitosan nanoparticle enhance the mucosal administration because of its tight binding with mucin, small size and the opening ability of tight junctions between epithelial cells.

Intra nasal administration of chitosan nanoparticle encapsulated mucosal vaccines against influenza, diphtheria, pertussis and hepatitis B virus (Illum et al. 2001; Pawar and Jaganathan 2014) stimulate the production of significant IgG and IgA responses in mice. Delated clearance time is also observed in the nasal mucosa with this vaccine and provokes both mucosal and humoral immune response (Pawar and Jaganathan 2014). Oral chitosan nanoparticles vaccine loaded with tetanus toxoid have also activated mucosal and humoral immunity (Barhate et al. 2014; Harde et al. 2014).

Mucoadhesive and osmotic properties of chitosan helps in adsorption and passage of protein peptides through the nasal epithelium (Dodane et al. 1992) and passage of macro molecules across the mucosal barrier (Huang et al. 2016). In recent studies glucomannan modified chitosan nanoparticles maintained biological activity of mediator molecules and blocked antigens. Glucuronidation based chitosan nanoparticles effectively induced systemic serum IgG, mucosal secretary IgA, cell mediated immune responses of IL-2 and IFN- γ (Harde et al. 2014).

N-2HACC and N,O-Carboxy methyl chitosan (CMC) is a vaccine adjuvant for New castle disease virus (NDV) and Infectious bronchitis virus (IBV). Nanoparticles containing NDV/IBV can enhance the proliferation of lymphocytes (Zhao et al. 2017) and induce intranasal inoculation of IgG and IgA antibodies. Chitosan nanoparticles along with plasmid DNA enhance antigen specific immune response (TaO et al. 2013). Some research studies have done on intranasal DNA vaccination (Torrieridramard et al. 2011).

Chitosan's immune stimulatory, mucoadhesive, negative zeta potential, poly cationic, non-reactive properties made it to use as adjuvant carrier for vaccines in the nanoparticle plat form.

17.5.5 CNPs in Drug Delivery

Chitosan is used as one of the important natural polymers with vast applications in drug delivery because of its solubility in the aqueous medium and its cationic amino groups function (Bellich et al. 2016). CNPs in the drug delivery are used to overcome the side effects of drugs, to maintain control rate of drug delivery and to ensure correctly the only targeted area is treated (Teare et al. 1995; Ewart et al. 2019).

The nanoparticles can pierce in to the infected cell (or) tissue due to the presence of larger junctions of epithelial cells. This piercing is of two types—passive targeting and active targeting. In actively targeting drug carrier system is conjugated to a tissue (or) specific cell ligand, whereas in passive targeting due to leaky junctions a nanoparticle reaches the target organ site (Varshosaz and Farzan 2015).

Desirable nanoparticles drug delivery system should reach, identify, bind and deliver its load to specific tissues and avoid drug induced harm to healthy tissues. Targeting ligands on the surface of nanoparticles should be in the form of peptides, antibodies, designed proteins, small molecules and nucleic acids (Liu et al. 2009; Friedman et al. 2013). Drugs which are encapsulated with chitosan nanoparticles can improve their absorption and bio-availability and allowing themselves to deliver gene and protein drugs and are effectively protected from enzyme degradation in vivo (Senapati et al. 2018). Cationic charges of chitosan when interact with anionic charges of nucleic acid molecules form poly electrolyte complex (PEC), this complex protects the nucleic acids from nuclease degradation (MacLaughlin et al. 1998). Bio distribution of chitosan can differ depending on the surface charge, size, molecular weight and hydrophobic nature of chitosan and its derivatives (He et al. 2010). Elimination of chitosan after the drug delivery is through renal clearance because of its solubility and low molecular size.

Mucosal membrane surfaces are common and easy routes for delivering drugs into the body system. Macromolecular drugs such as peptides and proteins are unable to cross the mucosal barriers as they are degraded by enzymes before reaching blood flow. To solve this problem, nanostructures based mucoadhesive polysaccharide chitosan is used (Amidi et al. 2010). Biotin (Vitamin H) has high

Table 17.2 Drug fabricated Chitosan nanoparticles for cancer therapy

S. No	Compound incorporated with CNPs	Treatment	References
1	5-fluorouracil (5-FU) encapsulated chitosan nanoparticles	Cancer therapy	Tıglı Aydın and Pulat (2012)
2	Photosensitizer tetraphenylchlorin Chitosan nanoparticles (TPC–CS NPs) loaded with mertansine (MRT) or cabazitaxel (CBZ)	Breast cancer cell lines	Pandya et al. (2020)
3	Copper-loaded chitosan nanoparticles (Cu-CNPs)	Osteosarcoma cancer	Jw and Liao (2017)
4	α-santalol functionalized chitosan nanoparticles (Sn-CNPs)	Breast cancer	Zhang et al. (2020)
5	Chitosan-PLGA based catechin hydrate nanoparticles (CS-CTH-PLGA-NPs)	Lung cancer	Ahmad et al. (2020)
6	Quercetin Loaded Chitosan Nanoparticles (Qu-CS NPs)	Colorectal cancer	Rashedi et al. (2019)
7	Curcumin-loaded Chitosan nanoparticles (Cu-CNPs)	Cancer therapy	Le et al. (2013)
8	Arg-Gly-Asp (RGD) peptide-labelled chitosan nanoparticles loaded with SiRNA (RGD-CH-NPs)	Ovarian cancer	Han et al. (2010)
9	Biotinylated chitosan nanoparticles (bio-CNPs)	Liver cancer	Cheng et al. (2017)
10	Alginic acid-coated chitosan nanoparticles (A.C. NPs)	Breast cancer	Liu et al. (2013)
11	Hyaluronic acid (HA)-decorated glycol chitosan (GC) nanoparticle conjugated to doxorubicin (DOX) and co-loaded celecoxib (CXB) (HA-GC-DOX/CXB)	Lung cancer	Lee et al. (2020)
12	Folate-Chitosan Nanoparticles Loaded with Ursolic Acid (FA-CS-UA-NPs)	Breast cancer	Jin et al. (2016)
13	Folic acid-conjugated temozolomide (TMZ)-loaded chitosan nanoparticles (CSTMZ-FLA-NP)	Lung cancer	Li et al. (2017)
14	Niclosamide loaded chitosan nanoparticles (Nic-Chi Np's)	Breast cancer Lung cancer	Naqvi et al. (2017)
15	Ketorolac-loaded chitosan nanoparticles	Cancer therapy	Venu et al. (2018)
16	Gemcitabine loaded fucoidan/chitosan nanoparticles	Breast cancer	Oliveira et al. (2018)

affinity for streptavidin and is used for conjugation with nanoparticles (Pramanik et al. 2016). It was reported certain tumor cells overexpress folate receptors than to normal cells, hence, folic acid (vitamin B9) used for targeting in several cancers treatments due to its immense affinity for folate receptors (Zhao et al. 2008). Similarly, the smart targeting of nanoparticles with specific carbohydrates, short peptides, antibodies and small molecules have been designed and studied (Friedman et al. 2013). Some of the recent studies on various cancers have revealed that CNPs have exceptional role in targeted drug delivery (Table 17.2).

17.6 Conclusion

Chitosan is an inexpensive biopolymer extracted from chitin, shows remarkable properties such as biocompatibility, biodegradability, non-antigenic, mucoadhesive, biological activity, cationic nature and low toxicity. A stupendous research has been done on chitosan and its functionalised derivatives to intend in wound healing, tissue engineering, drug delivery, antimicrobial and anti-tumour activity, anti-diabetic and a cholesterol reducing activity. Nanoparticles bio-fabrication transmits desirable functional characteristics to chitosan. Several effective methods are used for fabrication and characterization of CNPs. CNPs provide larger surface area volume to ratio, and their size ranges easily penetrate in to the cells and hence, used as a potential vehicle for delivery of several molecules including drugs, antibiotics, vaccines, genes, peptides, etc., and scaffolds for tissue engineering. However, all these CNPs based therapeutics are in preclinical stages and further extensive studies are required to reveal the safety and effectiveness for their applications.

Acknowledgments We are thankful to the Department of Genetics and Genomics, Yogi Vemana University, Kadapa, Andhra Pradesh, India for providing us the literature access.

References

Ahmad N, Ahmad R, Alrasheed RA et al (2020) A chitosan-PLGA based catechin hydrate nanoparticles used in targeting of lungs and cancer treatment. Saudi J Biol Sci 27 (9):2344–2357. https://doi.org/10.1016/j.sjbs.2020.05.023

Ahmadi F, Oveisi Z, Samani SM, Amoozgar Z (2015) Chitosan based hydrogels: Characteristics and pharmaceutical applications. Res Pharm Sci 10(1):1–16

Ali SW, Rajendran S, Joshi M (2011) Synthesis and characterization of chitosan and silver loaded chitosan nanoparticles for bioactive polyester. Carbohydr Polym 83(2):438–446. https://doi.org/10.1016/j.carbpol.2010.08.004

Amidi M, Mastrobattista E, Jiskoot W, Hennink WE (2010) Chitosan-based delivery systems for protein therapeutics and antigens. Adv Drug Deliv Rev 62:59–82. https://doi.org/10.1016/j.addr.2009.11.009

Aziz N, Fatma T, Varma A, Prasad R (2014) Biogenic synthesis of silver nanoparticles using Scenedesmus abundans and evaluation of their antibacterial activity. J Nanoparticles, Article ID 689419. https://doi.org/10.1155/2014/689419

Aziz N, Faraz M, Pandey R, Sakir M, Fatma T, Varma A, Barman I, Prasad R (2015) Facile algae-derived route to biogenic silver nanoparticles: synthesis, antibacterial and photocatalytic properties. Langmuir 31:11605–11612. https://doi.org/10.1021/acs.langmuir.5b03081

Aziz N, Pandey R, Barman I, Prasad R (2016) Leveraging the attributes of Mucor hiemalis-derived silver nanoparticles for a synergistic broad-spectrum antimicrobial platform. Front Microbiol 7:1984. https://doi.org/10.3389/fmicb.2016.01984

Baghdan E, Pinnapireddy SR, Strehlow B et al (2018) Lipid coated chitosan-DNA nanoparticles for enhanced gene delivery. Int J Pharm 535:473–479. https://doi.org/10.1016/j.ijpharm.2017.11.045

Bal-Ozturk A, Karal-Yilmaz O, Akguner ZP et al (2019) Sponge-like chitosan-based nanostructured antibacterial material as a topical hemostat. J Appl Polym Sci 136(19):47522. https://doi.org/10.1002/app.47522

Barhate G, Gautam M, Gairola S et al (2014) Enhanced mucosal immune responses against tetanus toxoid using novel delivery system comprised of chitosanfunctionalized gold nanoparticles and botanical adjuvant: characterization, immunogenicity, and stability assessment. J Pharm Sci 103 (11):3448–3456. https://doi.org/10.1002/jps.24161

Barth HG (ed) (1984) Modern methods of particle size analysis, vol 97. John Wiley and sons

Bellich B, D Agostino I, Semeraro S et al (2016) "The good, the bad and the ugly" of Chitosans. Mar Drugs 14:99. https://doi.org/10.3390/md14050099

Bermingham A, Derrick JP (2002) The folic acid biosynthesis pathway in bacteria: evaluation of potential for antibacterial drug discovery. Bioessays 24:637–648. https://doi.org/10.1002/bies.10114

Bhumkar DR, Pokharkar VB (2006) Studies on effect of pH on cross-linking of chitosan with sodium tripolyphosphate: a technical note. AAPS PharmSciTech 7(2):E138–E143. https://doi.org/10.1208/pt070250

Brunel F, Véron L, David L et al (2008) A novel synthesis of chitosan nanoparticles in reverse emulsion. Langmuir 24:11370–11377. https://doi.org/10.1021/la801917a

Cai N, Li C, Han C et al (2016) Tailoring mechanical and antibacterial properties of chitosan/gelatin nanofiber membranes with Fe3O4 nanoparticles for potential wound dressing application. Appl Surf Sci 369:492–500. https://doi.org/10.1016/j.apsusc.2016.02.053

Calvo P, Remuñán-López C, Vila-Jato JL, Alonso MJ (1997) Novel hydrophilic chitosan-polyethylene oxide nanoparticles as protein carriers. J Appl Polym Sci 63:125–132. https://doi.org/10.1002/(SICI)1097-4628

Carlstrom D (1957) The crystal structure of α-chitin (poly-N-acetyl-D-glucosamine). Int J Cell Biol 3(5):669–683. https://doi.org/10.1083/jcb.3.5.669

Chan HK, Kwok PCL (2011) Production methods for nanodrug particles using the bottom-up approach. Adv Drug Deliv Rev 63(6):406–416. https://doi.org/10.1016/j.addr.2011.03.011

Chandy T, Sharma CP (1990) Chitosan-as a biomaterial. Biomater Artif Cells Artif Organs 18:1–24. https://doi.org/10.3109/10731199009117286

Chen Y, Mohanraj VJ, Parkin JE (2003) Chitosan-dextran sulfate nanoparticles for delivery of an anti-angiogenesis peptide. Lett Pept Sci 10(5–6):621–629. https://doi.org/10.1007/BF02442596

Chen Y, Mohanraj VJ, Wang F, Benson HA (2007) Designing chitosan-dextran sulfate nanoparticles using charge ratios. Aaps Pharm Sci Tech 8(4):131–139. https://doi.org/10.1208/pt0804098

Chen F, Li X, Mo X et al (2008) Electrospun chitosan-P(LLA-CL) nanofibers for biomimetic extracellular matrix. J Biomat Sci Polym E 19:677–691. https://doi.org/10.1163/156856208784089661

Cheng M, Zhu W, Li Q (2017) Anti-cancer efficacy of biotinylated chitosan nanoparticles in liver cancer. Oncotarget 8(35):59068

Cheung RCF, Ng TB, Wong JH, Chan WY (2015) Chitosan: an update on potential biomedical and pharmaceutical applications. Mar Drugs 13(8):5156–5186. https://doi.org/10.3390/md13085156

Chien RC, Yen MT, Mau JL (2016) Antimicrobial and antitumor activities of chitosan from shiitake stipes, compared to commercial chitosan from crab shells. Carbohydr Polym 138:259–264. https://doi.org/10.1016/j.carbpol.2015.11.061

Choudhary RC, Kumaraswamy RV, Kumari S et al (2017a) Synthesis, characterization and application of chitosan nanomaterials loaded with zinc and copper for plant growth and protection. In: Prasad R et al (eds) Nanotechnology. Springer, Singapore. https://doi.org/10.1007/978-981-10-4573-8_10

Choudhary RC, Kumaraswamy RV, Kumari S et al (2017b) Cu-chitosan nanoparticle boost defense responses and plant growth in maize (Zea mays L.). Sci Rep 7:9754. https://doi.org/10.1038/s41598-017-08571-0

Ciro Y, Rojas J, Alhajj MJ et al (2020) Production and characterization of chitosan–polyanion nanoparticles by polyelectrolyte complexation assisted by high-intensity sonication for the modified release of methotrexate. Pharmaceuticals 13(1):11. https://doi.org/10.3390/ph13010011

Clifton LA, Skoda MW, Le Brun AP et al (2015) Effect of divalent cation removal on the structure of gram-negative bacterial outer membrane models. Langmuir 31(1):404–412. https://doi.org/10.1021/la504407v

Coma V, Martial-Gros A, Garreau S et al (2002) Edible antimicrobial films based on chitosan matrix. J Food Sci 67:1162–1169. https://doi.org/10.1111/j.1365-2621.2002.tb09470.x

Crini G (2006) Non-conventional low-cost adsorbents for dye removal: a review. Bioresour Technol 97(9):1061–1085. https://doi.org/10.1016/j.biortech.2005.05.001

Danielsen S, Vårum KM, Stokke BT (2004) Structural analysis of chitosan mediated DNA condensation by AFM: influence of chitosan molecular parameters. Biomacromolecules 5 (3):928–936. https://doi.org/10.1021/bm034502r

Dash M, Chiellini F, Ottenbrite RM, Chiellini E (2011) Chitosan-A versatile semisynthetic polymer in biomedical applications. Prog Polym Sci 36:981–1014. https://doi.org/10.1016/j.progpolymsci.2011.02.001

Deng QY, Zhou CR, Luo BH (2006) Preparation and characterization of chitosan nanoparticles containing lysozyme. Pharm Biol 44(5):336–342. https://doi.org/10.1080/13880200600746246

Denuzier A, Ferrier D, Damour O (1998) Chitosan–chondroitin sulfate and chitosan–hyaluronate polyelectrolyte complexes: biological properties. Biomaterials 19(14):1275–1285. https://doi.org/10.1016/S0142-9612(98)00036-2

Devlieghere F, Vermeulen A, Debevere J (2004) Chitosan: antimicrobial activity, interactions with food components and applicability as a coating on fruit and vegetables. Food Microbiol 21:703–714. https://doi.org/10.1016/j.fm.2004.02.008

Dodane V, Amin Khan M, Merwin JR (1992) Effect of chitosan on epithelial permeability and structure. Int J Pharm 182:21–32. https://doi.org/10.1016/S0378-5173(99)00030-7

Du WL, Niu SS, Xu YL et al (2009) Antibacterial activity of chitosan tripolyphosphate nanoparticles loaded with various metal ions. Carbohydr Polym 75(3):385–389. https://doi.org/10.1016/j.carbpol.2008.07.039

El-Shabouri MH (2002) Positively charged nanoparticles for improving the oral bioavailability of cyclosporin-A. Int J Pharm 249(1–2):101–108. https://doi.org/10.1016/S0378-5173(02)00461-1

Erbacher P, Zou S, Bettinger T et al (1998) Chitosan-based vector/DNA complexes for gene delivery: biophysical characteristics and transfection ability. Pharm Res 15:1332–1339. https://doi.org/10.1023/A:1011981000671

Ewart D, Peterson EJ, Steer CJ et al (2019) A new era of genetic engineering for autoimmune and inflammatory diseases. Semin Arthritis Rheum 49:e1–e7. https://doi.org/10.1016/j.semarthrit.2019.05.004

Felt O, Buri P, Gurny R (1998) Chitosan: a unique polysaccharide for drug delivery. Drug Dev Ind Pharm 24:979–993. https://doi.org/10.3109/03639049809089942

Friedman AD, Claypool SE et al (2013) The smart targeting of nanoparticles. Curr Pharm Des 19 (35):6315–6329. https://doi.org/10.2174/13816128113199990375

Gällstedt M, Hedenqvist MS (2006) Packaging-related mechanical and barrier properties of pulp–fiber–chitosan sheets. Carbohydr Polym 63(1):46–53. https://doi.org/10.1016/j.carbpol.2005.07.024

Ghannam S, Korayem H, Farghaly L, Hosny S (2018) The effect of chitosan nanosilver dressing versus mesenchymal stem cells on wound healing. J Afr Assoc Physiol Sci 6:23–31

Gortari MC, Hours RA (2013) Biotechnological processes for chitin recovery out of crustacean waste: a mini-review. Electron J Biotechnol 16(3):14–14. https://doi.org/10.2225/vol16-issue3-fulltext-10

Goy RC, Dd B, Assis OB (2009) A review of the antimicrobial activity of chitosan. Polímeros 19:241–247. https://doi.org/10.1590/S0104-14282009000300013

Grenha A (2012) Chitosan nanoparticles: a survey of preparation methods. J Drug Target 20:291–300. https://doi.org/10.3109/1061186X.2011.65412

Guo SA, DiPietro LA (2010) Factors affecting wound healing. J Dent Res 89(3):219–229. https://doi.org/10.1177/0022034509359125

Gupta AK, Naregalkar RR, Vaidya VD, Gupta M (2007) Recent advances on surface engineering of magnetic iron oxide nanoparticles and their biomedical applications. Nanomedicine 2(1):23–29. https://doi.org/10.2217/17435889.2.1.23

Gupta A, Mumtaz S, Li CH et al (2019) Combatting antibiotic-resistant bacteria using nanomaterials. Chem Soc Rev 48(2):415–427. https://doi.org/10.1039/C7CS00748E

Hajji S, Khedir SB, Hamza-Mnif I et al (2019) Biomedical potential of chitosan-silver nanoparticles with special reference to antioxidant, antibacterial, hemolytic and in vivo cutaneous wound healing effects. Biochim Biophys Acta 1863:241–254. https://doi.org/10.1016/j.bbagen.2018.10.010

Han LK, Kimura Y, Okuda H (1999) Reduction in fat storage during chitin-chitosan treatment in mice fed a high-fat diet. Int J Obes Relat Metab Disord 23:174–179

Han HD, Mangala LS, Lee JW et al (2010) Targeted gene silencing using RGD-labeled chitosan nanoparticles. Clin Cancer Res 16(15):3910–3922. https://doi.org/10.1158/1078-0432.CCR-10-0005

Harde H, Agrawal AK, Jain S (2014) Tetanus toxoids loaded glucomannosylated chitosan based nanohoming vaccine adjuvant with improved oral stability and immunostimulatory response. Pharm Res 32(1):122–134. https://doi.org/10.1007/s11095-014-1449-5

He C, Hu Y, Yin L et al (2010) Effects of particle size and surface charge on cellular uptake and biodistribution of polymeric nanoparticles. Biomaterials 31:3657–3666. https://doi.org/10.1016/j.biomaterials.2010.01.065

Helander IM, Nurmiaho-Lassila EL, Ahvenainen R et al (2001) Chitosan disrupts the barrier properties of the outer membrane of Gram-negative bacteria. Int J Food Microbiol 71 (2–3):235–244. https://doi.org/10.1016/S0168-1605(01)00609-2

Hoffman AS (2012) Hydrogels for biomedical applications. Adv Drug Deliv Rev 64:18–23. https://doi.org/10.1016/j.addr.2012.09.010

Hong W, Zeng J, Xie J (2014) Antibiotic drugs targeting bacterial RNAs. Acta Pharm Sin B 4:258–265. https://doi.org/10.1016/j.apsb.2014.06.012

Hosseinnejad M, Jafari SM (2016) Evaluation of different factors affecting antimicrobial properties of chitosan. Int J Biol Macromol 85:467–475. https://doi.org/10.1016/j.ijbiomac.2016.01.022

Huang H, Yuan Q, Yang X (2005) Morphology study of gold–chitosan nanocomposites. J Colloid Interface Sci 282:26–31. https://doi.org/10.1016/j.jcis.2004.08.063

Huang X, Sun Y, Nie J et al (2015) Using absorbable chitosan hemostatic sponges as a promising surgical dressing. Int J Biol Macromol 75:322–329. https://doi.org/10.1016/j.ijbiomac.2015.01.049

Huang Y, Li R, Chen J, Chen J (2016) Recent advances in chitosan-based nano particulate pulmonary drug delivery. Carbohydr Polym 138:114–122. https://doi.org/10.1039/C6NR03256G

Ichikawa S, Iwamoto S, Watanabe J (2005) Formation of biocompatible nanoparticles by self-assembly of enzymatic hydrolysates of chitosan and carboxymethyl cellulose. Biosci Biotechnol Biochem 69(9):1637–1642. https://doi.org/10.1271/bbb.69.1637

Illum L, Jabbal-Gill I, Hinchcliffe M et al (2001) Chitosan as a novel nasal delivery system for vaccines. Adv Drug Deliv Rev 51(1–3):81–96. https://doi.org/10.1016/S0169-409X(01)00171-5

Inamuddin, Ahamed MI, Prasad R (2021) Advanced antimicrobial materials and applications. Springer, Singapore. ISBN 978-981-15-7098-8. https://www.springer.com/gp/book/9789811570971

Ing LY, Zin NM, Sarwar A, Katas H (2012) Antifungal activity of chitosan nanoparticles and correlation with their physical properties. Int J Biomater 2012:632698. https://doi.org/10.1155/2012/632698

Jang MK, Kong BG, Jeong YI et al (2004) Physicochemical characterization of a-chitin, b-chitin and c-chitin separated from natural resources. J Polym Sci A Polym Chem 42:3423–3432. https://doi.org/10.1002/pola.20176

Jarmila V, Vavrikova E (2011) Chitosan derivatives with antimicrobial, antitumour and antioxidant 494 activities, a review. Curr Pharm Des 17(32):3596–3607. https://doi.org/10.2174/138161211798194468
Javid A, Ahmadian S, Saboury AA et al (2013) Chitosan-coated superparamagnetic iron oxide nanoparticles for doxorubicin delivery: synthesis and anticancer effect against human ovarian cancer cells. Chem Biol Drug Des 82:296–306. https://doi.org/10.1111/cbdd.12145
Jayasuriya AC (2017) Production of micro-and nanoscale chitosan particles for biomedical applications. In: Chitosan based biomaterials, vol 1. Woodhead Publishing, pp 185–209. https://doi.org/10.1016/B978-0-08-100230-8.00008-X
Jin H, Pi J, Yang F et al (2016) Folate-chitosan nanoparticles loaded with ursolic acid confer anti-breast cancer activities in vitro and in vivo. Sci Rep 6:30782. https://doi.org/10.1038/srep30782
Jingou J, Shilei H, Weiqi L et al (2011) Preparation, characterization of hydrophilic and hydrophobic drug in combine loaded chitosan/cyclodextrin nanoparticles and in vitro release study. J Colloid Interface Sci 83(1):103–107. https://doi.org/10.1016/j.colsurfb.2010.11.005
Jw A, Liao W (2017) Enhanced anticancer effect of copper-loaded chitosan nanoparticles against osteosarcoma cancer. RSC Adv 7:15971–15977. https://doi.org/10.1039/C6RA21648J
Kheiri A, Jorf SM, Malihipour A et al (2017) Synthesis and characterization of chitosan nanoparticles and their effect on Fusarium head blight and oxidative activity in wheat. Int J Biol Macromol 102:526–538. https://doi.org/10.1016/j.ijbiomac.2017.04.034
Kim TH, Jiang HL, Jere D et al (2007) Chemical modification of chitosan as a gene carrier in vitro and in vivo. Prog Polym Sci 32(7):726–753. https://doi.org/10.1016/j.progpolymsci.2007.05.001
Kmiec M, Pighinelli L, Tedesco MF (2017) Chitosan-properties and applications in dentistry. Adv Tissue Eng Regen Med Open Access 2(4):00035. https://doi.org/10.15406/atroa.2017.02.00035
Kong M, Chen XG, Xue YP et al (2008) Preparation and antibacterial activity of chitosan microshperes in a solid dispersing system. Front Mater Sci China 2:214–220. https://doi.org/10.1007/s11706-008-0036-2
Köping-Höggård M, Tubulekas I, Guan H et al (2001) Chitosan as a nonviral gene delivery system. Structure–property relationships and characteristics compared with polyethylenimine in vitro and after lung administration in vivo. Gene Ther 8(14):1108–1121. https://doi.org/10.1038/sj.gt.3301492
Kreuter J (1995) Nanoparticles as adjuvants for vaccines. In: Powell M, Newman M (eds) Vaccine design. Springer, New York, pp 463–472. https://doi.org/10.1007/978-1-4615-1823-5_19
Kulikov SN, Tikhonov VE, Bezrodnykh EA et al (2015) Comparative evaluation of antimicrobial activity of oligochitosans against Klebsiella pneumoniae. Russ J Bioorg Chem 41:57–62. https://doi.org/10.1134/S1068162015010100
Kumar MR, Muzzarelli RAA, Muzzarelli C et al (2004) Chitosan chemistry and pharmaceutical perspectives. Chem Rev 104(12):6017–6084. https://doi.org/10.1021/cr030441b
Kumar V, Bhatt PC, Rahman M et al (2017) Fabrication, optimization, and characterization of umbelliferone β-D-galactopyranoside-loaded PLGA nanoparticles in treatment of hepatocellular carcinoma: in vitro and in vivo studies. Int J Nanomedicine 12:6747. https://doi.org/10.2147/IJN.S136629
Kumar S, Ye F, Dobretsov S, Dutta J (2019) Chitosan nanocomposite coatings for food, paints, and water treatment applications. App Sci 9:2409. https://doi.org/10.3390/app9122409
Le TMP, Dang TML, La TH et al (2013) Preparation of curcumin-loaded pluronic F127/chitosan nanoparticles for cancer therapy. Adv Nat Sci-Nanosc 4(2):025001. https://doi.org/10.1088/2043-6262/4/2/025001
Lee R, Choi YJ, Jeong MS et al (2020) Hyaluronic acid-decorated glycol chitosan nanoparticles for pH-sensitive controlled release of doxorubicin and celecoxib in nonsmall cell lung cancer. Bioconjug Chem 31(3):923–932. https://doi.org/10.1021/acs.bioconjchem.0c00048
Li X, Feng X, Yang S, Fu G, Wang T, Su Z (2010) Chitosan kills Escherichia coli through damage to be of cell membrane mechanism. Carbohydr Polym 79:493–499. https://doi.org/10.1016/j.carbpol.2009.07.011

Li Z, Yang F, Yang R (2015) Synthesis and characterization of chitosan derivatives with dual-antibacterial functional groups. Int J Biol Macromol 75:378–387. https://doi.org/10.1016/j.ijbiomac.2015.01.056
Li K, Liang N, Yang H (2017) Temozolomide encapsulated and folic acid decorated chitosan nanoparticles for lung tumor targeting: improving therapeutic efficacy both in vitro and in vivo. Oncotarget 8(67):111318. https://doi.org/10.18632/oncotarget.22791
Li J, Rao J, Pu K (2018) Recent progress on semiconducting polymer nanoparticles for molecular imaging and cancer phototherapy. Biomaterials 155:217–235. https://doi.org/10.1016/j.biomaterials.2017.11.025
Liu S, Luo H, Li N (2001) Resonance Rayleigh scattering study of the interaction of heparin with some basic diphenyl naphthylmethane dyes. Anal Chem 73(16):3907–3914. https://doi.org/10.1021/ac001454h
Liu R, Kay BK, Jiang S (2009) Nanoparticle delivery: targeting and nonspecific binding. MRS Bulletin 34(6):432–440
Liu Z, Lv D, Liu S et al (2013) Alginic acid-coated chitosan nanoparticles loaded with legumain DNA vaccine: effect against breast cancer in mice. *PLoS One* 8(4):e60190. https://doi.org/10.1371/journal.pone.0060190
Liu H, Wang C, Li C et al (2018) A functional chitosan-based hydrogel as a wound dressing and drug delivery system in the treatment of wound healing. RSC Adv 8:7533–7549. https://doi.org/10.1039/C7RA13510F
Lustriane C, Dwivany FM, Suendo V, Reza M (2018) Effect of chitosan and chitosan-nanoparticles on post harvest quality of banana fruits. Plant Biotechnol J 45(1):36–44. https://doi.org/. https://doi.org/10.5010/JPB.2018.45.1.036
MacLaughlin FC, Mumper RJ, Wang J et al (1998) Chitosan and depolymerized chitosan oligomers as condensing carriers for in vivo plasmid delivery. J Control Release 56:259–272. https://doi.org/10.1016/S0168-3659(98)00097-2
Madhumathi K, Kumar PS, Abhilash S, Sreeja V et al (2010) Development of novel chitin/nanosilver composite scaffolds for wound dressing applications. J Mater Sci Mater Med 21:807–813. https://doi.org/10.1007/s10856-009-3877-z
Madihally SV, Matthew HW (1999) Porous chitosan scaffolds for tissue engineering. Biomaterials 20(12):1133–1142. https://doi.org/10.1016/S0142-9612(99)00011-3
Maitra A, Ghosh PK, De Tapas K (1997) Process for the preparation of highly monodispersed polymeric hydrophilic nanoparticles. https://patents.google.com/patent/US5874111A/en113
Manchanda R, Nimesh S (2010) Controlled size chitosan nanoparticles as an efficient, biocompatible oligonucleotides delivery system. J Appl Polym Sci 118(4):2071–2077. https://doi.org/10.1002/app.32508
Manni L, Ghorbel-bellaaj O, Jellouli K et al (2010) Extraction and characterization of chitin, chitosan, and protein hydrolysates prepared from shrimp waste by treatment with crude protease from Bacillus cereus SV 1. Appl Biochem 162(2):345–357. https://doi.org/10.1007/s12010-009-8846-y
Matica MA, Aachmann FL, Tøndervik A et al (2019) Chitosan as a wound dressing starting material: Antimicrobial properties and mode of action. Int J Mol Sci 20(23):5889. https://doi.org/10.3390/ijms20235889
Mi FL, Shyu SS, Wu YB, Lee ST et al (2001) Fabrication and characterization of a sponge-like asymmetric chitosan membrane as a wound dressing. Biomaterials 22:165–173. https://doi.org/10.1016/S0142-9612(00)00167-8
Mohammed MA, Syeda JTM, Wasan KM, Wasan EK (2017) An overview of chitosan nanoparticles and its applications in non-parenteral drug delivery. Pharmaceutics 9:53. https:/doi.org/10.3390/pharmaceutics9040053
Mohebbi S, Nezhad MN, Zarrintaj P et al (2019) Chitosan in biomedical engineering: a critical review. Curr Stem Cell Res Ther 14:93–116. https://doi.org/10.2174/1574888X13666180912142028

Mori M, Rossi S, Ferrari F et al (2016) Sponge-like dressings based on the association of chitosan and sericin for the treatment of chronic skin ulcers. I. Design of experiments–assisted development. J Pharm Sci 105:1180–1187. https://doi.org/10.1016/j.xphs.2015.11.047

Mudalige T, Qu H, Van Haute D (2019) Characterization of nanomaterials: tools and challenges. Nanomaterials Food App 2019:313–353. https://doi.org/10.1016/B978-0-12-814130-4.00011-7

Muzzarelli RAA (1996) Chitosan-based dietary foods. Carbohydr Polym 29:309–316. https://doi.org/10.1016/S0144-8617(96)00033-1

Naqvi S, Mohiyuddin S, Gopinath P (2017) Niclosamide loaded biodegradable chitosan nanocargoes: an in vitro study for potential application in cancer therapy. R Soc Open Sci 4 (11):170611. https://doi.org/10.1098/rsos.170611

Niwa T, Takeuchi H, Hino T (1993) Preparations of biodegradable nanospheres of water-soluble and insoluble drugs with D, L-lactide/glycolide copolymer by a novel spontaneous emulsification solvent diffusion method and the drug release behavior. J Control Release 25(1–2):89–98. https://doi.org/10.1016/0168-3659(93)90097-O

No HK, Young Park N, Ho Lee S, Meyers SP (2002) Antibacterial activity of chitosans and chitosan oligomers with different molecular weights. Int J Food Microbiol 74:65–72. https://doi.org/10.1016/S0168-1605(01)00717-6

No HK, Kim SH, Lee SH et al (2006) Stability and antibacterial activity of chitosan solutions affected by storage temperature and time. Carbohydr Polym 65:174–178. https://doi.org/10.1016/j.carbpol.2005.12.036

Oliveira C, Neves NM, Reis RL et al (2018) Gemcitabine delivered by fucoidan/chitosan nanoparticles presents increased toxicity over human breast cancer cells. Nanomedicine 13 (16):2037–2050. https://doi.org/10.2217/nnm-2018-0004

Onsosyen E, Skaugrud O (1990) Metal recovery using chitosan. J Chem Technol Biotechnol 49:395–404. https://doi.org/10.1002/jctb.280490410

Othman N, Masarudin MJ, Kuen CY et al (2018) Synthesis and optimization of chitosan nanoparticles loaded with L-ascorbic acid and thymoquinone. Nanomaterials 8(11):920. https://doi.org/10.3390/nano8110920

Pandya AD, Øverbye A, Sahariah P et al (2020) Drug-loaded photosensitizer-chitosan nanoparticles for combinatorial chemo-and photodynamic-therapy of cancer. Biomacromolecules 21(4):1489–1498. https://doi.org/10.1021/acs.biomac.0c00061

Patrick GL (2003) An introduction to medicinal chemistry. Oxford University Press, New York, NY

Pavinatto FJ, Caseli L, Oliveira ON Jr (2010) Chitosan in nanostructured thin films. Biomacromolecules 11(8):1897–1908. https://doi.org/10.1021/bm1004838

Pawar D, Jaganathan KS (2014) Mucoadhesive glycol chitosan nanoparticles for intranasal delivery of hepatitis B vaccine: enhancement of mucosal and systemic immune response. Drug Deliv 23:1–11. https://doi.org/10.3109/10717544.2014.908427

Pena A, Sánchez NS, Calahorra M (2013) Effects of chitosan on Candida albicans: conditions for its antifungal activity. Biomed Res Int 2013. https://doi.org/10.1155/2013/527549

Phan DN, Dorjjugder N, Khan MQ et al (2019) Synthesis and attachment of silver and copper nanoparticles on cellulose nanofibers and comparative antibacterial study. Cellul 26 (11):6629–6640. https://doi.org/10.1007/s10570-019-02542-6

Pillai CKS, Paul W, Sharma CP (2009) Chitin and chitosan polymers: chemistry, solubility and fiber formation. Prog Polym Sci 34(7):641–678. https://doi.org/10.1016/j.progpolymsci.2009.04.001

Pramanik AK, Palanimuthu D, Somasundaram K, Samuelson AG (2016) Biotin decorated gold nanoparticles for targeted delivery of a smart-linked anticancer active copper complex: in vitro and in vivo studies. Bioconjug Chem 27(12):2874–2885. https://doi.org/10.1021/acs.bioconjchem.6b00537

Prasad R, Pandey R, Barman I (2016) Engineering tailored nanoparticles with microbes: quo vadis. WIREs Nanomed Nanobiotechnol 8:316–330. https://doi.org/10.1002/wnan.1363

Prasad R, Pandey R, Varma A, Barman I (2017) Polymer based nanoparticles for drug delivery systems and cancer therapeutics. In: Kharkwal H, Janaswamy S (eds) Natural polymers for drug delivery. CAB International, London, pp 53–70

Qi L, Xu Z (2004) Lead sorption from aqueous solutions on chitosan nanoparticles. Colloids Surf A Physicochem 251(1–3):183–190. https://doi.org/10.1016/j.colsurfa.2004.10.010

Raafat D, Von Bargen K, Haas A, Sahl HG (2008) Insights into the mode of action of chitosan as an antibacterial compound. Appl Environ Microbiol 74:3764–3773. https://doi.org/10.1128/AEM.00453-08

Rabea EI, Badawy MET, Stevens CV et al (2003) Chitosan as antimicrobial agent: applications and mode of action. Biomacromolecules 4:1457–1465. https://doi.org/10.1021/bm034130m

Rahimi M, Ahmadi R, Samadi Kafil H, Shafiei-Irannejad V (2019) A novel bioactive quaternized chitosan and its silver-containing nanocomposites as a potent antimicrobial wound dressing: structural and biological properties. Mater Sci Eng C 101:360–369. https://doi.org/10.1016/j.msec.2019.03.092

Rampino A, Borgogna M, Blasi P et al (2013) Chitosan nanoparticles: preparation, size evolution and stability. Int J Pharm 455(1–2):219–228. https://doi.org/10.1016/j.ijpharm.2013.07.034

Rashedi J, Haghjo AG, Abbasi MM et al (2019) Anti-tumor effect of quercetin loaded chitosan nanoparticles on induced colon cancer in Wistar rats. Adv Pharm Bull 9(3):409. https://doi.org/10.15171/apb.2019.048

Regiel-Futyra A, Liskiewicz M, Sebastian V et al (2015) Development of non cytotoxic chitosan-gold nanocomposites as efficient antibacterial materials. ACS Appl Mater Interfaces 7 (2):1089–1099. https://doi.org/10.1021/am508094e

Saharan V, Mehrotra A, Khatik R et al (2013) Synthesis of chitosan based nanoparticles and their in vitro evaluation against phytopathogenic fungi. Int J Biol Macromol 62:677–683. https://doi.org/10.1016/j.ijbiomac.2013.10.012

Saharan V, Sharma G, Yadav M et al (2015) Synthesis and in vitro antifungal efficacy of Cu-chitosan nanoparticles against pathogenic fungi of tomato. Int J Biol Macromol 75:346–353. https://doi.org/10.1016/j.ijbiomac.2015.01.027

Sailaja AK, Amareshwar P, Chakravati P (2011) Different techniques used for the preparation of nanoparticles using, natural polymers and their application. Int J Pharm Pharm Sci 3:975–149

Sarkar K, Debnath M, Kundu PP (2013) Preparation of low toxic fluorescent chitosan-graft-polyethyleneimine copolymer for gene carrier. Carbohydr Polym 92(2):2048–2057. https://doi.org/10.1016/j.carbpol.2012.11.067

Senapati S, Mahanta AK, Kumar S, Maiti P (2018) Controlled drug delivery vehicles for cancer treatment and their performance. Signal Transduct Target Ther 3:7. https://doi.org/10.1038/s41392-017-0004-3

Simões D, Miguel SP, Ribeiro MP, Coutinho P et al (2018) Recent advances on antimicrobial wound dressing: a review. Eur J Pharm Biopharm 127:130–141. https://doi.org/10.1016/j.ejpb.2018.02.022

Soares RM, Rosangela MDA, Alviano DS et al (2000) Identification of sialic acids on the cell surface of Candida albicans. Biochimica et Biophysica Acta (BBA)-General Subjects 1474 (2):262–268. https://doi.org/10.1016/S0304-4165(00)00003-9

Sudarshan N, Hoover D, Knorr D (1992) Antibacterial action of chitosan. Food Biotechnol 6:257–272. https://doi.org/10.1080/08905439209549838

Takahashi T, Takayama K, Machida Y, Nagai T (1990) Characteristics of polyion complexes of chitosan with sodium alginate and sodium polyacrylate. Int J Pharm 61(1–2):35–41. https://doi.org/10.1016/0378-5173(90)90041-2

Takahashi T, Imai M, Suzuki I, Sawai J (2008) Growth inhibitory effect on bacteria of chitosan membranes regulated with deacetylation degree. Biochem Eng J 40:485–491. https://doi.org/10.1016/j.bej.2008.02.009

Tan Tang H, Hu J, Hu Y et al (2011) Controlled release of chitosan/heparin nanoparticle-delivered VEGF enhances regeneration of decellularized tissue-engineered scaffolds. Int J Nanomedicine 6:929. https://doi.org/10.2147/IJN.S18753

Tao W, Ziemer KS, Gill HS (2013) Gold nanoparticle-M2e conjugate coformulated with CpG induces protective immunity against influenza a virus. Nanomedicine 9:237–251. https://doi.org/10.2217/nnm.13.58

Teare J, Spedding C, Whitehead M et al (1995) Omeprazole and dry mouth. Scand J Gastroenterol 30:216–218. https://doi.org/10.3109/00365529509093266

Tıglı Aydın RS, Pulat M (2012) 5-Fluorouracil encapsulated chitosan nanoparticles for pH-stimulated drug delivery: evaluation of controlled release kinetics. J Nanomaterials 2012. https://doi.org/10.1155/2012/313961

Tiyaboonchai W (2003) Chitosan nanoparticles: a promising system for drug delivery. Naresuan University Journal: Science and Technology (NUJST) 11(3):51–66

Torrieridramard L, Lambrecht B, Ferreira HL et al (2011) Intranasal DNA vaccination induces potent mucosal and systemic immune responses and cross-protective immunity against influenza viruses. Mol Ther 19:602–611. https://doi.org/10.1038/mt.2010.222

Travan A, Pelillo C, Donati I et al (2009) Non-cytotoxic silver nanoparticle-polysaccharide nanocomposites with antimicrobial activity. Biomacromolecules 10:1429–1435. https://doi.org/10.1021/bm900039x

Tronchin G, Pihet M, Lopes-Bezerra LM et al (2008) Adherence mechanisms in human pathogenic fungi. Med Mycol 46:749–772. https://doi.org/10.1080/13693780802206435

Tsai GJ, Su WH (1999) Antibacterial activity of shrimp chitosan against Escherichia coli. J Food Prot 62(3):239–243. https://doi.org/10.4315/0362-028X-62.3.239

Tsigos I, Martinou A, Kafetzopoulos D, Bouriotis V (2000) Chitin deacetylases: new, versatile tools in biotechnology. Trends Biotechnol 18(7):305–312. https://doi.org/10.1016/S0167-7799(00)01462-1

Varma A, Deshpande S, Kennedy J (2004) Metal complexation by chitosan and its derivatives: a review. Carbohydr Polym 55:77–93. https://doi.org/10.1016/j.carbpol.2003.08.005

Varshosaz J, Farzan M (2015) Nanoparticles for targeted delivery of therapeutics and small interfering RNAs in hepatocellular carcinoma. World J Gastroenterol 21(42):12022–12041. https://doi.org/10.3748/wjg.v21.i42.12022

Venkatesan J, Kim SK (2010) Chitosan composites for bone tissue engineering-an overview. Mar Drugs 8(8):2252–2266. https://doi.org/10.3390/md8082252

Venu K, Mondal S, Mondal P (2018) Cytotoxic activity of ketorolac-loaded chitosan nanoparticles in SCC-29 cell lines. Egypt Pharmaceut J 17(1):53. https://doi.org/10.4103/epj.epj_1_18

Venugopal V (2008) Marine products for healthcare: functional and bioactive nutraceutical compounds from the ocean. CRC Press, Taylor and Fransis Group, Boca Raton, FL, pp 185–508

Wichai S, Chuysinuan P, Chaiarwut S et al (2019) Development of bacterial cellulose/ alginate/ chitosan composites incorporating copper (II) sulfate as an antibacterial wound dressing. J Drug Deliv Sci Technol 51:662–671. https://doi.org/10.1016/j.jddst.2019.03.043

Woo CH, Choi YC, Choi JS et al (2015) A bilayer composite composed of TiO2-incorporated electrospun chitosan membrane and human extracellular matrix sheet as a wound dressing. J Biomater Sci Polym Ed 26:841–854. https://doi.org/10.1080/09205063.2015.1061349

Xiang SD, Scholzen A, Minigo G et al (2006) Pathogen recognition and development of particulate vaccines: does size matter? Methods 40(1):1–9. https://doi.org/10.1016/j.ymeth.2006.05.016

Yan N, Chen X (2015) Sustainability: don't waste seafood waste. *Nature* 524(7564):155–157. https://doi.org/10.1038/524155a

Yang L, Lu W, Pang Y, Huang X et al (2016) Fabrication of a novel chitosan scaffold with asymmetric structure for guided tissue regeneration. RSC Adv 6:71567–71573. https://doi.org/10.1039/C6RA12370H

Younes I, Sellimi S, Rinaudo M et al (2014) Influence of acetylation degree and molecular weight of homogeneous chitosans on antibacterial and antifungal activities. Int J Food Microbiol 185:57–63. https://doi.org/10.1016/j.ijfoodmicro.2014.04.029

Yuan G, Lv H, Tang W et al (2016) Effect of chitosan coating combined with pomegranate peel extract on the quality of Pacific white shrimp during iced storage. Food Control 59:818–823. https://doi.org/10.1016/j.foodcont.2015.07.011

Zewde B, Ambaye A, Stubbs J III, Dharmara R (2016) A review of stabilized silver nanoparticles-synthesis, biological properties, characterization, and potential areas of applications. JSM Nanotechnol Nanomed 4:1–14

Zhang Y, Zhang M (2002) Three-dimensional macroporous calcium phosphate bioceramics with nested chitosan sponges for load-bearing bone implants. J Biomed Mater Res 61:1–8. https://doi.org/10.1002/jbm.10176

Zhang D, Zhou W, Wei B et al (2015) Carboxyl-modified poly (vinyl alcohol)- crosslinked chitosan hydrogel films for potential wound dressing. Carbohydr Polym 125:189–199. https://doi.org/10.1016/j.carbpol.2015.02.034

Zhang J, Wang Y, Li J et al (2020) α-Santalol functionalized chitosan nanoparticles as efficient inhibitors of polo-like kinase in triple negative breast cancer. RSC Adv 10(9):5487–5501. https://doi.org/10.1039/C9RA09084C

Zhao X, Li H, Lee RJ (2008) Targeted drug delivery via folate receptors. Expert Opin Drug Deliv 5 (3):309–319. https://doi.org/10.1517/17425247.5.3.309

Zhao K, Li S, Li W et al (2017) Quaternized chitosan nanoparticles loaded with the combined attenuated live vaccine against newcastle disease and infectious bronchitis elicit immune response in chicken after intranasal administration. Drug Deliv 24:1574–1586. https://doi.org/10.1080/10717544.2017.1388450

The manufacturer's authorised representative in the EU is Springer Nature Customer Service Centre GmbH, Europaplatz 3, 69115 Heidelberg, Germany. If you have any concerns regarding our products, please contact ProductSafety@springernature.com

Printed and bound by CPI Group (UK) Ltd, Croydon, CR0 4YY
30/09/2026
02225332-0012